THE NEANDERTHAL LEGACY

An Archaeological Perspective from Western Europe

THE NEANDERTHAL LEGACY

An Archaeological Perspective from Western Europe

Paul Mellars

**Department of Archaeology
Cambridge University, UK**

PRINCETON UNIVERSITY PRESS, PRINCETON, NEW JERSEY

Library of Congress Cataloging-in-Publication Data
Mellars, Paul.
   The Neanderthal legacy : an archaeological perspective from
western Europe / Paul Mellars.
      p.   cm.
   Includes bibliographical references and index.
   ISBN 0-691-03493-1 (cloth : acid-free paper)
   1. Neanderthals – Europe.   2. Paleolithic period – Europe.
3. Human evolution – Europe – Philosophy.   4. Behavior evolution – Europe.
5. Europe – Antiquities. I. Title.
GN285.M45   1995
936 – dc20   95-4300

Contents _____

v

List of Tables _____

List of Illustrations

Figure 13.15   *Bone artefacts and animal-tooth pendants from the Châtelperronian levels at Arcy-sur-*   415
                   Cure.
Figure 13.16   *Bifacial leaf points and associated tools from the Szeletian site of Vedrovice V*   417
                   (Czechoslovakia).

Preface

My colleague Chris Stringer warned me that writing a book about Neanderthals would be rather like ordering menus for College High Table dinners – there would be no way of pleasing everyone. For the past eighty years or so we seem to have been caught between two opposing camps – between those who see the Neanderthals as our immediate ancestors, only mildly different in at least their basic biological and behavioural patterns from fully 'modern' populations; and those who prefer to adhere to Marcellin Boule's original vision of the Neanderthals as representing an extinct side line of human evolution, with both behavioural patterns and most probably innate *capacities* for behaviour which were radically different from those of later populations. The recent controversies surrounding the origins and dispersal of anatomically modern populations have put the spotlight firmly onto the Neanderthals as arguably the last surviving representatives of the original 'archaic' lineages which populated the world prior to the dramatic demographic and behavioural dispersal of genetically modern populations.

While there have been many studies of the biological and anatomical aspects of Neanderthals, there have been surprisingly few systematic attempts to review their behavioural patterns. There has of course been no shortage of excavations of Middle Palaeolithic sites, combined with a deluge of analyses of the associated stone-tool industries and (to a lesser extent) faunal assemblages. My central belief in embarking on this book was that it was only by taking a very broad-ranging look at all aspects of the behaviour and organization of Neanderthal communities – their technology, subsistence patterns, spatial and demographic organization and patterns of communication and cognition – that we could hope to form any overall impression of their true behavioural capacities and, in particular, how far these may have differed from those of the ensuing anatomically modern populations. As far as I am aware, this is the first book which has attempted to bring all these issues together in a single study, focussed on a specific and exceptionally rich and well documented body of archaeological evidence.

The extent to which this book depends on the research of my French colleagues will be immediately apparent from dipping into any of the individual chapters. Following the pioneering research of François Bordes, André Leroi-Gourhan and others shortly after the Second World War, there has been an extraordinary upsurge in studies of the Middle Palaeolithic over the past twenty years, extending far beyond the traditional obsession with typological and taxonomic issues which dominated the field for so long. In southwestern France the work pioneered by Jean-Philippe Rigaud on the spatial distribution and organization of Mousterian sites has been systematically extended by a number of younger colleagues – most notably by Jean-Michel Geneste, Allain Turq and Christine Duchadeau-Kervazo in the Perigord region, and further south by Jacques Jaubert, Jean-Marie Le Tensorer, Liliane Meignen and others. Over the same period the research of Françoise Delpech, Jean-Luc Guadelli, Guy Laquay, Stephane Madelaine, Philip Chase and others has taken the analysis of Middle Palaeolithic faunal assemblages far beyond the traditional concern with the purely climatic and chronological aspects of the faunas, into a determined attempt to reconstruct the economic and carcase-processing strategies which lay behind the archaelogical bone assemblages. Above all, perhaps, the study of lithic technology has been revolutionized by the application of the 'chaîne opératoire' approach to technological analysis (involving the use of extensive refitting studies, experimental replication techniques and microscopic use-wear analyses), as well as by systematic studies of the sources and distribution patterns of the raw materials employed for tool production.

What is remarkable about this recent work is not merely its scientific quality, but the extent to which much of the critical data and analysis still remains embedded in a range of specialist site reports, conference volumes and (above all perhaps) privately circulated and only partially published doctoral dissertations, which are largely inaccessible to the majority of non-French workers. One of the primary aims of this book has been simply to bring together this extraordinary store of recently acquired information and to make it more readily available to a wider audience. Beyond this I have of course added my own material and analyses, gleaned in over thirty years of research into Middle Palaeolithic problems, and extending back to the PhD thesis I originally wrote at Cambridge (under the supervision of Charles McBurney) in the early 1960s. In one sense, this book is a personal odyssey to discover exactly what I believe has been learnt about Neanderthal behavioural patterns in the three decades or so since I wrote the original thesis.

It is hard to exaggerate the debt I owe to my French colleagues in sharing their ideas with me, providing information on the latest results of their fieldwork and laboratory analyses and allowing me to reproduce data and illustrations from their own publications. Jean-Philippe Rigaud in particular has been a tower of support, as he has been to many other British and American workers involved in the study of French prehistory. For similar help and cooperation in a variety of ways I am equally indebted to Eric Boëda, Jean-Pierre Chadelle, Jean-Jacques Cleyet-Merle, André Debénath, Françoise Delpech, Pierre-Yves Demars, Christine Duchadeau-Kervazo, Catherine Farizy, Jean-Michel Geneste, Jean-Luc Guadelli, Jacques Jaubert, Guy Laquay, Henri Laville, Michel Lenoir, François Lévêque, Stephane Madelaine, Liliane Meignen, Jacques Pelegrin, Denise de Sonneville-Bordes, Allain Tuffreau, Allain Turq and Bernard Vandermeersch. Without the cooperation of these and other French colleagues the book could never have been written. The same applies, from a slightly different perspective, to the help I have received from a number of north American colleagues involved in closely related studies of the European Middle Palaeolithic – above all to Harold Dibble, Nicholas Rolland, Philip Chase, Art Jelinek and Lewis Binford. Some of them may feel that their interpretations have received rather rough justice in some sections of the book, but I hope they will see the close attention I have paid to their work above all as a genuine mark of respect for their contributions. Lewis Binford in particular, I believe, has been the most positive and inspiring influence on European Palaeolithic studies over the past 30 years, and I am greatly indebted for his generosity not only in discussing his ideas with me at length (most notably during a memorable visit to Albuquerque in the summer of 1992) but also in allowing me to quote from his extremely important unpublished research at Combe Grenal.

The preparation of the book has benefited greatly from the encouragement and guidance of Bill Woodcock and Emily Wilkinson of Princeton University Press, and the artistic skills of John Rodford, who drew or redrew well over half the illustrations in the book. My colleagues in the Archaeology Department at Cambridge have provided various forms of intellectual stimulation – and restraint – over the past 15 years, while a succession of exceptionally bright and enthusiastic undergraduate and graduate students have done even more to stimulate and channel my thinking along new lines. The contributions of Nathan Schlanger, Paul Pettitt and Gilliane Monnier in particular will be apparent from the discussions in Chapters 3 and 4. The great inspiration throughout my own studies at Cambridge was provided by Professor (now Sir) Grahame Clark, and it is a pleasure to acknowledge the debt I owe to him. Two other colleagues at Cambridge – Nick Shackleton and Tjeerd Van Andel – were particularly helpful in reading through drafts of the environmental chapter and offering valuable comments.

Finally, I owe a special debt of gratitude to my wife. Anyone who embarks on writing a lengthy book knows that this demands endless tolerance for lost weekends, late nights, bouts of exasperation and no doubt other strains – backed up by endless promises that the end is just around the corner. As ever, she bore this with tolerance and good humour, and contributed greatly and in many ways to the final product.

Financial support towards the preparation of the book was provided mainly by the British Academy. In addition to travel grants for visits to France in 1990, 1991, 1992, and 1993, the Academy provided the initial stimulus and support for the work which lay behind the book, through the award of a two-year Research Readership from 1989 to 1991. For similar support – and for providing the ideal geographical and intellectual environment in which to write a book – I am indebted to the Master and Fellows of Corpus Christi College.

Paul Mellars
Cambridge
July 1995

To my wife and parents

CHAPTER 1

Introduction _____

The Neanderthals have always been something of an enigma. Since their initial discovery in the middle of the last century opinions have tended to polarize between two extremes: between those who saw the Neanderthals as standing directly astride the main course of human evolution, only slightly different in either their physical or mental capabilities from modern populations; and those who saw them, by contrast, as much more primitive figures, with behavioural and physical capacities radically different from those of later populations and almost certainly representing an extinct side branch of human evolution. According to one viewpoint the Neanderthals were our direct ancestors, while according to the other they were rather distant, and not very respectable, cousins. A spate of characterizations in media cartoons, as well as more thoughtful presentations in popular novels (such as William Golding's *The Inheritors*, and Jean Auel's *The Clan of the Cave Bear*) have served to enhance the mystique and uncertainty surrounding the true role of the Neanderthals in our own evolution.

Research and discoveries over the past ten years have tended to heighten rather than reduce these long-standing controversies over the place of the Neanderthals in human evolution. Recent research in molecular genetics has been interpreted to suggest that the Neanderthals may have made no direct contribution to the genetic ancestry of bio-logically modern populations in Eurasia, and indeed that the Neanderthals as a whole might well represent a separate biological species (Cann *et al.* 1987, 1994; Stoneking & Cann 1989; Stoneking *et al.* 1992; Stringer & Gamble 1993 etc.). Similarly, recent dating of a range of essentially modern anatomical remains at the sites of Skhul and Qafzeh in Israel, and at a number of sites in southern Africa, has shown that forms closely similar to ourselves had already emerged in several parts of the world long before their appearance, in a remarkably sudden and abrupt form, in the more western zones of Eurasia (Stringer & Andrews 1988; Stringer 1990, 1992, 1994; Stringer & Gamble 1993; Bräuer 1989; Vandermeersch 1989). These conclusions have been contested by proponents of the 'regional continuity' or 'multi-regional evolution' school, who argue that the entire framework of both genetic and anatomical evidence which has been used to support the demographic extinction of the Neanderthals is based on rank misrepresentations of the biological evidence, or at best on serious ambiguities in the interpretation of this evidence (Wolpoff 1989, 1992; Wolpoff *et al.* 1994; Thorne & Wolpoff 1992; Smith 1991, 1994). According to them European readers of this book are far more likely to have a strong component of Neanderthal genes in their direct ancestry than genes of a hypothetical intrusive modern population, from some exotic African or Asian source.

1

Similar debates have plagued recent inter-pretations of the archaeological records of Neanderthal behaviour. To many prehistor-ians the archaeological records of the Nean-derthals suggest a pattern of behaviour which is only radically and fundamen-tally different from that of the ensuing bio-logically modern populations but which indicates a fundamentally different structure of mind. Recent characterizations in this vein have suggested that the Neanderthals may have been incapable of hunting most of the larger species of animals; that they formed social groupings which were more akin to the sexually segregated foraging units of most primate communities than the fam-ily-based structure of modern human pop-ulations; that they lacked the capacity for long-range planning or organization of their economic and social activities; and that they almost certainly lacked complex, highly structured language (Binford 1989; Lieber-man 1989; Chase & Dibble 1987; Soffer 1994; Stringer & Gamble 1993). Opposing this view are those who see the general behaviour and cultural capacities of the Neanderthals as only marginally different from those of later populations, with the exception of a few post-Neanderthal embellishments in the form of representational art, more complex forms of bone and antler technology and a general predilection for manufacturing stone tools from more elongated and economical blade forms in preference to larger and heavier flakes (Clark & Lindly 1989; Lindly & Clark 1990; Clark 1992; Hayden 1993). The latter developments, it is argued, are more likely to represent a gradual, cumulative increase in the overall complexity of behavioural pat-terns over the course of later human evolu-tion than a radical transformation in the underlying intellectual and cultural capabil-ities of the populations involved.

So what exactly are the central issues in current studies of the Neanderthals? The question which lies at the heart of the present debate centres on the precise relationships of

the Neanderthal populations of Europe with the ensuing populations of anatomically and behaviourally modern humans, a transition which seems to have taken place in most regions of Europe between ca 40,000 and 35,000 years ago. Specifically, the major issues in this context can be reduced to three critical questions:

1. To what extent, if at all, did the Nean-derthals contribute to the genetic ancestry of later populations in Europe?

2. How far, and in what ways, did the behav-iour of Neanderthal populations contrast with that of the ensuing anatomically and behaviourally modern populations?

3. If we can document major contrasts between the behavioural patterns of Nean-derthal and modern populations, how should these contrasts be explained? Do they reflect simply a gradual, progressive increase in the overall complexity of dif-ferent behavioural systems over the course of time? Or do they represent a much more sudden and radical shift in behavioural patterns, which reflects an equally pro-found shift in the associated mental and cognitive capacities for behaviour of the populations involved?

These questions, addressed primarily to the archaeological records from western Europe, form the central focus and subject matter of the present book.

Who were the Neanderthals?

I shall make no attempt to discuss in any detail here the biological and anatomical fea-tures of the Neanderthals since this has already been dealt with comprehensively in two recent books – *In Search of the Nean-derthals* by Chris Stringer and Clive Gamble (1993), and *The Neanderthals: Changing the Image of Mankind* by Erik Trinkaus and Pat Shipman (1993). Both studies seem to agree that, in at least their major anatomical features, the Neanderthals form a reasonably distinctive and fairly well defined taxonomic grouping, even if the precise geographical and chronological limits of the grouping are more difficult to define. As in other fossil hominids, the most distinctive features of Neanderthal morphology are reflected in the skull and facial regions. In the case of the Neanderthals these include heavily enlarged supraorbital brow ridges, a generally low and flattened cranial vault with a strongly developed occipital 'bun', the heavily built structure of the jaws and teeth, with little trace of a chin, and a surprisingly large cranial capacity of around 1400–1600 cc implying an overall brain volume at least as large as that of modern populations. There seems to be equal agreement that at least some of these distinctive features of Neanderthal morphology can be seen as an adaptation to the specific environmental conditions of the more northern zones of Eurasia during the colder, glacial and sub-glacial episodes of the later Pleistocene. Thus the large noses and the generally inflated form of the facial region as a whole are often seen as an adaptation to accommodate the very large nasal channels that were essential to warm the cold, dry air of these exceptionally harsh climates (Howell 1957; Coon 1962; Wolpoff 1980 but see Trinkaus 1989b for a different view). Similarly, the generally short, heavy body structure typical of the Neanderthals is usually seen not only as an adaptation to a very active and strenuous life style which demanded considerable physical strength, but also as an adaptation to conserve body heat in severe, seasonally fluctuating climates (Trinkaus 1983, 1989b; Trinkaus & Shipman 1993; Smith 1991). Whether all the distinctive features of Neanderthal morphology can be explained in these terms is more controversial. Smith and others, for example, have suggested that the large and heavily built form of the Neanderthal face may have been related more to the stresses involved in

the habitual use of jaws and teeth for various 'paramasticatory' activities (i.e. using the jaws as tools) than to any cold-climatic adaptation (Smith & Paquette 1989).

To put exact limits on the Neanderthals in a time and space framework is more difficult. However, there seems to be reasonable agreement that most of the distinctive features of Neanderthal anatomy can be traced across a broad arc of Europe and western Asia, extending from the Atlantic coasts of France and the Iberian peninsula to the western parts of the Middle East and central Asia – for example at Tabun, Amud and Kebara in Israel, at Shanidar in Iraq, and as far eastwards as Teshik Tash in Uzbekistan (Smith 1991; Stringer & Gamble 1993; Trinkaus & Shipman 1993). Whether anything distinctively Neanderthal can be identified to the south of this zone (for example, some of the North African fossils such as those from Jebel Irhoud and Dar-es-Soltan in Morocco) remains more controversial. The main geographical range corresponds, in other words, to the more western zones of Eurasia, and predominantly to those areas which experienced recurrent episodes of sharply colder climate during the middle and later stages of the Pleistocene.

In a chronological sense, most of the well dated and 'classic' Neanderthal forms belong to the earlier stages of the last glacial period, between ca 110,000 and 35,000 BP. Both Trinkaus & Shipman (1993) and Stringer & Gamble (1993), however, have argued that many distinctive Neanderthal traits can be traced back into the period of the penultimate glacial and perhaps, as for example in the remains from Biache-Saint-Vaast in northern France, to the period of isotope stage 7, around 200–250,000 BP. Earlier forms such as the hominids from Swanscombe in England, Steinheim in Germany, Petralona in Greece and Tautavel in France tend to be regarded as 'pre' or 'proto' Neanderthal forms, anatomically transitional between the late *Homo erectus/Homo heidelbergensis* populations of

Europe and the succeeding Neanderthal forms (Stringer *et al.* 1984; Stringer & Andrews 1988; Stringer & Gamble 1993). Clearly, if one accepts that the Neanderthals are indeed the direct descendants of the earlier *erectus* populations in Europe it would be unrealistic to expect to recognize a sharp line of demarcation between these taxa in anatomical or evolutionary terms.

Similar problems are encountered in attempts to define the precise chronological limits of what conventionally has been defined as the 'Middle Palaeolithic' phase of technological development. It should be stressed that it is not the aim of this book to look at the problem of the technological origins or emergence of Middle Palaeolithic technology from the earlier patterns of Lower Palaeolithic technology. Any serious study of this question would not only require a book in its own right but would be seriously handicapped by the extremely patchy, coarse-grained and above all very poorly dated records of human technological development prior to the last 250,000 years. As a working definition I am happy to conform to what has now become the conventional practice of regarding the prime hallmark of Middle Palaeolithic technology as being the emergence of more complex and sophisticated patterns of prepared-core flaking, classically illustrated by the various Levallois and allied techniques discussed in Chapter 2. As several workers have recently stressed, the emergence of these techniques could be seen as a major turning point not only in a purely technological sense but also as a potential watershed in the whole conceptual and cognitive basis of lithic technology, implying a much greater degree of forward planning, time depth, and capacity for strategic problem-solving in the working of stone resources (Roebroeks *et al.* 1988; Rolland 1990; Mellars 1991; Klein 1989a). Whether or not this viewpoint is accepted, it is now clear that remarkably complex and varied forms of prepared-core techniques were being practised

in several parts of the Old World by the time of the penultimate interglacial period between ca 200,000 and 250,000 BP (for example at Biache-Saint-Vaast and the Grotte Vaufrey in France and Maastricht-Belvédère in Holland) and probably at roughly the same time at a range of sites in western Asia and Africa (see Chapter 4, and Klein 1989a; Bar-Yosef 1992). Whether it is entirely coincidental that these complex stone-working techniques appeared in Europe at roughly the same time as the rapid increase in cranial capacities which is one of the most distinctive features of the Neanderthals, is an interesting point for speculation.

The chronological scope of this book therefore coincides essentially with the period from the final disappearance of the Neanderthals in western Europe around 30–35,000 years ago to the period of oxygen-isotope stage 7 around 250,000 BP. The surviving archaeological records of human behaviour within this time range are very unevenly distributed and are far more abundant, more fully documented and more chronologically fine-grained during the later stages of the Middle Palaeolithic than during its earlier stages. The book should therefore be seen primarily as a study of the archaeological evidence of human behaviour during the earlier stages of the last glacial period, between ca 115,0000 and 35,000 years ago. It is this period for which I will reserve the term 'Mousterian' throughout the book. In my view it is only for this period that we have a sufficient quantity of well documented evidence, and sufficient control over the quality and resolution of this evidence, to present any really secure and convincing reconstructions of human behavioural patterns within the Neanderthal time range.

The archaeological perspective

It soon became clear in planning this book that to attempt a general survey of the archaeological evidence for Neanderthal behaviour across the whole geographical range discussed above – i.e. extending from the Atlantic coast of Europe to the Middle East – would not only be a daunting task in terms of the amount of material and data to be considered, but could become a rather questionable exercise. There are two principal reasons for this. First, there is the problem of knowing exactly where in the archaeological records of western Eurasia we see the products of Neanderthal populations. One of the most significant facts to emerge during the last decade is that we can no longer assume that all archaeological assemblages conventionally classified as 'Middle Palaeolithic' in a purely technological sense were indeed the products of Neanderthals, or indeed other archaic forms of hominids. The discoveries at both Skhul and Qafzeh in Israel and possibly at Staroselje in the Crimea have demonstrated that in certain contexts technologically Middle Palaeolithic industries were produced by hominids who in most anatomical respects were much closer to biologically modern populations than to Neanderthals – and at the remarkably early date of around 100,000 BP (Vandermeersch 1989; Bar-Yosef 1992; Stringer & Gamble 1993). Even in parts of central and eastern Europe there is still debate as to whether some of the skeletal remains recovered from Middle Palaeolithic contexts (such as Krapina in Croatia, or Kulna in Moravia) can be confidently attributed to Neanderthals as opposed to anatomically modern populations (Smith 1984, 1991). These observations raise critical questions concerning the ultimate relationships between Neanderthal and anatomically modern populations in these areas, which will be pursued further in the final chapter of this book. The fact remains, however, that it is only in the extreme western zones of Europe that we have a sufficiently large, well documented and consistent association between technologically 'Middle Palaeolithic' industries and taxonomically 'Neanderthal' remains to make any reasonably confident

correlations between the archaeological and skeletal records. While we can never be sure that every Middle Palaeolithic industry in western Europe was produced by a Neanderthal, we can be far more confident in making this correlation in this region than in any other part of Eurasia.

My second reason for choosing a more restricted geographical focus for this book is more pragmatic and relates to the way in which we approach the analysis and interpretation of the archaeological evidence. The importance of adopting a specifically regional approach in this context is now widely recognized (e.g. Gamble 1984, 1986). It would make little sense, for example, to make direct comparisons between the patterns of animal exploitation or the relative frequencies of different species of animals exploited, even in two areas as geographically close as southwestern France and northern Spain, where the overall patterns of climate, vegetation, topography etc. are likely to have been significantly different. The same applies with equal force to comparisons between, say, the relative use of cave versus open-air sites in different areas, the overall patterns of settlement and mobility of the human groups, or the patterns of procurement and distribution of different raw materials (Féblot-Augustins 1993). Only by adopting a regional approach to these issues can we hope to provide a coherent reconstruction of the detailed patterns of behavioural adaptation of Neanderthal communities to the rapidly oscillating climatic and ecological conditions of the later Pleistocene. The choice in the present context was therefore fairly stark: either to focus the present study primarily on the evidence from one specific and well documented region; or to range more widely, and inevitably more superficially, over the evidence from many different regions and run the risk of failing to deal adequately with the situation in any one region.

When viewed in these terms the rationale

for focusing primarily on the evidence from western Europe, and largely on the evidence from the so-called 'classic' region of south-western France, is self-evident. This is not merely the classic region for Middle Palaeolithic studies in a purely historical sense (e.g. Lartet & Christy 1864; de Mortillet 1869, 1883) but it has also produced a wealth and concentration of hard archaeological evidence for Middle Palaeolithic behavioural patterns which, by any criteria, is much richer, more detailed and better documented than that from any other area of comparable size in either Europe or Asia. From south-western France alone we now have a total of over fifty cave and rock shelter sites with relatively substantial and well documented evidence for Middle Palaeolithic occupation, in many cases in the form of long and detailed occupation sequences. The majority of these sites have produced rich and well preserved faunal assemblages, and from many of them we now have detailed information not only on the technological features of the stone-tool assemblages but also on the character and geological sources of the raw materials from which they were made (Geneste 1985, 1989a; Turq 1989; 1992b). All this research has been combined with a meticulous concern with the geological and climatic associations of the human occupations, reflected in a wealth of published data on the sedimentological, palaeobotanical and faunal associations of the different sites.

The point of these observations, needless to say, is not to imply that the patterns of Neanderthal behaviour in the extreme western zones of Europe are inherently more important or more interesting than those in other regions of Eurasia, still less to suggest that any patterns or regularities observed in this region can be extrapolated automatically and uncritically to other regions. Whether such extrapolations can be made must remain one of the prime targets of future research. The point to be emphasized is that the archaeological and behavioural records from these rich and well studied regions of western Europe cover an impressive span of time and reflect the behaviour and adaptations of Neanderthal populations in sharply contrasting climatic and environmental conditions. If we can discern any general patterns or regularities in behavioural patterns over this time range, these should not be dismissed lightly. Wherever possible and appropriate in the following chapters I have attempted to make specific comparisons with the evidence from other regions of Europe, especially where these seem to hint at a pattern of behaviour significantly different from that reflected in the western European data. Overall, however, the arguments for focusing any study of this kind primarily on the evidence from one specific and well studied region are largely self evident and I make no apology for choosing the region where the available archaeological records of Neanderthal behaviour are exceptionally abundant and well documented. Given the specific theoretical orientation of this study, there is also something to be said for choosing an area where archaeological records for the behaviour of the succeeding anatomically modern populations are equally rich and clearly defined.

Theoretical perspectives

My aim in this book is not to adopt any specific *a priori* theoretical stance towards the analysis and interpretation of the archaeological record beyond what I would describe as a simple 'rationalist' one – i.e. to focus on a number of specific, clearly defined questions relating to particular aspects of Neanderthal behaviour and to evaluate a range of alternative answers to these questions from whichever aspects of the archaeological record seem most directly relevant. There are, nevertheless, a number of themes which run through many of the chapters and which should be recognized at the outset.

First, it is clear that much of the following discussion will be seen in many quarters as an explicit reaction against some of the more extreme 'hyperfunctionalist' approaches which have dominated much of the literature on the Middle Palaeolithic over the past two decades – as I would see it largely as an overzealous extrapolation of the more modest forms of functionalism which characterized the 'processualist' archaeology of the 1960s and 1970s. As a product and great admirer of many aspects of the New Archaeology of the mid-1960s, I regard myself, in many respects, as a committed processualist, in the sense that I believe that a close analysis of the potential functional interrelationships between the different aspects of behavioural patterns represents the obvious starting point for any productive analysis of human behaviour in the past. I also admire the fact that functionalist or processualist approaches place the primary emphasis on those aspects of human behaviour and adaptation which leave the most direct traces in the archaeological record – such as technology, subsistence, settlement patterns, human-environment relationships etc. In these respects I believe that a broadly 'functionalist' viewpoint forms a natural and logical starting point for any constructive study of human behavioural patterns and adaptations in the past.

What disturbs me is the lengths to which functionalist interpretations have been carried in some of the studies of Neanderthal behaviour over the past twenty years. I detect an indication that the pursuit of functional interpretations has become not merely a reasonable point of departure for the analysis of behavioural patterns, but something approaching an *a priori credo* to be pursued and defended against almost any alternative interpretation. As several authors have recently pointed out, the limitation of this approach is not merely that it largely excludes the possibility of any explicitly cultural component in the behaviour of earlier Palaeolithic communities but that it virtually ignores the entire dimension of socially constructed behaviour and learning processes which must inevitably have shaped the behavioural patterns of all communities in the past, extending back to our primate origins (e.g. Mithen 1994; Mellars 1992a, 1994; Bar-Yosef & Meignen 1992; McGrew 1993). These and related issues are discussed more fully in Chapter 10.

The second point is to some extent interrelated with the first and concerns the importance of viewing the organization and behaviour of past human communities within a demographic perspective, that is, in terms of networks of communities linked by varying degrees of social interraction. This approach assumes in other words that human populations do not form uniform, homogeneous networks extending indefinitely across the landscape, but that they are divided into smaller and more restricted social and breeding units which interact in specific and often closely prescribed ways (Wobst 1974, 1976; Dunbar 1987; Foley & Lee 1989; Rodseth *et al.* 1991). This approach to the analysis of demographic and behavioural patterns is commonplace in studies of other animal and primate communities but has so far received remarkably little attention in studies of the behaviour and organization of Palaeolithic communities. As discussed further in Chapter 10, this question has direct implications not only for understanding how Neanderthal communities may have been organized in purely *social* terms but also for issues such as the viability of these communities in conditions of rapid climatic and ecological change, as well as for any associated population shifts or displacements during these periods of environmental or demographic change. All these questions have direct and important implications for studies of Neanderthal technological patterns, as the discussions in Chapter 10 will attempt to make clear.

The third dimension which I believe has

been surprisingly neglected in earlier studies
of the Middle Palaeolithic is the question of
changes in behavioural and adaptive patterns over the course of time. Even if we
define the Middle Palaeolithic in its most
restricted sense (i.e. as coinciding with the
earlier stages of the last glaciation) it is now
clear that this covers a span of at least 80,000
years – that is almost three times as long as
the entire time-span of the Upper Palaeolithic
succession in Europe. If we extend the Middle Palaeolithic to include all the manifestations of Levallois and related prepared core
technologies reaching back to the time of
isotope stage 7, the time range expands to
over 200,000 years. It is now clear that this
period witnessed some of the most rapid and
extreme shifts in climatic and ecological patterns during the whole of the Pleistocene
period, with conditions in western Europe
shifting repeatedly from dense deciduous
forest to almost treeless tundra or steppe (e.g.
Zagwijn 1990; Behre 1990; Jones & Keen 1993;
Dansgaard *et al*. 1993). It is inconceivable that
this period could have passed without radical shifts or adaptations in both human technological patterns and in the specific geographical and ecological ranges occupied by
different human groups. Admittedly, in the
past, studies of these patterns of technological and ecological adaptation have been
seriously hampered by the long-running
debates over the relative and absolute chronology of many of the most important Middle
Palaeolithic sequences in western Europe.
Nevertheless, the importance of understanding and controlling the chronological dimension of human behavioural patterns is no less
self-evident in the Middle Palaeolithic than
in any of the later periods of prehistory. The
attempt to ignore or side-step this dimension
of human behavioural patterns has, I believe,

led to some of the more serious errors and
confusions in studies of the Middle Palaeolithic archaeological record during the last
fifty years.

The final theme of the book is that the true
picture of Neanderthal behaviour, and
behavioural capacities, almost inevitably lies
somewhere between the two positions
described at the start of this introduction.
There seems to be an irresistible urge to
polarize scientific debate into extreme positions. The truth is rarely that simple. I believe
that there is ample evidence that many of the
most basic behavioural patterns of the Neanderthals were significantly different from
those of the ensuing anatomically modern
populations, that the cognitive capacities of
the Neanderthals for certain behavioural patterns *may* have been different from our own
(though this remains difficult to prove) and
that, at least in the western zones of Europe,
the Neanderthals probably did become
extinct (Mellars 1989a,b, 1991, 1992a). But
this is a far cry from some of the more
extreme scenarios which see the Neanderthals as having virtually no capacity for strategic or symbolic thought, no language and
social patterns which were more akin to those
of the great apes than to behaviourally and
anatomically modern populations. I prefer
not to see my own position as reflecting 'a
typically British spirit of compromise' (Haddenham 1980: 48), but as adopting a more
realistic approach to the behaviour and cognitive capacity of populations who had
brains just as large as ours and who were the
product of at least five million years of
intense socially and ecologically competitive
evolution from their closest primate ancestors. If this is seen as a characteristically
Anglo-Saxon compromise, so be it!

CHAPTER 2

The Environmental Background to Middle Palaeolithic Occupation _____

The Middle Palaeolithic spans a period of dramatic climatic and ecological change, coinciding broadly with the later stages of the Pleistocene period. The Pleistocene period as a whole was characterized by a long succession of climatic oscillations in which conditions shifted repeatedly between periods of very cold, 'glacial' climate, leading to major expansions of the ice sheets in the northern and southern hemispheres, and intervening 'interglacial' episodes in which the climate returned to conditions broadly similar to those of the present day. In all, at least ten of these glacial/interglacial cycles can now be documented during the past one million years. Exactly what triggered these major changes in world-wide climate is still the subject of lively debate amongst climatic specialists. Most climatologists, however, now subscribe to a version of the Milankovitch 'astronomical hypothesis' in which subtle but significant variations in the pattern of the earth's rotation around the sun, with regular and predictable cycles of around 23,000, 41,000 and 100,000 years, are assumed to have provided the initial stimulus for the majority of documented climatic shifts (see Imbrie *et al*. 1989, 1992; Bradley 1985). A clear understanding of these climatic changes and their effects on vegetation patterns, animal populations, annual temperature regimes etc. is an essential prerequisite to any study of the patterns of human occupation by Neanderthals in Europe.

The period which is of most concern to this study is that of the last glacial and interglacial cycle which spans the past 130,000 years. Most of the evidence for climatic and environmental conditions over this time range comes from two main sources: first, from studies of the changing oxygen-isotope ($^{18}O/^{16}O$) ratios in deep-sea sediments (Fig. 2.1), which provide a fairly direct record of the total quantity of sea water contained in the world's oceans and therefore of the total amount of water locked up on land in the form of ice sheets (see Shackleton 1977, 1987; Bradley 1985); and second, from some of the remarkably long and detailed pollen sequences recently documented from a number of sites in western and southern Europe, such as La Grande Pile and Les Echets in eastern France (Figs 2.6, 2.7), Tenaghi Phillipon in Greece, Valle di Castiglione in Italy and Padul in southern Spain (Woillard & Mook 1982; de Beaulieu & Reille 1984 etc.). Other important sources of information have come from studies of some of the shorter pollen sequences recorded from sites in northern Europe (Brørup, Amersfoort, Hengelo etc.) (Fig. 2.9), from studies of the changing surface temperatures of the world's oceans based on studies of the varying frequencies of particular planktonic species of marine organisms preserved in sea-bed sediments (Fig. 2.11) and from more general records of climatic and temperature changes derived from studies of Arctic and Antarctic ice cores

9

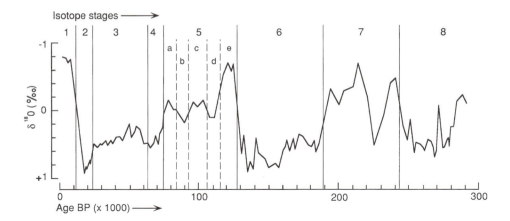

Figure 2.1 *Pattern of oxygen-isotope ($^{18}O/^{16}O$) fluctuations in deep-sea sediments over the past 300,000 years, based on studies of five separate ocean cores, with the divisions into major climatic stages indicated. The high points in the curve reflect warm periods with reduced glaciation, while the troughs represent glacial episodes, when the lighter ^{16}O isotope was selectively removed from the oceans (in the form of water vapour) to form continental ice sheets. Stage 5e represents the peak of the last interglacial. The time scale is derived from the 'orbital forcing' calculations of Martinson* et al. *1987.*

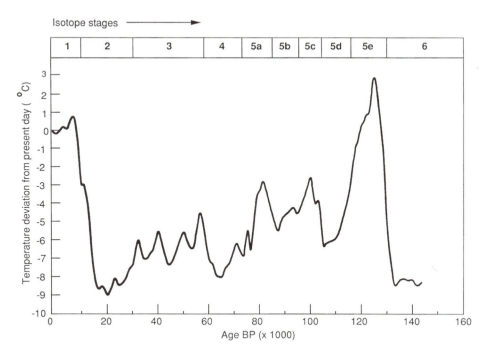

Figure 2.2 *Temperature variations over the past 140,000 years estimated from studies of oxygen-isotope and deuterium ratios in the Voztok ice core from eastern Antarctica. The inferred correlation with deep-sea core isotropic stages is shown at the top of the graph. After Jouzel* et al. *1987.*

(Figs 2.2, 2.14) (Zagwijn 1990; Behre 1990; Behre & Plicht 1992; Bond et al. 1993; Dansgaard et al. 1993; Jouzel et al. 1987). Through a combination of these different lines of evidence we can now present a picture of changing climatic and environmental conditions during the later stages of the Pleistocene period with a level of detail and clarity which would have been unimaginable twenty, or even ten, years ago.

In the following sections attention will be focused first on some of the more general patterns of climatic and environmental change which can be documented in western Europe over the course of the later Pleistocene period, and second, on some of the more specific evidence from sites in the region which will provide the main case study throughout the remainder of this book, i.e. the Perigord region and adjacent areas of southwestern France.

The last interglacial period

The last interglacial period extending from ca 126,000 to 118,000 BP is one of the most intensively studied and best documented periods of the Pleistocene with relatively detailed records available from deep-sea cores, ice cores, high sea-level stands and many detailed pollen sequences (Watts 1988; Bradley 1985; Bowen 1978; West 1977; Shackleton 1969, 1977; LIGA 1991a; Sejrup & Larsen 1991). The period is known by different names in different parts of Europe – the 'Eemian' in northern Europe, the 'Ipswichian' in Britain, the 'Riss-Würm' in France, and the 'Mikulino' in Russia. In the isotope records of deep-sea cores (Fig. 2.1) it is now generally agreed that the last interglacial in the strict sense should be restricted to the period of isotope stage 5e (Shackleton 1987; Chappell & Shackleton 1986; Watts 1988; LIGA 1991a; Sejrup & Larsen 1991).

Perhaps the most striking feature is the abruptness with which the last interglacial started. Immediately preceding the interglacial there was a period of exceptionally cold climate (represented by isotope stage 6 in the deep-sea core records) during which temperature conditions probably fell to levels similar to those attained at the peak of the last glaciation (around 18,000 BP) and the continental ice sheets expanded to almost their maximal positions in northern Europe. The abrupt termination of this glaciation seems to have been caused by a sharp increase in solar radiation at around 130,000 BP. Within a period of ca 5000 years year-round temperatures rose by at least 10–15°C, the ice sheets retreated to near to their present positions and world-wide sea levels rose to 5–6 metres higher than those of today (Fig. 2.17) (Martinson et al. 1987; Chappell & Shackleton 1986; Imbrie et al. 1989; LIGA 1991a).

The most detailed records of climatic and ecological conditions during the last interglacial are provided by the numerous pollen sequences which have been recorded from well over a hundred separate locations in different parts of northern, western and eastern Europe (see for example Watts 1988; Zagwijn 1990; Woillard 1978; de Beaulieu & Reille 1984; Bowen 1978; West 1977; LIGA 1991a). Most of these show a similar pattern of forest development, with only minor variations in different parts of the continent (see Figs 2.3, 2.7). Typically, the vegetational sequences commence with a brief period of birch and pine dominated forest during the opening stages of the interglacial (the so-called 'pre-temperate' or 'protocratic' phase) which is then followed by a consistent succession of warmth-demanding deciduous species ranging successively from elm and oak, through to alder, hazel, yew and hornbeam. During the later stages of the interglacial there was a reversion to more coniferous species such as spruce, silver fir and pine. Only during the very earliest and latest stages of the interglacial would there have been any significant areas of open vegetation to break the monolithic dominance of these dense forests which characterized the

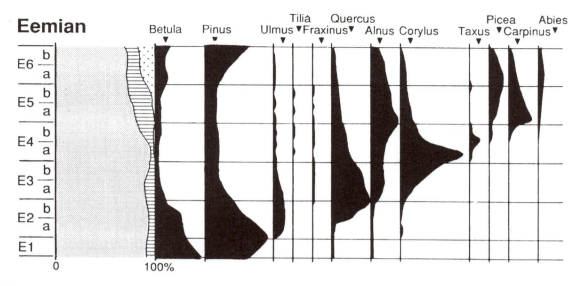

Figure 2.3 *Typical vegetational succession recorded in pollen sequences spanning the last interglacial period ('Eemian' interglacial) in northern Europe. Similar patterns of vegetational development are recorded over large areas of Europe during this period. The Eemian interglacial is generally agreed to correlate with isotope stage 5e, from ca 126–118,000 BP. After Zagwijn 1990.*

whole of the last interglacial throughout most regions of Europe.

Exactly how climatic conditions varied during the interglacial is still open to some dispute. Most authors are agreed that the initial rise in temperatures at the start of the interglacial must have been very rapid and that the warmest temperatures were probably attained during the earlier stages of the interglacial, around 120–125,000 BP (Fig. 2.4) (Watts 1988). At that time, world-wide sea levels seem to have risen to at least 5–6 metres above present-day levels (implying a significant reduction in the total volume of the continental ice sheets) and the temperature records in the Arctic and Antarctic ice cores suggest that average year-round temperatures were at least 2–3°C higher than those in the same regions today (Figs 2.2, 2.14, 2.17) (Shackleton 1987; Jouzel *et al.* 1987; Lorius *et al.* 1985; Dansgaard *et al.* 1993; LIGA 1991a; Sejrup & Larsen 1991). There is equally

strong evidence for a clear climatic optimum at this time in the vegetational records of northern Europe (Fig. 2.3). As Watts (1988) points out, several species such as *Stratiotes* (water aloe) and *Hydrocharis* (frog bit) extended further to the north during the peak of the Eemian interglacial than at any time during the present postglacial period and there is evidence for the presence of such semi-tropical species as hippopotamus and the European pond tortoise in the faunal records of northern Europe (see also West 1977; Stuart 1982; Bowen 1978; Zagwijn 1990; LIGA 1991a). Watts argues that the earlier stages of the interglacial were probably characterized by a relatively continental pattern of climate, (i.e. with substantially warmer summers than at present and rather cooler winters) which became more oceanic, with less marked seasonal contrast, during the middle and later stages of the interglacial.

Most of the debate in the literature has

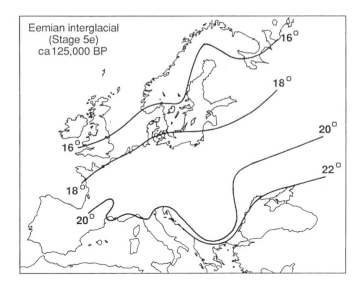

Figure 2.4 *Estimated mid-summer temperatures across Europe at the peak of the last (Eemian) interglacial, reconstructed from vegetational data by Zagwijn 1990.*

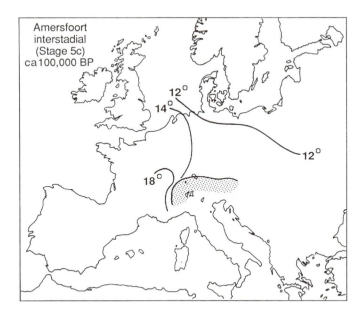

Figure 2.5 *Estimated mid-summer temperatures in Europe at the time of the Amersfoort interstadial (= isotope stage 5c, ca 100,000 BP) as reconstructed from vegetational data by Zagwijn 1990. Note the much sharper temperature gradients from southwest to northeast than those during the Eemian interglacial (Fig. 2.4), presumably due to the presence of an ice sheet over northern Scandinavia.*

centred on the precise causes of the changes in vegetational patterns during the closing stages of the interglacial – the so-called 'post-temperate' or 'telocratic' phases (Watts 1988). As noted above, this period is characterized by an increase in the frequencies of various coniferous species of pine, spruce and silver fir and by a general expansion in more open vegetation at the expense of closed woodland (see Fig. 2.3). One early suggestion put forward by Iversen (1958) and others was that these changes could have been caused purely by progressive deterioration in soil conditions during the course of the interglacial rather than by any significant change in climate. It now seems more likely, however, that there was indeed a significant deterioration in climatic conditions (at least in summer temperatures) during the later part of the interglacial (Watts 1988), as reflected, for example, in the temperature records of ice cores, in the gradual increase in global ice volumes reflected in oceanic oxygen–isotope ratios and in the various faunal indications of sea-surface temperatures in the North Atlantic (Figs 2.1, 2.2, 2.11, 2.14) (Jouzel et al. 1987; Imbrie et al. 1989; Sancetta et al. 1973). A similar deterioration in climate during the later stages of the interglacial also seems predicted from theoretical models of solar radiation implied by the Milankovitch hypothesis (Watts 1988). Thus, all evidence suggests that the last interglacial period came to an end in a much more gradual way than it began and was characterized by a progressive shift towards increasingly glacial conditions rather than by any sudden climatic event.

The Early Glacial period, ca 118,000–75,000 BP

It is now clear that the earlier stages of the last glacial period were characterized by a complex but clearly defined pattern of climatic oscillations. In the isotope records of deep-sea cores (Fig. 2.1) this period comprised four main stages, of which two, stages 5d and 5b, were quite clearly glacial in character while the other two, 5c and 5a, were much warmer 'interstadial' episodes marked by a sharp increase in world-wide temperature conditions and a corresponding reduction in the overall extent of glaciation (Shackleton 1977, 1987). Collectively, this sequence of major climatic fluctuations occupies a period of around 40,000 years, from ca 118,000 to ca 75,000 BP.

The character of climatic and ecological conditions during the so-called 'early glacial' or 'early Weichselian' is now well documented in many parts of western and central Europe from the long and continuous pollen sequences recorded at La Grande Pile (Fig. 2.6), Les Echets (Fig. 2.7), Padul, Tenaghi Pillipon, Valle di Castiglione and several other localities (Woillard 1978; de Beaulieu & Reille 1984; Reille & de Beaulieu 1990; Florschuz et al. 1971; Wyjmstra et al. 1990; Follieri et al. 1989). Following much debate there is now effective agreement as to how these sequences should be correlated with the oxygen–isotope records in ocean cores. The two major warm periods represented in the Grande Pile and Les Echets diagrams (i.e. St Germain interstadials I and II) clearly correlate with the major periods of deglaciation represented in isotope stages 5c and 5a, while the intervening colder periods (Melisey stages I and II) correlate with the major phases of glacial advance in stages 5d and 5b (Fig. 2.6) (Woillard & Mook 1982; de Beaulieu & Reille 1984; Zagwijn 1990; Behre 1990; Behre & Plicht 1992; LIGA 1991a).

Support for this correlation has been provided by the analysis of sea-bed sediments just to the west of the Spanish coast, where a characteristic series of pollen-zone assemblages can be correlated directly with parallel fluctuations of the $^{18}O/^{16}O$ record of the deep-sea sediments themselves (Turon 1984). In the more northern parts of Europe there is now equal agreement that the well known

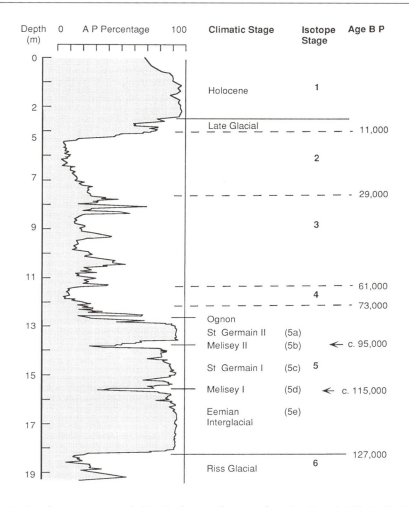

Figure 2.6 *Vegetational sequence recorded in the long pollen core from La Grande Pile in the Vosges region of northeastern France, covering the period of the last interglacial and glacial cycle. The dates for the upper part of the sequence (post 70,000 BP) are based on radiocarbon measurements, while those for the lower part are based on the inferred correlations with the isotopic stages in deep-sea cores. The names given to the major stadial and interstadial phases in the sequence are indicated in the central column. After Woillard & Mook 1982.*

succession of Amersfoort, Brørup and Odderade interstadials can be correlated closely with the vegetational sequences at La Grande Pile, Les Echets and elsewhere, with the Amersfoort and Brørup intervals corresponding essentially with the earlier and later parts of the St Germain I interstadial (see below) and the Odderade interstadial

with the major warming during St Germain II (Figs 2.9, 2.10) (Zagwijn 1990, de Beaulieu & Reille 1984; Behre 1990; Behre & Plicht 1992). Several interstadial deposits recorded in southern Britain (at Chelford, Wretton and elsewhere) most probably correlate with the period of St Germain I (Bowen 1990; Simpson & West 1958; West *et al.* 1974).

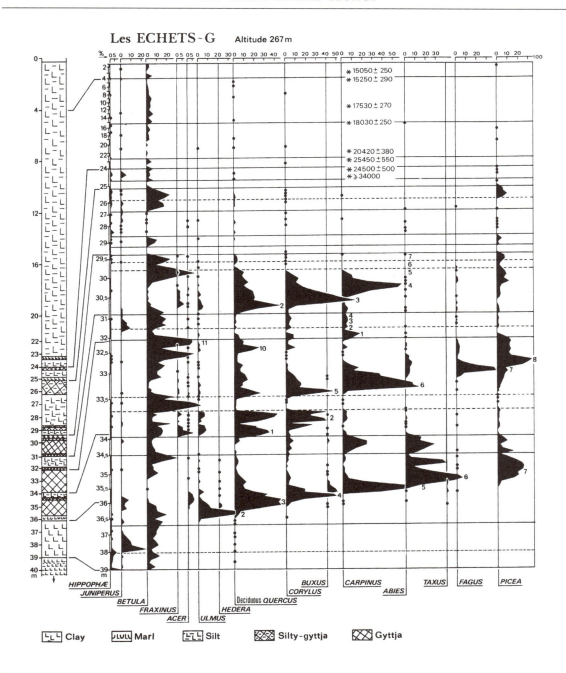

Figure 2.7 *Vegetational sequence covering the last glacial/interglacial cycle recorded at Les Echets in the northern Rhône valley (eastern France). Correlations with the main stadial and interstadial phases defined in the Grande Pile sequence (Fig. 2.6) are shown on the right. After de Beaulieu & Reille 1984.*

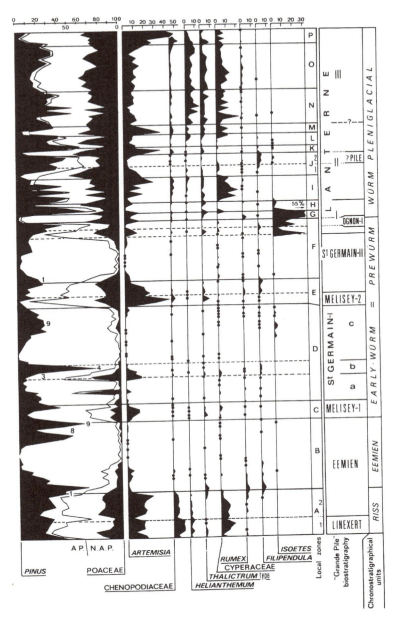

Figure 2.7
continued

On the basis of this information we can present a reasonably detailed reconstruction of climatic and ecological fluctuations over this time range, covering a wide area of western and central Europe (see for example Behre 1990; Zagwijn 1990; LIGA 1991a; Sejrup & Larsen 1991; Behre & Plicht 1992).

There is now no doubt that the two major cold periods, i.e. isotope stages 5d and 5b, were glacial in character, in the sense of involving a major expansion of the continental ice sheets from their previous interglacial limits (Shackleton 1977, 1987) (Fig. 2.1). The oxygen–isotope records leave no

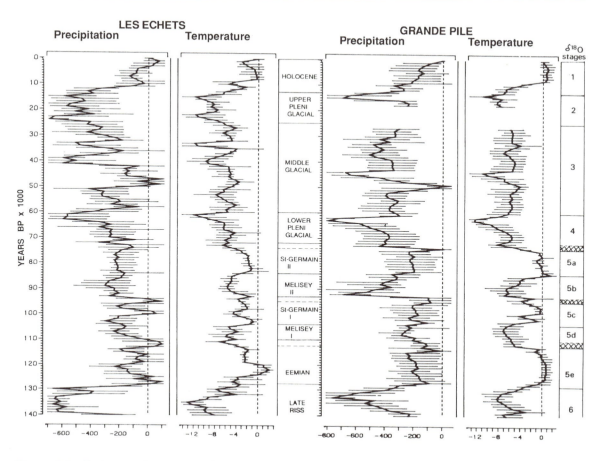

Figure 2.8 *Estimates of mean annual temperatures and precipitation over the past 140,000 years derived from climatic analyses of the vegetational patterns in the Les Echets and La Grande Pile pollen cores in eastern France (see Figs 2.6, 2.7). The main climatic phases defined in the Grande Pile sequence are shown in the central column; correlations with isotope stages in deep-sea cores are shown on the right. The horizontal bars indicate the confidence limits of the temperature and precipitation estimates. After Guiot* et al. *1989, 1993.*

doubt that in global terms the extent of the ice sheets formed during these periods must have attained almost half of the volume of those attained during the period of the last glacial maximum, though of course the growth of these ice sheets could have taken place at rather different rates in different areas of the world – with the growth of the North American ice sheets perhaps preceding by several thousand years those in northern Europe. In northern Europe there is still no clear evidence for the total extent of gla-

ciation during either of these periods, since the direct geological evidence for these ice advances has been obscured by later glaciations in the same areas. The pattern of sharp climatic gradients recorded from north to south and to some extent from east to west across the continent, however, (see below) leaves little doubt that some very substantial glaciation must have built up in the northern parts of Europe during both stages 5d and 5b (Zagwijn 1990; Behre 1990: Sejrup & Larsen 1991).

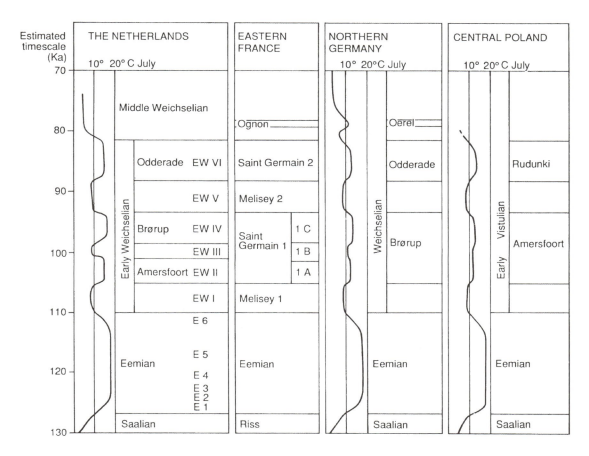

Figure 2.9 *Principal stadial and interstadial periods recognized in different areas of northern and western Europe during the earlier part of the last glaciation, with associated summer temperature estimates. After Zagwijn 1990.*

The essentially periglacial character of the climate during these two periods is indicated equally clearly by the composition of the pollen records from many parts of northern, western and central Europe (Figs 2.6, 2.7, 2.10). In pollen diagrams ranging from southern Denmark to the Rhone valley the vegetation during both stages 5d and 5b seems to have been open in character, characterized predominantly either by grasses, sedges or other tundra-like plants, or, further to the south and east, by high frequencies of more steppic species such as *Artemisia, Thalictrum* and *Chenopodiaceae* (Zagwijn 1990; Behre 1990). Nevertheless, in all these areas there is evidence that some localized patches of more hardy trees such as pine, birch and willow were able to survive during these periods, suggesting that climatic conditions never became quite as severe as those experienced during the later full glacial periods of isotope stages 2 and 4. The vegetation over the different zones of northern and western Europe during both stages 5d and 5b has been char-

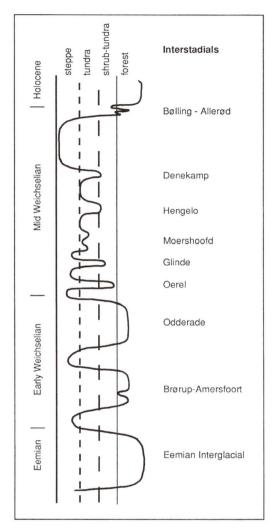

Figure 2.10 *Reconstructed vegetational sequence throughout the last interglacial/glacial cycle in Holland and northern Germany, based on combined evidence from several sites. The major interstadials recognized in this area are shown on the right. After Behre 1990.*

acterized by Zagwijn (1990: 62) as essentially 'open with patches of birch-pine forest', though with an increasing component of woodland further to the south and east across the continent.

The precise implications of these vegetational patterns in climatic terms are more difficult to reconstruct. Zagwijn (1990) has suggested that over a zone of northern Europe extending from the Netherlands through northern Germany to central Poland, mid-summer temperatures during both stages 5d and 5b can hardly have risen above ca 8°C – implying an overall depression in temperatures (compared with those of the present day in the same regions) of at least 8–10°C (Fig. 2.9). Further to the south, the scale of temperature reduction seems to have been much less. From various multivariate analyses of vegetational patterns at La Grande Pile and Les Echets, Guiot *et al.* (1989) have estimated an average depression in year-round temperatures during these two periods of around 5–6°C (Fig. 2.8). As noted earlier, these and other lines of evidence indicate a much sharper pattern of climatic gradients from north to south and from east to west across Europe than those recorded during the preceding interglacial period (Sejrup & Larsen 1991).

Exactly what caused these episodes of sharp climatic deterioration during the earlier stages of the last glacial period is still a matter for debate. Almost certainly, a major part of this climatic cooling was due to a sharp reduction in the intensity of solar radiation around 118,000 BP as the Milankovitch hypothesis would predict (Martinson *et al.* 1987; Imbrie *et al.* 1989; LIGA 1991a,b). But in the more northern zones of Europe there was probably a major feedback factor in the accumulation of major ice masses in the Scandinavian region, combined with a major southward extension in the flow of polar waters from the north Atlantic region, into the central Atlantic zone (Zagwijn 1990). According to the reconstructions by Ruddiman and MacIntyre (1976; MacIntyre *et al.* 1972, 1975), this polar front extended southwards to the latitude of ca 52°N (i.e. approximately to the latitude of southern England) during both stages 5d and 5b (Fig. 2.12), deflecting the warmer waters of the Gulf Stream far to the south, and providing a further source of extremely cold air masses to much of the northern and western zones of Europe (Lamb 1977; LIGA 1991b).

In some ways the most striking feature of the early glacial period is not the severity of the colder periods but the relative warmth of the intervening 'interstadial' periods represented by stages 5c and 5a of the oxygen–isotope records. The implications of the ocean core isotope ratios (Fig. 2.1) are that the total volume of ice masses during these periods was reduced to approximately half the volume attained during the colder episodes of stages 5d and 5b – although still much greater than those which existed during fully interglacial periods (Shackleton 1977, 1987). In the same way there is clear evidence that the position of the polar front in the North Atlantic region retreated several hundred miles to the north of the earlier glacial episodes, from around the latitude of 50°N to above 60°N (Fig. 2.12) (MacIntyre et al. 1972, 1975; Ruddiman & MacIntyre 1976).

The most detailed reflection of these major interstadial periods is provided once again in the long pollen sequences recorded at La Grande Pile, Les Echets and elsewhere (Figs 2.6–2.8). As noted earlier, all these diagrams show two major periods of climatic warming (generally referred to as St Germain phases I and II) which can be correlated unambiguously with stages 5c and 5a respectively of the deep-sea core sequences (Zagwijn 1990; Woillard & Mook 1982; de Beaulieu & Reille 1984; Behre 1990) and with only slightly more reservations with the well known sequence of Amersfoort/Brørup and Odderade interstadials within the more fragmentary pollen records of northern Europe (Fig. 2.9) (LIGA 1991a; Sejrup & Larsen 1991; Behre & Plicht 1992). As both Zagwijn (1990) and Behre (1990) have recently emphasized, it is now clear that the exact ways in which these two interstadials were reflected in the local vegetational and climatic records varied sharply in

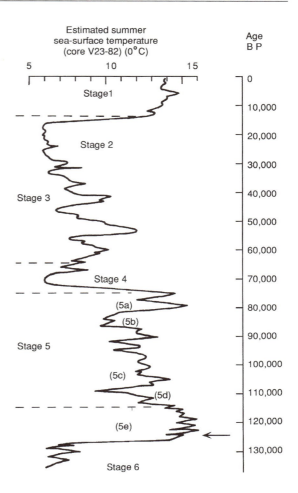

Figure 2.11 *Estimated mid-summer sea-surface temperatures in the north Atlantic region over the past 130,000 years, based on analyses of the varying frequencies of temperature-sensitive surface-living organisms in deep-sea core V23–82. After Sancetta* et al. *1973.*

the different regions of Europe. In the northern zones of the Netherlands, Britain, Denmark and northern Germany, for example, both of these interstadials (the Amersfoort/Brørup and Odderade) were characterized by the migration of dense pine and birch woodland into areas which had previously been

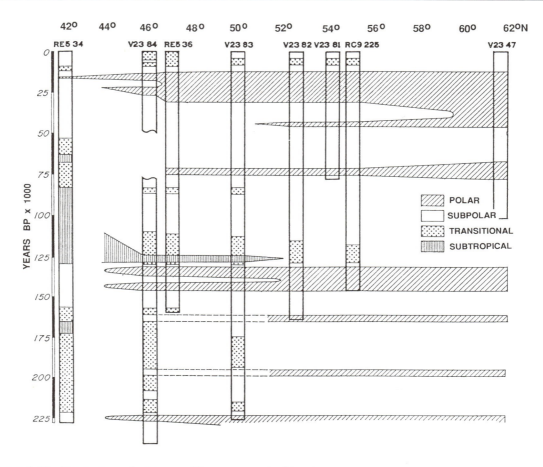

Figure 2.12 *Reconstructed patterns of fluctuations of cold and warm sea-water currents in the north Atlantic region, based on faunal evidence from eight deep-sea cores at latitudes ranging from 42°N to 62°N. During the coldest glacial phases currents of 'polar' sea water extended southwards along the European coastline to ca 44°N – i.e. to the latitude of northern Spain. During the last interglacial, 'subtropical' currents extended northwards as far as the southern coast of Britain. After McIntyre* et al. *1972.*

occupied by open or shrub–tundra communities. Further to the south in the Rhone valley and the western flanks of the Alps, however, the same episodes were characterized by a succession of more warmth-demanding deciduous species, which shows an overall pattern of ecological succession remarkably similar in many respects to that recorded in the same areas during the different stages of the last interglacial – i.e. a succession from birch and pine, through to

elm, oak, hazel and hornbeam forests and then back to pine, spruce and other cold-tolerant species (Woillard 1978; de Beaulieu & Reille 1984). Both Zagwijn and Behre see this as a further indication of the very steep climatic and ecological gradients which existed over northern, western and central Europe during the earlier stages of the last glaciation, resulting partly from the continuing presence of some land ice in Scandinavia, partly from the presence of extremely cold sub-polar waters in the northern parts of the Atlantic ocean and partly from the increased level of continentality of the climate in north-western Europe caused by the general reduction in world-wide sea levels.

The climatic reconstruction provided by Zagwijn (1990) for these periods is illustrated in Fig. 2.5. He suggests that July temperatures ranged from around 12°C (approximately 6°C below present-day values) in southern Scandinavia and northern Germany, to around 18–20°C (1–2°C below present values) in the southern parts of France and along the Atlantic coast. Essentially the same patterns are confirmed by the recent multivariate analyses of climatic and vegetational patterns proposed by Guiot *et al.* (1989, 1993) for the pollen sequences at La Grande Pile and Les Echets (Fig. 2.8). In the case of Les Echets, the peak of warm conditions during both the St Germain I and II episodes are calculated as being ca 1–2°C cooler than the present day, whilst at La Grande Pile, 250 km to the north-east of Les Echets in the foothills of the Vosges mountains, it would seem that average annual temperatures could have been almost identical to those of the present in the same region (Sejrup & Larsen 1991).

One of the intriguing details to emerge from the study of these early interstadial deposits in Europe is the evidence for a bipartite climatic sequence during the earlier of the two interstadials, St Germain I. In the pollen diagrams from both Grande Pile and Les Echets (Figs 2.6, 2.7) this is marked by a very short episode, perhaps only 1,000–2,000 years in length, during which overall tree-pollen frequencies were sharply reduced and there was a temporary re-establishment of essentially open vegetation clearly implying a major climatic cooling (Woillard 1978; de Beaulieu & Reille 1984). The same episode is probably indicated by the similar reduction in tree-pollen ratios which separates the twin peaks of the Amersfoort and Brørup interstadials in northern Europe (Figs 2.9, 2.10) and in similar fluctuations recorded in many

other contemporaneous pollen sequences extending as far to the south and east as Grenada in Spain and Tenaghi Phillipon in Greece (Zagwijn 1990; Behre 1990). Taking all the relevant data into account, Zagwijn (1990) suggests that during this interval overall climatic conditions in Europe may have fallen to levels similar to those experienced during the glacial episodes of isotope stages 5d and 5b. The explanation for this brief but very sharp climatic oscillation remains uncertain. For the human groups who were occupying the various regions of Europe, however, the impact of this climatic event must have been just as significant as that of the more long-term climatic and ecological deteriorations which characterized stages 5d and 5b of the isotope records.

The Middle Glacial period, ca 75,000–25,000 BP

The end of the early glacial period was marked by the onset of much more rigorous climatic conditions which define stage 4 of the oxygen–isotope sequence. This period between ca 75,000 and 60,000 BP (Martinson *et al.* 1987), clearly represents the main glacial maximum of the first half of the last glaciation. In the oxygen–isotope records this period is marked by a rapid decrease in the $^{18}O/^{16}O$ ratios (Fig. 2.1) indicating a major expansion of the continental ice sheets far beyond the limits reached during the earlier cold periods of stages 5d and 5b (Shackleton 1977, 1987; Chappell & Shackleton 1986). The same period is equally apparent in the various faunal indicators of sea-surface temperatures in the North Atlantic region (Fig. 2.11) which suggest that surface temperatures in this region fell to at least 3–4°C lower than those attained during any of the preceding stages of the glaciation (Sancetta *et al.* 1973; Imbrie *et al.* 1989). According to the reconstructions of MacIntyre *et al.* (1975) it would seem that the position of the extremely cold

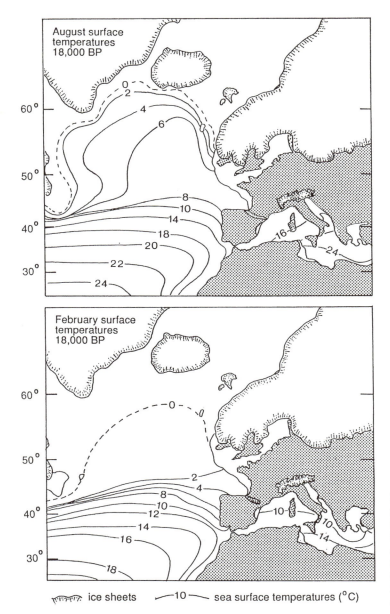

Figure 2.13 *Estimated mid-summer and mid-winter sea-surface temperatures in the Atlantic region at the time of the overall maximum of the last glaciation, ca 18,000 BP. The approximate positions of the continental ice sheets and coastlines of Europe at this time are also shown. After CLIMAP 1981.*

polar front in the North Atlantic area extended to the latitude of ca 45°N, that is, at least 500 miles to the south of the positions attained during isotope stages 5d and 5b and almost as far south as that reached during the overall maximum of the last glaciation at around 18,000 BP (Fig. 2.12; see also Ruddiman & MacIntyre 1976).

In Europe there are clear indications of the character of ecological and climatic conditions during this period in most of the long vegetational sequences discussed earlier. In Holland, Denmark and northern Germany it seems that vegetation during this period consisted entirely of open tundra communities with perhaps some areas of almost barren 'polar desert' during the coldest climatic phases (Figs 2.9, 2.10) (Behre 1990; Zagwijn 1990). Mid-summer temperatures in this region have been estimated by Zagwijn (1990) as around 5°C, with average temperatures which must have remained well below freezing point for large parts of the year. A similar pattern is reflected in the pollen records from La Grande Pile and Les Echets. Here it seems that conditions were more steppic than tundra-like in character (reflected in high frequencies of species such as *Artemisia* and *Chenopodiaceae*, implying generally drier and more continental climates in these more southerly areas) and included a slightly higher component of some of the hardier tree species such as birch and pine (Woillard 1978; Woillard & Mook 1982; de Beaulieu & Reille 1984). Nevertheless, the overall temperatures in these regions are thought to have fallen, once again, to at least 3–4°C lower than those experienced at any earlier stage in the last glaciation (Fig. 2.8) (Guiot *et al.* 1989). There is now no question that the period of isotope stage 4 witnessed some of the most severe climatic conditions of the last glaciation – in most respects probably only slightly less severe than those which characterized the period of the overall maximum of the last glaciation at around 18,000 BP (Figs 2.13, 2.31).

Exactly how far the ice sheets expanded in northern Europe during isotope stage 4 is still uncertain. It is clear from the isotope records of deep-sea cores that the ice sheets were generally much less extensive during stage 4 than during the period of the glacial maximum at ca 18,000 BP (Shackleton 1987; Chappell & Shackleton 1986) so that most of

the potential geological traces of the precise limits of the earlier glacial advances have been largely obliterated from the geological record. In Britain, Bowen (1990) has suggested that the ice sheets during stage 4 probably extended to around the latitude of the Isle of Man; and in northern Germany it has been suggested that the partially over-ridden Stetten end moraine may have formed during this period (Butzer 1972). For the present, however, most of these suggestions remain highly speculative. Overall, there is no doubt that stage 4 must have been characterized by fairly massive and extensive glaciation over the northern parts of Europe, but the exact limits of this glaciation remain unclear.

Isotope stage 3

The ensuing period of isotope stage 3, from ca 60,000 to 25,000 BP, is one of the most enigmatic parts of the last glaciation. In broad terms this was a period of predominantly mild climate, in which the extent of global glaciation was substantially reduced (see Fig. 2.1 Shackleton 1977, 1987; Martinson *et al.* 1987). The difficulties of reconstructing climatic and environmental patterns during this interval stem from the sharply *oscillatory* nature of climatic events, which can be seen clearly in all the associated climatic and palaeoenvironmental records.

The best records of these rapid climatic oscillations can be seen in the recent, high-resolution studies of two long ice cores (the so-called 'GRIP' and 'GISP2' cores) from central Greenland (GRIP 1993; Dansgaard *et al.* 1993; Bond *et al.* 1993; Grootes *et al.* 1993; Boulton 1993; Kerr 1993). Through detailed studies of the oxygen–isotope ratios in the cores it has been possible to identify at least a dozen significant climatic oscillations between ca 25,000 and 60,000 BP, in which temperatures over the area of the ice sheet itself seem to have risen by between 5 and 8°C, often within remarkably short periods of

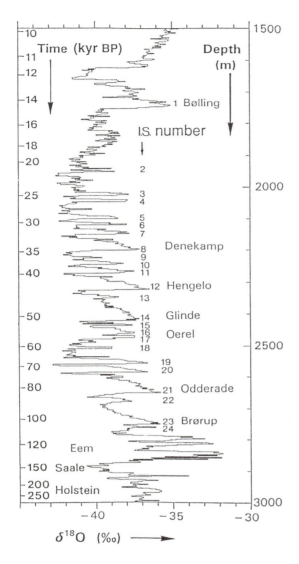

Figure 2.14 *Fluctuating oxygen-isotope ($^{18}O/^{16}O$) ratios recorded through the recent 'GRIP Summit' ice core from Greenland. The various 'interstadials' recognized in the sequence are indicated by numbers (in the central column), and correlated tentatively with those recognized in pollen and other stratigraphic sequences in northwestern Europe. After Dansgaard* et al. *1993. (N.B. More recent work on the core has cast doubt on the reality of the apparent climatic fluctuations during the period of the Eemian interglacial: see Grootes* et al. *1993.)*

less than 50 years or so (Dansgaard *et al.* 1993; Grootes *et al.* 1993). Most of these warmer oscillations were short lived, of the order of ca 1,000 years. At four points in the sequence, however, there is evidence for periods of more prolonged warming of 2,000–4,000 years and with temperature peaks (usually at the start of each interval) slightly higher than those of the shorter intervals. The clearest illustration of these patterns which has been published so far is shown in Fig. 2.14.

It is now apparent that many of the shorter interstadials within stage 3 were either too brief, or not sufficiently marked, to show up in many of the other records of climatic change preserved, for example, in most of the deep-sea core oxygen–isotope sequences (but see Bond *et al.* 1993 for a possible exception) or in most pollen sequences from sites in northern and western Europe. The shorter intervals may have been simply too brief for trees to migrate into northern Europe from their refuge areas further south. The longer and more major interstadials, however, *are* clearly represented in many of these sequences, including for example the pollen records from La Grande Pile and Les Echets in eastern France (Woillard 1978; de Beaulieu & Reille 1984; Guiot *et al.* 1989), in some shorter pollen sequences from northern Europe (Behre 1990; Zagwijn 1990; Behre & Plicht 1992), in oxygen–isotope records from the Mediterranean and north Atlantic (Labeyrie 1984; Paterne *et al.* 1984), and in some of the more detailed and high resolution records of sea–surface temperatures (Sancetta *et al.* 1973; Bond *et al.* 1993) (see Figs 2.6–2.11, 2.15). Unfortunately the close spacing of several of these intervals combined with the poor resolution of most absolute dating methods within this time range make an exact correlation of the individual climatic oscillations within the different sequences extremely difficult (see Fig. 2.16).

Perhaps the main point to emphasize is that the ecological effects of these various interstadial episodes within isotope stage 3

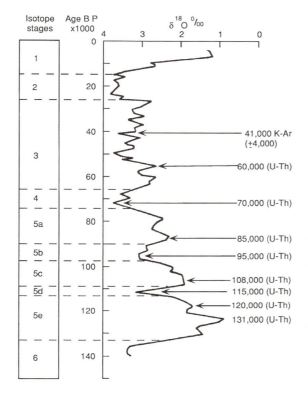

Figure 2.15 *Changing oxygen-isotope ratios over the past 140,000 years recorded in core KET-8004, from the north Mediterranean. In this area, variations in the oxygen-isotope ratios are likely to be much more directly controlled by temperature variations than in the major ocean basins. Absolute dates for the major stadial and interstadial phases obtained by Potassium-Argon and Uranium-Thorium methods are indicated on the right. After Labeyrie 1984 (see also Paterne et al. 1986).*

full coniferous forest (Zagwijn 1990; Behre 1990; Behre & Plicht 1992; Van der Hammen *et al.* 1967: see Figs 2.9, 2.10). Similarly in the pollen records from Grande Pile and Les Echets, the various interstadial episodes of stage 3 were characterized by arboreal pollen percentages of only ca 30–50 percent (consisting almost entirely of birch and pine) while the earlier St Germain I and II interstadials had been marked by tree frequencies of around 90 percent, and consisting of full deciduous forest not very different from that of the preceding interglacial (Figs. 2.6, 2.7). Most of these ecological and climatic records from Europe converge on figures of around 5–7°C for the overall scale of the temperature rises during the major interstadial episodes of isotope stage 3 (see Figs 2.2, 2.8, 2.9, 2.11). Even if climatic conditions during stage 3 were highly oscillatory, therefore, they never seem to have attained the degree of climatic amelioration which characterized the more major interstadials during the earlier stages of the last-glaciation.

The exact causes of these rapid and sudden climatic oscillations during isotope stage 3 have recently generated lively controversy among Quaternary climatologists (e.g. Dansgaard *et al.* 1993; Bond *et al.* 1993; Kerr 1993). The most significant discovery is that the most pronounced periods of rapid cooling seem to have coincided with periods when large numbers of icebergs broke away from the Laurentide ice sheet in the North Atlantic region (leaving clearly defined layers of detrital material in the contemporaneous deep-sea sediments) which in turn would have reduced the salinity of the surface waters in the North Atlantic, and triggered off movements in ocean currents which would have caused an influx of much warmer waters – a pattern which has recently been described as 'Bond cycles' and related 'Heinrich events' (Bond *et al.* 1993; Kerr 1993). Other climatologists, however, are inclined to see most of these climatic oscillations as a possible direct result of various astronomical orbital-forcing

were much less marked than those which characterized the earlier, more pronounced interstadials of isotope 5a and 5c. In the pollen records from northern Europe, for example, the Denekamp and Hengelo interstadials were marked simply by a shift from open tundra to shrub tundra communities, whereas the earlier, much longer, Amersfoort/Brørup and Odderade interstadials had been characterized in the same areas by

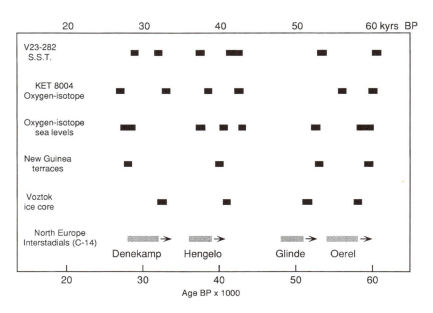

Figure 2.16 *Comparison of age estimates for various interstadial episodes during the course of isotope stage 3, derived from the various climatic sequences shown in Figs 2.2, 2.11, 2.15, & 2.17. Note that the dates for the major interstadials recognized in northern Europe (at the base of the diagram) are based mainly on radiocarbon measurements, and may therefore understimate the true age of these intervals by up to 3,000 years (cf. Bard* et al. *1990a).*

mechanisms of the kind discussed earlier (Kerr 1993).

Leaving these debates aside, the pattern of highly variable, rapidly fluctuating climatic and ecological conditions during the period of isotope stage 3 is of particular interest from an archaeological and behavioural standpoint, since it coincides with the final stages of the Middle Palaeolithic occupation of Europe and the ensuing cultural and demographic transition to the Upper Palaeolithic. As we shall see in a later chapter, the character of these ecological oscillations may well have a critical bearing on the specific demographic and other mechanisms by which anatomically modern populations were able to compete with and apparently eventually replace the local Neanderthal populations within the different regions of Europe.

Sea-level fluctuations

As noted earlier, fluctuations in world-wide sea levels during the course of the later Pleistocene period can now be reconstructed with a remarkable degree of accuracy. The evidence comes from two major sources (Fig. 2.17): from the oxygen–isotope records in deep-sea cores, which can be used to provide an estimate of the total volume of sea water in the world's oceans (Shackleton 1977, 1987); and from the more direct evidence of varying sea-level positions preserved in the form of coral terraces and other forms of raised beach deposits in various areas of the world (Aharon & Chappell 1986; Chappell & Shackleton 1986). Estimates of sea levels from oxygen–isotope records need to be handled with caution, since some specific allowance must be made as to the particular component in

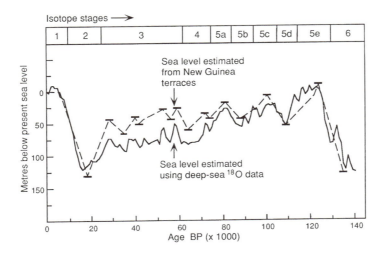

Figure 2.17 *Estimates of world-wide sea level fluctuations over the past 140,000 years derived from studies of oxygen-isotope ratios in surface and deep-water ocean sediments (continuous line) and dated coral terraces in New Guinea (dashed line). After Shackleton 1987.*

these changing isotope ratios which is attributable directly to temperature changes, rather than to the removal of isotopically light sea water from the oceans to form the continental ice sheets (Shackleton 1987). In principle the direct records of coral reefs and terraces are simpler to interpret, although in these cases some allowance must be made for the effects of local tectonic displacements in the recorded altitudes of the individual terraces. The most detailed and extensively studied traces of coral terraces are those preserved in the Huon Peninsula of New Guinea, from which large numbers of precise and internally consistent radiocarbon and uranium-series dates have been obtained (Aharon 1983; Aharon & Chappell 1986; Chappell & Shackleton 1986). Similar terrace sequences are documented from Barbados, Timor and the New Hebrides, all of which have yielded patterns comparable with those documented from the New Guinea terraces (Aharon & Chappell 1986).

The general picture of these 'eustatic' sea-level fluctuations which emerges for the later stages of the Pleistocene period is summarized in Fig. 2.17. The period of maximum sea level can now be shown to have coincided closely with the period of oxygen–isotope stage 5e (i.e. the peak of the last interglacial), and to have reached a level around 6–10 metres higher than that of the present day. There is clear evidence for this in many parts of the world not only in the coral terraces preserved in New Guinea and elsewhere but also in the form of beach deposits located at altitudes of ca 6–10 metres in many parts of the world (for example in northern France, southern Britain and the Mediterranean) (Bowen 1978, 1990). Following this period of interglacial high sea level, there was a series of complex fluctuations during the earlier stages of the last glacial period. During the initial cold stages of isotope stages 5d and 5b it seems that world-wide sea level fell to around 50–60 metres below present levels (Shackleton 1987). During the much warmer interstadial episodes of stages 5c and 5a, by contrast, sea levels rose sharply to levels only perhaps 10–20 metres below those of the present day.

Clearly defined episodes of relatively high sea-level stands during these major interstadial episodes are well documented in several parts of the world (especially New Guinea and Barbados) and have been dated by direct uranium-series measurements to around 105–110,000 and 80–85,000 BP (Aharon & Chappell 1986). Sea levels fell to their early glacial minimum during the period of isotope stage 4, to around 60–80 metres below present levels. A succession of rapid fluctuations then occurred during the period of isotope stage 3 (between ca 60,000 and 30,000 BP) before the massive reduction in sea levels, down to around −130 metres below modern levels, occurred during the overall maximum of the glaciation at around 18,000 BP (Shackleton 1987).

The effects of these sea-level fluctuations on the coastal geography of Europe are reasonably easy to reconstruct from the patterns of submarine contours around various parts of the coastline. The major reduction of sea level during isotope stage 4 had the effect of exposing large areas of the current North Sea basin and English Channel as dry land, effectively integrating southern Britain, northern France, the Low Countries and southern Scandinavia as part of a greatly extended North European plain. At the same time the coastal plain along the Atlantic coast of France was extended by at least 30–40 km towards the west. The much higher sea levels which occurred during the warmer interstadial episodes brought the coastlines much closer to modern patterns. Even so, the highest sea levels recorded during the major interstadials of the last glaciation would have had a major impact on the geography of the areas bordering the North Sea basin and would have left Britain an integral part of the main European land mass throughout the whole of the last glacial period. It was only during the period of the last interglacial that Britain was fully isolated from the continent, with coastlines broadly similar to those of the present day.

Loess deposition

One of the most conspicuous geological effects of periods of glacial climate in the northern zones of Europe was the formation of thick deposits of wind-blown loess. The precise origins and mode of formation of these deposits are still open to some debate (Butzer 1972; Bowen 1978; Wintle 1990). The deposits were evidently laid down by heavy wind action and were presumably derived from areas where extensive spreads of fine-grained sand or silt deposits were exposed to the effects of wind erosion without the protection of a continuous vegetation cover. But how far they were derived from river or beach deposits and how far from various forms of glacial outwash deposits is more debatable, and no doubt varied between different locations.

In western Europe, thick deposits of loess can be traced in a broad arc from the western coasts of Brittany and Normandy through Picardy and the Paris basin into Belgium and northern Germany. Southwards, occasional deposits of loess can be traced into the Loire valley and into at least the northern parts of Burgundy (Fig. 2.18). The thickness attained can be impressive; in parts of northern France, for example, it is estimated that up to 10–15 metres of loess may have built up during the various stages of the last glaciation.

The major problem has always been precise dating of loess deposits. Wintle et al. (1984) have suggested that the formation of loess may be a rare event during the Pleistocene period as a whole, perhaps occupying only 10 percent or less of Pleistocene time. The only form of absolute dating which can be applied directly to the loess deposits is that of thermoluminescence (TL) dating – based on the presumed bleaching of the TL signal in quartz grains by the action of sunlight as particles were transported through the air (Wintle 1990). By applying this technique to loess profiles in Normandy, Wintle et al. (1984) have suggested that the principal

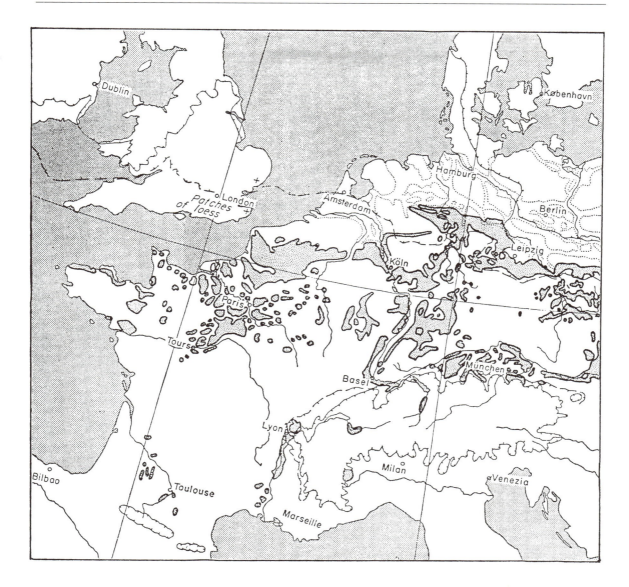

Figure 2.18 *Distribution of loess and related wind-blown deposits in western Europe. The deposits are likely to have formed during a number of different glacial episodes over the past 500,000 years. After Flint 1957.*

phases of loess deposition during the last glaciation were concentrated mainly during isotope stages 2 and 4, with perhaps an earlier, brief period of accumulation during stages 5b or 5d (see also Parks & Rendell 1992). A similar pattern has been suggested recently by Van Vliet-Lanoë (1989, 1990) based on the overall stratigraphy of the loess deposits and associated soil profiles (Fig. 2.19). The intervening, warmer episodes of the Pleistocene were characterized by heavy weathering of the existing loess deposits and the formation of a series of complex soil profiles, such as the well known Warneton or Rocourt soil com-

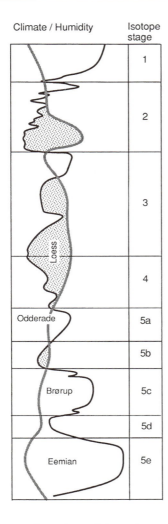

Climate / Humidity

Isotope stage

Loess

Odderade

Brørup

Eemian

Figure 2.19 *Chronology of the major periods of loess deposition in northern France during the last glacial period, as proposed by Van Vliet-Lanoë (1990). Loess deposition in this area is thought to have been favoured by a combination very cold, dry climatic conditions.*

these extreme northern fringes of Europe, the accumulation of loess sheets must have been a crucially important environmental factor – temporarily reducing large areas of the North European Plain to virtual deserts and presumably creating almost intolerable conditions during periods of heavy winds. As Bordes (1954a) and others have emphasized, traces of human occupation are effectively unknown from within the loess deposits themselves, and are confined to the various weathering and soil horizons which separate the major phases of loess accumulation. The problem of accurately dating and correlating these individual soil horizons in different loess profiles remains one of the most difficult and controversial issues in current Pleistocene studies in Europe (Wintle 1990).

Climatic and environmental patterns in southwestern France

The evidence for climatic and environmental conditions in the Perigord and adjacent regions of southwestern France derives almost entirely from one rather specialized geological source, namely, from the various stratigraphic sequences preserved within the cave and rock-shelter sites of the region. These sequences can be studied from several different perspectives by studies of the detailed sedimentology; geochemistry and micro-morphology of the deposits, by analyses of the preserved pollen content, and by studies of the associated faunal assemblages, including both macro-faunal and micro-faunal remains. All these approaches have been applied extensively within southwestern France over the past 30 years, providing an impressive battery of detailed information from many different sites. In particular, the studies of Laville and others on the sedimentology of the cave and rock-shelter and sequences, Paquereau, Leroi-Gourhan and Leyroyer on the palynology of the deposits, and Prat, Delpech, Guadelli, Laquay and others on the faunal assemblages, have been

plexes, which separate the deposits of Younger and Older loess and which evidently formed during the various episodes of isotope stage 5 (Van Vliet-Lanoë 1990).

The present uncertainty over the precise chronology of loess deposition is unfortunate. For the human groups who occupied crucially important. We have far more information on these aspects of the cave and rock shelter sequences within southwestern France than we do for any other region of Europe at the present time.

In practice, there are a number of problems in studying and interpreting climatic and palaeoenvironmental data from cave and rock shelter sequences which have been widely debated in the recent literature. The use of sedimentological data in particular has come under close scrutiny, mainly owing to the sparsity of controlled studies of the precise effects of varying climatic regimes on different types of bedrock formations and the difficulty of assessing the impact of local or even humanly induced factors on the specific climatic and environmental conditions of the individual sites (e.g. Farrand 1975, 1988; Texier 1990). The interpretation of pollen samples from cave and rock-shelter deposits is also problematical, partly owing to the possibility of downwashing of pollen grains in relatively coarsely textured sediments and partly to the danger of selective destruction of particular pollen taxa in the highly calcareous, alkaline environments encountered in most sites (Turner 1985; Turner & Hannon 1988; Sánchez Goñi 1991, 1993). Even the study of faunal assemblages is not without problems: the challenge here is to differentiate between the effects of purely climatic and ecological changes on the composition of local faunal communities and the role of human selection in the economic exploitation of different species (see Chapter 7 below). In addition one must correlate and synchronize individual stratigraphic sequences recorded in the different sites and recognize significant depositional or erosional hiatuses within the deposits (Laville 1988; Laville et al. 1986; Farrand 1975, 1988).

Whether or not these problems are as serious in practice as in theory is debatable. Most of the cave and rock-shelter sequences which have been systematically studied within southwestern France present a generally coherent and internally consistent record of climatic and environmental events during the later stages of the Pleistocene in which the majority of the parallel and complementary lines of evidence lead to similar conclusions. The aim of this section is to focus on the most complete and fully documented of these sequences and see how far they can be used to construct a general climatic and environmental framework for the associated records of Middle Palaeolithic occupation in the region.

The Combe Grenal sequence

By far the most detailed and complete record of climatic and environmental conditions throughout the Middle Palaeolithic period in southwestern France is provided by the long sequence excavated by François Bordes at the site of Combe Grenal between 1953 and 1965. The site at present consists of a small cave behind a rock overhang, which lies on the flanks of a small dry valley one kilometre to the south of the Dordogne river (Bordes 1955, 1972; Laville et al. 1980). Over the course of its geological history the position of the rock shelter receded gradually towards the north (as a result of progressive weathering and erosion of the bedrock limestone deposits) leaving the major phases of the geological infilling standing on a series of separate rock platforms, which are progressively younger with increasing height above the adjacent valley (Fig. 2.20). The depth of the filling is almost 13 metres. All this sequence has now been studied in detail from the perspectives of sedimentology, pollen analysis and the composition of the associated faunal assemblages (Bordes et al. 1966; Laville 1975; Laville et al. 1980; Bordes & Prat 1965; Delpech et al. 1983; Laquay 1981; Guadelli 1987; Guadelli & Laville 1990) (Figs 2.21, 2.22). As a record of changing climatic and ecological conditions throughout the greater part of the Middle Palaeolithic time range, this sequence is unique within southwestern France.

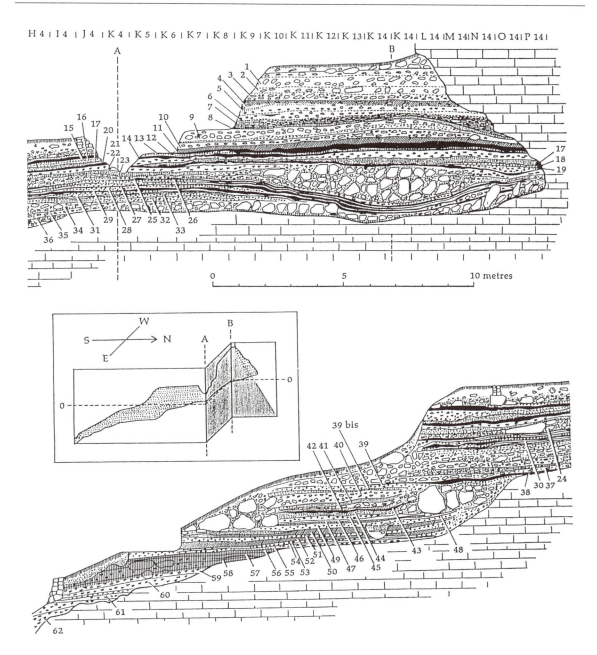

Figure 2.20 *Stratigraphy of the rock-shelter deposits at Combe Grenal (Dordogne, southwestern France), showing the upper (above) and lower (below) parts of the sequence. (Note that the section shown here is folded, and therefore includes both lateral and transverse components of the stratigraphy, as indicated in the central diagram.) Note how the deposits lie on a series of erosional platforms in the underlying limestone, which become progressively younger in age as the rock shelter recedes towards the north. The deposits span a total of over 13 metres, and cover the period from the end of the penultimate ('Riss') glaciation (layers 56–65) to around the middle of the last glaciation (see Fig. 2.23). After Bordes 1972.*

Briefly, the main features of the sequence can be summarized as follows:

1. The basal part of the filling (layers 65–56) consists of a series of stony *'eboulis'* deposits which extend over a depth of ca 2.5 metres and lie on the lowermost erosional step in the bedrock formation (Fig. 2.20). All available evidence indicates that these levels were formed under extremely severe climatic conditions – possibly the coldest conditions in the whole of the Combe Grenal sequence. The percentages of arboreal pollen recorded in these levels (Fig. 2.21) remain consistently below 15 percent and in the majority of levels pine is the only tree species represented (Bordes *et al.* 1966). The same pattern is reflected in the character of the associated sediments, which also point to a period of extremely cold, dry climate with intensive frost action and little evidence for any contemporaneous chemical weathering of the deposits (Bordes *et al.* 1966; Laville 1975; Laville *et al.* 1980). The composition of the associated faunal assemblages provides further support for this conclusion (Fig. 2.22); throughout all levels remains of reindeer account for between 92 and 97 percent of total ungulate remains with only sporadic traces of other species such as red deer and horse (Bordes & Prat 1965).

All the available evidence therefore points to exceptionally rigorous conditions during the formation of the lower levels, which are assumed to correlate with the period of extremely cold, full glacial conditions represented by the later stages of oxygen–isotope stage 6 (Fig. 2.23) (Laville *et al.* 1983, 1986; Guadelli & Laville 1990; Mellars 1986a, 1988). Archaeologically, these levels produced a series of late Acheulian industries characterized by rather crudely made hand axes and low frequencies of Levallois technique (Bordes 1971a).

2. The second major climatic episode in the Combe Grenal sequence is represented by a period of heavy weathering and associated soil formation which affected virtually all deposits formed during the preceding 'Rissian' period to a depth of at least 1.0 metres. According to Laville (1975; Laville *et al.* 1980), it is unlikely that any entirely new deposits were laid down during this interval and it would seem that the upper levels of the associated soil horizon were truncated and eroded by the effects of heavy humidity during the later stages of this period. As a result there are no direct records of either pollen, faunal remains or associated archaeological assemblages for this interval. Nevertheless, it is clear that this period represents a prolonged episode of very warm, humid climate which must have persisted for several thousand years. All authors are now agreed that this can be correlated with the peak of the last interglacial period (i.e. isotope stage 5e of the deep-sea core sequence) and accordingly dated to around 120–125,000 BP (Fig. 2.23) (Laville *et al.* 1983, 1986).

3. The third major phase in the Combe Grenal sequence is represented by the block of deposits between levels 55 and 38 spanning approximately 2.5 metres in depth and lying, once again, on a separate erosional step in the bedrock formation. The combination of the sedimentological and pollen evidence shows that these deposits formed during a succession of four major climatic oscillations comprising two major cold episodes and two prolonged episodes of much warmer climate. The cold episodes, represented in layers 55–53 and 49–44, were characterized by severe climatic conditions leading to intensive frost action in the formation of the associated sediments and reducing tree-pollen frequencies in the local habitat to around 12–15 percent (Fig. 2.21). Nevertheless, it is significant that arboreal pollen frequencies in these levels are consistently higher than those recorded in the majority of underlying 'late Rissian' levels and show a continuous record of species such as juniper alder and hazel in addition to the hardier species of

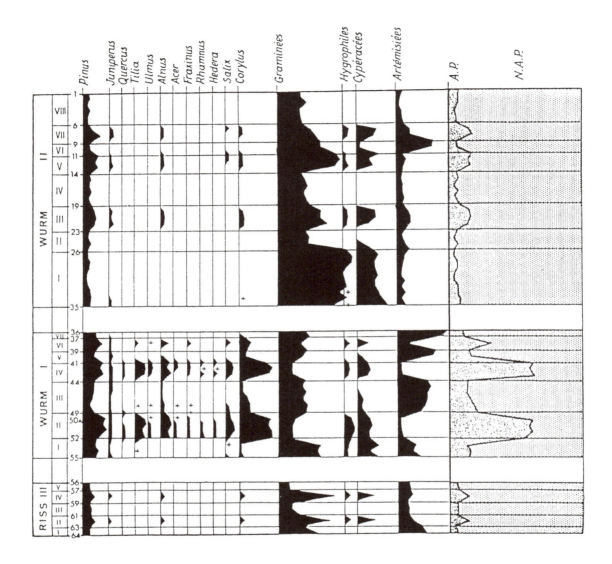

Figure 2.21 *Pollen succession recorded through the Combe Grenal sequence. After Bordes* et al. *1966.*

birch and pine (Bordes *et al.* 1966), which suggest that climatic conditions never approached the severity of those recorded during the preceding cold phase of isotope stage 6.

The two warm episodes in the sequence (in layers 52–50A and 43–41) reflect a dramatic change from these conditions. In these levels overall arboreal pollen ratios rise to between 60 and 70 percent and now include substantial frequencies of a wide range of warmth-demanding deciduous species such

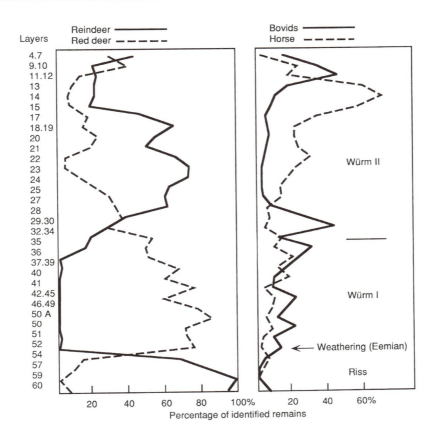

Figure 2.22 *Fluctuating frequencies of the four principal faunal taxa recorded throughout the Combe Grenal sequence, based on counts of numbers of identified specimens (NISP) of the different taxa. The most striking feature is the shift from the heavily red-deer dominated faunas in the earlier, 'Würm I' levels of the last-glacial sequence to the mainly reindeer-dominated faunas in the later, 'Würm II' levels. In the Würm I levels, reindeer are represented by only seven identified specimens (from layers 52 and 40) which could well be derived from the underlying 'Rissian' levels, in which reindeer remains are abundant. After Bordes & Prat 1965.*

as elm, lime, oak, alder and hazel in addition to the hardier species such as pine and juniper (Fig. 2.21). The same pattern is reflected in the associated sedimentological data which indicate a sharp increase in both temperature and humidity and some degree of soil formation (Laville 1975; Laville *et al.* 1980). In certain respects, therefore, climatic conditions in these levels approach those of fully interglacial conditions. Paquereau, however, emphasizes that the character of the associated pollen spectra – still domi-

nated mainly by pine and with relatively low percentages of oak – is by no means fully 'interglacial' in character and implies that climatic conditions during these intervals were significantly cooler than those experienced either during the preceding Eemian interglacial or at the present day (Bordes *et al.* 1966: 12, 16). The evidence suggests, in other words, relatively mild interstadial conditions during the formation of these levels.

Despite these climatic fluctuations the composition of the faunal assemblages recor-

ded throughout these so-called 'Würm' I levels in the Combe Grenal sequence remain
remarkably stable throughout the entire
sequence of levels 55–38 (Fig. 2.22). In all
levels remains of red deer are clearly dominant, accompanied by more sporadic remains
of horses, bovids, wild pig and roe deer
(Bordes & Prat 1965; Laquay 1981). How one
should explain this relative stability of the
faunal assemblages in the face of rapidly
changing climatic and vegetational conditions raises interesting issues which will be
discussed more fully in Chapter 7. In general
terms, however, the character of the faunal
remains provides further evidence that climatic conditions during the formation of
these levels never attained the extremely rigorous, full glacial conditions reflected in both
the earlier Rissian levels of the Combe Grenal
sequence and in the overlying Würm II levels
on the site.

From the general character and stratigraphic position of the various climatic
episodes discussed above there is now effectively unanimous agreement as to how these
levels should be correlated with the overall
climatic sequence for the Upper Pleistocene
(Laville *et al.* 1983, 1986; Guadelli & Laville
1990; Mellars 1986a, 1988 etc.). Since these
deposits rest directly on top of a presumed
last interglacial soil horizon and immediately
below a long period of much colder climate
(discussed   below)   they   are   generally
assumed to correlate with the well documented succession of colder and warmer episodes represented by stages 5d–5a of the
deep-sea core sequence (Fig. 2.23). Indeed, it
could be argued that the patterns suggested
by both the sedimentological and pollen evidence for this part of the Combe Grenal
sequence coincide almost exactly with what
one would predict from the vegetational and
climatic sequences recorded at La Grande
Pile and Les Echets (Figs 2.6, 2.7). This
implies a dating for these levels in the region
of ca 115,000–75,000 BP. The only ambiguity
lies in the dating of the initial cold episode

represented by levels 55–53 of the Combe
Grenal sequence. Whilst the obvious correlation of these levels would be with stage 5d of
the ocean core sequence (i.e. the first cold
phase which effectively initiated the last glaciation), it is conceivable that these levels
could correspond with the much briefer cold
episode which occurred mid-way through
the course of the St Germain I interval (i.e.
isotope stage 5c) at Grande Pile, Les Echets
and elsewhere (i.e. the so-called 'Montaigu'
episode) (cf. de Beaulieu & Reille 1984: 125;
Reille & de Beaulieu 1990). If this correlation
were adopted it would imply that any deposits formed during the very earliest stages of
the last glacial period (during stage 5d and
the earlier part of stage 5c) had been removed
or truncated by erosion in the early stages of
the climatic deterioration. But this remaining
uncertainty has little effect on the overall
correlation of these Würm I levels in the
Combe Grenal sequence with the basic pattern of climatic oscillations during isotope
stages 5a–5c.

4. A more dramatic shift in the climatic history of the Combe Grenal sequence is reflected between layers 37 and 35. This episode
has been described by Laville *et al.* (1986: 38)
as a major threshold ('*seuil*') in the climatic
succession and is marked by the appearance
of climatic and vegetational conditions which
are far more rigorous than those reflected in
any earlier levels in the last-glacial sequence
in the site. In the first place, overall arboreal
pollen ratios fall to levels substantially and
consistently lower than those recorded in any
of the earlier Würmian levels in the site and
are marked by the total disappearance of all
species other than pine – most notably the
warmth-loving hazel which had persisted
throughout all of the preceding Würm I levels in the sequence (Fig. 2.21) (Bordes *et al.*
1966). Second, Laville *et al.*(1986) draw attention to a more general change in the character
of the non-arboreal pollen in these levels
marked especially by the sharp increase in

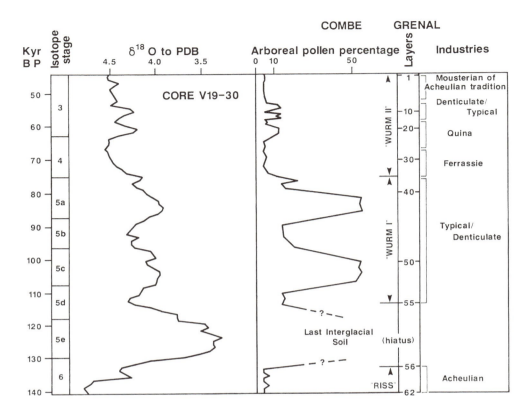

Figure 2.23 *Proposed correlation between the climatic and vegetational sequence at Combe Grenal and the sequence of oxygen-isotope stages in deep-sea cores (reproduced from Mellars 1986a). The main succession of industries on the site is shown on the right.*

species such as *Ephedra, Galium, Armeria* and *Poterium* which appear to mark the onset not only of more rigorous but much drier and more steppic conditions. Third, this period is marked by a dramatic increase in the frequency of reindeer in the associated faunal assemblages, which rise from effectively zero percent throughout layers 55–37 up to ca 20 percent in layer 35 and then to between 60 and 70 percent in the overlying layers 28–23 (Fig. 2.22) (Bordes & Prat 1965; Guadelli 1987). According to Laville (1975) the associated sedimentological data point equally to the onset of much colder and drier conditions, commencing with layers 36–37 and

continuing throughout most of the later levels in the sequence.

There can be little doubt about the correlation of these levels with the climatic sequence in deep-sea cores (Fig. 2.23). All the French workers now agree that the dramatic deterioration in climatic conditions reflected between levels 38 and 35 must coincide with the transition from stage 5 to stage 4 of the isotope record, dated closely in the ocean cores to around 73,000 BP (Laville *et al.* 1983, 1986; Guadelli & Laville 1990; Martinson *et al.* 1987). Whether this should be regarded as a very abrupt climatic transition is more open to debate. In the vegetational records at La

Grande Pile and Les Echets, for example, the transition from stage 5a to stage 4 is characterized by a sequence of rapid climatic oscillations (the so-called 'Ognon' oscillations) which preceded the emergence of fully 'glacial' conditions at the peak of isotope stage 4 (Figs 2.6, 2.7). A similar pattern may be reflected in the Combe Grenal sequence by the briefly warmer and probably wetter episodes represented in layer 38 and in the clear weathering horizon which separates layers 36 and 35. It could be that the overall climatic sequence at Combe Grenal has been partially truncated over this interval by a phase of erosion between layers 36 and 35. Nevertheless, there can be little doubt as to the general correlation of this climatic transition with the transition from stage 5 to stage 4 in the deep-sea core sequence and there seems no reason to postulate any major break or hiatus at this point in the Combe Grenal sequence.

5. The extremely cold climatic conditions which first appeared in levels 37–35 persisted throughout the great part of the ensuing sequence on the site, i.e. throughout the majority of levels 35–1. In most of these levels overall tree-pollen frequencies remain at extremely low levels (mostly between 5 and 10 percent) and in the majority, pine is the only tree species represented (Fig. 2.21). More significant variations can be detected in some of the non-arboreal species. Thus, relative frequencies of grasses versus composites and heliophiles fluctuate substantially between different levels in the sequence, though perhaps with a general shift from predominantly grass-dominated vegetation in the lower levels (layers 35–26) to mainly composite-dominated spectra in the later levels (layers 25–1) (Bordes *et al.* 1966).

The major question hinges on how far one can identify clear traces of the onset of more temperate climatic conditions in the upper part of the Combe Grenal sequence, coincid-

ing with the shift to the generally milder, 'interstadial' conditions of isotope stage 3. There is apparently pollen evidence for three relatively short-lived episodes of slightly milder climate (represented respectively in levels 22–20, 13–11, and 8–7), in each case reflected by an increase in overall tree-pollen frequencies to around 15–20 percent and by the temporary reappearance of species such as hazel, alder and juniper (Fig. 2.21). The same episodes were also marked by an increase in local humidity reflected in both the sedimentological data and a sharp increase in species such as *Cyperaceae* and other hygrophiles. Significantly, there is also evidence for a major shift in the composition of the associated faunal assemblages in these upper levels (Fig. 2.22). This is marked by a sharp decrease in the overall frequency of reindeer, which falls from ca 60 percent in levels 19–18 to around 20 percent in levels 15–8, and by a corresponding increase in red deer, horses and bovids (Bordes & Prat 1965; Guadelli 1987). All these shifts can probably be taken to reflect a significant improvement in climatic conditions in the uppermost levels of the Combe Grenal sequence, which almost certainly correlates with the transition from isotope stage 4 to stage 3 (Guadelli & Laville 1990; Mellars 1986a, 1988).

Any attempt at precise dating of the upper levels in the Combe Grenal sequence is difficult from the present evidence. The problems of identifying, separating and correlating the complex succession of separate interstadial episodes which characterizes stage 3 of the ocean core sequence have already been emphasized and any attempt at specific correlation of the events registered at Combe Grenal in these terms might be premature. As discussed below, the evidence from some other sites in southwestern France (especially Le Moustier) suggests that the stratigraphic sequence at Combe Grenal comes to an end well before the end of the overall Mousterian succession in this region, probably by around 50,000 BP (Mellars 1986a, 1988). For more secure evidence of these later stages in the climatic sequence we must rely on findings from other, complementary sites discussed further below.

The question of the absolute chronology of the Combe Grenal sequence has been left to the end of this discussion since this has recently generated a lively controversy. At present there is a series of six thermoluminescence (TL) dates for the sequence produced by the British Museum Laboratory based on samples of burnt flint collected during the excavations by François Bordes in 1958–63 and covering the range from 44,000 to 113,000 BP (Bowman & Sieveking 1983). As several workers have pointed out, if these dates are accepted at face value they not only conflict with the overall chronology and climatic correlations proposed above (i.e. with the correlations now accepted by apparently all French workers) but also imply a pattern of climatic fluctuations within the Combe Grenal sequence which is in obvious conflict with the patterns documented in all other sources of climatic information (i.e. deep-sea cores, ice cores, sea-surface temperatures, long pollen sequences etc.) covering the same time range (Laville et al. 1986; Laville 1988; Mellars 1986a,b, 1988). For example, the dates of ca 105,000 and 113,000 BP obtained for layer 60 at the extreme base of the sequence imply that the most severe climatic conditions recorded during the entire Combe Grenal sequence occurred during the period of isotope stage 5d – which emerges in this dating as much colder than isotope stage 4. The dates obtained for layers 49–55 and layer 20 (ca 61,000–68,000 BP and 44,000 BP respectively) similarly reflect the reverse of the climatic patterns expected over this time range, implying exceptionally warm conditions, with full deciduous forest, during the period of isotope stage 4, and very much colder conditions during the interstadial period of isotope stage 3 (Fig. 2.23). As Laville et al. (1986: 38–40) have indicated, the existing series of TL dates for the Combe

Grenal sequence is difficult to reconcile either with any reasonable interpretation of the overall climatic and environmental sequence at Combe Grenal itself, or any other climatic sequences documented over this time range.

As Laville and others have pointed out, there are a number of obvious problems with both the provenance and the measurement of the burnt flint samples employed for the TL dating. The six samples employed for the TL measurements were collected during excavations carried out on the site over thirty years ago, were probably exposed to strong sunlight at the time of collection and were stored for nearly 20 years with the archaeological collections at the University of Bordeaux before dating by the laboratory. In addition, it would seem that the levels of background radioactivity in two of the layers involved in the dating were not measured directly (Bowman & Sieveking 1983; Laville et al. 1986; Laville 1988). As Aitken and others have emphasized, these are hardly ideal conditions for TL dating of burnt flint samples (Aitken 1985, 1990; Aitken et al. 1986). Perhaps more to the point, a series of parallel TL measurements undertaken by the same laboratory on samples from the nearby site of Pech de l'Azé IV produced results which the laboratory itself described as 'too young to be acceptable' and yielded a date for one of the upper (but not final) Mousterian layers of 19,600±1600 BP – i.e. less than half the known age of the sample (Bowman et al. 1982). In the light of these results it would be unwise to place any strong reliance on the existing TL measurements for the Combe Grenal sequence. Given the close correspondence which can be demonstrated between the overall climatic sequence at Combe Grenal and similar patterns in deep-sea cores, ice cores and the long vegetational sequences at La Grande Pile, Les Echets and elsewhere, it seems most reasonable to accept these correlations as providing the best overall chronological framework for the Combe

Grenal sequence, until a new and more tightly controlled programme of absolute dating (preferably employing a combination of several dating techniques) can be applied to the site.

Climatic patterns in other southwestern French sites

No other site in southwestern France has produced a climatic and environmental sequence for the earlier stages of the Upper Pleistocene which can compare either in length and completeness, or in the level of detail and resolution, with that recorded at Combe Grenal. There are, however, a number of sites which can be used either to corroborate and reinforce the general climatic implications of the Combe Grenal sequence or to add some new elements to the final stages of the climatic sequence which are not well represented in the Combe Grenal deposits. The two main sites in this connection are Pech de l'Azé II, located close to the small valley of the Enea only 6 km to the north of Combe Grenal, and the lower shelter at Le Moustier located in the Vézère valley, some 22 km to the north-west (see Fig. 8.1). The importance of these sites is that they have both been dated recently by a long and internally coherent series of absolute dates, which can be used to provide direct control over the correlation of the climatic and environmental sequences with the more general climatic sequences recorded in deep-sea cores (Valladas et al. 1986; Grün et al. 1991; Mellars & Grün 1991).

The sequence recorded at Pech de l'Azé site II (summarized in Figs 2.24–2.26) can be described briefly, since this appears to provide in most respects a close parallel for the climatic, vegetational and faunal sequence documented in the earlier part of the last-glacial sequence at Combe Grenal (Bordes 1971a, 1972; Laville 1975; Laville et al. 1980; Paquereau 1969; Bordes & Prat 1965; Grün et al. 1991). The major interstadial episode rep-

resented in layers 4A–4C (characterized by overall arboreal pollen ratios of around 50–60 percent and including many warmth-demanding deciduous tree species: Fig. 2.24) is almost certainly the equivalent of the similar warm episode reflected in levels 43–41 at Combe Grenal and has been dated by electron-spin-resonance (ESR) measurements on associated animal teeth to around 70–85,000 BP (Fig. 2.26) – i.e. clearly coinciding with isotope stage 5a (Grün et al. 1991). The subsequent, much less marked, climatic ameliorations in levels 2G' and 2G most probably correlate with the similar milder intervals recorded in layer 38 and between layers 35 and 36 at Combe Grenal. After this point the entire character of the sedimentary, pollen and faunal sequence reflects a sharp transition to much colder and apparently much drier conditions, clearly coinciding with the climatic transition from isotope stages 5a to 4. The ESR dates for these upper levels centre on ca 60–70,000 BP, although with a degree of dispersal in the individual measurements which could cover almost any period from the beginning of isotope stage 4 to the earlier part of stage 3 (Fig. 2.26). Perhaps the most striking aspect of this climatic sequence is reflected in the associated faunal assemblages (Bordes & Prat 1965; Laquay 1981). As at Combe Grenal, the lower part of the last-glacial sequence (layers 4E-2G) is marked by a heavy predominance of red deer and other temperate species such as wild pig, roe deer and aurochs, with only very sporadic remains of reindeer. Reindeer first appear clearly in level 4A2 and rise rapidly to dominance in the upper part of layer 2, apparently associated with the early glacial maximum of isotope stage 4 (Fig. 2.25). The entire sequence at Pech de l'Azé II therefore provides in most respects a close parallel for that documented in the lower and middle parts of the Combe Grenal sequence. The Pech de l'Azé II sequence is particularly important in confirming the overall chronology proposed for this part of the climatic succession at

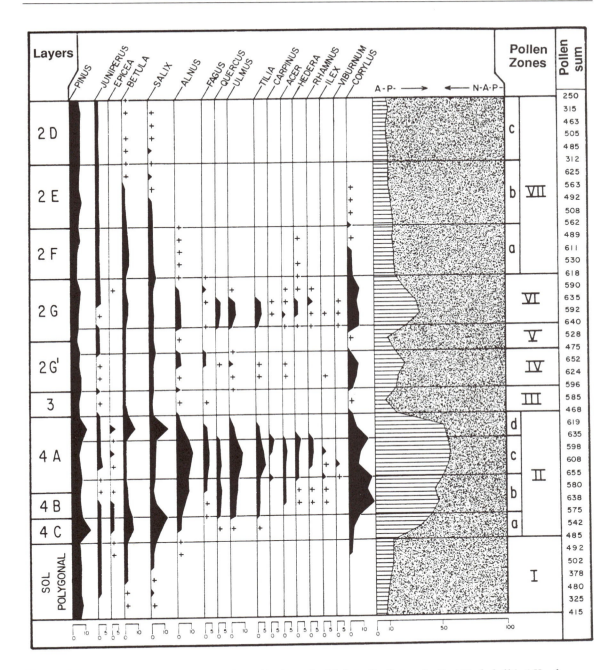

Figure 2.24 *Pollen sequence through the earlier last-glacial deposits (layers 2–4) at Pech de l'Azé II, after Paquereau 1969. The proposed chronology for the sequence is indicated in Fig. 2.25.*

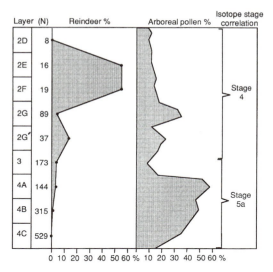

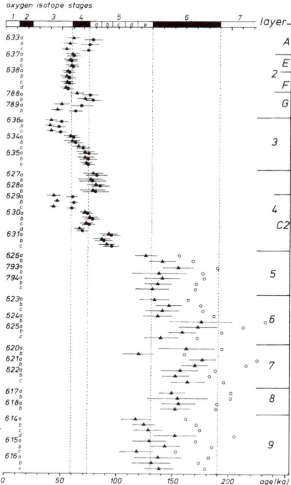

Figure 2.25 *Proposed correlation of the faunal and vegetational sequence at Pech de l'Azé II with isotope stages 5a and 4 in deep-sea cores. As shown in Fig. 2.26, ESR dates for layer 4 centre on ca 75–80,000 BP, while dates for layer 2 and 3 range from 55–72,000 BP. From Grün et al. 1991.*

Figure 2.26 *Electron-spin-resonance (ESR) dates obtained on animal teeth for the sequence at Pech de l'Azé II. Dates based on the early-uptake (EU) model of uranium uptake in the teeth are indicated by triangles; the slightly earlier dates based on the linear-uptake (LU) model are indicated by circles. The lower levels in the sequence (layers 5–9) contain Acheulian industries and evidently date from the period of isotope stage 6. After Grün et al. 1991.*

Figure 2.27 *Photograph of the stratigraphic section in the lower shelter at Le Moustier (see also Fig. 2.28).*

Combe Grenal and, in particular, reinforcing the specific correlations proposed with the general climatic succession in deep-sea cores.

The stratigraphic and climatic sequence recorded in the lower shelter at Le Moustier (Figs 2.27–2.29) is important because it complements the evidence recorded at Combe Grenal and Pech de l'Azé II by adding much more detail to the pattern of climatic events during the later stages of the Mousterian sequence between ca 55,000 and 40,000 BP

(Peyrony 1930; Laville 1975; Laville *et al.* 1980; Paquereau 1975; Valladas *et al* 1986; Mellars & Grün 1991). As noted earlier, these final stages in the Mousterian sequence are not well represented at Combe Grenal and cannot be dated with any confidence or precision in absolute terms. By contrast, we now have what would seem to be an internally coherent and apparently secure chronology for the stratigraphic sequence in the lower shelter at Le Moustier, based on long and detailed sequences of both TL measurements

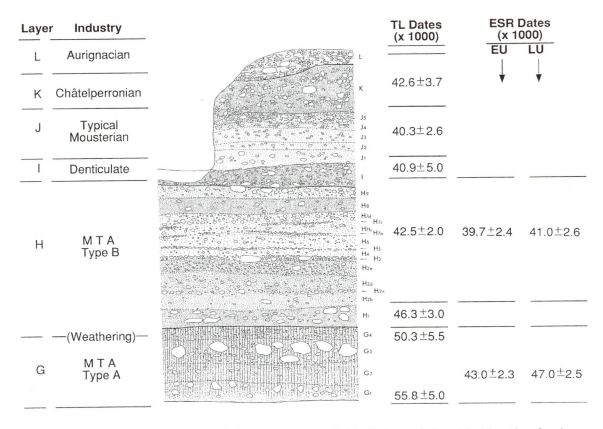

Layer	Industry		TL Dates (x 1000)	ESR Dates (x 1000)	
				EU	LU
L	Aurignacian			↓	↓
K	Châtelperronian		42.6±3.7		
J	Typical Mousterian		40.3±2.6		
I	Denticulate		40.9±5.0		
H	M T A Type B		42.5±2.0	39.7±2.4	41.0±2.6
			46.3±3.0		
—(Weathering)—			50.3±5.5		
G	M T A Type A			43.0±2.3	47.0±2.5
			55.8±5.0		

Figure 2.28 *Stratigraphic and archaeological sequence recorded in the lower shelter at Le Moustier, showing the results of TL dating of burnt flint samples carried out by Valladas* et al. *(1986) and ESR dating of animal teeth by Mellars & Grün (1991). Note that the ESR dates are calculated according to two different models of uranium uptake in the teeth (the 'early-uptake' and 'linear-uptake' models), which yield slightly different results; on both models the ESR dates suggest a slightly younger age for layer G (43–47,000 BP) than that suggested by the TL dating (50–55,000 BP). After Mellars & Grün 1991 (see also Fig. 6.19).*

of burnt flint samples (Valladas *et al.* 1986) and by a parallel series of ESR measurements on animal teeth from the same levels (Figs 2.28, 6.19) (Mellars & Grün 1991; see also Mellars 1986a). At present therefore the sequence recorded in the lower shelter of Le Moustier must be regarded as by far the best and most securely dated sequence of stratigraphic and archaeological levels so far documented from the later stages of the Mousterian succession in southwestern France.

The general climatic pattern recorded in the dated part of the stratigraphic sequence at Le Moustier, (i.e. between layers G and J) is summarized in Fig. 2.29. As will be seen, the pollen record throughout the greater part of the sequence indicates a succession of predominantly cold climatic conditions characterized by generally low frequencies of arbo-

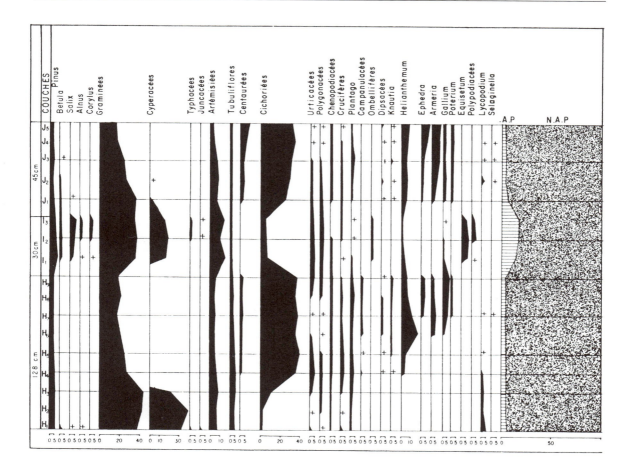

Figure 2.29 *Pollen sequence recorded by Paquereau (1975) through the upper levels (layers H-J) at Le Moustier. For details of the stratigraphy and dating of the sequence, see Fig. 2.28.*

real pollen, mostly between 2 and 5 percent, and with pine as the only tree species represented in the majority of the levels (Paquereau 1975). At three separate points in the sequence, however, there are indications of phases of much milder climate. In the pollen records, traces of apparent interstadials can be recognized in level G2 and in levels I1–I3, in each case marked by a sharp rise in the overall frequency of tree pollen from 2–5 to around 15–20 percent and with the appear-

ance of more warmth-demanding deciduous species such as birch, alder, hazel and willow. In the sedimentary record there is evidence for an additional and more major climatic amelioration coinciding with the interface between levels G and H (Laville 1975; Laville *et al.* 1980). In this case there is no direct pollen evidence for the climatic amelioration – since the phase was evidently marked more by weathering and erosion of the existing deposits than by the accumula-

tion of new sediments – but the presence of a
major warm interval is strongly indicated by
the presence of a heavily weathered soil horizon which penetrates for a depth of almost
60 cm into the underlying deposits of layer G
(Fig. 2.28). In the earlier literature this level
was generally taken to represent the position
of the main 'Würm I/II' interstadial within
the Le Moustier succession, before the recent
redating of the site by Valladas *et al*. (see
Laville 1975, 1988; Laville *et al*. 1980: 174–181;
1986; Mellars 1986a, 1988). In all we can
probably recognize a succession of at least
three separate interstadial episodes within
the later stages of the Le Moustier sequence
separated by phases of much colder climate
and all lying within the time range of
55–40,000 BP (Figs 2.28, 2.29). In archaeological terms, these levels were associated
with typical Mousterian of Acheulian tradition (MTA) industries (of both Type A and
Type B forms: Bourgon 1957) and with a succession of faunal assemblages heavily dominated by remains of large bovids (i.e. either
Bos or *Bison*) (Peyrony 1930; S. Madelaine,
personal communication). All these climatic
oscillations fall within the time range of isotope stage 3 of the deep-sea core sequences,
which can be seen again as a period of relatively complex, sharply fluctuating climate.
To attempt a more specific correlation of
these interstadials with those recorded in
some of the other climatic sequences summarized in Fig. 2.16, however, might be premature for the reasons already discussed.

The precise dating and correlation of the
various climatic fluctuations recorded in the
basal part of the Le Moustier sequence
(between levels A and F) remain for the
present much more debatable. As Laville *et
al*. (1980: 174–7) have emphasized, these
deposits represent an entirely separate episode in the geological history of the Le
Moustier site, dominated by a series of fluviatile deposits laid down during successive
periods of flooding by the adjacent Vézère
river. There is almost certainly a major strati-

graphic and erosional hiatus between the
formation of these levels and that of the
overlying layers G–J which, in the absence of
any direct dating evidence, could relate to
almost any point within the last glacial
sequence. The pollen evidence from these
levels reveals two major warm episodes
which might well equate with those recorded
in levels 50–52 and 41–43 at Combe Grenal
(i.e. coinciding with isotope stages 5c and 5a)
(Paquereau 1975; Laville *et al*. 1980: 181; Laville 1975: 186–7). In the absence of any independent chronological control for this part of
the Le Moustier sequence, however, it might
be unwise to press these correlations any
further.

Although not within the Middle Palaeolithic time range, it should be noted that there
is evidence for a further, well defined interstadial during the opening stages of the
Upper Palaeolithic sequence in southwestern
France, coinciding broadly with the earlier
stages of the Châtelperronian. This can be
seen in the detailed pollen sequences from
both Saint-Césaire and Quinçay and is
strongly hinted at in the contemporaneous
pollen and sedimentological sequences from
Les Cottés, Les Tambourets, La Ferrassie,
Trou de la Chèvre and, further to the north-
east, at Arcy-sur-Cure (Fig. 2.30) (Leroi-
Gourhan 1984; Leroyer 1986, 1988, 1990;
Paquereau 1984; Leroi-Gourhan & Renault-
Miskovsky 1977; Renault-Miskovsky &
Leroi-Gourhan 1981; Laville *et al*. 1980). The
existence of a major interstadial at this point
in the early Upper Palaeolithic is now so
widely recognized that it has been formally
designated by Leroi-Gourhan and Renault-
Miskovsky (1977; also Renault-Miskovsky &
Leroi-Gourhan 1981) as the 'Les Cottés' interstadial. Indeed, in the latest stratigraphic
schemes of Laville, this has been elevated to
the most significant interstadial in the whole
of the last-glacial succession in southwestern
France and accordingly designated the 'interstade Würmien' (Laville *et al*. 1986; Laville
1988; Guadelli & Laville 1990). The exact

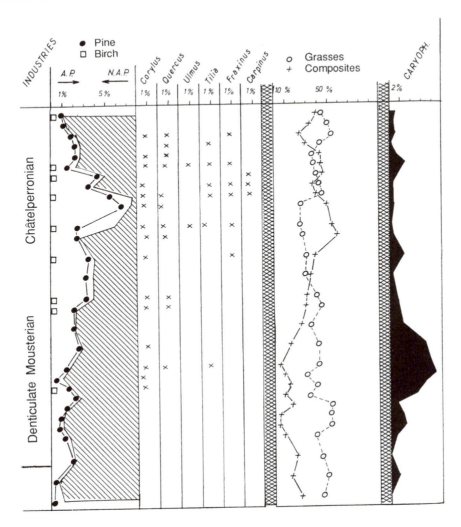

Figure 2.30 *Pollen sequence recorded through the later Mousterian and Châtelperronian levels at Arcy-sur-Cure (central France), showing evidence for a major climatic amelioration during the formation of the Châtelperronian levels. This interval is generally referred to as the 'Les Cottés' interstadial in France, and assumed (more tenuously) to correlate with the Hengelo interstadial in the Netherlands. After Leroi-Gourhan 1988.*

dating of the interstadial remains ambiguous, mainly owing to the imprecision of most radiocarbon dates in this time range, but is generally estimated around 36,000 BP (Leroi-Gourhan & Renault-Miskovsky 1977; Renault-Miskovsky & Leroi-Gourhan 1981). Certainly, the amelioration is known to have come to an end by ca 34,000 BP when there is clear evidence for the re-establishment of extremely cold full-glacial conditions coinciding with the appearance of the earliest Aurignacian industries in the Perigord region (Movius 1975; Farrand 1975, 1988; Renault-Miskovsky & Leroi-Gourhan 1981).

The obvious correlation, therefore, would be to regard this interval as the equivalent of the so-called 'Hengelo' interstadial, now documented widely from many other parts of northwestern Europe and similarly dated by radiocarbon to around 36–38,000 BP (Figs 2.10, 2.16: Van der Hammen *et al.* 1967; Leroi-Gourhan & Renault-Miskovsky 1977; Zagwijn 1990). In absolute terms we should probably regard this interval as lying closer to ca 39–41,000 BP, if we take into account the accumulating evidence for an approximately 3,000-year displacement in the radiocarbon time-scale over this period (Bard *et al.* 1990a). In any event, this well defined interstadial during the earliest stages of the Upper Palaeolithic sequence must be seen as a further major component within the overall sharply oscillating climatic succession of oxygen–isotope stage 3.

Southwestern France as a human habitat

One of the most striking and widely recognized features of the archaeological record of the Perigord and adjacent areas of southwestern France is the sheer wealth and abundance of the evidence for Palaeolithic occupation. This pattern is well documented for both the Middle and Upper Palaeolithic periods and is reflected not only in the overall totals of occupied sites – running into several hundred for both the Middle and Upper Palaeolithic (see Chapter 8) – but also in the density and concentration of the archaeological remains recovered from the sites (Mellars 1985). Making allowance for various possible forms of bias in the survival and documentation of the archaeological evidence (such as the long history of Palaeolithic research in the region and the favourable conditions for preservation of occupation deposits in cave and rock shelter sites) no one would seriously question that the southwestern French region has yielded the richest and most concentrated record of human occupation during the various stages of the last glacial sequence so far documented in Europe (Sackett 1968; Laville *et al.* 1980; David 1985).

No doubt part of the explanation for this wealth and intensity of Palaeolithic occupation lies in some of the basic topographic and geological features of the Perigord region – including the abundance of naturally protected cave and rock shelter sites, the obvious protection offered by these sheltered valley habitats against harsh glacial climates and the relative abundance and widespread distribution of high quality flint supplies for tool manufacture (see Chapter 5). It is arguable, however, that the most critical features which attracted and supported a high level of human occupation in this region throughout the last glacial period were specific local climatic and ecological conditions. As I have discussed more fully elsewhere (Mellars 1985) the main features of these conditions are as follows:

1. All attempts to map the distribution of ecological zones throughout Europe under the various 'glacial' regimes of the Pleistocene period indicate that the southwestern zone of France must have supported the most southern areas of essentially open tundra or steppe-like vegetation within the European continent (e.g. Butzer 1972: Fig. 51; Iversen 1973: 16–17). The explanation for this is related simply to the highly maritime character of the climate in this extreme western zone of Europe, which keeps summer temperatures to a much lower level than those experienced in more continental zones of central and eastern Europe and provides correspondingly less favourable conditions for tree growth. As shown in Fig. 2.31, the latitudinal margin of forest growth tends to shift progressively towards the north as one moves from west to east across Europe into regions of increasingly continental climate and correspondingly warmer summers. The important implication is that these extreme

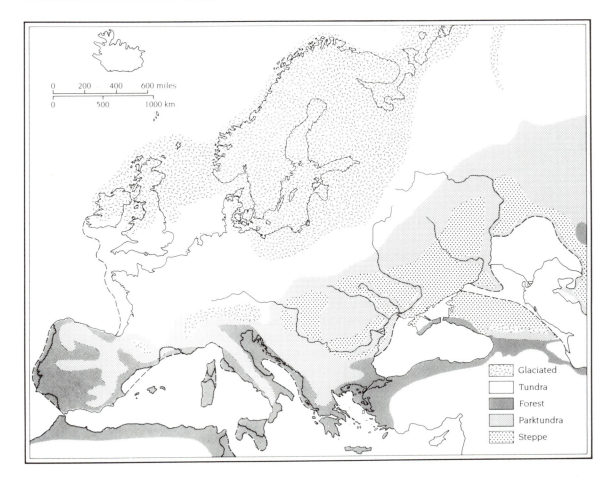

Figure 2.31 *Reconstruction of the major vegetation zones, ice sheets, and coastlines in Europe at the time of the last glacial maximum, ca 18–20,000 BP, according to Iversen 1973. Note how the zone of open tundra and steppe vegetation extends much further to the south in western Europe than in central and eastern Europe.*

southern tundra and steppe landscapes received exceptionally high levels of solar radiation (due to their latitude) and consequently would have supported some of the highest levels of plant productivity within the European continent (Butzer 1972: 463). As a habitat for many open-living herbivorous species, including reindeer, horse, bison etc., these tundra and steppe zones of southwestern France were probably unique within the last-glacial landscapes of Europe.

2. A second factor which would almost certainly have had a major influence on the overall carrying capacity of southwestern France for animal populations was the relative oceanicity of the climate and the corresponding mildness, by glacial standards, of the winters. Accurate temperature estimates are notoriously difficult to derive from palaeoecological evidence, but one estimate based on botanical data has suggested an annual temperature range in the Perigord

region around the time of the last glacial maximum from ca 12–15°C in summer to 0°C in winter (Wilson 1975: 185). Temperatures would have risen a few degrees higher than this during the major interstadial phases of the last glaciation.

The comparative mildness of the winters would have benefited animal as well as human populations in a variety of ways, most notably by reducing the depth and duration of snow cover and by ameliorating some of the worst effects of the punishing weather conditions which must have posed a major obstacle to human and animal survival in some of the central and eastern zones of Europe (e.g. Gamble 1983). Possibly the most important consequence of these milder winters, however, would have been to extend the duration of the growing season for most species of plants and therefore to increase substantially the quantities of forage which were available to animal populations throughout the winter season. This in turn would have had a major impact on the overall carrying capacity of the region for animal herds, since it is now generally recognized that it is the availability of *winter* forage, especially during the late winter and earliest spring months, that represents the most critical factor in determining the overall density and biomass of animal populations which can be supported in any region on a long-term basis (Moen 1973: 404–13).

3. Whilst the abundance and concentration of local animal populations was no doubt one of the critical factors in supporting high human population densities, it is arguable that the most important single factor was the sheer variety and diversity of local ecological conditions within this region (Jochim 1983; Mellars 1985). The Perigord region is characterized by two contrasting types of habitat: the very open, exposed environments on the extensive limestone plateaux; and the sheltered habitats within the major river valleys which dissect these plateaux. At present these areas are characterized by significantly different micro-habitats – with generally warmer climatic conditions in the valleys associated with greater protection from wind exposure, and richer, deeper and more fertile soils than on the higher, more exposed plateaux (Laville 1975; Duchadeau-Kervazo 1982).

This complex interdigitation of plateau and valley habitats helps to explain the ecological diversity reflected within both the palaeobotanical and faunal evidence from the southwestern French sites throughout the last glacial period. As Paquereau (1979) has pointed out, the major river valleys in the Perigord and adjacent areas almost certainly supported some tree growth throughout the last glacial succession. During the coldest periods this would have been reduced to the hardier species such as pine and birch, perhaps confined mainly to the deeper and more sheltered valleys. During the milder interstadial periods, however, this range of species expanded to include a variety of specifically warmth-loving species such as oak, elm, lime, hazel and alder. As Paquereau (1979) and others have emphasized, we should visualize vegetational conditions in the Perigord and adjacent areas as a mosaic of contrasting communities with the zones of woodland extending mainly along the more sheltered valleys and predominantly open, tundra or steppe-like vegetation flourishing over the higher and more exposed plateau areas.

This element of ecological diversity is equally reflected in the composition of the faunal assemblages from Middle and Upper Palaeolithic sites (Jochim 1983; Mellars 1985; Delpech 1983). As discussed in Chapter 7, many of the sites show a predominance of one particular species (normally either reindeer, red deer, horse or large bovids) which could indicate either a deliberate selection for the exploitation of these particular species by the human groups themselves, or alternatively, the specific character of the local ecological conditions within the immediate environment of each site. However, the most remarkable aspect of the faunal evidence is the exceptional range of different animal species represented in most sites and the ways in which their remains normally occur, side by side, within the same occupation levels.

In economic terms the crucial importance of ecological diversity lies in the degree of security which this provides against periods of occasional failure of particular economic resources (Drury 1975). Even if the populations of some animal species, such as reindeer, may occasionally have suffered rapid declines as a result of short-term ecological fluctuations – or even overexploitation by the human groups themselves (Mithen 1989, 1993) – it is likely that the human communities would still have had access to many other, complementary species of animals to tide them over these periods of temporary resource failure. It is this aspect of diversity in economic resources which helps to explain the capacity of southwestern France to support dense and concentrated human populations and, apparently, to provide a high degree of economic security for them over long spans of time.

4. The fourth aspect of ecological conditions in southwestern France which has been emphasized in several studies (e.g. Drury 1975; Mellars 1985; Raynal & Guadelli 1990) is the marked compression or steepening of ecological zones along an east-to-west axis which characterized this region throughout the last glacial period. The general topography and relief of the Perigord and adjacent areas is such that there is a natural and fairly rapid succession of topographic and environmental zones as one moves progressively westwards from the higher elevations of the Massif Central towards the coastal Atlantic Plain. Under glacial conditions, however, these ecological gradients were made even steeper by the presence of local glaciers in some of the higher elevations of the Massif

Central (Fig. 2.32) and by sharply reduced temperatures within these upland areas. As Raynal and Guadelli (1990) have pointed out, there appears to be evidence for this in some of the recent pollen records from the Massif and adjacent areas which point to a rapid change in vegetational and associated climatic patterns between the Massif Central and the Atlantic coast.

The direct effect of this steepening of ecological gradients from east to west was to add a further dimension of potential ecological diversity to the overall range of environments available to the human communities. The patterns of animal migration within this region are still subject to some controversy (e.g. Bouchud 1966; Spiess 1979; Gordon 1988; Boyle 1990; Pike-Tay 1991, 1993; Burke 1993) but there can be little doubt, on purely ecological grounds, that some of the major migratory movements of species such as reindeer, and perhaps horse and red deer, were along an essentially west–east trajectory from the more low-lying and sheltered valley areas of the Perigord and adjacent areas during the winter months towards the higher elevations of the Massif Central and its immediate foothills during the summer. Bouchud (1966) and others, however, have argued that any migrations along this axis are likely to have been on a limited scale, and unlikely to have extended more than perhaps 80–100 km between winter and summer ranges. With the sharp compression of climatic and ecological gradients along this east–west axis during the coldest glacial episodes, it could be argued that any migrations along this axis were even further reduced. The implication is that these migratory animal populations were probably never very far from the Perigord region during any part of the annual cycle, and presumably were accessible for exploitation by the local human groups within at most a few days of travel either towards the middle or upper foothill zones of the Massif Central, or perhaps into the coastal Atlantic Plain (Mellars 1985).

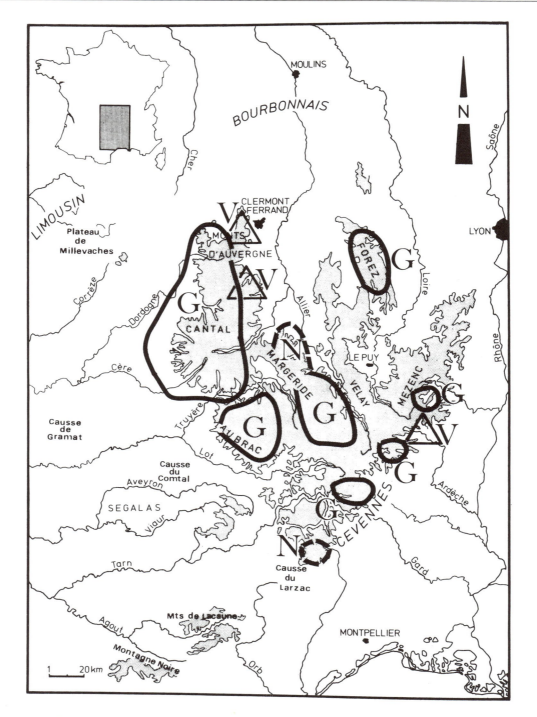

Figure 2.32 *Estimated extent of glaciers in the Massif Central region of central France at the time of the last glacial maximum, ca 18–20,000 BP (after Daugas & Raynal 1989). The presence of these glaciers would inevitably have created much sharper east-west gradients in climatic and ecological conditions across the southwestern French region than at the present day.*

5. The final point concerns the probable role of the major river valleys which traverse the Perigord region as the major, habitual migration trails of species such as reindeer. This has been discussed in several earlier studies (e.g. Bouchud 1966; Spiess 1979) and may be critical to understanding many aspects of the detailed distribution of both Middle and Upper Palaeolithic sites in this region. The point is that by locating settlements or hunting locations directly astride these major migration trails it was possible for human groups to intercept animal populations deriving from relatively large territories within southwestern France – i.e. the combined summer and winter ranges – at a single location (Mellars 1985: 280). As an explanation for the remarkable concentration of Middle and Upper Palaeolithic sites at particular locations in the Perigord region (for example in certain sections of the Vézère and Dordogne valleys: see Figs 8.1, 8.2)

this is probably the most important single factor.

The factors discussed above are very general features of the ecological and environmental conditions within southwestern France, and their precise character would have fluctuated sharply and repeatedly during the different chronological and climatic phases of the Upper Pleistocene. We are still, unfortunately, very poorly equipped with information on some of the more specific aspects of environmental patterns, most notably reliable estimates of temperature conditions, varying snow-fall regimes and the extent of seasonal contrasts in climate. It is against this background that we will examine the archaeological records of the behaviour and organization of Neanderthal communities within these western fringes of Europe.

CHAPTER 3

Stone Tool Technology ————————————

Studies of stone-tool technology have always occupied a central position in approaches to the Middle Palaeolithic. The reasons for this are evident. Here, as in the rest of the Palaeolithic, stone-tool assemblages provide by far the most durable and complete record of human development with a degree of continuity and fine-scale resolution which is much better than that of the associated faunal assemblages and far more complete than that of the skeletal remains of the populations involved. Not surprisingly, the intricacies of changes in stone tool flaking techniques and the fine, chronologically patterned changes in the forms of stone tools have always provided the principal framework both for constructing regional chronological sequences and for documenting divergences in patterns of technological and cultural development in different regions.

As a result of more sophisticated approaches to analysis developed over the past two decades, however, it is now clear that sympathetic approaches to the study of lithic technology can go much further. As the subsequent chapters will show, a correct understanding of the nature and structure of stone-tool technology can shed important light on the patterns of movement of human groups over the landscape; the particular economic and technological activities carried out in different sites, and potentially on the mental and cognitive processes which lay behind the production of the tools. There is

little doubt that these and other approaches to the analysis of stone-tool technology will continue to provide a central focus of research into the behaviour and organization of Palaeolithic communities well into the future.

The analysis of stone-tool technology can be carried out at many different levels, each posing its own particular problems of methodology and interpretation and each providing rather different insights into the structure and organization of the activities which lay behind the production of stone-tool residues. The most basic aspect, which forms the focus of this chapter, is the strategies by which the available sources of flint or other raw materials were systematically reduced into various flakes, blades or other blank forms either for immediate use or for subsequent modification into a range of retouched tool forms. Second, there is the question of how these initial blank forms were reduced into more regular, extensively retouched implement types and the significance in functional, stylistic or other terms which can be attached to these morphologically or typologically distinct tool categories (discussed in Chapter 4). Equally significant are the strategies and procedures by which the original sources of raw materials were exploited by Middle Palaeolithic groups and subsequently distributed across the landscape for use or for further reworking on eventual occupation or special-activity sites (Chapter 5). Finally, and currently most controversial, there is the question of precisely what significance should be attached to the bewildering technical and typological variations which can now be documented over the wide space and time range of the Middle Palaeolithic. Since the latter issues have effectively monopolized much of the literature on the Middle Palaeolithic over the past thirty years, they will be raised and discussed in some detail in Chapter 10.

Primary flaking technology

Studies of the primary flaking techniques by which available supplies of flint or other raw materials were systematically reduced into suitable blanks or pre-forms for tool production are fundamental to any study of Palaeolithic technology. This is an area in which extensive research has been carried out over the past ten years and in which some of the most impressive advances have been made. Credit for these developments belongs mainly to the French school of Palaeolithic studies, stimulated initially by the pioneering research of François Bordes (Bordes 1947, 1950a,b, 1953a,b, 1954a, 1961a, 1980, 1984; Bordes & Bourgon 1951). It was Bordes who introduced the notion of controlled quantitative approaches to the study of Lower and Middle Palaeolithic technology and thereby laid the foundations for all later systematic studies of technological variation and development (Bordes 1950a, 1961a). It was Bordes, too, who was responsible for some of the earliest experimental approaches to lithic technology, involving not only a personal mastery of different Palaeolithic techniques but also controlled experiments to investigate the effects of different flaking strategies on both the forms of finished tools and the kinds of flaking debris resulting from their manufacture (Bordes 1947).

More recent studies have developed these approaches in a variety of ways. Further experiments in the systematic replication of Palaeolithic flaking techniques have been undertaken by Jaques Tixier, for example, and more recently by a number of his students (most notably Eric Boëda and Jacques Pelegrin) working in the Laboratory for Prehistory and Technology at Meudon (Tixier 1978; Tixier et al. 1980; Boëda 1982, 1986; Pelegrin 1986, 1990; Boëda & Pelegrin 1983). Similar approaches have been developed by Bruce Bradley (1977), Harold Dibble and others in the United States, by Mark Newcomer (1971) and Peter Jones (1981) in Britain, and by Jean-Michel Geneste (1985), Alain Turq (1989b, 1992a, 1992b), Liliane Meignen and several other workers in France. Equally important advances over the same period have come from systematic studies of the detailed spatial distribution of lithic artefacts and flaking debris over occupation surfaces, combined with painstaking reconstructions of the products of knapping debris (Fig. 3.2), to allow insights into the overall sequence and underlying strategies of the flaking procedures involved (Tuffreau & Sommé 1988; Rigaud 1988; Geneste 1988; Révillion 1989; de Heinzelin & Haessaerts 1983; Roebroeks 1988; Schlanger 1994). All these approaches are directly relevant not only to understanding the various stages of production of lithic artefacts, but to investigating the basic conceptual and mental processes which lay behind the sequence of planning and manufacture of stone tools.

The central concept which underlies all of these recent studies of lithic technology is generally referred to in the French literature as the chaîne opératoire approach, and in the English literature as 'lithic reduction' studies (e.g. Tixier 1978; Geneste 1985, 1991; Boëda 1986; Boëda et al. 1990; Bradley 1977). All these approaches are based on the recognition that the entire process of production and shaping of stone artefacts is essentially a reductive procedure which passed through several discrete and separate stages (see Tables 3.1, 3.2; Fig. 3.1). The process started with selection of suitable nodules or other blocks of material which were assessed,

Table 3.1

*Principal stages of flake and tool production, use and discard
recognized in the reduction sequence scheme* (chaîne opératoire) *of Geneste
(1985: 179).*

Acquisition stage

Stage 0:   Extraction and testing of nodule

Production stages

Stage 1:   Decortication of nodule
Initial shaping of core
Preparation of striking platforms

Stage 2:   Production of primary flake blanks (flakes, blades etc.)

Shaping/retouching stage

Stage 3:   Retouching of tools

Utilization stage

Stage 4:   Use of retouched and/or unretouched pieces
Resharpening/reworking of tools

Discard stage

Stage 5:   Breakage
Terminal edge-wear/damage
Discard

The major products generated during the different reduction stages are
listed in Table 3.2.

usually by one or two trial blows, as suitable for the flaking procedures envisaged. In most cases it then proceeded through a stage of systematic 'decortication' of the nodule to remove the outer covering of irregular cortex or skin. Once this procedure was completed, there was usually a phase of shaping the core into a more or less regular pre-form from which flakes of a preselected shape, size or regularity could be removed. Subsequent stages in the sequence included successive removals of these preferred flake forms, usually accompanied by intermittent episodes of reshaping or correcting the core to allow further, controlled flake removals. The final stages in the reduction sequence involved selecting specific flakes for use either in the unretouched state or for systematic shaping by retouch into regular implement forms. The process could be extended by subsequent reworking or resharpening of the edges of the tools as they became blunted or damaged in the course of use. Finally the use-life of the tool was effectively exhausted and the piece was discarded to form part of the accumulating refuse on the site. This is an idealized sequence and in particular contexts certain stages of the reduction sequence were either attenuated or even omitted – for example in some of the simpler, non-Levallois techniques where the initial phase of systematic decortication of the nodules was often by-passed. Nevertheless, as a basic conceptual framework for analysing the different

Table 3.2

Principal types of products recognized by Geneste in his reduction sequence
(chaîne opératoire) *scheme for the production of Levallois flakes*

Technological phase	Product no.	Type of product
0	0	Unworked/tested nodule
	1	Cortical flake (>50%)
1	2	Partial cortical flake (<50%)
	3	Natural backed knife
2A	4	Ordinary flake/point
	5	Ordinary blade
	6	Atypical Levallois flake or blade
	7	Levallois flake
	8	Levallois blade
	9	Levallois point
	10	Pseudo-levallois point
2B	11	Discoid core
	12	Diverse core types
	13	Levallois flake or point core
	14	Levallois blade core
	15	Edge-removal flake
	16	Crested flake/blade
	17	Core fragment
2C	18	'Kombewa' type flake core
	19	Truncated and thinned flake
	20	'Kombewa' flake
	21	Unidentifiable non-cortical flakes
3	22	Biface-retouch flake
	23	Other retouch/resharpening flakes
Miscellaneous	24	Flaking debris >30 mm
	25	Flaking debris <30 mm
	26	Small flakes/fragments <30 mm

After Geneste (1988): Table 1.

technological strategies of stone tool production and the various kinds of flaking debris encountered on archaeological sites, the *chaîne opératoire* approach has now become central to all modern studies of lithic technology (see Boëda *et al.* 1990).

These recent analytical approaches to Middle Palaeolithic technology are still in their early stages and have as yet been applied to a limited number of specific sites and assemblages. So far, detailed refitting studies have been applied to only a handful of sites, for example at Maastricht-Belvédère in Holland (Roebroeks 1988; Schlanger 1994) and Saint-Valéry-sur-Somme, Seclin and Grotte Vaufrey in France (de Heinzelin & Haesaerts 1983; Révillion 1989; Rigaud 1988; Geneste 1988) (e.g. Fig. 3.2). Nevertheless, the results of these studies are sufficient to demonstrate the range and diversity of different flaking techniques employed in the Middle Palaeolithic, and the complexity and apparent

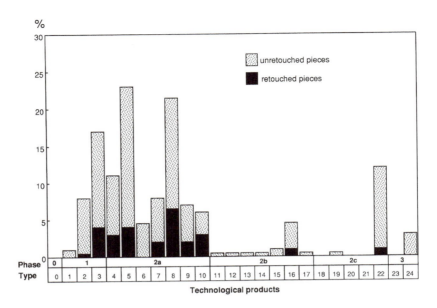

Figure 3.1   *Analysis of the retouched and unretouched pieces in the assemblage from layer VII of Grotte Vaufrey, according to the 'chaîne opératoire' scheme of Geneste (1985). (See Table 3.1 for a list of the numbered technological products and stages.) After Geneste 1988.*

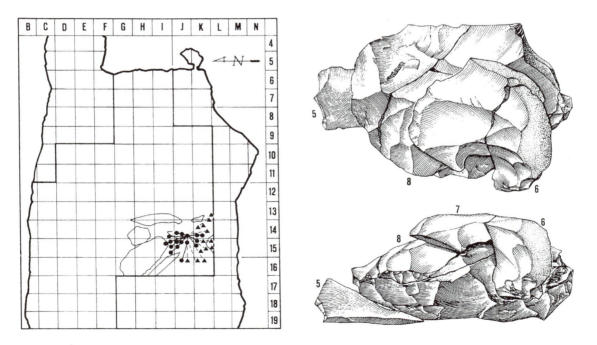

Figure 3.2   *Refitted block of flakes from level VIII of the Grotte Vaufrey. The spatial distribution of the refitted pieces within this level is shown on the left. After Geneste 1988.*

sophistication of these strategies in both conceptual and technological terms. The aim of the following sections is to illustrate this with particular case studies. For convenience and by historical convention these have been divided into the broad categories of 'Levallois' methods, 'non-Levallois' methods and 'blade techniques'. However, this categorization is far from rigid and the inherent fluidity and flexibility of the different flaking strategies in many cases cross-cuts these conventional technological categories (Schlanger 1994).

Levallois techniques

Levallois techniques have been recognized as one of the most distinctive hallmarks of Middle Palaeolithic technology since the original definition of the Mousterian by Gabriel de Mortillet in the late nineteenth century (de Mortillet 1883: 240, 255; see also Commont 1909: 122). The central and diagnostic feature of these techniques has always been seen as the attempt to control or predetermine the overall shape and size of the intended flakes by means of careful preparation of one face of the parent core. The classic definition of the Levallois concept is embodied in the widely quoted definition of Levallois flakes proposed by François Bordes: i.e. 'un éclat à forme prédéterminée par une préparation spéciale du nucléus avant enlèvement de cet éclat' (Bordes 1961a: 14; 1980: 45). Later confusions have arisen mainly from the wide variety of forms of flakes and associated core types which can be accommodated within this definition. As Bordes himself was at pains to emphasize (e.g. 1950a, 1961a, 1980 etc.), Levallois techniques – defined by his criteria – could be used to produce a remarkable variety of different flake forms ranging from 'classic' forms of broad, oval flakes showing distinctively converging 'centripetal' patterns of core preparation (Figs 3.3, 3.4) through to more elongated, tapering forms

which could be variously described as either 'Levallois points' or 'Levallois blades' depending on the particular morphological or metrical criteria employed (Figs 3.7, 3.8, 3.18). With this range of diversity, it is hardly surprising that the literature has been plagued by debates on the exact implications and definitions of Levallois flaking. (See for example, Boëda (1993b) for a general review of these debates and Copeland (1983) for a discussion of some of the confusions over the use of the term Levallois in the literature on Middle Eastern industries.)

Recently, several attempts have been made to gain new insight into the essential character of Levallois flaking techniques, partly to resolve some of these long-standing terminological confusions and partly to clarify the technological strategies which underlay these procedures (e.g. Bradley 1977; Shchelinskii 1974, 1983 (in Plisson 1988); Geneste 1985; Perpère 1989; Van Peer 1991, 1992; Delagnes 1990, 1992; Schlanger 1994). The most detailed studies which have been applied to European industries so far are those carried out by Eric Boëda in the course of his doctoral research at the University of Paris (Boëda 1982, 1984, 1986, 1988a,b,c, 1993a,b; Boëda et al. 1990; Beyries & Boëda 1983). Boëda has identified what he refers to as a basic 'Levallois concept', which to him represents the unifying element behind all flaking techniques to which the term Levallois can properly be applied. This is defined by a basic division in the initial stages of preparation and shaping of Levallois cores into two main components: first, the preparation of a continuous striking platform extending around most of the perimeter of the selected nodule – normally produced by successive blows delivered more or less vertically on the upper face of the core and extending over a substantial part of the lower face; and second, by the systematic shaping of the upper surface of the core by blows delivered from various points around the perimeter of this prepared striking platform

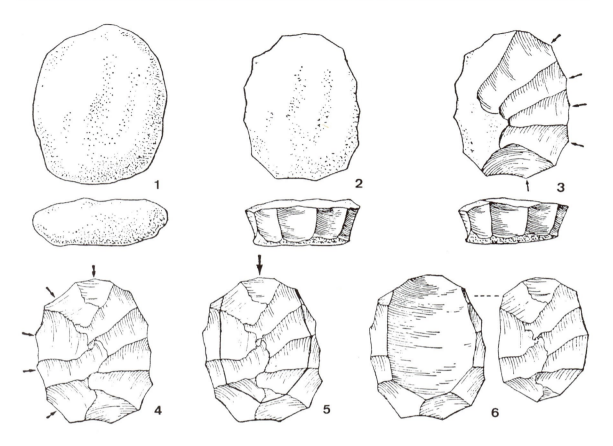

Figure 3.3   *Stages of production of a 'classic' Levallois core, according to Bordes 1961b.*

(Boëda 1988a; Boëda *et al.* 1990). As he points out, this defines effectively, in both an operational and a conceptual sense, two major components of the core form and at the same time effectively restricts all the subsequent primary flake removals to a prescribed and delimited area of the core surface (Figs 3.5, 3.9). Boëda emphasizes that this also defines the potential productivity of the core in a 'volumetric' sense by restricting the effective volume of flake production to the uppermost part of the core surface.

As Boëda points out, however, this still leaves scope for many different flaking strategies within the general heading of his Levallois concept and allows a surprising latitude in the form and character of the flakes produced by different techniques. Boëda (1988a, 1993b; Boëda *et al.* 1990) divides these different strategies into two main groups, which he refers to respectively as 'lineal' and 'recurrent' Levallois techniques. Within each a number of more specific strategies of core preparation led to a variety of flake products. Briefly, the main features of these different strategies can be summarized as follows:

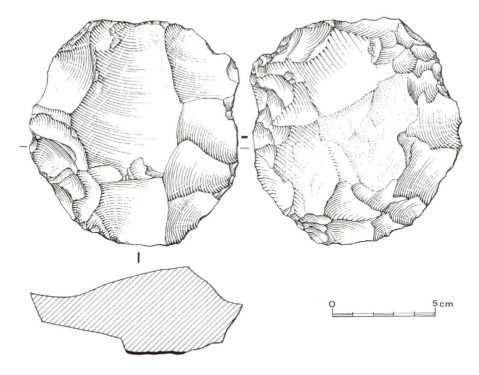

Figure 3.4 *'Classic' Levallois core from the site of La Borde (Lot). After Jaubert* et al. *1990.*

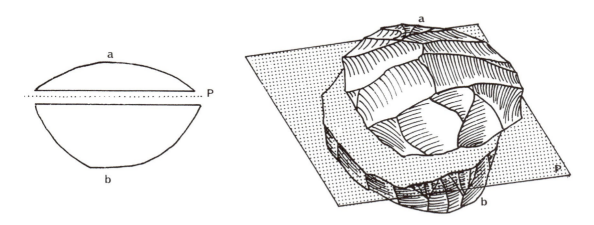

Figure 3.5 *Schematic representation of the basic 'Levallois concept' as defined by Eric Boëda. The core is divided into two major 'surfaces' (a and b) and two corresponding 'volumes'. The lower face is used to prepare the continuous striking platform around the perimeter of the core, while the upper face is carefully prepared for the production of one or more Levallois flakes The intersection of these two surfaces is defined by the plane 'P'. After Boëda 1988c.*

Lineal Levallois techniques

In terms of Boëda's definition lineal Levallois techniques are strategies designed essentially for production of only a single major flake removal from the prepared core surface. In this sense the definition coincides with the classic definition of Levallois flakes presented in most of the earlier textbooks (Fig. 3.3). In Boëda's research, these techniques were best represented at the site of Bagarre (Pas-de-Calais) in northern France which appears to date from a stage of the penultimate glaciation (Tuffreau & Zuate y Zuber 1975). The basic flaking strategy here coincides precisely with his overall definition of the Levallois concept – i.e. involving the initial preparation of a continuous striking platform extending around the perimeter of the nodule followed by careful shaping of the upper surface to define a single, clearly predetermined and apparently preconceived shape of flake from the prepared surface (Boëda 1982, 1984, 1988a, 1993b). The main point emphasized by Boëda, however, is that the precise character and orientation of this preparatory flaking was manipulated in different ways to produce a variety of contrasting forms in the finished flakes (Fig. 3.6).

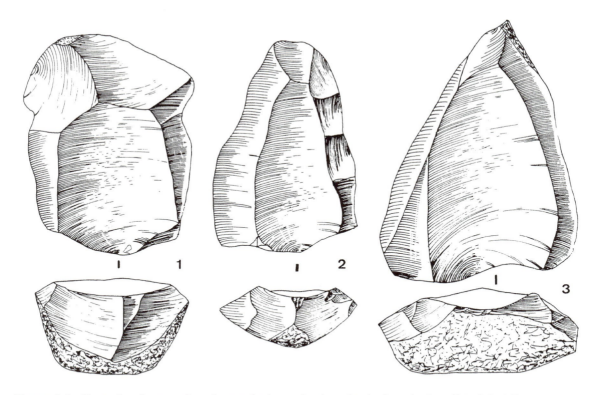

Figure 3.6 *Examples of cores oriented towards the production of a single major Levallois flake (illustrating Boëda's 'lineal Levallois' strategy) from the site of Bagorre (northern France). By different patterns of core preparation, flakes ranging in shape from rectangular (no. 1) to triangular (nos 2, 3) could be produced. After Boëda 1988a.*

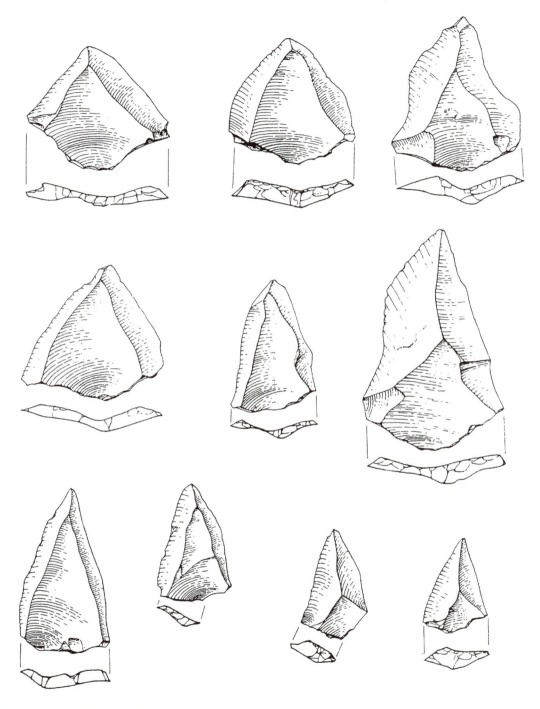

Figure 3.7 *Typical 'Levallois points' and similar forms from the later Mousterian levels of the Kebara cave (Israel). After Bar-Yosef & Meignen 1992.*

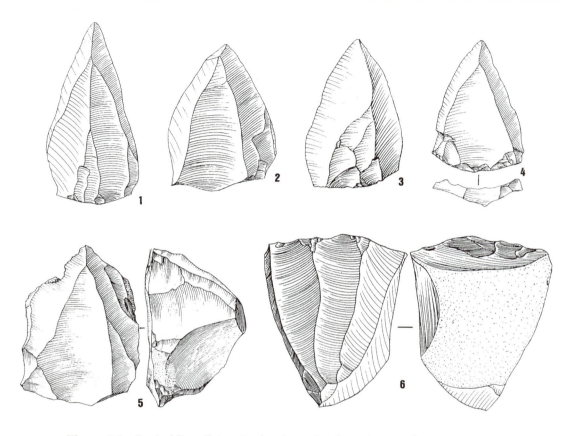

Figure 3.8 *Typical 'Levallois points' and associated core forms. After Bordes 1961a.*

Where the preparatory flaking was carried out in a predominantly radial or centripetal fashion from various points around the core perimeter, the resulting flakes are oval or rectangular in outline and show clear traces of this radial flaking on their dorsal surfaces. In other cases, however, the preparatory flaking was oriented either primarily or exclusively from either one or both ends of the core, leading to the production of flakes with a more elongated or triangular form (Fig. 3.6). At least some of the cores from Bagarre seem to have been intended for the production of typical, symmetrical 'Levallois points' of the kind illustrated in Figs 3.7 and 3.8. As

Boëda (1988a) stresses, the basic strategy of lineal Levallois flaking was therefore a highly flexible procedure which could be used to produce a variety of clearly differentiated forms of flakes simply by differing approaches in the initial stages of core preparation.

The notion of lineal Levallois techniques does not imply that the total capacity of a core was restricted to the production of only a single flake. As in all variants of Levallois techniques there was the opportunity, once the initial flake (or flakes) had been removed, to repeat the cycle of special preparation of the upper surface of the core to allow for the removal of further flakes from successively deeper levels in the core interior. In practice, sequences of up to four or five of these stratified flake removals could be produced in this way (Bradley 1977; Geneste 1985; Shchelinskii in Plisson 1988). The essence of the lineal procedure was simply that following the removal of each of these major flakes, a separate phase of preparation and reshaping of the core surface would be needed to allow for further flake removals.

Recurrent techniques

The definition of 'recurrent' Levallois techniques in Boëda's terms lies in the clear intention, from the initial stages of core preparation, to produce not one but a repeated succession of flakes of predetermined shape and size from the same, carefully prepared upper face of the core (Boëda 1988a) (e.g. Fig. 3.9). In this sense they are more economical in terms of flaking effort than the lineal techniques described above, requiring much less systematic reshaping or modification of the core between the successive flake removals. Boëda uses two main sites to illustrate these techniques – those of Biache-Saint-Vaast in the Pas-de-Calais (dating from around the middle of the penultimate glaciation, with thermoluminescence dates of ca 175,000 BP) and the later site of Corbehem (also in the Pas-de-Calais) dating from an early stage of the last glaciation (Boëda 1982, 1986, 1988c; Tuffreau 1979; Tuffreau & Sommé 1988). The detailed character of the flaking and core-preparation strategies recorded in these two sites show some major differences which led, inevitably, to corresponding differences in the resulting flake and core forms.

The principal flaking strategy employed at Biache-Saint-Vaast was intended for the production of long, narrow flakes and therefore depended on the production of elongated cores, usually of roughly rectangular form (Boëda 1988c). The initial stages in the prepa-

ration and shaping of these cores were effectively identical to those involved in the lineal or classical Levallois techniques and required the initial preparation of a continuous striking platform extending around the perimeter of the core, followed by a succession of centripetal flake removals to block out the domed, upper surface (Fig. 3.9). Once this initial phase of centripetal flaking was completed, however, the subsequent sequence of Levallois flake removals was carried out exclusively from either one or both ends of the core with the aim of producing a series of clearly elongated flakes. The principal feature which distinguishes it from the lineal procedures was that for each of the major sequences of flake removals a succession of either two or three flakes was removed from immediately adjacent points on the core surface without any intervening phase of deliberate preparation (or re-preparation) of the core surface (Fig. 3.10). In other words, whilst the first major flake removed in this sequence (i.e. the 'primary' flake) might be closely similar if not identical to that produced in the lineal techniques, the ensuing sequence of flake removals from the same surface will inevitably show the negative scars of the earlier removals along one or more of the margins of the resulting flake (see Fig. 3.11). In this way Boëda has been able to define a hierarchy of three separate types of adjacent, overlapping flake removals (which he refers to simply as first, second and third flakes) all removed in direct succession from immediately adjacent points on the core surface, and all showing distinctive scar patterns on their dorsal surfaces (Fig. 3.11). Where these sequences of adjacent flakes were removed from both ends of the core (i.e. in his 'bipolar' as opposed to 'unipolar' sequences) the resulting dorsal scar patterns are more complex but again can be categorized into a succession of primary, secondary and tertiary removals (see Fig. 3.10). The whole procedure seems to have been designed to produce a complex succession of elongated flakes of

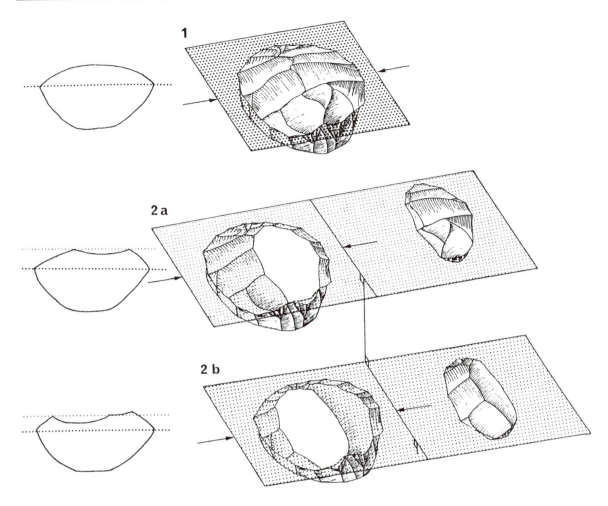

Figure 3.9 *Schematic representation of the 'recurrent unipolar' Levallois technique as defined by Boëda. In this case the upper face of the core is prepared to produce two immediately successive, overlapping Levallois flakes, with distinctive scar patterns on the surfaces of both the detached flakes and the residual core. After Boëda 1988a.*

variable, but predictable outlines which eliminated the need for individual preparation of each of these intended flakes. In this sense the technique is more economical in terms of time and flaking effort than the lineal. Classified in morphological terms, the flakes thus produced can vary from essentially classic, oval Levallois flakes to other forms normally described as either Levallois blades, or elongated forms of Levallois points (Boëda 1988a,c).

Although grouped together by Boëda under the same general heading of recurrent Levallois methods, the flaking technique documented at Corbehem is much simpler than that represented at Biache-Saint-Vaast

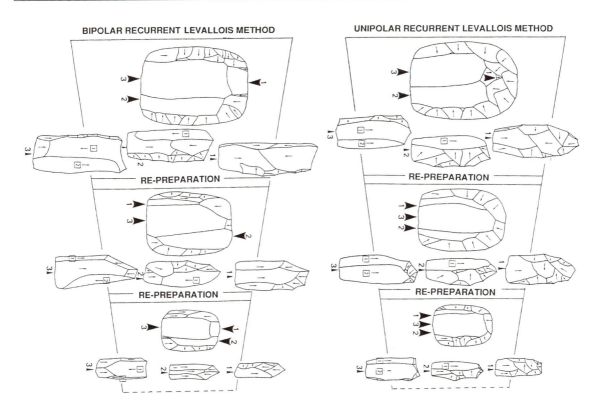

Figure 3.10 *Illustrations of the 'recurrent bipolar' and 'recurrent unipolar' Levallois techniques as represented at the site of Biache-Saint-Vaast in northern France. In both strategies the upper face of the core can be repeatedly re-prepared to produce a succession of up to three adjacent and overlapping flakes, detached from either one end of the core (in the 'unipolar recurrent' technique) or from both ends (in the 'bipolar' technique). The arrows indicate directions of flake removals. After Boëda 1988a.*

and is, in essence, more like that of the classical Levallois (lineal) techniques, described earlier (Boëda 1982, 1986, 1988a, 1993b). Here, the whole of the flaking strategy seems to have been organized in a radial fashion; flakes were removed successively from all directions on the core perimeter (Fig. 3.12), in contrast to the exclusively unipolar or bipolar patterns of flaking documented at Biache. The critical feature which distinguishes this strategy from that documented in the lineal Levallois methods is simply that the removal of a succession of radial or centripetal flakes

was carried out, apparently in close succession, without any major, intervening phase of deliberate reshaping or remodelling of the main face of the core. Thus, the form of the flakes detached in this strategy was controlled almost entirely by the patterns of earlier flake removals from the same surface, without any intermediate phase of trimming or re-shaping of the core surface (see Fig. 3.12).

One might ask how far the technique documented at Corbehem differs from the conventional notion of 'disc-core' technique as

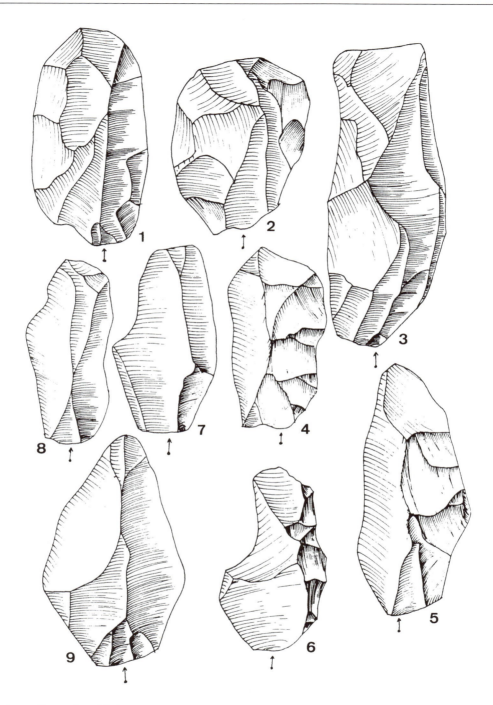

Figure 3.11 *Examples of flakes produced by the 'recurrent unipolar' and 'recurrent bipolar' techniques at Biache-Saint-Vaast. Nos 1–3 are 'first-order' removals; nos 4–6 'second order'; and 7–9 'third order – each characterized by distinctive scar patterns on the dorsal surfaces. After Boëda 1988a.*

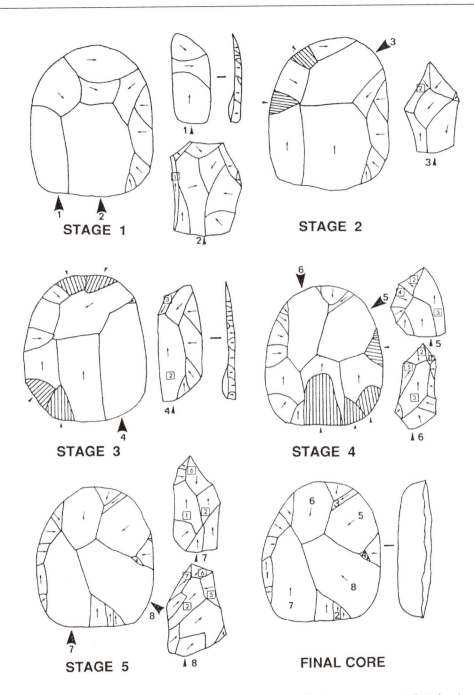

STAGE 1

STAGE 2

STAGE 3

STAGE 4

STAGE 5

FINAL CORE

Figure 3.12 *Illustration of the 'recurrent centripetal' Levallois technique as represented at the site of Corbehem in northern France. The diagrams illustrate how a sequence of 8 or more successive Levallois flakes can be detached from different parts of the core perimeter, with only a limited amount of intervening re-preparation of the core surface (indicated by shading) between the successive flake removals. After Boëda 1988a.*

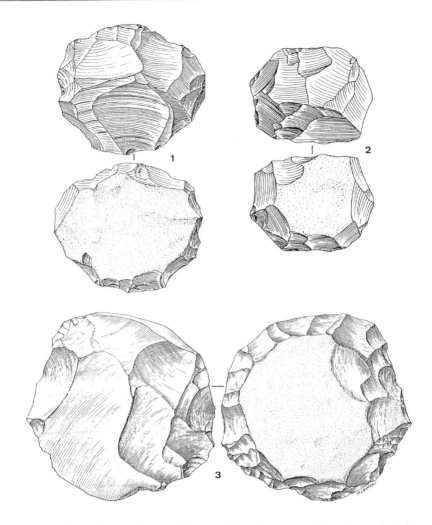

Figure 3.13 *Examples of 'disc' cores. Many of these are likely to represent heavily worked-down versions of what were initially either 'recurrent centripetal' or other forms of Levallois cores. After Bordes 1961a.*

defined by Bordes (1950a, 1961a) and many earlier workers. Ultimately, perhaps, this is a matter of semantics. As discussed below, few workers now would dispute that most aspects of conventional disc-core techniques are essentially Levallois in all basic techno-logical and conceptual respects. Also, as Boëda stresses, virtually all the flakes pro-duced in the course of the core reduction sequences documented at Corbehem seem to conform to normal definitions of Levallois

flakes (including the sense defined by Bordes: 1980) when assessed in terms of over-all patterns and complexity of negative scar patterns on the dorsal surfaces of flakes. Even if the cores produced in the course of these strategies have many elements in common with those of conventional disc-core tech-niques (Fig. 3.13), the flakes produced show most of the distinctive morphological fea-tures usually regarded as fully Levallois techniques.

Non-Levallois techniques

Levallois techniques were only one of the major strategies of primary flake production employed by Middle Palaeolithic groups. The existence of many alternative flaking strategies has been recognized since the earliest stages of research into the Middle Palaeolithic (e.g. Bourlon 1906, 1910, 1911), and it is now evident that these alternative techniques were often employed alongside Levallois methods or, in some cases, to the total exclusion of Levallois techniques. The special case of blade production techniques will be discussed in a later section. Other techniques, however, were clearly designed, as in the Levallois method, for the production of relatively broad, substantial flakes but apparently with much less conscious effort at systematic shaping of the core prior to the main sequence of flake removals. Despite the obvious variety of these techniques, they are usually grouped together collectively under the heading of non-Levallois techniques.

The question of how far disc-core techniques (Bordes 1961a: 72–3; Pigeot 1991; Boëda 1993a) can be separated from the broader grouping of Levallois methods has been raised in the preceding section. The point to emphasize is that these disc-core techniques were reliant on precisely the same basic sequence of core preparation as that in the classic Levallois techniques, involving the initial preparation of a continuous striking platform around the perimeter of this nodule, followed by successive removals of flakes from the upper surface of this nodule (Bordes 1950a, 1961a). The only criterion for differentiating between the two techniques (Levallois on the one hand, and disc-core on the other) seems to lie in the varying degrees of special preparation applied to the upper face of the core. My own inclination is to see this distinction as a matter of degree rather than of kind (see Boëda 1993a for a contrasting view). Even authors who support this basic distinction (e.g. Bordes 1961a: 85) have

pointed out that the effective operation of disc-core techniques frequently involved the removal of occasional flakes to regularize the main flaking face of the cores – and thereby to allow better control over the shape and size of subsequent flake removals. In many contexts, examples of cores have been found which are apparently intermediate between typical Levallois and conventional disc-core forms (Boëda 1986; Mellars 1964). Many typical disc cores are likely to represent the heavily reduced end products of relatively intensive flaking strategies in which cores were progressively worked down from larger and more complex to smaller and simpler forms (Bordes 1961a: 16, 73) (Fig. 3.13). This would explain the relatively frequent cases where typical Levallois flakes have been recovered from sites where associated parent cores are apparently totally lacking.

More interesting technologically are the strategies which have little in common with the Levallois techniques and which reflect entirely different methodological and conceptual approaches to primary flake production. These have attracted less attention than the more impressive forms of Levallois technology and are less well documented in the literature. The most detailed studies of these strategies have been undertaken recently by Alain Turq (1988b, 1989b, 1992a,b), in the course of his studies of a number of classic Quina-Mousterian industries. As an illustration of the potential complexity and in some ways sophistication of these non-Levallois techniques, the results of Turq's recent work are worth examining closely.

Quina Mousterian assemblages are by definition non-Levallois in character and depended for almost all aspects of tool manufacture on the production of simple forms of flakes, most of which are comparatively thick in cross-section and often retain large amounts of cortex on their dorsal surfaces (Bordes 1953a, 1968a, 1981, 1984: 158–60; Turq 1989b, 1992a). These pieces were used predominantly for the production of various side-scraper forms, usually characterized by fairly heavy, invasive retouch along one edge of the flake which defines a steep, heavily convex working edge. In some cases there is evidence that tools were successively resharpened during use, in the course of which the worked edge bit progressively deeper into the central and thicker parts of the flakes (Lenoir 1986; Meignen 1988; see below).

Turq (1989b, 1992a) has argued that the apparently simple or even crude forms of the flakes employed in Quina industries were not simply the result of unstructured or haphazard flaking strategies but were deliberately selected as the most appropriate forms of flake blanks for producing distinctively Quina tools. He argues that the primary hallmark of these Quina flakes was the location of the maximum thickness of the flake directly opposed to the main working edge of the eventual tool, partly to assist in the handling and grip of the tool and partly to allow the maximum scope for successive resharpening in the course of use (Fig. 3.14).

By selecting these thick, triangular-section flakes it was possible to continue the process of progressively resharpening the tool edges through several different phases of tool use and still maintain a substantial rear edge to the piece which could be either held in the hand or, conceivably, attached to a wooden haft. Turq describes a number of flaking strategies by which these thick, triangular-section flakes could be produced – based partly on his studies of material from a number of typical Quina–Mousterian assemblages in the Dordogne region, and partly on his own experiments in replicating similar techniques (Turq 1988b, 1989b, 1992a, b). As he points out, the range of possible alternative strategies is flexible, and appears to have varied in response to such factors as the different flaking qualities of each raw material, and the specific forms of the nodules that were locally available. In essence, however, two basic strategies of core reduction can be identified:

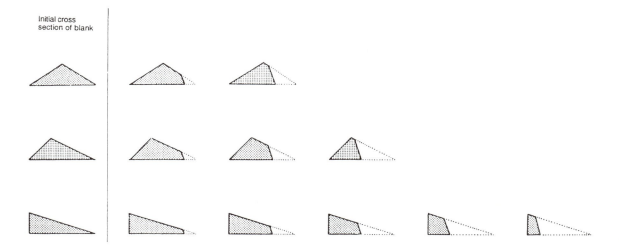

Figure 3.14 *Diagram to illustrate the potential of flakes with varying cross-sections for repeated episodes of edge-resharpening. The triangular-section flakes shown at the base are particularly characteristic of the Quina-Mousterian. After Turq 1992a.*

1. The first has generally been referred to in the earlier literature as the 'salami slice' or 'citrus slice' technique and has been recognized as one of the distinctive hallmarks of typical Quina-type industries since the early years of this century (e.g. Bourlon 1906, 1910; see also Cheynier 1953; Bordes 1981: 79 etc.). As the name suggests, the technique consists of removing a sequence of successive, transverse slices through an elongated flint nodule, much as one might slice through a lemon or a link of salami (Fig. 3.15). The resulting flakes inevitably retain a strip of cortex around a large part of the edge, and could be transformed into characteristic Quina-type racloirs simply by applying heavy retouch to one of the longer margins of the flake (Turq 1989b, 1992a). In some of the later stages of flaking the patterns of negative scars on the dorsal surfaces of the flakes might become slightly more complex if some of the earlier salami-slice removals extended through only part of the section of the nodule. Turq stresses, however, (1989b) that this technique is only really practicable where the available raw material occurs in the form of elongated, narrow nodules and appears to be lacking in certain Quina assemblages where other forms of raw materials (e.g. tabular flint or more irregular nodules) were employed.

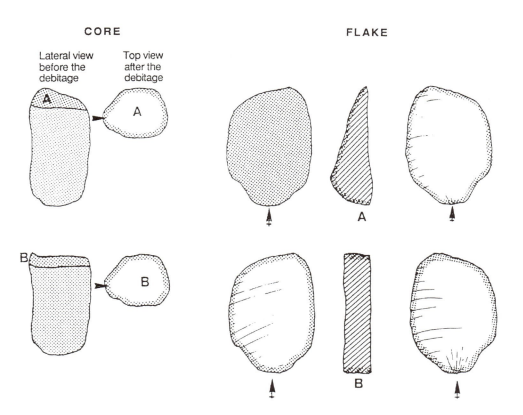

CORE

Lateral view before the debitage

Top view after the debitage

A

A

B

B

FLAKE

A

B

Figure 3.15 *'Salami slice' technique of flake production, as found in many Quina-Mousterian industries. After Turq 1992a.*

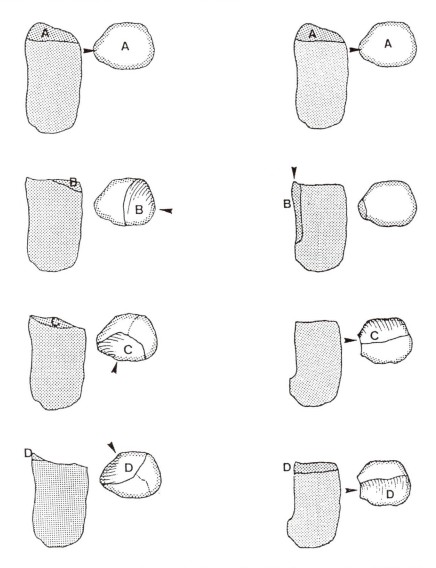

Figure 3.16 *More complex patterns of core reduction employed for the production of thick, triangular-section flakes typical of Quina-Mousterian assemblages. After Turq 1992a.*

2. The second strategy is more complex and involved additional stages in the initial preparation and flaking of the nodules (Fig. 3.16). Typically, the flaking commenced by removing one major preparatory flake from one end of the nodule and then employing this initial flake surface as a prepared striking platform for the removal of one or more elongated flakes extending vertically down the face of the nodule (Turq 1989b). Several of the flakes produced in this way could have been used directly as blanks for typical Quina tools

especially where the flakes retained a sub-stantial, naturally blunted back formed by the original cortex retained along one edge of the flake. In the subsequent stages of flaking these elongated, longitudinal flake removals could then be used to provide a prepared striking platform for further flake removals extending transversely across the breadth of the core in much the same way as in the simpler salami-slice techniques. In certain respects, therefore, this technique was a slightly more complex variant of the salami-slice method but with two significant differ-ences; first, the majority of the flakes would show clear traces of a simple prepared strik-ing platform as opposed to purely cortical platforms in the case of the salami-slice pro-cedures; second, the flakes were removed exclusively from a single, predetermined direction through the nodule, rather than from a succession of different points extend-ing around the perimeter of the nodule (Turq 1988b, 1989b). Several of the flake removals, therefore, would tend to show a simple, lon-gitudinal pattern of flake scars on their dorsal surfaces and be more elongated or rectan-gular than the flakes produced from the sim-pler salami-slice techniques.

As Turq points out, there were several other variants but all involved essentially minor modifications of the two basic strate-gies described above. His argument is that all these strategies were apparently deliberate and preconceived in the sense that they were all oriented towards the production of flakes which were especially suitable for the pro-duction of distinctively Quina tools – i.e. flakes in which the maximum thickness was located directly opposite the eventual, retou-ched edge of the finished tool (Fig. 3.14). In many ways the flaking strategies are much simpler than those involved in the conven-tional forms of Levallois technology, re-quiring fewer steps in the initial stages of preparing both the dorsal surfaces of the cores and the associated striking platforms.

But as Turq points out, this hardly reduces the level of intentionality or predetermina-tion of the strategies which lay behind the flake production or the relative efficiency of the techniques for the purposes for which they were intended. Turq (1989b) argues that the level of success achieved in this kind of flaking, in the sense of producing flakes which were immediately suitable for retouching into tools, seems to have been substantially higher than that achieved in most Levallois techniques. According to his estimates, these Quina strategies seem to have produced in the region of 60–75 percent of immediately usable flakes, as opposed to around 15–25 percent in most Levallois methods (cf. Geneste 1985: 257–9). He con-cludes that these procedures were not merely deliberate and consciously structured but also highly economical in producing specific forms of flake blanks suitable for a particular range of retouched tool forms from particular varieties of available raw materials.

Blade technology

One of the most significant developments over the past few years has been the recogni-tion that the repertoire of primary flaking techniques employed in the European Mid-dle Palaeolithic involved not only the con-ventional forms of Levallois or non-Levallois flake production but a surprisingly strong component of deliberate and highly special-ized blade production. This has been recog-nized for more than 40 years in the Middle Palaeolithic sequence of the Near East (nota-bly in the various occurrences of the so-called 'pre-Aurignacian' or 'Amudian' industries: Rust 1950; Bordes 1955b; Garrod 1956; Jelinek 1990), but until recently remained less securely documented in the European indus-tries. Hints of the existence of this technology had been available for some time in early discoveries at sites such as Crayford in Eng-land (Spurrell 1880; Cook 1986) and Coquelles in northern France (Lefèbvre 1969;

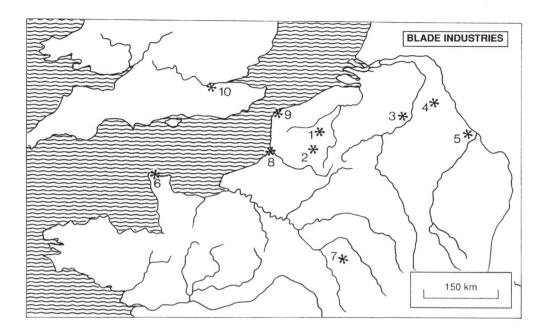

Figure 3.17 *Map of Middle Palaeolithic sites with heavily blade-dominated industries in northern Europe. The sites are as follows: 1 Seclin; 2 Riencourt-lès-Bapaume; 3 Rocourt; 4 Rheindalen; 5 Tönchesberg ; 6 Saint-Germain-des-Vaux; 7. Vallée de la Vanne; 8 St-Valéry-sur-Somme; 9 Coquelles; 10 Crayford. After Ameloot-van der Heijden 1993b.*

Tuffreau 1971) but they remained poorly documented in the literature. Recent discoveries at sites such as Seclin, Port Racine, Saint-Valéry-sur-Somme and Riencourt-lès-Bapaume in France (Révillion 1989; de Heinzelin & Haesaerts 1983; Tuffreau 1992, 1993), Rocourt in Belgium (Cahen 1984), Rheindalen and Tönchesberg in western Germany (Bosinski 1973) and Piekary in Poland (Morawski 1976) (see Fig. 3.17) have now confirmed not only the existence of these specialized blade-producing technologies at a surprisingly early date in the Middle Palaeolithic sequence but also the complexity and variety of the different flaking strategies involved (see Conard 1990).

A detailed review of these early blade technologies in western Europe has been pro-vided by Eric Boëda (1988b) as part of his general analysis of Middle Palaeolithic flaking strategies. Despite the limited number of sites in which these techniques have been identified, Boëda suggests that they can be divided into three separate groups, each involving significantly different concepts in the basic approaches to core reduction and each leading to the production of recognizably different forms of cores and associated flaking débitage. In terms of his own definitions, two of these techniques can be regarded as essentially Levallois in a conceptual sense, whilst the third seems to be effectively identical in most respects to the techniques documented in the much later Upper Palaeolithic industries. Following his criteria, these can be summarized as follows:

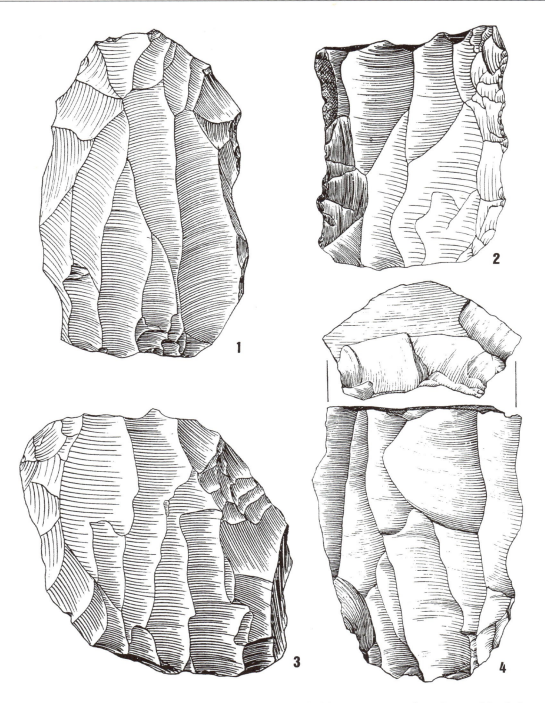

Figure 3.18 *Examples of cores used to produce 'Levallois blades' from sites in northern France. Nos 1–3 are based on the 'classic Levallois' technique of core preparation, while no. 4 conforms to Boëda's definition of the 'specialized Levallois' blade technique (see Fig. 3.19). After Bordes 1961a, Boëda 1988b.*

'Classical Levallois' blade technology

The fact that certain forms of essentially typical Levallois cores were intended for the production of elongated, blade-like flakes has been recognized sporadically in the literature throughout the present century (e.g. Commont 1909, 1913; Breuil & Kozlowski 1932; Bordes 1961a: 72). The distinctively Levallois aspect of these cores lies in the deliberate preparation of a continuous striking platform extending around the circumference of the core, from which a series of initial, preparatory flakes were struck from all parts of the perimeter, converging towards the centre (Fig. 3.18). In every sense, therefore, this technique seems to be entirely Levallois in its basic approach to core reduction. The adaptation of these cores specifically for blade production was achieved in the later stages of core reduction by the detachment of a succession of elongated, narrow flakes extending down the greater part of the length of the core and struck from specially prepared platforms at either one or both ends of the core. As shown in Figure 3.18, the repetition of this flaking over a single, prepared surface led to the production of cores with regular and clearly defined scar patterns running vertically down the face of the core and consequently to the production of elongated, parallel-sided blades. In terms of Boëda's definitions (1988a), these procedures would be grouped broadly under his heading of recurrent Levallois techniques. Nevertheless, the flaking strategy was clearly designed specifically for the production of elongated blade-like forms and can be regarded in this sense as an explicitly blade-producing technology.

'Specialized Levallois' blade technology

In this case Boëda recognizes that the concept of Levallois technology is being stretched beyond its conventional definitions and concedes that the blade techniques which he describes under this heading approach more closely to those documented in fully Upper Palaeolithic industries (Boëda 1988b). Two main criteria are used to distinguish these methods from those described above (Fig. 3.19). First, the flaking surfaces of these specialized blade cores are usually more

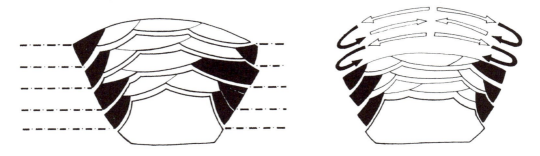

Figure 3.19 *Schematic representation of Boëda's 'specialized Levallois' technique of blade production, as illustrated at the site of Seclin (northern France). In this technique a succession of elongated blades was removed from successively deeper levels of one face of the core, assisted by the detachment of 'edge-removal' flakes ('éclats débordants') from the two lateral edges of the core. After Boëda 1988b*

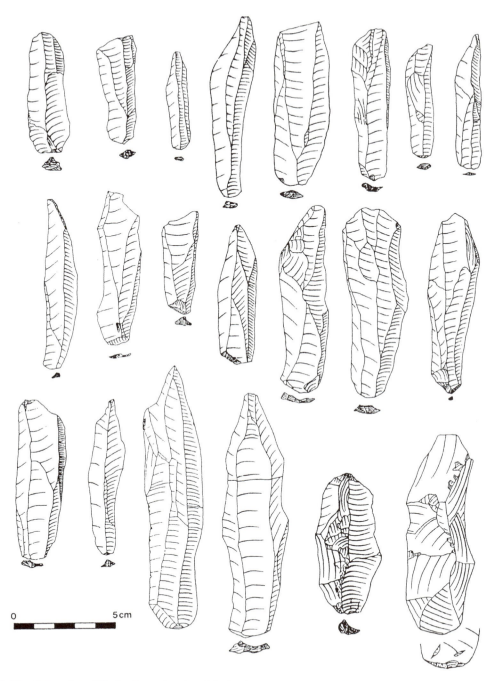

Figure 3.20   *Examples of blades from the site of Seclin (northern France) dated by thermoluminescence to ca 90,000 BP. The two pieces on the bottom right-hand-side are examples of typical crested blades ('lames-à-crête') used to initiate the sequence of blade detachments from the core, and closely resembling techniques used on Upper Palaeolithic sites. After Révillion 1989.*

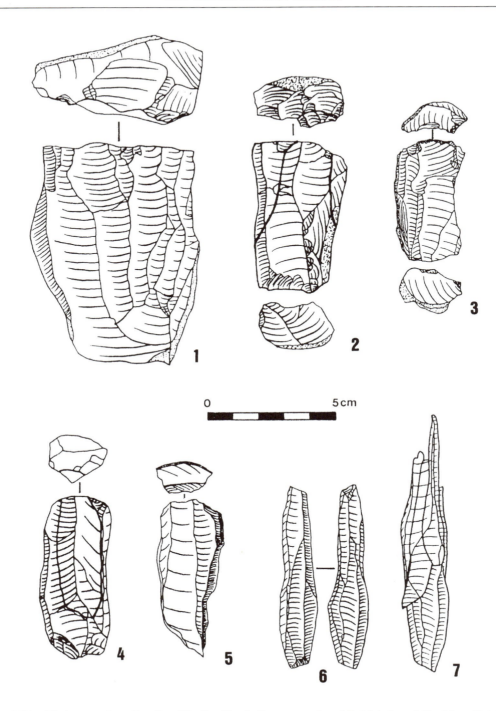

Figure 3.21 *Blade cores from the site of Seclin. Nos 1–5 are examples of Boëda's 'specialized Levallois' blade cores (see Fig. 3.19), while nos 6–7 are examples of fully prismatic cores, closely similar to Upper Palaeolithic forms. After Révillion 1989.*

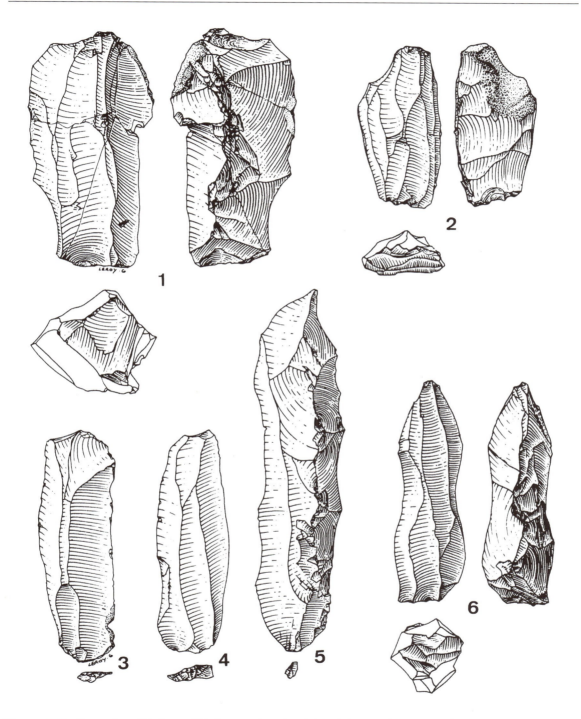

Figure 3.22 *Examples of blades and blade cores from level CA of the site of Riencourt-lès-Bapaumes in northern France. Nos 1, 5 & 6 illustrate the use of the 'lame-à-crête' technique. After Tuffreau 1992.*

markedly convex than in the case of normal Levallois cores, and as a result allowed the removal of blades over a greater proportion of the total core surface than the conventional, more flat-faced Levallois cores. In this sense the technique can be seen as more economical or more productive than in the classical Levallois methods, allowing the production of a much greater number of blades from a given volume of raw material. Second, the lateral edges of these specialized cores were prepared not by initial, centripetal flaking around all the core perimeter, as in the case of classical Levallois methods, but instead by the removal of two major edge-preparation flakes (*éclats débordants*) extending vertically down each side of the core (cf. Beyries & Boëda 1983). Once the basic form of the core had been roughed out in this way, a succession of elongated, regular blade removals could be detached in direct succession from prepared striking platforms at either or both ends of the core, i.e. by means of either unipolar or bipolar flaking (Fig. 3.19).

The clearest illustration of this technique has been documented at the site of Seclin (Pas-de-Calais), excavated by Alain Tuffreau and dated to the earlier stages of the last glaciation, around 90,000 BP, according to the results of TL dating of burnt flint samples (Tuffreau *et al.* 1985). A detailed account of the blade technology at Seclin has been published by S. Révillion (1989) based largely on a series of reconstructions of groups of conjoining flakes and associated cores. The basic blade-production strategies described by Révillion correspond reasonably closely with those described by Boëda under his heading of 'specialized Levallois' strategies discussed above. Révillion, however, is more inclined to stress the distinctively Upper Palaeolithic character of some aspects of the blade production on the site. He points out, for example, that many of the blade cores show blade removals extending around a substantial part of the core surface and in some cases around

the whole of the core to produce a fully prismatic form (Fig. 3.21). He argues that in 'volumetric' terms, these cores are more akin to Upper Palaeolithic than to Middle Palaeolithic forms. He also stresses the relatively standardized character of the blades, mostly with length-over-breadth ratios ranging between 2.0 and 2.5, though admitting that these are generally less elongated than those documented from most Upper Palaeolithic sites and generally show more complex patterns of preparatory scar facets on their dorsal surfaces (Fig. 3.20). More significantly, perhaps, he also points out that the great majority of these blade forms show clearly faceted striking platforms and seem invariably to have been struck with a 'hard' rather than a 'soft' hammer – both of which would usually be regarded as much more typically Middle than Upper Palaeolithic features. Overall, therefore, the assemblage could be said to show an interesting mixture of both Middle and Upper Palaeolithic forms.

Non-Levallois blade techniques

This final strategy of blade production is categorized by Boëda as entirely 'non-Levallois' in character, and is regarded by him in most respects as almost identical to that documented in fully Upper Palaeolithic industries. The best documented example discussed by Boëda is from the site of Saint-Valéry-sur-Somme where a small but closely associated series of conjoining flakes (Figs 3.23, 3.24) was recovered from deposits provisionally and rather tenuously attributed to the early or middle stages of the penultimate glaciation, possibly equivalent to isotope stages 7 or 8 of the ocean-core sequence (de Heinzelin & Haesaerts 1983). Despite the small size of the assemblage, the results of the refitting studies appear to demonstrate a strategy of highly specialized and standardized blade production which involved the use of carefully prepared, elongated cores struck from two opposed striking platforms.

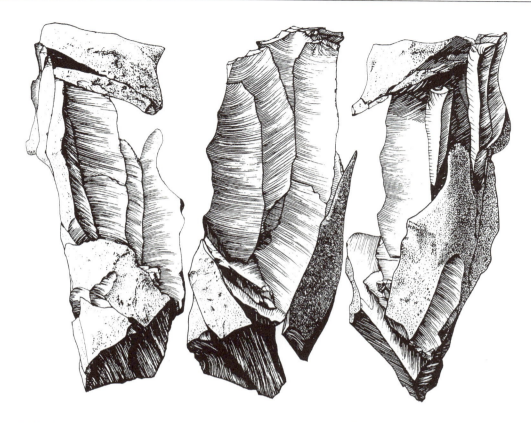

Figure 3.23 *Refitted group of blades and striking-platform preparation flakes from St-Valéry-sur-Somme (northern France). After de Heinzelin & Haesaerts 1983.*

The critical feature which distinguishes this strategy from that documented in the various Levallois techniques discussed earlier is that the active face of the core, from which the various blade removals were struck, appears to extend around the greater part of the circumference of the core rather than being restricted to one clearly delimited flaking surface. As Boëda (1988b) points out, this technique allowed virtually continuous flaking and successive reduction of the greater part of the core surface and accordingly allowed the maximum possible production of blade removals from the available nodules of raw material (Fig. 3.25). In terms of Boëda's concepts, it is this capacity for almost exponential reduction of the core volume which sets this technique apart from all recognized Levallois strategies and which characterizes the flaking strategy as fully Upper Palaeolithic in both technology and concept.

How far the blade technique documented at Saint-Valéry-sur-Somme can be separated from that documented at Seclin discussed above is more debatable. Although the majority of the blade cores recovered from Seclin were apparently flaked over only a single major surface (and in this sense con-

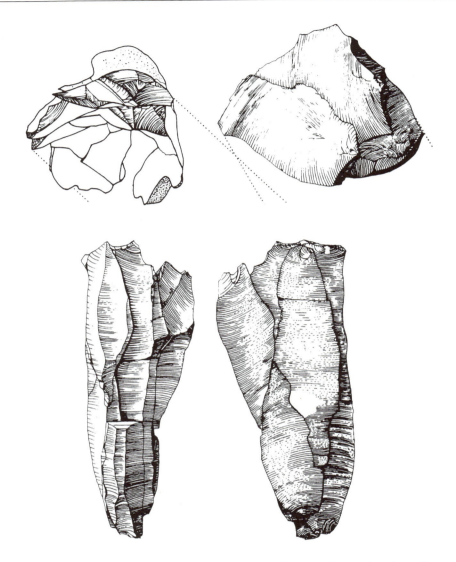

Figure 3.24 *Two groups of refitted blades and flakes from St-Valéry-sur-Somme (northern France). After de Heinzelin & Haesaerts 1983.*

form to Boëda's notion of the basic Levallois concept) the recent analysis of Révillion shows that in some cases this flaking was extended continuously around all the core circumference, leading to the production of cores that were fully prismatic (Révillion 1989). Thus, at least some of the cores at Seclin would appear to reflect an equal if not greater capacity for exponential reduction of the core volume than those documented at Saint-Valéry (Fig. 3.21). Equally significant is that the blade technology employed at Seclin

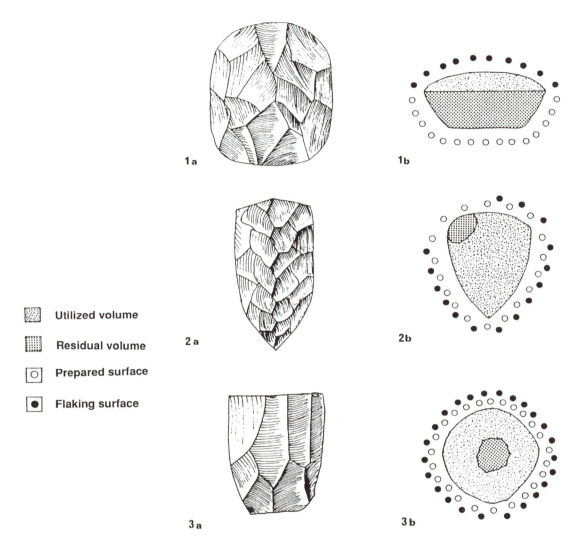

Utilized volume

Residual volume

Prepared surface

Flaking surface

Figure 3.25 *Diagram to illustrate the differing 'volumetric' use of raw materials achieved by typical Levallois techniques (upper) compared with two forms of blade techniques (middle, lower). By detaching blades from most of the core circumference, blade techniques generally allow the utilization of a much larger proportion of the original core volume than Levallois techniques. After Boëda 1988b.*

(as well as in the recently excavated nearby site of Riencourt-lès-Bapaume: Ameloot-van der Heijden 1993a,b) involved the use of typically crested blade (*lame-à-crête*) technique to initiate the sequence of blade detachments from the core, in a way which closely mimics Upper Palaeolithic techniques

(see Figs 3.20, 3.22). In general, therefore, the recent studies by Révillion bring the blade technology documented at Seclin more closely into line with those documented in fully Upper Palaeolithic contexts than the earlier study by Boëda had implied.

Discussion

Following this lengthy review of Middle Palaeolithic flaking strategies, what general patterns can be discerned? The main features can be summarized as follows:

1. Arguably the most impressive feature is the wide diversity and flexibility of the primary flaking strategies which can now be documented in different Middle Palaeolithic contexts. Under the broad heading of Levallois techniques, for example, we can identify at least five or six clearly differentiated strategies, all oriented towards the production of different forms of primary flakes and all involving different strategies in the various stages of preparation and flaking of the cores (Bordes 1980; Boëda 1986, 1988a, 1993b; Boëda *et al.* 1990). Whether we define these techniques in terms of the forms of the resulting flakes, the character of the residual cores or the detailed sequence and patterning of the preparatory flake removals, the variety is remarkable. In addition to the various Levallois techniques there is a range of taxonomically non-Levallois techniques which, even if basically simpler, nevertheless seem to have been equally deliberately oriented towards the production of specific kinds of flake blanks (Turq 1989b, 1992a). There are also various blade techniques which now can be documented from a surprisingly early stage of the Middle Palaeolithic and which in some cases approach closely those documented in fully Upper Palaeolithic contexts (Boëda 1988b).

2. The second impressive feature is the inherent technological complexity of some of the individual flaking strategies. In the various Levallois techniques, for example, the total sequence of preparation, production and eventual use of the flakes must have involved at least five or six clearly separate stages, all of which were apparently planned and preconceived from the initial stages of the flaking sequences (cf. Boëda 1988c; Boëda *et al.*

1990). This is illustrated most clearly in some of the recurrent Levallois strategies documented at sites such as Biache-Saint-Vaast, Grotte Vaufrey and Maastricht-Belvédère (Tuffreau & Sommé 1988; Boëda 1988a, 1993b; Rigaud 1988; Roebroeks *et al.* 1988; Schlanger 1994) in each case apparently dating from well before the end of the penultimate glaciation. At Biache, for example, the total sequence of operations involved in the production of various forms of elongated and pointed flakes must have involved six separate technological operations ranging from initial selection and decortication of the original flint nodules through various stages in the preparation of the peripheral striking platforms and dorsal surfaces of the cores to the successive removal of three distinctive forms of primary flakes (Fig. 3.10) (Boëda 1988c). If we add that some of these primary flakes seem to have been used selectively for the production of equally specific forms of retouched tool forms (and eventually inserted into wooden hafts: cf. Beyries 1988b), then the total sequence of successive technological procedures becomes even more impressive. As will be discussed further in Chapter 12, the remarkable complexity and elaboration of these flake and tool production strategies must have implications for the overall cognitive capacities and structures of Neanderthal populations.

3. One of the most intriguing and currently controversial issues raised by these recent technological studies centres on the factors or constraints which may have influenced the selection and use of different flaking strategies in different geographical or behavioural contexts (cf. Dibble 1991a; Dibble & Rolland 1992). As noted earlier, a striking feature of Middle Palaeolithic technology is the way in which these different flaking strategies vary not only between different chronological and geographical contexts of the Middle Palaeolithic, but often between separate, closely adjacent levels in the same sites (Rolland

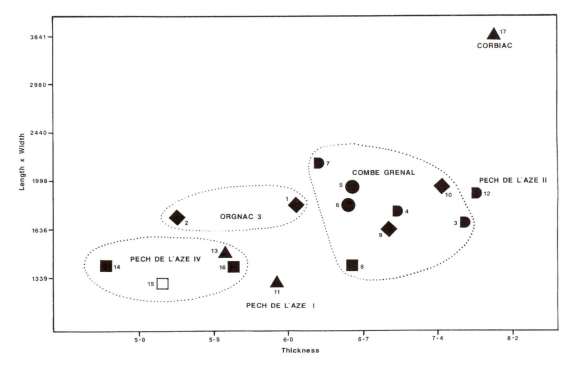

Figure 3.26 *Relationship between the thickness and surface area of Levallois flakes recorded in a number of French Middle Palaeolithic assemblages, after Dibble 1985. Dibble suggests that the differences between the flake dimensions are related mainly to variations in the size and quality of the local raw material supplies employed on the different sites.*

1988b). Explaining this variability remains one of the most challenging issues in current studies of the Middle Palaeolithic.

As discussed in Chapter 10, there is little doubt that some of these variations are related to the character and quality of the lithic raw material supplies available in different geographical contexts. Several workers have pointed out that some of the most frequent and elaborate occurrences of various Levallois techniques occur mainly in areas where local flint supplies are both relatively abundant and available in the form of relatively large, high quality nodules (Fish 1981; Bordes 1947, 1950b, 1954a, 1984: 169; Turq 1989a; Van Peer 1992). This can be seen in many of the industries from the flint-rich

areas of northern France, in many parts of the Middle East and North Africa, and in some of the more localized occurrences of high quality materials in areas such as the Bergerac region of southwestern France and parts of western Provence (Fig. 3.26). But as Bordes emphasized (e.g. 1968a: 138), it is difficult to see this factor as more than a partial explanation for the variable occurrence of different flaking techniques. Bordes himself pointed out that in several contexts relatively large and elaborate forms of Levallois flakes can be shown to have been produced from fairly coarse-grained, apparently intractable raw materials, such as quartzite or even fine-grained sandstone (Green 1984). There are equally frequent examples of the use of relatively sophisticated and abundant Levallois techniques in areas where the available flint supplies are either scarce or present in the form of small, irregular nodules – as for example the Grotte Vaufrey, Abri Caminade, Pech de l'Azé and other sites in the Dordogne valley or at Le Moustier, La Rochette and Fonseigner in the valleys of the Vézère and Dronne. None of these areas are characterized by particularly high quality flint supplies (Dibble 1985; Geneste 1985; Bordes 1968a: 138). Most striking of all, perhaps, is the dramatic way in which the relative use of various Levallois techniques can often be seen to vary between different occupation levels within precisely the same occupation sites – for example between the different stratigraphic levels at Combe-Grenal, at the Roc-de-Marsal and at Pech de l'Azé sites II and IV (Rolland 1988b: 165–9). Any suggestion that these different flaking techniques were related in a simple, direct and spontaneous way to the varying availability and flaking qualities of local raw materials would therefore seem to be contradicted directly by the evidence from many Middle Palaeolithic sites. Evidently, some factors other than the simple character and accessibility of local raw materials were involved in the selection and variable use of different flaking techniques.

4. Finally, some useful insights into the character and operation of different primary flaking strategies have been provided by recent experimental approaches to the replication of Middle Palaeolithic techniques. The essential aims of these studies are to reproduce as accurately as possible strategies of flake production documented in particular archaeological contexts, and in this way to investigate specific problems encountered in dealing with different varieties of raw materials, and the characteristic forms of *débitage* generated during the different stages of flake and core production. In other words, these studies aim to clarify the conceptual planning strategies involved in the design and operation of different flaking techniques and to generate specific predictions as to how these different flaking strategies may be reflected in the resulting lithic residues.

At present this research is still at an early stage and virtually all experimental studies so far reported in any detail in the literature have been focused on the various forms of Levallois techniques. The most fully published results are those applied by Geneste (1985: 203–70) to approximately 40 different nodules, comprising several different varieties of flint from southwestern France. Although Geneste's results were published before the recent attempts by Boëda to redefine and reclassify major variants of Levallois techniques, it would seem that most of Geneste's experiments were oriented towards the production of centripetally prepared Levallois flakes, coinciding essentially with Boëda's (1988a) categories of either lineal or recurrent centripetal techniques (Fig. 3.27). The studies by Shchelinskii, based on a total of around 60 experimental cores, have been published more briefly but were oriented towards the production of both radial and more elongated convergent or laminar flakes, corresponding broadly with Boëda's categories of unipolar and bipolar recurrent techniques (Shchelinskii 1974, 1983, summarized in Plisson 1988). Despite some apparent differences in the precise design and organization of the flaking experiments, a number of fairly clear and consistent patterns seem to emerge from both of these studies (see also Bradley 1977; Bradley & Sampson 1986; Van Peer 1992):

(a) One point emphasized by both Geneste and Shchelinskii is that Levallois techniques, however tightly controlled and designed, can only produce a limited component of typically Levallois flake products in relation to total quantities of flake débitage generated during different flaking sequences. Geneste (1985: 253) reports that in his experiments the overall percentages of Levallois, as opposed

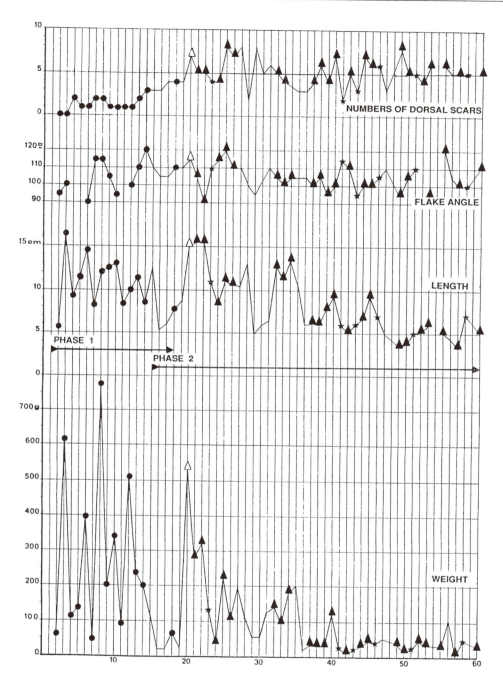

Figure 3.27 *Metrical parameters of a succession of flakes produced by Geneste during the experimental replication of Levallois flaking techniques. Phase 1 represents the initial phase of core preparation and decortication, while phase 2 represents the main succession of flake removals. After Geneste 1985.*

to non-Levallois, flakes covered a fairly wide range but centred on an average figure of 18 percent. The results reported by Shchelinskii point broadly to the same conclusion. Assessed in terms of potentially 'usable' flakes, he reports that the different experimental sequences tended to yield an average of around 20 percent of fully Levallois flakes in the case of both radially prepared and more convergent techniques – though the ranges recorded in individual cores could vary from less than 15 percent to as high as 40 percent (see Plisson 1988). Interestingly, he reported a rather lower success rate of around 11–16 percent in attempting to produce more elongated flake blanks from longitudinally prepared cores. The implication of these studies is that no attempt at producing typically Levallois flake blanks from original, unworked nodules is likely to produce more than approximately one-fifth of technically Levallois products among the overall range of knapping debris generated on the flaking sites (Fig. 3.27).

(b) Similar regularities emerge in the production of various forms of cortical flakes in the course of different core reduction sequences. Again Geneste reports that these frequencies can vary within fairly wide limits but seem to centre on an average value of around 30 percent over the flaking experiments as a whole. He goes on to suggest that a systematic comparison of the relative frequencies of cortical flakes, compared with the frequencies of much smaller débitage flakes, can be used to predict fairly accurately the kind of nodules employed in the use of different flaking strategies (Geneste 1985: Fig. 73). One particularly interesting result reported by Geneste is that the relative frequencies of cortical as opposed to fully Levallois flakes do not seem to have been significantly influenced (in his studies) by the specific flaking qualities of the different types of flints employed in the experiments – comprising, as noted above, at least three distinct vari-

eties of flint from separate outcrops in the Perigord region (Geneste 1985: 254). He stresses, however, that much more extensive experiments would be needed to assess the overall potential impact of different raw material types on the quality and efficiency of production of different flaking strategies.

(c) Finally, Geneste reports some rather more predictable results relating to the overall size and character of the flakes produced during his experiments. The conclusion, not surprisingly, is that typically Levallois flakes tend to be substantially larger than those produced during the various preparatory stages of core flaking and also tend to show more complex patterns of flake scars on their dorsal surfaces (Figs 3.27, 3.28) (Geneste 1985: 259–68). Both these variables must inevitably be related to some extent to the sizes of the original nodules, as well as to the relative intensity of the deliberate preparation applied before the removal of the individual flakes. Nevertheless, his own experimental figures of an average of 7.4 dorsal flake scars per flake – rising for the largest flakes to around 11–16 scars – provides a useful guide to the patterns which can be achieved by carefully controlled use of Levallois techniques. He goes on to point out that equally complex patterns of dorsal-scar preparation are reflected in the material recovered from several Middle Palaeolithic sites in both Europe and the Middle East, for example in certain levels of the Tabun Cave in Israel, and at Le Tillet and Pech de l'Azé IV in France (Table 3.3: Geneste 1985: 266–7). The implication is that many of the Middle Palaeolithic populations in Eurasia achieved no less highly controlled mastery of Levallois techniques than those attained in the recent experimental studies of modern flint knappers!

All these recent experimental approaches to the replication and analysis of Middle Palaeolithic flaking techniques could be carried much further. It would be particularly interesting to know how far variations in the

Table 3.3

Average numbers of dorsal flake scars recorded on samples of Levallois flakes in different Middle Palaeolithic assemblages in France and the Middle East, according to Dibble (1983): 58

Site/layer	Industry	Average no. of dorsal scars
Pech de l'Azé IV ca F4	MTA	5.78
Pech de l'Azé IV ca I2	Typical Mousterian	5.63
Pech de l'Azé IV ca J3b	Asinipodian	6.21
Combe Grenal ca 35	Ferrassie Mousterian	4.98
Combe Grenal ca 50	Typical Mousterian	5.57
Corbiac ca M1	MTA	5.81
Le Tillet	MTA	7.33
Tabun Ensemble I	Levallois-Mousterian	6.77
Tabun Ensemble I	Levallois-Mousterian	5.04
Tabun Ensemble X	Levallois-Mousterian	5.03
Tabun Ensemble XI	Amudian	3.45
Geneste (1985) – Experimental Levallois techniques		7.38

The numbers of scars observed on the experimental series of Levallois flakes produced by Geneste (1985) are indicated at the base (after Geneste 1985: 268).

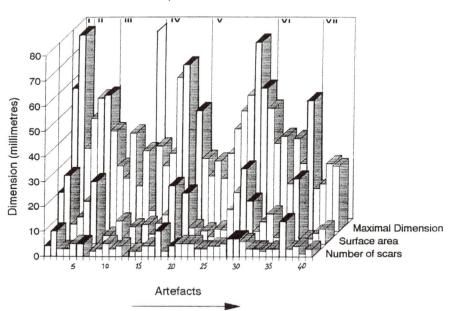

Figure 3.28 *Metrical parameters of a sequence of 41 refitted Levallois and other flakes from the penultimate-glacial site of Maastricht-Belvédère in Holland (from Schlanger 1994).*

shape and size of flint nodules could influence the choice of alternative flaking techniques and the overall productivity or
efficiency of the different techniques in terms
of the total numbers of Levallois or other
flakes which can be produced from specific
quantities of raw material. It would be
equally interesting to know how variations in
the inherent flaking qualities of the different
materials (e.g. the use of fine-grained flint
versus coarser-grained chert or quartzite)
could influence some of the basic morphological features of the flakes produced, such
as size, shape, thickness, complexity of dorsal
scar patterns etc. There is scope for more
work in all these areas, which could shed
important light on several of the critical
debates currently surrounding these aspects
of variability in Middle Palaeolithic technology.

CHAPTER 4

Tool Morphology, Function and Typology

Few topics have generated greater debate over the past few years than the basic 'typology' of Middle Palaeolithic industries. Inevitably, these questions are intertwined with more general issues of the significance of industrial variability in lithic industries and ultimately cannot be separated from them (see Chapter 10). Recently these issues have been brought into sharp focus by the publications of Harold Dibble, Nicholas Rolland and others which take a strongly 'reductionist' line (in both a literal and metaphorical sense) on the whole question of morphological and typological structure in Middle Palaeolithic industries (e.g. Rolland 1977, 1981; Dibble 1987a, 1988a, 1991a, b; Rolland & Dibble 1990; Dibble & Rolland 1992).

Reduced to basic terms the principal issues to have emerged in the recent literature concerning the morphology of Middle Palaeolithic tools can be summarized as follows:

1. To what extent can we identify discrete morphological 'types' in Middle Palaeolithic industries – in the sense of retouched tool forms which clearly have significance not only in the minds of modern, typologically oriented archaeologists but also in the minds, and related manufacturing procedures, of the individuals and groups who produced the tools?

2. If we can isolate morphologically and by implication conceptually discrete types, how far can they be correlated with specific economic or technological functions? In other words, is there any simple correlation between form and function in Middle Palaeolithic tools?

3. To what extent can the documented variation in the forms of Middle Palaeolithic retouched tools be attributed not to any intention or design in the minds of the original artisans but simply to systematic resharpening or reworking of the tools as they became worn or damaged through use?

4. What was the effect of using different raw materials on the detailed form and character of retouched tools? Can we identify correlations between specific forms of retouched tools and specific raw materials, and if so how much did variations in the character or quality of different raw materials influence the eventual forms of the tools?

5. Can we recognize any patterning in the forms of Middle Palaeolithic tools which goes beyond their purely functional or technical aspects and which appears to reflect an explicit visual or symbolic component, i.e. what I (Mellars 1989b, 1991) and others have described as deliberately 'imposed' form in tool production patterns?

Any discussion of these issues must involve the long-debated significance of 'types' in stone-tool industries, which generated a formidable literature in the early years of the New Archaeology (e.g. Spaulding 1953; Sackett 1966, 1968, 1973; Movius *et al.* 1968; Clarke 1968; Doran & Hodson 1975). With hindsight, many of these debates were perhaps reduced more to questions of semantics than to real issues of the cultural and behavioural significance of variations in stone-tool production although some significant and enduring issues were raised. At one level any discussion of artefact typology raises the basic problem of 'emic' versus 'etic' approaches to past human behaviour – how far we can extrapolate from our own perceptions of the morphology and taxonomy of different artefact forms to those of the individuals and groups who produced the tools. On the other hand there is the problem of applying notions of types across loosely related industries covering a broad chronological and geographical span. Thus, how far can we hope to identify or define notions of discrete 'types' which apply not only to the products of a single archaeological assemblage (i.e. a single human group over a short span of time) but to the products of many different communities, which may have had significantly different manufacturing norms (or even significantly different functional requirements) in tool production? Neither issue is simple, in either theoretical or procedural terms, and it is hardly surprising that many of the debates over the past 20–30 years have failed to come to any simple or generally accepted conclusions.

This discussion will adopt a more pragmatic stance. The aim is to concentrate on some of the most widely recognized and intuitively obvious retouched tool forms which have conventionally been recognized in Middle Palaeolithic industries and to ask how far these apparently discrete types can be substantiated on the basis of the combined morphological and, wherever possible, functional attributes of the tools. Some of the more general issues relating to the broader functional or symbolic components of tool production patterns will be discussed in the final section of the chapter and pursued further in Chapters 10 and 12.

Side scrapers

Side scrapers (alternatively known by the French term *racloirs*) have always been seen as one of the most distinctive retouched tool forms in Middle Palaeolithic industries and in the earlier literature were often regarded as a diagnostic feature of these industries (e.g. de Mortillet 1883). In fact, typical side-scraper forms, effectively identical to those encountered in the Mousterian and related industries, are known to occur throughout the greater part of the Lower Palaeolithic sequence extending back to the earlier Middle Pleistocene. Nevertheless, high frequencies of typical side-scraper or racloir forms are one of the most pervasive features of typically Middle Palaeolithic technology not only in Europe but in most parts of Africa and in at least the western parts of Asia (Klein 1989a).

In their most basic forms, side scrapers are very simple, almost elemental tools, characterized by two basic features (Bordes 1961a; Mellars 1964): first by a single, major retouched edge almost invariably located along one of the longest margins of the original flake blank; and second by the application of retouch obviously intended to produce a regular, sharp working edge along the retouched margin of the tool (Figs 4.1, 4.2). All the microscopic use-wear studies which have so far been carried out on typical side-scraper forms confirm that the zones of use were confined largely to these retouched edges (Semenov 1964; Keeley 1980; Beyries 1986, 1987, 1988a,b; Plisson 1988; Anderson-Gerfaud 1990). The specific functions of the tools seem to have been more variable, ranging from use as either cutting or slicing tools

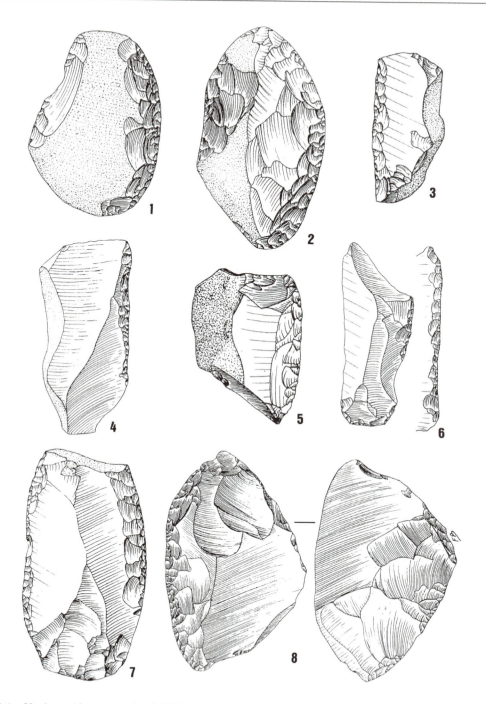

Figure 4.1 *Various side-scraper (racloir) forms, classified according to the categories of François Bordes: 1–5 single-edged lateral forms; 6 'bulbar face' form; 7 double-edged form; 8 racloir with thinned back. After Bordes 1961a.*

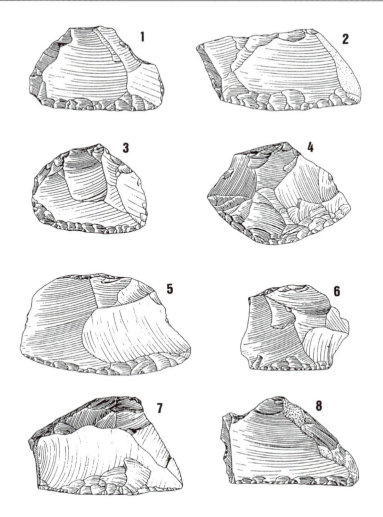

Figure 4.2 *Typical transverse racloir forms – characterized by the presence of a retouched edge located opposite the bulb of percussion and striking platform of the parent flake. After Bordes 1961a.*

(variously applied to wood, meat or skin) to more heavy-duty scraping of hide or bone. Clearly, the functions of these tools were much more flexible than their conventional English and French names would imply.

When defined in these terms there is scope for relatively wide variation in both the precise forms of side scrapers and in the character and treatment of the retouched edges. The

clearest reflection of this can be seen in the complex system of typological divisions for various racloir forms proposed in the standard typology of François Bordes (1961a). Leaving aside the more complex forms of convergent and *déjeté* racloirs, Bordes advocated 17 separate divisions for the simpler racloir forms (see Table 6.1). The major distinctions in Bordes' system are based on the position of the retouched edges in relation to the main flaking axis of the tool and on the overall shape of the retouched edge itself. Thus a basic distinction is made between pieces where the retouched edge is aligned parallel to the main axis of the flake (defined as 'lateral racloirs') and those where the retouched edge is oriented transversely across this axis ('transverse racloirs'). Further subdivisions of these types are based on the forms of the retouched edges – whether straight, convex or concave. In addition to these basic forms Bordes recognized separate categories of tools retouched either exclusively or partially on the ventral surface of the flakes (*racloirs sur face plane* or *à retouche alterne*) and other forms with extensive retouch on both the dorsal and ventral surfaces (*racloirs à retouche biface* and *racloirs à dos aminci*). Tools with characteristic retouch on two separate edges are grouped together collectively as double racloirs, and further subdivided into six categories based on various potential permutations in the curvature of the two retouched edges (see Table 6.1; Figs 4.1, 4.2).

From these distinctions it is clear that the manufacture of even relatively simple racloir forms left Middle Palaeolithic artisans with much flexibility in their choice of both overall design (i.e. the number and location of the retouched edges) and the precise patterns of retouch applied to the worked edges. Recent debates on the significance of these formal variations in racloir morphology have focused on two basic issues:

1. How far was the retouch applied to racloirs a deliberate feature applied in the initial stages of tool production and how far was it simply a result of *ad hoc* resharpening and reworking of the edges as they became progressively worn and blunted in the course of use (Rolland 1977, 1981; Dibble 1984a, 1987a; Rolland & Dibble 1990; Dibble & Rolland 1992)?

2. What significance can be attached to docu-

mented variations in the forms of the tools – either in terms of function or specific design norms or 'mental templates' that lay behind the conception and production of the tools (Dibble 1987a,b,c, 1989; Dibble & Rolland 1992; Kuhn 1992a)?

As discussed earlier, the question of tool reduction and resharpening has emerged as a central issue in Middle Palaeolithic technology with radical implications not only for the morphology of individual tool forms but also for the more general issue of inter-assemblage variation in the Mousterian (see Chapter 10). In a series of publications over the past 15 years Nicholas Rolland and Harold Dibble have presented a stark alternative to conventional perspectives on side-scraper typology by suggesting that these tools may never have been planned as retouched tool forms but may simply have acquired their characteristic retouch in the course of progressive resharpening of their edges during use (Rolland 1977, 1981, 1988a, 1990; Dibble 1984a,b, 1987a,b,c, 1988a,b, 1989, 1991a,b; Rolland & Dibble 1990; Dibble & Rolland 1992). They envisage that the great majority of conventional racloir forms started their use-lives simply as unretouched flakes, which were only systematically retouched as their originally sharp edges became progressively worn and damaged through heavy use. The logical extension of this is that effectively all the documented variation in the relative frequencies of racloirs in different Middle Palaeolithic assemblages can be attributed largely to variations in the degree to which these tool-resharpening processes were carried out in the different sites. Thus taxonomically Denticulate Mousterian industries, in which the overall frequency of racloirs is low, are seen as lightly reduced industries in which few raw flakes were transformed into retouched racloir forms, while the various Ferrassie and Quina-type industries (in which overall racloir frequency is high) are seen as heavily reduced indus-

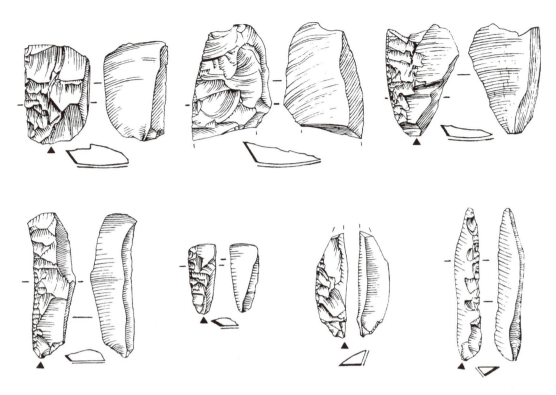

Figure 4.3 *Resharpening flakes detached from the edges of racloirs or similar tools, from the penultimate-glacial levels of the Cotte de St Brelade (Jersey). After Cornford 1986.*

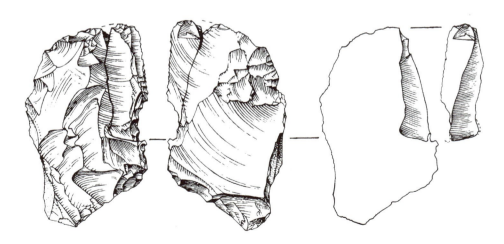

Figure 4.4 *Resharpening spall refitted to the edge of the original tool, from the Cotte de St Brelade (Jersey). After Cornford 1986.*

tries in which a large proportion of the available flakes were transformed into systematically resharpened racloir forms (Rolland 1977, 1981, 1988a; Rolland & Dibble 1990; Dibble & Rolland 1992).

These patterns in turn are thought to be closely related either to environmental factors (reflecting the relative abundance and accessibility of local raw material supplies) or to varying degrees of duration and intensity of occupation in the different sites (Dibble & Rolland 1992; Dibble 1991a,b). These issues will be discussed in more detail in Chapter 10 in the context of more general models of inter-assemblage variation in the Mousterian. The assumptions of these models, however, are directly relevant to the present issue of how far tool-reduction models can be invoked to account for the patterns of production of racloir forms.

There is no doubt that Rolland and Dibble have identified an important potential aspect of variation in Middle Palaeolithic technology which is in line with the findings of much recent research. Recent results of microscopic use-wear studies, for example, leave no doubt that many specimens of unretouched flakes in Middle Palaeolithic industries were subjected to heavy use, apparently for a range of different functions. Amongst others, the studies carried out by Sylvie Beyries on the assemblage from Corbehem in northern France (Beyries & Boëda 1983) and similar studies by Patricia Anderson-Gerfaud (1990) on assemblages from several southwestern French sites leave no doubt about this (see also Beyries 1986, 1987, 1988a,b, 1990; Keeley 1980). There is also now clear evidence from several sites that the retouched edges of typical racloir forms were systematically resharpened by the removal of deliberate resharpening spalls (Fig. 4.3). Typical specimens of these resharpening flakes have been recovered, for example, from Combe Grenal (Lenoir 1986), Marillac (Meignen 1988), La Cotte de Saint-Brelade (Cornford 1986) and La Micoque (Schlanger 1989), and in some

cases have been refitted directly to the parent tools (e.g. Fig. 4.4). These discoveries leave no doubt that systematic resharpening of blunted or damaged edges of side scrapers was carried out commonly on a range of Middle Palaeolithic sites.

The central issue here is the relative *scale* on which resharpening was carried out in Middle Palaeolithic sites. Can we use these resharpening models to argue that typical side-scraper forms were rarely if ever manufactured as deliberate, *a priori* tool forms (as Rolland and Dibble imply) or was resharpening applied purely as a secondary technological device to extend the use-life of tools conceived and produced from the outset as retouched tool forms? The main points are as follows:

1. The evidence from many sites does not support easily the notion that systematic reduction of raw flakes into characteristic racloir forms was a product of either intensive patterns of site occupation or the scarcity of local raw materials. There are now several examples of sites where high frequencies of typical racloirs coincide with locations where local raw materials were both of high flaking quality and apparently available in almost limitless supplies. The most obvious example is the site of Combe Capelle, where a major outcrop of high quality flint occurs directly on the occupation site (Peyrony 1943; Bourgon 1957). In this, as in other sites such as Champlost in Burgundy (Farizy 1985) and Biache-Saint-Vaast in northern France (Tuffreau & Sommé 1988), there would have been an ample capacity for producing fresh, unretouched flakes for tool use, without any need for the intensive resharpening of flakes as these became worn and damaged through use. Why it should have been necessary in these contexts to apply highly 'economizing' strategies in the use of available flint supplies is by no means clear.

2. Similarly, there are now many sites where a high overall frequency of racloirs (in rela-

tion to other tool forms) can be seen to coincide with an equally high frequency of unretouched flakes. At Biache-Saint-Vaast, for instance, an industry comprising over 60 percent of retouched racloir forms occurs in the context of an assemblage which is dominated by unmodified Levallois flakes (Tuffreau and Sommé 1988). A similar pattern is reflected at Champvoisy (Marne: Tuffreau 1989b) and again at Champlost in Burgundy (Farizy 1985: 406). In these and other cases it is difficult to see why intensive retouching and resharpening of tools should have been needed when there were large numbers of unretouched flakes immediately accessible and readily available for use on the occupation sites.

3. Even more problematic in terms of the racloir-reduction models is the available data on the relative sizes of side scrapers in different types of Mousterian assemblages. For example, several studies have shown that in assemblages comprising the lowest overall percentages of racloirs (i.e. those of the Denticulate variant) the average lengths of the side scrapers present in the assemblages are shorter than those recorded in some of the most heavily reduced industries, such as those of the Ferrassie and Quina Mousterian variants (Fig. 4.5; Table 10.1). This has been clearly documented, for example, by Rolland himself for the various levels of Denticulate, Quina and Ferrassie Mousterian at Combe Grenal (Rolland 1988b: 173, Table 9.4B) and is equally apparent in similar assemblages from the Abri Chadourne, La Quina, Hauteroche, Arcy-sur-Cure and elsewhere. Exactly how these observation can be reconciled with the hypothesis that the extent of systematic resharpening and reduction of tools is actually much greater in the Quina and Ferrassie industries than in those of the Denticulate variant (i.e. how systematic reduction can somehow make the tools larger) has yet to be explained.

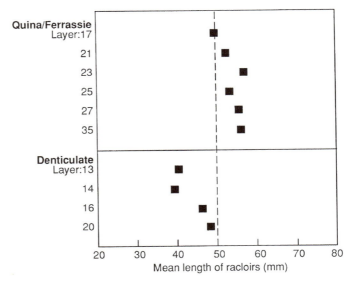

Figure 4.5 *Mean lengths of racloirs recorded in various levels of Ferrassie, Quina and Denticulate Mousterian at Combe Grenal, as documented by Rolland (1988b: Table 9.4b). The larger sizes of the racloirs recorded in the Ferrassie and Quina Mousterian levels clearly conflicts with the notion that these are heavily 'reduced' versons of the tools encountered in Denticulate industries.*

4. Finally, it should be emphasized that there are strong *a priori* reasons for assuming that in many contexts racloirs would have been produced as an essential and quite deliberate policy in general tool-making strategies. All studies of racloir morphology have recognized that two major objectives lie behind the production of typical racloir forms: first, to secure the maximum possible length of working edge from the available flake blanks, and second, to impose a regular, smooth form on this worked edge (Bordes 1961a; Mellars 1964; Dibble 1987a). A moment's reflection will show that the only way to achieve these objectives simultaneously on a high proportion of flake blanks is by applying systematic retouch to the edges of the tools (Fig. 4.6). In cases where primary flaking strategies produced elongated, regular flakes (for example in some of the Levallois strategies discussed earlier) then substantial numbers of flakes with naturally long, regular edges would have been readily available for use on the sites. But in most of the simpler flaking strategies (such as those in many Quina or disc-core strategies) it was only by systematically modifying the edges of the original flakes that these two objectives could be effectively achieved (Turq 1988b, 1989b, 1992a). Direct illustrations of these strategies can be seen in the detailed patterns of retouch applied in the shaping of racloirs in several sites. At Biache-Saint-Vaast it can be seen that the retouch on the edges of many of the racloirs was applied in a very sparing, discontinuous way apparently intended either to remove some obvious irregularity on the original edge of the flake or to extend the effective length of the working edge down the maximum possible length of the tool (Tuffreau and Sommé 1988). In these and many other cases it seems that retouch was applied not merely to rejuvenate heavily worn and damaged edges but as a deliberate policy to maximize the inherent potential of the available flakes for the specific functions envisaged.

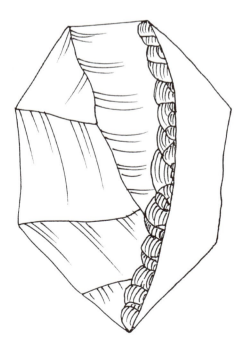

Figure 4.6 *Schematic illustration to show how the application of retouch can substantially increase the effective length of the working edge in the production of a typical 'racloir' form.*

None of this is meant to deny either the extensive use of totally unretouched flakes in many Middle Palaeolithic industries nor the reality of systematic resharpening of certain tools as a way of extending their natural use-lives. As noted above, both features can now be documented in the archaeological material from several sites. But to suggest that all the abundant and highly varied racloir forms encountered in Lower and Middle Palaeolithic industries can be dismissed as opportunistic end-products of these resharpening sequences would be not only in conflict with several specific features of the archaeological evidence but contrary to most reasonable expectations of tool production strategies.

Variations in side-scraper forms

Even if we accept that most side-scraper forms were produced from the outset as retouched tools, this still leaves much variation to be explained within the overall side-scraper range – i.e. the contrast between lateral and transverse types, single versus double-edged forms, variations in the shapes and treatment of the retouched edges (Figs 4.1, 4.2). What significance can be attached to this variation, either in terms of deliberate design norms in the initial production of the tools or in terms of subsequent use and resharpening?

Debates on the significance of varying frequencies of transverse versus lateral racloir forms in different Middle Palaeolithic industries provide a classic illustration of the kind of issues generated not only by studies of side-scraper forms but by studies of Middle Palaeolithic tool morphology in general. The distinction between these two forms rests strictly on the location of the retouched edge in relation to the main flaking axis of the tools – i.e. essentially parallel to the flaking axis in the case of lateral racloirs and at right angles to this axis in the case of transverse forms (Bordes 1961a). As Bordes (1961b, 1968a,

1984) and others have pointed out, the relative frequencies of these two forms show some striking variations between different industrial variants of the Mousterian. In particular, very high frequencies of transverse racloirs seem to be especially characteristic of the classic Quina-type industries in western Europe (Fig. 6.12), and of the Yabrudian industries in the Middle East (Bordes 1955b, 1984). In most other industries typically transverse forms are relatively rare and normally account for only about 5–10 percent of racloir forms in general.

Most workers in the past (including Bordes) have assumed that the production of transverse as opposed to lateral racloirs represented a deliberate decision on the part of the flint workers controlled largely by the forms of the original flake blanks selected. Where available flakes were relatively long in relation to their breadth (as for example in many Levallois industries) the longest working edges on the tool could usually be obtained along the main, longitudinal flaking axis of the original flake (Bordes 1961b: 806, 1968a: 101, 1977: 38, 1981: 78–9, 1984: 164; Bordes & de Sonneville-Bordes 1970: 61; Turq 1989b; Mellars 1967, 1992a). By contrast, where the original flake blanks were relatively broad in relation to their length (for example in most of the Quina-type flaking strategies discussed in the preceding chapter) then it was more often possible to obtain the maximum length of working edge by retouching the flakes on the transverse margin, directly opposite the striking platform. The distinction between lateral and transverse racloirs represented a simple technological decision dictated by the form of flakes immediately available for tool manufacture (see Fig. 4.7).

A sharply conflicting interpretation of the transverse/lateral racloir distinction has been put forward by Dibble in the context of his general tool-reduction models in the Mousterian (Dibble 1984a,b, 1987a,b,c; Dibble & Rolland 1992 etc.). His hypothesis is that

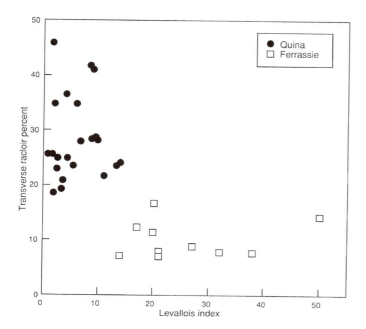

Figure 4.7 *Relationship between the relative frequencies of transverse versus lateral racloirs, and the variable utilization of Levallois flaking techniques (as reflected in the Levallois Index) recorded in Ferrassie and Quina Mousterian industries in southwestern France. The data suggest strongly that the production of transverse as opposed to lateral racloir forms was dictated mainly by the shapes of the original flake blanks produced by the different flaking strategies. From Mellars 1992a: Fig. 2.2, with additions.*

most, if not all, transverse racloirs started life initially as more elongated lateral racloir forms (either with or without deliberate retouching of the utilized edges) and were only transformed gradually into typical transverse forms as a result of repeated resharpening of the worked edges as they became progressively damaged or blunted by use. According to this model, the orientation of the worked edges shifted progressively from a lateral to a transverse orientation during successive phases of reworking and reduction. Dibble's own representation of this transformation is illustrated in Fig. 4.8.

In support of this hypothesis Dibble advances a number of observations on the metrical features and retouch characteristics of lateral and transverse racloir forms drawn from sites in both southwestern France (La Quina, Combe Grenal, Pech de l'Azé) and the Middle East (Bisitun) (Dibble 1987a,b,c). He argues that transverse racloirs tend to show larger striking platforms than those on most lateral racloirs together with higher ratios of flake-

thickness to total tool-area and generally heavier retouch along the worked edges (Fig. 4.9) (Dibble 1987a: Fig. 1, Table 1). He argues that these features are consistent with the idea that transverse racloirs were initially much larger in their original form than in their final, discarded, form and appear to show direct evidence for this heavy reduction of the tools in the invasive character of the retouch along the worked edges.

As in the more general racloir-reduction arguments discussed above there may well be an element of truth in Dibble's arguments. There is no doubt that many typical specimens of transverse racloirs do show evidence of apparently heavy and perhaps repeated resharpening of the worked edges (e.g. Lenoir 1986; Meignen 1988) and it may well be that in progressive resharpening some racloirs were occasionally transformed from lateral to transverse forms. The question, again, is the *scale* on which this technological transition occurred. Does this transformation account for almost all documented trans-

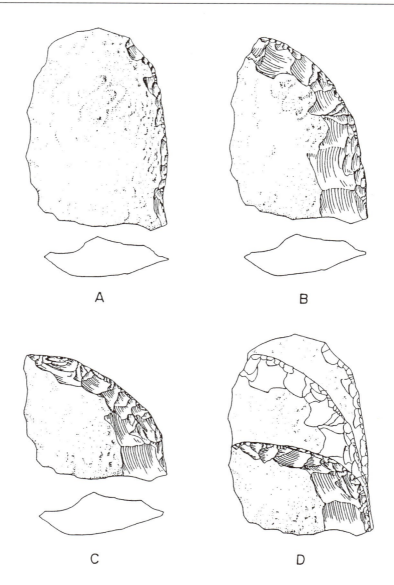

A

B

C

D

Figure 4.8 *Dibble's hypothetical reconstruction of the transformation from lateral to transverse racloir forms, in the course of successive episodes of resharpening the edges of the tools. Note how the overall length of the tool reduces progressively in the course of this transformation. From Dibble 1987a.*

verse racloir forms, as Dibble apparently implies, or was it a relatively rare and atypical occurrence which accounts for only a small percentage of the documented trans-

verse racloir forms in most Middle Palaeolithic assemblages? The relevant issues are as follows:

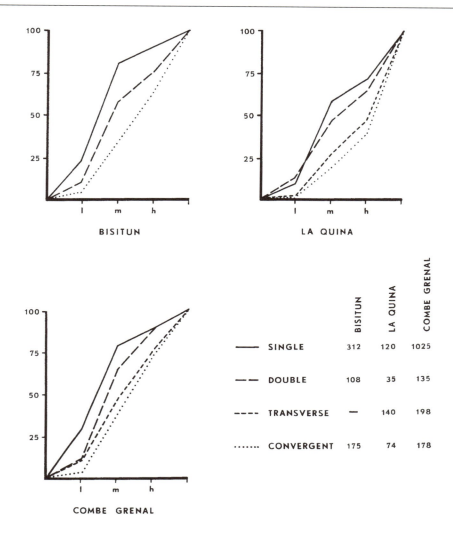

Figure 4.9 *Variable intensity of edge retouch recorded on different forms of racloirs (single, double, transverse and convergent) in the Mousterian assemblages from Bisitun (Iran), and La Quina and Combe Grenal (France), according to Dibble 1987a. Dibble argues that in all these sites the relative intensity of the retouch conforms with his hypothetical model of the variable degrees of edge resharpening and reduction represented by the different racloir forms (l = light retouch; m = medium retouch; h = heavy retouch).*

1. It can be seen immediately from a simple visual examination of the tools that many characteristic transverse racloirs can never have started life as conventional lateral forms. The tool edges are frequently oriented at almost 90° to the main flaking axis of the original flakes, and it is often clear that these retouched margins could never have extended around the lateral edges of the flakes in the way visualized in Dibble's hypothetical tool reduction sequences (see Fig. 4.2). This point has recently been emphasized by Turq

(1989b) who argues that in most of the Quina-type industries of southwestern France there is a clear morphological break between typical lateral and typical transverse racloir forms, with relatively few pieces which are intermediate between the two. Clearly therefore, the tool-reduction arguments cannot be applied to all, or even the great majority of documented transverse racloir forms in the classic southwestern French industries.

2. There is now evidence from several contexts that typical transverse racloir forms were manufactured on flake blanks which were very different from those used for lateral racloirs (cf. Meignen 1988; Bordes 1961b: 806, 1984: 164). This is best documented in the Quina Mousterian assemblages where, according to Turq (1989b, 1992a), there was not only a clear selection of certain specialized forms of blanks for the production of transverse, as opposed to lateral, racloirs, but perhaps even a deliberate attempt to produce them by special flaking strategies (see Chapter 3; also Lenoir 1986, Meignen 1988). As Turq points out, there can be no question in

these cases of a progressive transformation of lateral into typical transverse racloir forms.

3. There is evidence that despite the massive resharpening and progressive reduction envisaged for the production of transverse racloirs in Dibble's models the actual dimensions of typically transverse racloir forms are remarkably similar to those of typically lateral racloirs in most Mousterian industries (Fig. 4.10; see also Turq 1989b: Fig. 2). To explain this in terms of progressive tool reduction models it is necessary to assume that the original (unreduced) forms of the transverse racloirs were initially much larger than those of the lateral racloirs. According to Dibble's own reconstruction they needed originally to have been almost twice the size of those documented in their final discarded form (see Fig. 4.8). Even if this observation could be accommodated in principle within the general flake-reduction model, it is doubtful whether flakes of these dimensions could ever have been produced with the particular flaking techniques and raw material supplies documented in the relevant archaeological sites.

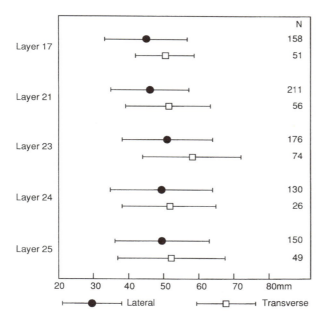

Figure 4.10 *Comparison of the maximum dimensions of lateral and transverse racloir forms recorded in different levels of Quina Mousterian at Combe Grenal, according to Pettitt (forthcoming). It will be seen that in all the levels transverse racloirs show consistently larger average lengths than the lateral forms, making it highly unlikely that the former types represent heavily reduced versions of the latter.*

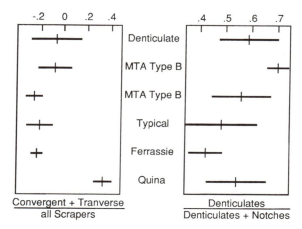

Figure 4.11 *Relative frequencies of transverse and convergent racloir forms, and the corresponding ratios of denticulates to notches, recorded in different industrial variants of the southwestern French Mousterian, according to Dibble 1988a: Fig. 10.6. Note how the relative frequencies of the two main racloir groups are broadly similar in all the industrial variants, with the exception of the Quina variant, in which transverse racloir frequencies are exceptionally high.*

4. There are similar problems in applying this tool-reduction model across the range of Mousterian assemblages in western France. As Dibble has documented (1988a: Fig. 10.6) the relative frequency of transverse to lateral racloirs remains essentially stable across the spectrum of Mousterian variants in this region, with the notable exception of those recorded in the Quina-type assemblages; in all other industrial variants (Ferrassie, Typical, Denticulate and Mousterian of Acheulian tradition (MTA) types) the relative proportions of transverse to lateral racloirs generally lie within the range of 0–15 percent, and only rise to higher levels (ca 20–30 percent) in the fully Quina industries (Fig. 4.11; see also Fig. 6.12). This is difficult to reconcile with the notion that these other, non-Quina variants embrace a wide spectrum of reduction-intensity pressures from virtually no reduction in the case of the Denticulate industries, to very heavy reduction in the case of the Ferrassie industries (cf. Dibble & Rolland 1992; Rolland 1977, 1981, 1988a). In terms of Dibble's arguments one would expect to observe much higher ratios of transverse to lateral racloirs in the case of heavily reduced industries (notably those of the Ferrassie variant) than in the hypothetically unreduced industries of the Denticulate and MTA Type B variants; in fact the ratios of transverse to lateral racloirs appear to be somewhat higher on average in the latter industries than in those of the heavily reduced Ferrassie group (Fig. 4.11) (Dibble 1988a: Fig.10.6).

5. Finally, it is now clear that all the specific metrical observations that Dibble (1987a) has advanced in support of the intensive reduction model for the production of transverse racloir forms are open to much simpler interpretation. The evidence from direct knapping experiments shows that production of broad, short flakes which seem to have been used mainly for manufacturing transverse racloirs almost inevitably entail the use of large striking platforms and will tend to yield flakes that are thick in relation to their length and total surface area (Turq 1989b). Thus, the large striking platforms and high ratios of thickness to surface area which Dibble has documented for many transverse racloir forms are almost inevitable products of the original flaking strategies, and certainly need not imply that the original length of the parent flakes was very much greater than that of the final reduced length of the tools.

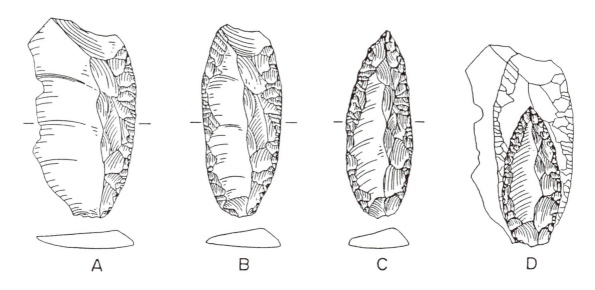

Figure 4.12 *Dibble's hypothetical reconstruction of the transformation from single, through double-edged, to convergent racloir forms, in the course of repeated resharpening and reduction of the edges of the tools. After Dibble 1987a.*

Similarly, it is clear that where retouching is applied to the transverse edges of these broad, thick flakes it will need to be relatively heavy and invasive (Fig. 4.9) to shape and control effectively the overall form of the retouched edge (cf Turq 1989b; Lenoir 1986; Meignen 1988). These features are inherent in flaking strategies applied to relatively short, thick flakes and need not imply massive deliberate resharpening and reduction in the eventual form of the tools. Even if these observations are consistent with the racloir-reduction models they can hardly provide any specific support for this argument as opposed to a more conventional interpretation of the basic form and technology of typically transverse racloirs.

These debates have been presented in some detail because they epitomize more general issues in the application of tool-reduction models in the Middle Palaeolithic. As noted earlier, the aim is not to suggest that pro-

gressive tool resharpening and reduction did not occur in certain Middle Palaeolithic contexts nor to deny that some variation in the forms of tools might be explicable in these terms. The issue is simply whether these models can be accepted as an all-embracing explanation for the wide variation documented in Middle Palaeolithic tool inventories or whether they account, in practice, for only a small component of this variation.

Pointed forms

In conventional conceptions of Middle Palaeolithic technology various forms of points are generally seen, along with side scrapers, as amongst the most distinctive and characteristic retouched tool forms (e.g. de Mortillet 1883). The diagnostic features are two carefully and more or less symmetrically retouched edges, usually shaped by flat, invasive retouch, which converge towards the distal end of the flake to form a fairly sharply

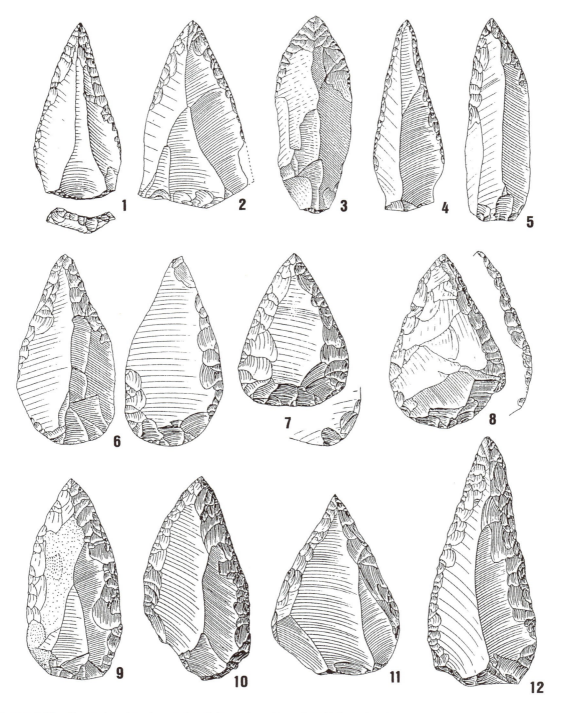

Figure 4.13 *Examples of various pointed forms from French Middle Palaeolithic sites. After Bordes 1961a.*

defined point (Fig. 4.13). The traditional notion was that these pieces functioned primarily as the hafted tips of hunting spears; as discussed below, however, this interpretation would not now be accepted for more than a small proportion of conventional pointed forms.

The classic taxonomy of these pointed forms was formulated in the typology of François Bordes (1961a) (Table 6.1). According to Bordes at least three main variants can be differentiated, representing two major functional categories: first, 'true' Mousterian points, subdivided into ordinary and elongated variants, which he felt probably (along with his category of retouched Levallois points) represented true, hafted missile heads; and second, a variety of forms of convergent and *déjeté* racloirs, which he believed were functionally separate from true points and represented specialized variants of his racloir category. The distinctions between these categories of true points and convergent or *déjeté* racloirs has always been one of the more controversial aspects of the Bordes typological system and the one most difficult to apply in practice. Bordes could give only general and subjective guidelines for distinguishing between the different forms. Thus true points were defined essentially by their overall symmetry, regularity and relative pointedness while convergent racloirs were generally thicker, less regular and less intuitively appropriate for use as missile tips (Bordes 1961a). The well known cartoon by Pierre Laurent (1965: 39) epitomizes the dilemma faced by many workers applying these distinctions to the wide spectrum of pointed forms documented in many Middle Palaeolithic industries. The one major variant in Bordes' taxonomy which can be identified in a fairly objective way is the category of *déjeté* racloirs, which is defined not so much by the overall form of the tools as by the distinctively angled orientation of the retouched edges in relation to the main flaking axis of the parent flake (Bordes 1961a).

The main problems posed by the analysis and interpretation of these pointed forms are similar to those involved in the interpretation of various racloir forms discussed in the preceding section:

1. To what extent do these represent intentional, deliberate forms imposed on the tools in the initial stages of manufacture, either in terms of specific design norms or intended function?

2. How far can we identify significant distinctions within the category of pointed forms – either in morphological terms or in terms of intended tool function?

Both issues have again been brought into focus by the recent tool reduction models of Dibble and others (e.g. Dibble 1984a,b, 1987a,b,c, 1988a,b, 1989, 1991a,b; Rolland & Dibble 1990; Dibble & Rolland 1992; Rolland 1990; Jelinek 1976; Holdaway 1989). As in the case of the various side-scraper forms discussed above, Dibble (1984a,b, 1987a,b,c) has argued that very few, if any, of the conventional forms of Mousterian points and convergent or *déjeté* racloirs were conceived and manufactured from the outset as such but represent heavily reduced versions of much simpler tool forms that were successively resharpened and remodelled in the course of use. Specifically, he envisages a sequence in which certain forms of flake blanks were transformed from simple, single-edged racloirs, to more heavily reduced double-edged forms and finally into forms in which the two retouched edges eventually converged to produce a point (Fig. 4.12). According to this model there is no significant distinction, either conceptually or functionally, between the various categories of pointed forms and most of the simpler racloir types. Nor, in his view, can we make any distinction between the heavily reduced forms of convergent and *déjeté* racloirs and pieces which Bordes and others have always regarded as 'true' points in the sense of functionally specialized missile heads (Dibble 1989; also Holdaway 1989).

As in the analogous models of lateral/transverse racloir reduction sequences, Dibble cites two specific features in support of these interpretations (e.g. 1984a, 1987a,b): first, the fact that the majority of pointed forms carry significantly heavier patterns of retouch than do the simpler, single and double-edged racloir types; and second, that these pieces tend to show smaller ratios of tool surface area in relation to the sizes of the striking platforms of the parent flakes (Fig. 4.9). Both observations, he argues, are consistent with the view that points have been subjected to much heavier resharpening and reduction than have the morphologically simpler racloir types.

My own impression is that while Dibble may well have identified a significant component of variation in some of the documented point and convergent racloir forms, this is not sufficient to dismiss the whole status of point types as an intentional tool form in the Middle Palaeolithic. The features I would emphasize in this context are as follows:

1. It seems evident from a visual examination of the tools that not all of the forms of convergent racloirs and Mousterian points can be dismissed as simple end-products of racloir-reduction sequences of the kind envisaged by Dibble. As Bordes and others have emphasized, many of these pieces show an overall regularity and clear bilateral symmetry in form (as reflected for example in the length, shape and overall treatment of the retouched edges) which seems to argue strongly for some kind of design-norms or mental templates in the minds of the flint workers (Bordes 1961a; Mellars 1964; Callow 1986a). Arguably, the most significant feature of the tools, however, is the specific treatment applied to the tips and bases of the implements. In the case of Bordes' categories of true Mousterian points and retouched

Levallois points, for example, the retouch was often applied in a highly selective, discontinuous way which was clearly intended to improve the sharpness and regularity of the point of the tool rather than to modify, let alone resharpen, the lateral edges (see Fig. 4.13). Similar observations can be made on the treatment applied to the bases. Here the retouching often extends around the base in such a way that the original striking platform of the flake has been totally removed. In other cases there is evidence for extensive, invasive flaking on the ventral surface which usually has the effect of either reducing or obliterating the original bulb of percussion (see Figs 4.13, 4.14). It is difficult to see how any of these features can be attributed to an opportunistic *ad hoc* resharpening of the edges of simple side-scraper forms. As discussed further below, these features are more likely to reflect deliberate attempts to accommodate the bases of the tools to specific hafting procedures.

2. Recent research seems to reveal some clear patterns in the deliberate selection of certain specific forms of flake blanks for the production of particular forms of pointed tools. The best evidence has come from the studies of Eric Boëda (1988c) on the material from level IIA at the site of Biache-Saint-Vaast. Thus Boëda points out that virtually all the typical specimens of convergent racloir and Mousterian point forms represented in this industry seem to have been manufactured on one particular form of flake blank which was produced at one specific point in the overall core reduction sequence – namely, the third-order removals in his strategy of 'bipolar recurrent' Levallois techniques (see Chapter 3, Figs 3.10, 3.11). All the simpler forms of single- and double-edged racloirs, by contrast, show no obvious selectivity, and seem to have been manufactured fairly indiscriminately from a wide variety of flake blanks (Boëda 1988c: 211, Table 5). The obvious conclusion, as Boëda points out, is that these

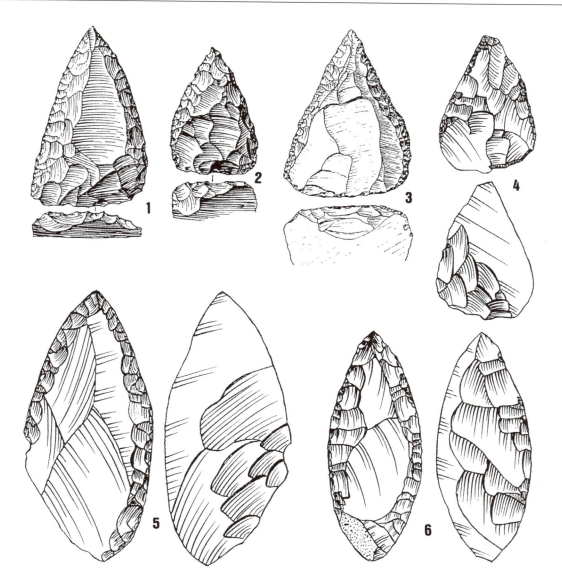

Figure 4.14 *Pointed forms with basal trimming of the bulbar surfaces. From Bordes 1961a and other sources. It is most likely that the thinning of the bases of the tools was related to some form of hafting procedures for these pieces (see Fig. 4.15).*

specific forms of flake blanks were either produced, or at least selected, deliberately with production of the convergent racloir forms in mind.

3. These conclusions are further reinforced

by results of recent micro-wear studies. The clearest data come from Sylvie Beyries' studies of the convergent racloirs from Biache-Saint-Vaast, and Patricia Anderson-Gerfaud's similar studies of the material from the open-

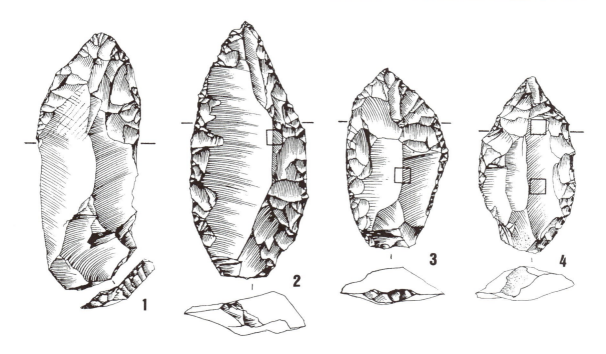

Figure 4.15 *Convergent racloir forms showing micro-wear traces of hafting, from the site of Biache-Saint-Vaast in northern France. After Beyries 1988b. The horizontal lines shown against each piece indicate the extent of the hafting traces along the length of the tools (see Beyries 1988a: Fig. 12.6).*

air site of Corbiac in southwestern France. At both sites it was found that virtually all specimens of typically convergent racloirs and Mousterian points showed evidence of deliberate hafting, in the form of distinctive patterns of abrasion and polishing confined exclusively to the lower, proximal parts of the tools and visible on both the dorsal and ventral surfaces (Fig. 4.15) (Beyries 1988a,b; Anderson-Gerfaud 1990). Anderson-Gerfaud argues that these seem to reflect hafting of tools in a relatively loose form which allowed some movement between the tool and associated haft during use. The functions documented showed more variation, ranging from the whittling of wood to the scraping of bone or skins, but it is significant that all these use-wear traces were normally confined to the distal ends of the tools which

extended beyond the encasing hafts (Fig. 4.15) (Anderson-Gerfaud 1990). None of this evidence easily supports the notion that all the retouch applied to these tools was due to the resharpening of utilized edges, since in most cases the retouched edges extended well below the parts encased in the hafts. In any event it is significant that with the exception of these typically convergent forms very little evidence for deliberate hafting has so far been documented on any other forms of Mousterian tools (Beyries 1987, 1988a; Anderson-Gerfaud 1990). The evidence provides strong support for the idea that many examples of convergent racloir forms were intended for a specific economic function involving the use of a haft, and that the overall form and design of the tools was dictated largely by this anticipated function.

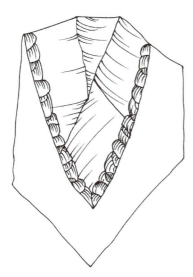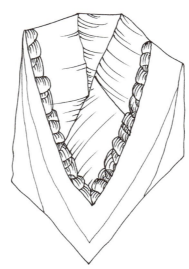

Figure 4.16 *Diagram to illustrate how the deliberate production of typical pointed forms (shown on the left) can result in identical patterns of edge retouch to those resulting from the progressive 'reduction' of different racloir forms, as envisaged in the tool-reduction models of Dibble (see Fig. 4.12).*

4. It is difficult to see how the specific metrical features documented by Dibble (1984a, 1987a,b,c) for convergent scrapers and Mousterian points can be used to argue for his racloir-reduction model rather than for the alternative interpretations suggested above (Fig. 4.9). As Dibble (1987a) points out, both the major features which he has documented (i.e. relatively heavy retouch on the worked edges and apparently reduced areas of tool surface in relation to the sizes of the associated striking platforms) can be seen essentially as a reflection of the amount of deliberate shaping (or in his terms reduction) applied to the tools. But both these features are equally consistent with the hypothesis that deliberate retouching and reduction of tools was designed not simply to resharpen damaged edges but to allow a specific overall form to be imposed on the finished tools. If convergent racloirs and points were produced to a large extent according to an explicit design, then this would inevitably involve in many cases a substantial degree of reduction of the original flake blanks, with consequently heavy retouching on certain parts of the tool edges, to achieve this overall,

preconceived form (Fig. 4.16). Dibble's observations on these features are interesting and relevant but hardly allow one to choose objectively between the alternative interpretations discussed above.

The final question of whether there were significant morphological or functional distinctions within the broad category of pointed forms is by far the most difficult issue to resolve. The problems stem partly from the scarcity of detailed studies of the overall range of morphological variation of different pointed forms in particular industries and partly from the sparsity of detailed use-wear studies. The long-standing debate over the existence of spear heads in the Middle Palaeolithic seems impossible to resolve with complete certainty. Both Anderson-Gerfaud (1990) and Beyries (1987, 1988a) have stressed that in their recent studies of microwear patterns there is as yet no unambiguous evidence for the existence of missile points – although it is arguable that the use of Mousterian points for this function might well leave few discernible traces in the microwear data (cf. Jelinek 1988b). John Shea (1989,

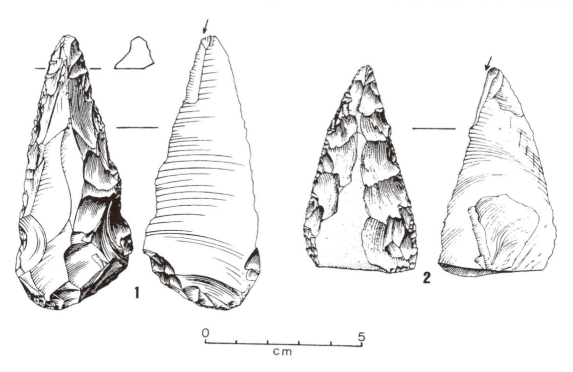

Figure 4.17 *Apparent impact fractures resulting from the use of Mousterian points as missile or weapon tips, from the penultimate-glacial Middle Palaeolithic levels at the Cotte de St Brelade (Jersey). After Callow 1986a.*

1993) by contrast, has claimed much more positive microwear evidence for the use of both Mousterian and Levallois points as missile heads in several Near Eastern sites. The strongest argument for the existence of spear points in the European Middle Palaeolithic has been advanced by Paul Callow (1986a) based on the evidence for apparently distinctive impact-fracture patterns on several of the typically pointed forms from the penultimate-glacial occupation levels at La Cotte de Saint Brelade in the Channel Islands (Fig. 4.17). If these interpretations of the fracture patterns are reliable then this could be seen as a major breakthrough in understanding not only the typology but also the functional and economic implications of Middle Palaeolithic technology. Further studies of the kind carried out at the Cotte de Saint Brelade – combined no doubt with more detailed experimental studies of impact fracture patterns – are needed to resolve this issue convincingly.

Notched and denticulated tools

Various forms of notched and denticulated tools are generally seen as some of the least impressive products of Middle Palaeolithic technology, both visually and in terms of the knapping skills and technology involved in their production (Fig. 4.18). As Bordes has pointed out (1961a: 35, 1963: 43), these pieces were not generally recognized as deliberate retouched tool forms until well into the present century and were often either ignored or collected in a highly selective way in the earlier excavations. The fact that these forms can represent the most dominant element in certain Mousterian industries was first

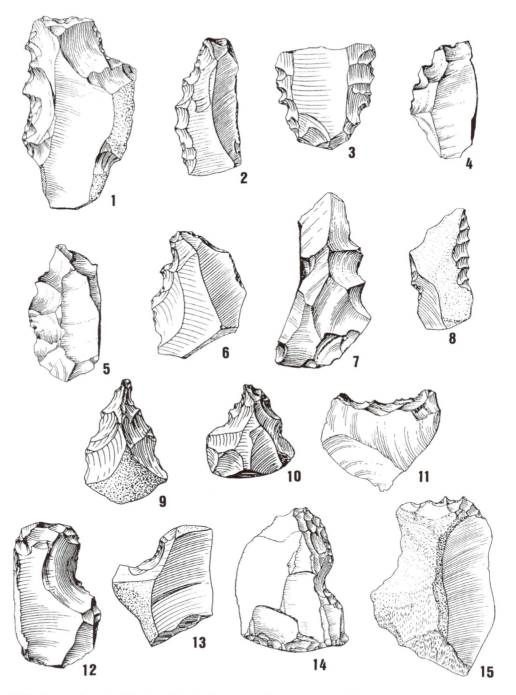

Figure 4.18 *Examples of notched and denticulated tools from French Middle Palaeolithic sites. Nos 12–14 would be classified by Bordes as 'Clactonian notches' and nos 9–10 as 'Tayac points'. After Bordes 1961a; Girard et al. 1975; Callow 1986a.*

clearly recognized in the excavations of Bour-rinet and Darpeix at the Sandougne rock-shelter in 1928 (Darpeix 1936). Over the past 40 years, principally owing to the publica-tions of François Bordes, the notion of highly specialized Denticulate variants of the Mousterian has emerged as one of the most enigmatic and challenging aspects of indus-trial variability in the Middle Palaeolithic (see especially Bordes 1963, 1984, and Chapter 10).

As defined in the Bordes typology (1961a) two major forms are involved: simple notch forms characterized by a single major inden-tation worked on the edge of a flake, and denticulated forms characterized by a series of two or more closely juxtaposed notches located along the same edge. In practice this leaves scope for a wide range of variation in both overall forms and techniques of manu-facture. In particular, Bordes and other workers have pointed out that the individual notches on both simple notched forms and more complex denticulated forms can be pro-duced by either a single major blow (gen-erally referred to as 'Clactonian notches') or by a succession of lighter, repeated blows at the same point ('irregular notches'). Simi-larly, the numbers of individual notches on the edges of denticulated forms can vary from as few as two or three to as many as ten or twelve. Further distinctions can be made according to the position of the notched/denticulated edges in their relation to the long axis of the flake or to the presence of two or more separately denticulated edges on the same tool (see for example Jaubert 1984, 1990). Certain more complex forms in which two denticulated edges converge towards a thick point were classified separately by Bordes as 'Tayac points' (Fig. 4.18) (Bordes 1961a).

The functional interpretation of notched and denticulated forms has caused much speculation. Bordes was apparently inclined to regard most of these tools as planes or spokeshaves for working wood, but admitted the possibility of other interpretations

(Bordes 1963: 47). Binford initially interpre-ted denticulates as playing a poorly specified role in the processing of plant materials (Bin-ford & Binford 1966: 256, 259, 1969: 79) but later suggested that at least some tools might have functioned as knives for hacking or slicing strips of meat to assist in drying or preservation processes. Dibble and Rolland are less specific but suggest that both notches and denticulates may have been more appro-priate for processing plant or woody ma-terials than for use in butchery or animal processing activities (Dibble & Rolland 1992; Rolland 1981: 26–31, 1990: 371–3).

Recent applications of micro-wear analyses have gone some way towards resolving these issues but still allow different functions for individual tools. The most positive results have come from studies of simple 'Clactonian notch' forms. From material from five dif-ferent Middle Palaeolithic sites, Beyries (1987, 1988a) has claimed that almost all these tools were used essentially as planes or scrapers for shaping wooden stakes or shafts. More complex denticulated forms in the same industries appear to have been used predominantly for similar wood-working activities, although a few pieces showed evi-dence for use on either skins or meat. Similar results have emerged from the studies of Anderson-Gerfaud (1981, 1990) on several assemblages belonging to the MTA variant in southwestern France. In these assemblages she found that all except one of the denticu-lates had been used for either scraping, cut-ting or planing wood. The one exception, a concave-edged denticulate from the later MTA levels at Pech de l'Azé IV, appeared to have been used in the processing of 'soft plant material'.

The specific functions of notched and den-ticulated tools, therefore, still remain prob-lematic although the processing of plant, (especially wood), as opposed to animal materials seems to be indicated by the cur-rent use-wear data. At the same time the working of wood and plant materials is more likely to produce detectable use-wear traces than softer meat or animal tissue as Anderson-Gerfaud (1990), Jelinek (1988b: 221), Beyries (1990) and others have pointed out. One issue not yet resolved by micro-wear or systematic morphological studies is how far simple notches and more complex denticulate forms can be separated into distinct tool categories. Not surprisingly Dibble and Rolland (1992) have suggested that most forms of denticulates are repeatedly re-used and resharpened notch forms. Whilst this may be plausible for some of the simpler denticulated forms it is more difficult to accept for some of the more elaborate forms characterized by multiple regular and evenly spaced denticulations aligned along a single edge (see Fig. 4.18). What is not in doubt is that both notched and denticulated forms were deliberate tools which evidently played a major economic role in certain Middle Palaeolithic assemblages. As discussed in Chapter 10, the dichotomy between these heavily denticulate-dominated industries and those dominated by various side-scraper or pointed forms presents the greatest challenge of industrial variability in the Middle Palaeolithic.

Backed knives

Typical backed knives provide a fourth distinctive category of retouched tool forms in the Middle Palaeolithic. The basic technological features are clear cut: first, one sharp, regular and normally unretouched edge running along one lateral margin; and second, the presence of steep, abrupt retouch applied to much of the opposite edge (Fig. 4.19). The general functional orientation of the tools seems equally clear; the unretouched, sharp edge of the flake is generally assumed to represent a knife edge whilst the blunting applied to the opposite edge was presumably intended to allow pressure to be applied to the back of the tool whilst in use (Bordes 1961a: 32; Binford & Binford 1966: 244).

Whether the tools were held in the hand or attached to a haft has not yet been clearly documented by use-wear studies.

Within this category most of the obvious formal variation is related to the different forms of the original flake blanks selected for production (Monnier 1992). As shown in Fig. 4.19, these can vary from broad, fairly heavy flakes, often retaining a strip of cortex along the backed edge, to thinner, more elongated blade-like forms. Normally the retouch applied to the backed edge is relatively light and adheres closely to the original outlines. Other pieces, however, show much heavier retouch apparently intended to impose a more regular, convex form on the blunted edge. How far some of the pieces could be said to exhibit clearly 'imposed form' is an interesting point for debate (see below; Mellars 1989b; 1991). In any event the pieces illustrated in Fig. 4.19 show that individual shapes of backed knives are highly variable and reflect little obvious attempt at morphological standardization beyond the combination of the two basic features (one naturally sharp edge and one deliberately blunted edge) (see also Monnier 1992).

At present we have very little direct use-wear data on the specific functions of backed-knife forms. The samples analysed by Beyries (1987, 1988a) included only one specimen of a retouched backed knife, which showed some signs of use on bone. Most backed knives analysed in her samples belonged to Bordes' category of 'naturally backed knives' characterized by natural blunting extending along one edge of the flake (usually as a strip of cortex) but carry no deliberate retouch. Of 32 pieces in this category analysed by Beyries the functions were divided evenly between use on bone (nine examples), on wood (nine examples) and on meat (ten examples). Whether or not these results have any direct relevance for interpreting typical (i.e. retouched) backed-knife forms is an open question. Provisionally, the results could suggest that backed-knife forms had many functions

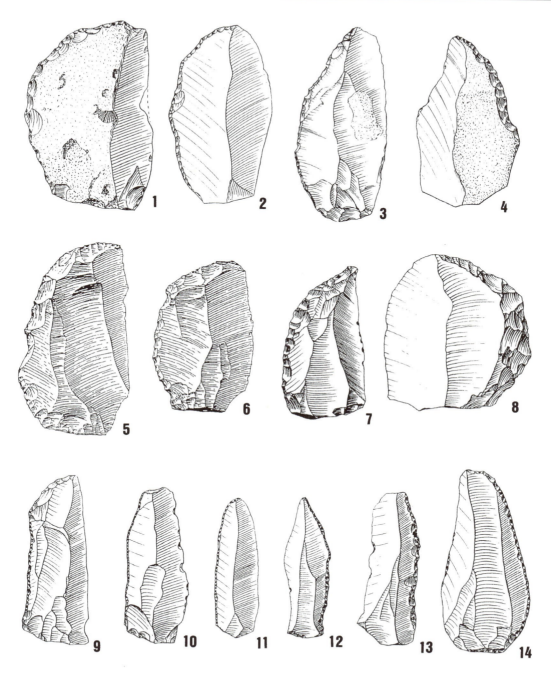

Figure 4.19 *Examples of typical backed knives from MTA sites in southwestern France. All of the tools are characterized by the presence of a single, continuously retouched and blunted edge, located opposite a naturally sharp, regular, and unretouched edge. While in most of the pieces the retouch is confined to the immediate edges of the tool, in other cases (e.g. nos 5–8) it cuts more deeply into the central and thicker parts of the parent flakes. After Bordes 1961a; Lalanne & Bouyssonie 1946; Delporte 1962.*

and were used on wood or plant materials as well as in the processing of animal products.

Arguably the most interesting feature of typical backed-knife forms is their strong association with industries of the MTA group (see Bordes 1961a: 33, 1984: 137–49; also Peyrony 1920, 1930; Bourgon 1957). As discussed in Chapter 6, all the industries from western France in which the frequency of these tools exceeds ca 1–2 percent of the total tool inventories can be attributed either on typological or closely related stratigraphic grounds to the MTA variant – in some cases with overall frequency of 20–30 percent (Fig. 6.4). Similar associations between backed knives and hand axes have been recorded in some of the open-air MTA sites in northern France (Bordes 1954a; Tuffreau 1971; Farizy & Tuffreau 1986) and apparently in much earlier industries such as the typically Acheulian industry from the Atelier Commont in the Somme valley (Bordes & Fitte 1953). Whether interpreted in 'cultural' or other terms, the reality of this association is now well documented in the French industries.

Rarer forms

The various forms of retouched tools described above represent the dominant morphological and typological categories within the Middle Palaeolithic industries of western Europe and account for the overwhelming majority of retouched tool forms. The basic type list compiled by Bordes (1961a) includes several other categories of minor types including such forms as piercers, planes (rabots), becs, truncated flakes, hachoirs etc. Assessment of the significance of these very rare tool forms is extremely difficult. As Dibble has documented (1988a: Figs 10.3, 10.4) most occur in a sporadic, patchy fashion in all the major variants of the Mousterian, almost invariably in frequencies of less than 1 percent and apparently without a clear pattern of association with other technological or typological features of the industries. Granted this, and in view of the very simple character of most of the forms in question, it is debatable how far any of these forms can be recognized as discrete, deliberate types. Many of these pieces could represent accidental products, as a result of unfinished or unsuccessful attempts at manufacturing other, well defined tool forms. For example, some truncated flakes could represent unfinished or partially retouched attempts at backed knives; other forms such as becs, percoirs or rabots might represent similarly incomplete or unsuccessful attempts at notches, denticulates or Tayac points. Above all as Bordes emphasized (e.g. 1963: 48) it is likely that some of these pieces were simply the products of purely taphonomic processes such as the effects of human trampling on the sites, or the effects of cryoturbation or other geological crushing processes (see also Dibble & Rolland 1992: 5). To accord any of these forms the status of discrete, conceptually defined types would certainly be premature from the present evidence.

Possibly more significant is the existence of end scrapers and burins as discrete, intentional types in Middle Palaeolithic industries. Occasional examples of both types have been reported from many Middle Palaeolithic contexts in western Europe, and in some cases the appearance of these pieces (especially in illustrations) appears convincing (Fig. 4.20). The problem is that these pieces normally occur in such low frequencies in individual tool assemblages that their status as intentional products requires careful scrutiny. Again there is the question whether these are deliberate products or whether they could represent unfinished or discarded attempts at some of the commoner types. Patricia Anderson-Gerfaud (1990) has claimed that almost all the examples of end-scrapers examined in her micro-wear analyses of MTA industries in southwestern France would be classified as 'atypical' in a morphological sense. More significantly, she

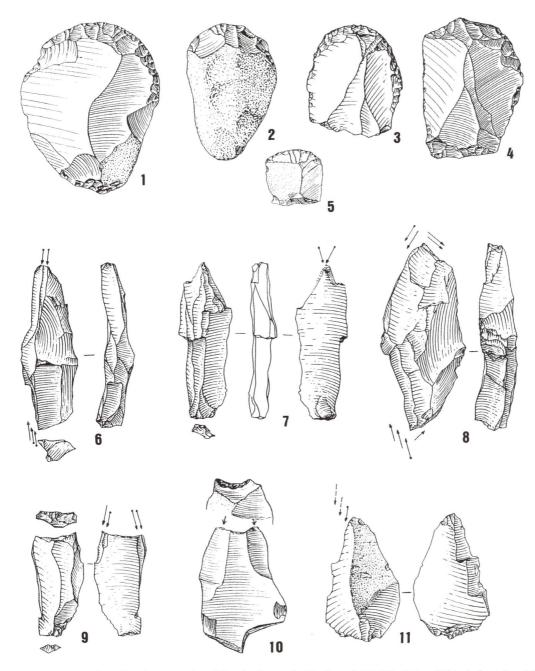

Figure 4.20 *Examples of 'end scrapers' and 'burins' recorded in French Middle Palaeolithic industries. While some of the end-scraper forms could represent atypical or unfinished examples of other tool types (such as backed knives), the series of burins recorded from the early last-glacial site of Riencourt-les-Bapaume in northern France (nos 6–9) is highly typical, and remarkably similar to Upper Palaeolithic forms. After Bordes 1961a, and Ameloot-van der Heijden 1993a.*

argues that from a functional standpoint almost all these pieces seem to group with the other Mousterian forms of woodworking tools, rather than with skin-working tools. In this sense at least there seems to be a fundamental distinction between the morphologically 'end-scraper' forms documented in Middle Palaeolithic assemblages and the typical end-scraper forms encountered in the majority of Upper Palaeolithic industries (Anderson-Gerfaud 1990).

The status of supposedly 'burin' forms identified in many Middle Palaeolithic industries in Europe is equally debatable. As several workers have pointed out, apparently typical burin-spall removals can be produced either deliberately or accidentally in many stages of tool manufacture, and sometimes as a simple by-product of producing other tool forms. Again it is significant that burins of either typical or atypical form rarely account for more than 1–2 percent of the total tool inventories in well documented assemblages from European sites. Interestingly, this is in sharp contrast to the situation in the Middle East where typical burin forms have been recorded in frequencies of up to 10–20 percent in some Middle Palaeolithic contexts (for example at Rosh ein Mor in Israel: Crew 1976: 99–105). By far the most impressive burins from west European contexts are those reported recently from the early last-glacial site of Riencourt-les-Bapaume in northern France, some of which are remarkably similar to Upper Palaeolithic types (Fig. 4.20: Ameloot-van der Heijden 1993a, 1993b).

Bifacial tools

Bifacially worked tools in the Middle Palaeolithic make up a broad and heterogeneous group which evidently comprises a number of discrete and sharply differentiated forms. Since these forms have been fully described in earlier literature (e.g. Bordes 1961a, 1984) and pose rather fewer interpretative prob-

lems than some of the other types, they will be discussed here more briefly.

Forms conventionally described as 'hand axes' are distinguished essentially by four basic features: first, by a continuous pattern of bifacial trimming usually extending over most of both faces of the tools; second, by a clear pattern of bilateral symmetry centred around the long (i.e. vertical) axis of the tools; third, by a pronounced pattern of *asymmetry* around the horizontal axis, generally defining a more pointed or convergent upper part of the tool and a more obtuse or rounded lower part; and fourth, by a largely continuous edge running around the greater part of the tool perimeter, formed by the intersection of the flake removals from the upper and lower faces.

Defined in these terms, the various forms of hand axes encountered in Middle Palaeolithic industries appear to extend the long tradition of hand-axe manufacture from the earlier stages of the Pleistocene. The precise shapes and technology of these hand axes vary in reasonably well documented ways in different stages of the Lower and Middle Palaeolithic succession. In some of the last-glacial industries of central Europe, for example, the various forms of 'Micoquian' hand axes (Fig. 6.8) perpetuate some of the sharply pointed hand-axe forms which, in other contexts, can be traced back well into the Lower Palaeolithic sequence (for example, at La Micoque itself or at sites such as Swanscombe and Hoxne in southern Britain: Bordes 1961a, 1984; Roe 1981). In western Europe, by contrast, these more elongated and pointed hand-axe types (Fig. 4.21) are restricted largely to contexts earlier than the last glaciation, and are generally regarded as hallmarks of fully 'Acheulian' industries (Bordes 1984). Whether or not these Acheulian and Micoquian forms persisted into the last-glacial industries in, for example, northern or western France is still a largely open question (see for example Tuffreau 1971; Bordes 1984; Farizy & Tuffreau 1986).

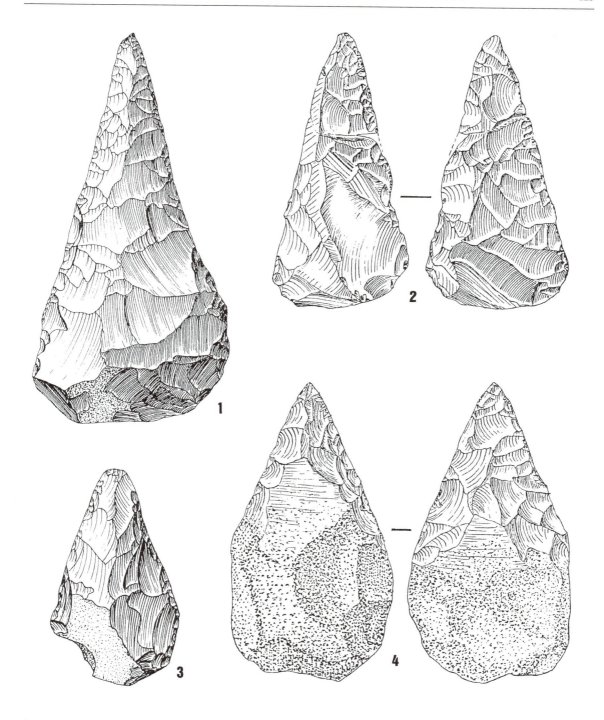

Figure 4.21 *Examples of elongated, pointed 'Micoquian' type hand axes from the sites of La Micoque, France (nos 1, 3) and Bocksteinschmiede, Germany (nos 2, 4). After Bordes 1961a; Bosinski 1967.*

Figure 4.22 *Typical 'cordiform' and related biface forms, characteristic of the French MTA industries. From layer 4 of Pech de l' Azé site I (Dordogne). After Bordes 1954–55.*

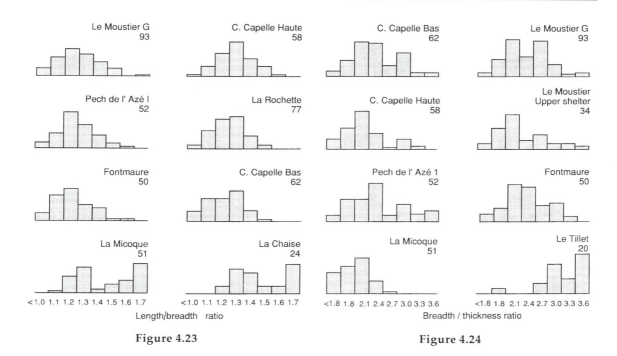

Figure 4.23   **Figure 4.24**

Figure 4.23   *Length-over-breadth ratios of bifaces recorded for various MTA industries from southwestern French sites, compared with those for the earlier 'Acheulian' industries from La Micoque and La Chaise. The graph shows that elongated, pointed hand-axe forms are effectively lacking from the MTA industries. From Mellars 1967.*

Figure 4.24   *Breadth-over-thickness ratios of bifaces recorded for various MTA industries in southwestern France, compared with those for the 'Micoquian' assemblage from La Micoque, and the northern French site of Le Tillet ('café au lait' series). The clearly bimodal distribution apparent in most of the graphs almost certainly reflects the manufacture of bifaces from two different forms of blanks – nodules on the one hand, as opposed to large flakes on the other. The bifaces from the northern French site of Le Tillet would appear to be manufactured entirely from large flakes, while those from La Micoque were presumably manufactured from nodules. From Mellars 1967.*

The classic hand-axe forms encountered within the last-glacial industries of western Europe conform largely to the broad grouping of 'cordiform' types (Fig. 4.22) – and in this form are generally regarded as the diagnostic hallmarks of the MTA variant (Peyrony 1920, 1930; Bordes 1953a, 1984 etc.). The relative uniformity and standardization of these tools in a morphological sense has been emphasized in many earlier studies (e.g.

McBurney 1950; Bordes 1961a: Fig. 7; Mellars 1967). The distinctively broad, squat forms of the tools can be differentiated quite easily from the more elongated Acheulian or Micoquian hand axes on the basis of simple length-over-breadth measurements alone (Fig. 4.23). The precise forms of the tools show rather more variation. While the term cordiform or heart-shaped defines the central shape tendency of the majority of the tools (Fig. 4.22),

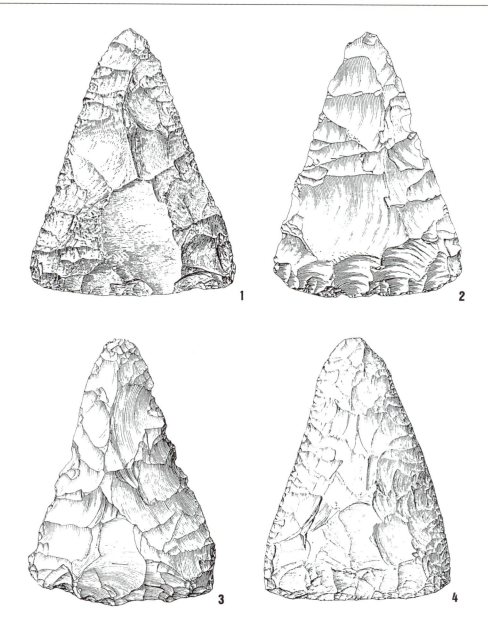

Figure 4.25 *Sharply triangular hand-axe forms, recorded from various surface contexts in western France. After Bordes 1961a, Turq 1992b.*

individual hand axes can vary from almost oval, or sometimes nearly circular, to more sharply angular outlines, including distinctive triangular (Fig. 4.25) and so-called *bout-coupé* (Fig. 4.26) forms (Bordes 1961a; Shackley 1977; Roe 1981). As Bordes frequently emphasized (e.g. 1954a: 441, 1961a: 78; 1984) the latter are particularly characteristic of

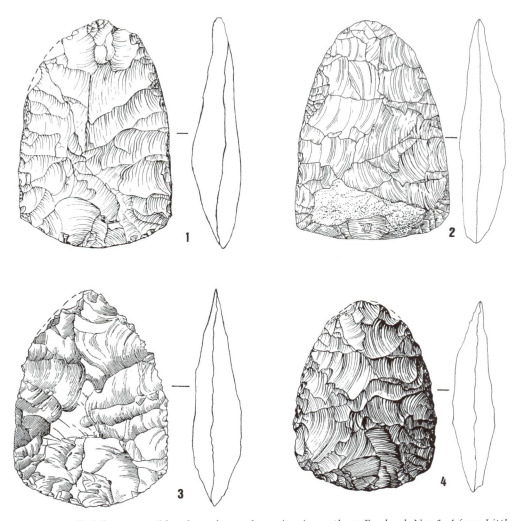

Figure 4.26   *So-called 'bout-coupé' hand-axe forms, from sites in southern England. Nos 1, 4 from Little Cressingham (Norfolk); no. 2 from Bournemouth (Hampshire); no. 3 from St Neots (Cambridgeshire). After Lawson 1978, Coulson 1986, Paterson & Tebbutt 1947.*

some of the earlier MTA industries from the loess regions of northern France (see also Tuffreau 1971). It can be seen from both visual inspection of partially retouched pieces and simple measurements of breadth-over-thickness ratios (Fig. 4.24) that the tools could be made from either complete nodules of raw material or from large flakes (Mellars 1967). The sizes are similarly variable and may range from up to 20 cm in length to very much smaller forms less than 5 cm in length. To judge by the available stratigraphic data there seems to be a general tendency for the overall sizes of the hand axes to decrease continuously throughout the chronological development of the MTA industries within

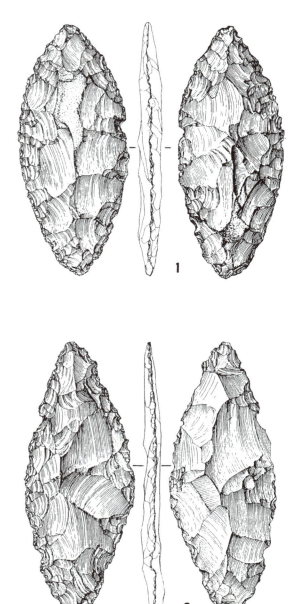

Figure 4.27 *Bifacial leaf points from the later Mousterian levels of the Mauern cave (southern Germany). No. 2 shows typically 'plano-convex' patterns of retouch, marked by the initial detachment of very broad flake removals from the lower face, followed by the finer trimming of the upper, more convex face. After Bohmers 1951.*

western France (Mellars 1965, 1967). Regardless of all variations in shape and size there can be no serious doubt as to the highly distinctive character of these cordiform hand-axe types within the French industries nor to their close association with industries of the classic MTA group. As Bordes often stressed (e.g. 1961b, 1968a; Bordes & de Sonneville-Bordes 1970) typical examples of cordiform hand axes are effectively lacking from all other distinctive industrial variants of the Mousterian (Quina, Ferrassie, Denticulate etc.) within southwestern French sites.

Other forms of fully bifacial tools are not nearly as well represented within the industrial sequence in western Europe. The broad category of bifacial leaf points is one of the distinctive hallmarks of some later Mousterian industries of central and eastern Europe (Figs 4.27, 6.9) and is as yet virtually unknown in the extreme western zones of Europe (Bordes 1968a, 1984). Leaving aside one or two very rare, isolated and generally atypical occurrences (for example the single pieces illustrated by Bordes (1954–55: Figs 14, 17; 1961a: Fig. 49) from Pech de l'Azé site I and Fontmaure) the most typical specimens of characteristic leaf points recorded from the French sites seem to be those described by de Lumley from the later Rissian levels at the Baume Bonne cave in Provence (de Lumley 1969c: 258–61). Whatever the particular origins and functions of these tools, they would seem to have little in common with the much more widespread and characteristic forms of cordiform hand-axes documented within the classic MTA industries of western Europe.

A final distinctive, if rather enigmatic, form is the so-called 'Vasconian' flake-cleaver (Fig. 4.28) recorded from a range of sites in the Pyrenees and the adjacent Cantabrian region of northwest Spain (Abri Olha, Castillo, Cueva Morín etc.: Bordes 1953a, 1984: 166, 206–9; Cabrera Valdés 1988). Although sometimes linked with broadly cordiform hand-axe forms, these tools stand as a separate morphological and technical

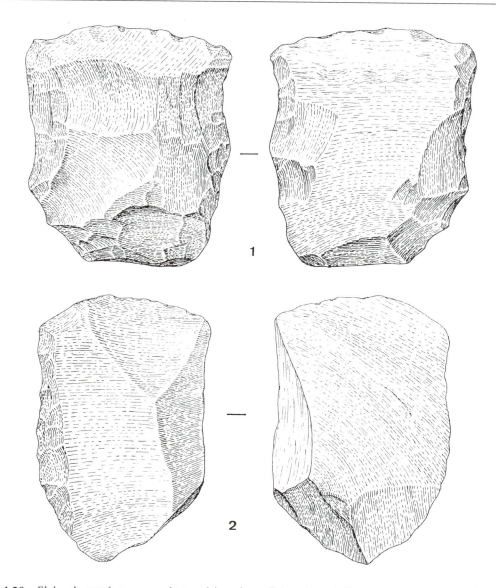

Figure 4.28 *Flake-cleaver forms manufactured from large flakes of quartzite or similar rocks from the so-called 'Vasconian' Mousterian levels of El Castillo, Cantabria. After Bordes 1961a.*

group. As the name suggests, the tools consist of large flakes in which the retouching is confined largely to the lateral margins on either one or both faces, and in which the presumed working edge is formed by one of the original sharp and unretouched edges of the parent flake (Fig. 4.28) (Bordes 1961a: Figs 75–6, 82, 1984: 206–9). Most of these tools were manufactured from large side-struck flakes apparently produced selectively from some of the local fine-grained rocks such as ophite or quartzite (Cabrera

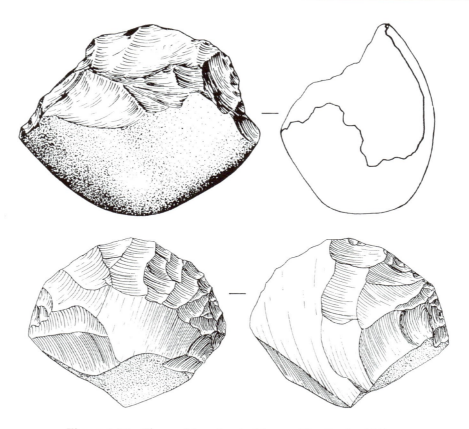

Figure 4.29 *Chopper/chopping-tool forms. After Bordes 1961a.*

Valdés 1988). Only in a few rare instances (as at Abri Olha) are these forms apparently associated with typical cordiform hand axes (Bordes 1953a, 1984: 166).

How far the various forms of so-called choppers or chopping-tools should be included within the general grouping of bifacial tools is more open to debate (Fig. 4.29). The distinguishing feature of the great majority of these tools is that they are manufactured predominantly from rounded pebbles of various coarsely textured rocks (usually quartz or quartzite, very rarely from flint) on which the flaking is confined entirely to one end of the parent pebbles to define a rather jagged and irregular cutting edge

(Bordes 1961a). In many cases there is a real problem in deciding whether these forms represent intentional tools or whether they are simply discarded, partially worked cores. In certain contexts, however, these pieces do show a degree of regularity and systematization in the flaking which leaves little doubt that they were intended as deliberate tool forms (e.g. Figs 7.30, 7.31). Within the French industries these forms are best represented in some of the MTA industries (as at Fonseigner and Les Ourteix: Geneste 1985) and in some of the open-air occurrences of taxonomically Denticulate industries at sites such as Mauran, Coudoulous and La Borde (Figs 7.30, 7.31) (Jaubert 1984, 1990; Girard *et al.* 1975).

Discussion

Following this discussion of the various retouched tool forms encountered in Middle Palaeolithic industries, what kinds of general patterns can be discerned? What is the inherent character of this morphological variation in Middle Palaeolithic tools, and what does it tell us about the broader behavioural and conceptual patterns which lay behind tool production?

The first point now effectively established is that there are indeed several categories of retouched tool forms in Middle Palaeolithic contexts which almost certainly existed as distinct concepts in the minds of the groups who produced them. The overall range of these discrete forms is very much smaller than that represented in Upper Palaeolithic industries (Mellars 1989b) and almost certainly much smaller than that implied in the classic typology of François Bordes. Nevertheless, the basic range discussed in the preceding sections represents the irreducible minimum of morphologically and apparently conceptually discrete tool forms which must be recognized to account for the total documented variation within Middle Palaeolithic tools.

Many other related issues such as the effect of repeated resharpening on the overall forms of retouched tools and the specific functions of different types remain more enigmatic and require fuller investigation than they have received so far. Whatever weight one may attach to the tool-reduction models of Dibble and Rolland these can only account at best for a very limited component of the total documented variations discussed above. The question of purely functional determination of tool forms presents a wider range of interpretative issues which will be pursued in Chapter 10. The only point which can be made with any confidence at present is that the functions of many of the apparently discrete and morphologically separate forms seem to have overlapped in a complex and

poorly defined way. Certainly no simple one-to-one correlations between form and function have been revealed by recent applications of micro-wear analyses (e.g. Beyries 1987, 1988a; Anderson-Gerfaud 1990).

How these different morphological types were visualized by the Middle Palaeolithic flint workers poses equally intriguing questions. The various attributes which define and characterize the different types make up a complex combination of both 'technical' features (i.e. relating to the basic flaking techniques and procedures of production) and more specifically morphological features, (i.e. relating to the particular location, form and character of the retouch applied at different points around the edges of the tools). All the basic distinctions between, e.g. racloirs, points, denticulates, backed knives etc. are defined essentially in these simple 'morpho-technical' terms. Leaving aside the specific functional interpretations of these different forms the central question is whether we can identify any other significant component of conceptual patterning in the production of Middle Palaeolithic tool forms which goes beyond this simple combination of technical and morphological features.

As I have discussed (e.g. Mellars 1989a,b, 1991) one of the central issues is whether the production of Middle Palaeolithic tools involved any explicit component of 'imposed form' of the kind which can be documented for many if not the majority of Upper Palaeolithic tools. The notion of imposed form resides essentially in the idea that a deliberate attempt was made to influence and control the overall shapes of the retouched tools which went beyond their immediate functional requirements. Typically, this involves large-scale reduction of the original flakes in a way which influences not only the active working edges of the finished tools but also their appearance (Fig. 4.30) and ensures the production to relatively distinctive and morphologically standardized forms.

For the majority of retouched tool forms in

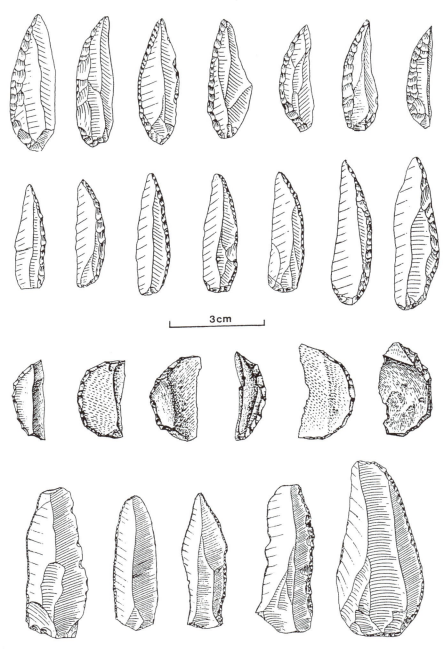

Figure 4.30 *Comparison of 'backed-knife' forms in early Upper Palaeolithic and Mousterian industries. In the case of both Châtelperron points (upper two rows) and Uluzzian crescents (middle row) it can be seen that much of the original flake blank has been chipped away to produce a relatively high degree of standardization and 'imposed form'. In the case of the MTA backed knives (bottom row) it is apparent that even where these tools are manufactured from blades, the retouch usually adheres closely to the original outlines of the blade, and has little effect on the overall form of the tool (see also Fig. 4.19). After Mellars 1989a, Bordes 1961a.*

Middle Palaeolithic industries it can be argued that this kind of imposed form is largely, and in many cases conspicuously, lacking. The various kinds of retouch seem to have been designed to enhance the functional aspect of the tools, usually by shaping and controlling their main working edges (for example racloirs, notches, denticulates etc.) or in other cases (as in typical backed knives) by improving the capacities of the tools either to be held comfortably in the hand or attached to wooden hafts. But beyond this attention to the immediate edges there is little indication that the flint workers went to any great lengths to control the overall, visual form.

One of the clearest illustrations of this contrast can be seen by comparing the forms and strategies of production of the various Mousterian backed-knife forms (Fig. 4.19) with those of the broadly analogous blunted-back forms encountered in early Upper Palaeolithic industries – for example in the French Châtelperronian or in the roughly contemporaneous Uluzzian industries of Italy (Fig. 4.30) (Mellars 1989b: 344–8). In the Mousterian backed-knife group the flint worker often invested much effort in blunting the dorsal edges of the flakes, but this retouch is usually restricted to the immediate edges of the flakes and rarely involved the extensive removal or reduction of large areas of the original flake surface (Monnier 1992). The combination of this strategy with the use of a range of different blank forms for tool production leads to the highly diverse appearance of the tools shown in Fig. 4.19. Whatever criteria are employed, the tools could hardly be said to reflect a great degree of either imposed form or standardization in their finished forms.

The contrast with the early Upper Palaeolithic forms is immediately apparent from examples of both Châtelperron points and Uluzzian crescents illustrated in Fig. 4.30. In both, the overall forms of the different tools show a relatively standardized appearance which in most cases was the result of large-scale reduction of the original flake blanks to achieve these repeated and tightly controlled forms. That this standardization is not due simply to the use of more uniform blades rather than flake blanks for Upper Palaeolithic tools is demonstrated by the fact that the majority of the Uluzzian tools are manufactured from flakes rather than blades (Goia 1990).

Similar observations can be applied to the majority of the other retouched tool forms encountered in Middle Palaeolithic industries, including racloirs and even more obviously to the morphologically simpler notched and denticulated forms (see for example Dibble 1989) (Figs 4.1, 4.2, 4.18). Focusing exclusively on these major categories of flake tools, one could maintain that imposed form was effectively lacking within the documented repertoire of Middle Palaeolithic tool forms.

The obvious exceptions to this generalization are provided by some forms of fully bifacial tools discussed in the immediately preceding section – most notably some of the more distinctive forms of hand axes recorded in the French MTA industries and bifacial leaf points which characterize the later Mousterian industries of central and eastern Europe. Imposed form in the production of hand axes and leaf points is effectively inherent in the definition of these tools, and is apparent in the specimens illustrated in Figs 4.21, 4.22, 4.25, 4.26 and 4.27. As discussed earlier, the morphology of these tools is characterized by three major features: first, by a high degree of bilateral symmetry in the overall shapes and patterns of retouch; second, by the standardized appearance of the majority of the tools, reflected in the distinctively cordiform shapes of most of the MTA hand axes (Fig. 4.22), or the laurel-leaf forms of the majority of leaf points (Fig. 4.27); and third, by the existence of certain forms which show highly idiosyncratic outlines (such as the sharply triangular form of some of the earlier MTA hand axes or the even more distinctive *bout-coupé* forms: Figs 4.25, 4.26) which were evidently imposed on the tools in a deliberate, repeated and premeditated way (Bordes 1961a, 1984). Whatever emphasis one may place on some of the inherent technological constraints involved in the manufacture (such as particular varieties of raw materials or the use of specific flaking strategies: cf. Dibble 1989) it is impossible to see this as more than a limited and partial factor involved in the overall form of the finished tools (Wynn & Tierson 1990). Certainly, no appeal to raw material constraints can explain why the specific shapes of hand axes encountered in the French MTA industries are so strikingly and consistently different from those encountered in the earlier Acheulian or Micoquian industries from the same regions (Figs 4.21, 4.23: e.g. Bordes 1971a, 1984; Wynn & Tierson 1990). However the data are interpreted, the specific shapes of these fully bifacial tools points to the existence of clear morphological norms or conceptual templates in the minds of the individuals who manufactured them (see Chapter 12).

Whether an element of imposed form can be identified in any other varieties of Middle Palaeolithic tools remains debatable. It could be argued that some of the more extensively worked examples of points or convergent racloirs exhibit an overall regularity and symmetry in outline which hints at the existence of preconceived intentionality in the overall shapes of the finished tools (see Figs 4.13, 4.14). However, it is difficult to evaluate whether this was dictated by an *a priori* conceptual interest in the shapes or by the underlying functional requirements of tools intended specifically for hafting either as double-edged knife forms or spear tips (see above and Beyries 1988a,b; Anderson-Gerfaud 1990; Callow 1986a). Similar ambiguities surround some other distinctive forms encountered in certain Middle Palaeolithic industries including, for example, some of

the regular and symmetrical *limace* forms documented in many of the Quina and Ferrassie-type assemblages (Fig. 6.7).

The specific morphology and relative degree of standardization of Middle Palaeolithic tool forms require more specific and detailed documentation. After almost a century of research, methodological approaches to the morphology, technology and indeed intended functions of Middle Palaeolithic tools remain at a surprisingly rudimentary level. From present data it is evident that the overall morphological patterning of Middle Palaeolithic tool forms is much simpler than that documented in Upper Palaeolithic industries, and shows much less explicit patterning at different times and places throughout the Middle Palaeolithic universe (Mellars 1973, 1989a,b). As discussed in Chapter 12 these contrasts must have some clear implications for both the basic conceptual patterning which lay behind tool production and the degree to which this may reflect significant social, demographic or cultural divisions within Middle Palaeolithic populations. However, the systematic study of this patterning is still at an early stage and this is one of the areas where more sharply oriented and structured research is urgently required.

Raw material patterns in tool production

The question of apparent links or relationships between particular kinds of raw material and particular forms of retouched tools in Middle Palaeolithic contexts has been raised several times in the literature (e.g. Tavoso 1984; Geneste 1985: 526–37, 1988: 460, 487–9; Otte *et al.* 1988: 96–8; Dibble 1991a; Dibble & Rolland 1992; Binford 1992). As yet, only limited attention has been devoted to these patterns and systematic analytical data to support the claimed correlations between tool morphology and varying raw material types remain relatively sparse. Some of the potential correlations and explanations advanced to account for these patterns can be summarized as follows:

1. The pattern which has attracted the most comment in the earlier literature is the tendency for some of the more morphologically complex and extensively shaped tool forms (most notably various forms of racloirs, points and bifacial hand-axe forms) to be manufactured preferentially from the more fine-grained and high-quality raw materials, and for the morphologically simpler and generally smaller tool forms (principally notches, denticulates and related tools) to be manufactured predominantly from poorer-quality materials (e.g. Tavoso 1984; Geneste 1985: 527, 1988: 460; Otte *et al.* 1988: 96–8; Dibble 1991a: 35; Binford 1992). The available data are sparse, but as summarized in Fig. 4.31 provide some support for this suggestion.

Insofar as this pattern has been discussed at any length in the literature the assumption has generally been that the variable use of different raw materials most probably reflects the particular economic or social contexts in which the different tool forms were employed. Geneste (1985) has suggested that the use of poorer quality materials for the manufacture of notches and denticulates may reflect the essentially *ad hoc* nature of these tools, generally made from what lay immediately to hand during the course of very brief episodes of activity on particular sites and subsequently discarded either at or very close to the point where they were made. By contrast he argues that the use of better quality materials for the various forms of racloirs, points and bifaces may well imply that these were extensively transported tools – that is, tools made from the best available raw mate-

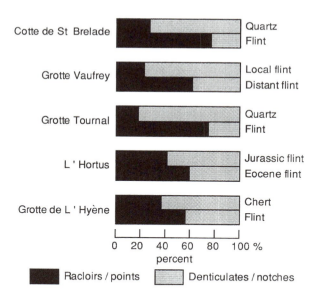

Figure 4.31 *Relative frequencies of racloirs versus notched and denticulated tools manufactured from different raw materials in a range of Middle Palaeolithic industries. Note how in all the assemblages racloirs tend to be manufactured selectively from the better quality raw materials (usually obtained from more distant sources) while notches and denticulates tend to be manufactured from poorer materials, usually derived from sources closely adjacent to the sites. Data from Callow & Cornford 1986 (Cotte de St Brelade); Geneste 1988 (Grotte Vaufrey); Tavoso 1984 (Grotte Tournal); de Lumley 1972 (Grotte de l'Hortus); Girard 1978 (Grotte de l'Hyène, Arcy-sur-Cure).*

rials at some distance from the point where they were eventually discarded and probably deliberately carried over the landscape with these future uses specifically in mind (Geneste 1985: 526–37; 1988: 487–90). Binford (1992 and personal communication) has toyed with a broadly similar model for the variable use and production of racloirs versus notched/denticulated tools. His view, in essence, is that the notches and denticulates were probably used in the course of very short-range foraging activities – mainly for collecting and processing plant foods and probably carried out by the females in the Neanderthal groups. As a result materials tend to be from the most immediate flint outcrops rather than better quality materials from more distant outcrops. He suggests that racloirs, points and bifaces by contrast were probably used for much wider ranging activities related mainly to the procurement and butchery of animal carcasses, which allowed access to more distant and often much better quality materials. For both Geneste and Binford, therefore access to varying types and quality of raw material supplies is a direct reflection of the specific activities for which the tools were employed and the extent to which the different economic activities involved varying degrees of mobility.

2. I would suggest that there is a simpler and more pragmatic explanation for this pattern, which relates the use of different types of raw material more directly to the particular technological constraints involved in tool production. Where a flint worker had to manufacture large and more complex forms of tools there would inevitably be a strong incentive to select the best quality raw materials available, to allow maximum control over the precise form and relative finesse of the finished tools. The manufacture of most forms of notches and denticulates, by contrast, imposed far fewer demands on the skill of the flint worker or the quality of raw materials employed. A notched or denticu-

lated tool can be produced easily and effectively from almost any kind of raw material. More simply, if a flint knapper has access to two or more varieties of raw material of sharply varying flaking quality, he/she will almost inevitably select the better quality material for the production of the technologically more demanding forms and presumably reserve the poorer quality material for the morphologically simpler, technologically less demanding forms. This assumption need not imply any great foresight or planning on the part of the Neanderthal flint workers and merely assumes that any experienced craftsman will inevitably have some astute appreciation of the variable qualities of different raw materials for particular technological procedures. To use a woodworking analogy, one would not expect a carpenter to manufacture survey pegs or clothes props out of expensive, fine-grained mahogany when cheaper and more readily available pine or other softwoods would suffice equally well for the task in hand.

3. A similar interpretative dichotomy can be envisaged for some of the other correlations between the relative complexity of retouched tool forms and the selective use of different raw materials which have sometimes been claimed in the literature. The apparent tendency for the more complex forms of multiple edged tools, such as double, convergent and *déjeté* racloirs, to be manufactured more often from the best quality raw materials than the simpler forms of single edged racloirs (e.g. Niederlender *et al.* 1956: 225–6; Tavoso 1984) can be interpreted in at least two different ways. One is to invoke essentially the same reasoning as in the preceding paragraph. A second view would be to invoke the notion of repeated resharpening of retouched edges along the lines suggested by Dibble and others. In this case one could argue that the best quality raw materials might not only be more amenable to effective, repeated resharpening of the

retouched edges, but that the greater economic or conceptual 'value' attached to the better quality materials might serve as a further incentive to extend the use life of the material for as long as possible (see Meignen 1988). In either case the incentive to retouch the maximum number of different edges on the tools was to some extent dictated by the character and quality of the raw material itself.

These speculations could no doubt be pursued further. The implication is that whilst certain broad correlations between the use of different raw materials and the overall morphology of the associated tools can be documented in Middle Palaeolithic industries, the precise mechanisms for these correlations are by no means self-evident. In addition to the points discussed above, there is of course the intervening factor of the effect of varying raw material quality, as well as the shape and size of the parent nodules, in influencing the selection of alternative flaking strategies for the production of the primary flake blanks from which the retouched tools were made (see Chapter 3; Dibble 1985). Any raw material effects of this kind could impinge directly on several aspects of the overall forms of the eventual tools by influencing, for example, the relative frequencies of elongated versus broader flakes, the relative thickness of the flakes available for tool manufacture or the relative frequencies of cortical and non-cortical flakes (Mellars 1964: 231; Kuhn 1992a). The study of these questions of the specific relationships between tool typology and technology and the varying character and

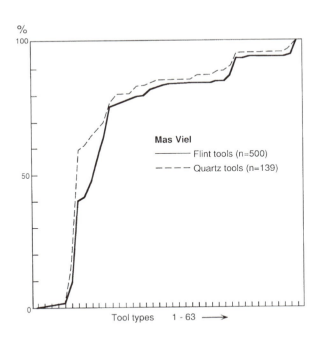

Figure 4.32 *Cumulative graphs of the tool assemblages manufactured from flint (continuous line) versus quartzite (dashed line) in the Quina-Mousterian assemblage from Mas-Viel (Lot). The only significant difference between the two assemblages lies in the higher ratio of double to single-edged racloir forms in the series manufactured from the better-quality flint. From Niederlender et al. 1956.*

quality of available raw material supplies is still in its infancy and clearly deserves more systematic attention and analysis.

Finally, most of the illustrations and examples discussed above derive from contexts where the stone workers had access to two or more different raw materials and where the central issue concerns how they were selected and used for different forms of tools. Whether these patterns imply that different raw materials automatically demand or dictate the production of significantly different tool forms is an entirely separate and far more debatable issue. Bordes addressed this issue on several occasions and was adamant that variability of raw material could account for only an extremely limited part of the total range of technological and typological variation documented within the Mousterian complex, pointing out that effectively all the different Mousterian variants documented within the French sites can be shown to have been produced from widely differing materials. Arguably one of the most striking illustrations of this point was provided by Bordes' own analysis of the typical Quina-Mousterian assemblage from the site of Mas-Viel in the Lot (Niederlender *et al.* 1956).

Here, Bordes was able to demonstrate that although the assemblage was manufactured in roughly equal proportions from two different raw materials (i.e. high quality flint versus more coarse-grained quartzite) the composition and frequency of the tools made in the two materials were effectively identical and in each case conformed closely to the classic definition of the Quina variant (Fig. 4.32). The only significant point of contrast lay in the lower percentages of more complex forms of double, convergent and *déjeté* racloirs, as opposed to single edged racloirs, made from the poorer quality quartzite, than in the series made from the better quality flint Niederlender *et al.* 1956: 225–6). The possible explanation for this has been discussed above, but as Bordes stressed, this particular feature has no effect whatever on the overall diagnosis of the assemblage as corresponding to the typical Quina-Mousterian group. While these patterns of raw material variability are of considerable interest, therefore, there is no reason to think that they take us more than a small way towards explaining the bewildering degree of variation in tool production patterns which characterizes the Mousterian complex.

CHAPTER 5

The Procurement and Distribution of Raw Materials

One area in which impressive advances have been made during the last few years is in studying patterns of procurement and distribution of raw materials in Middle Palaeolithic contexts. These studies are of critical importance in several respects. First, they can provide a direct insight into patterns of movement of human groups over the landscape, and potentially on their contacts or relationships with neighbouring groups. Second, the patterns of procurement and utilization of raw materials reflect the organization which lay behind the production and use of stone tools, and the kind of cognitive planning processes involved. Third, it is now clear that the character, quality and local abundance of raw material supplies in some cases had a significant effect on the form of Middle Palaeolithic industries, influencing not only the character of primary flaking strategies employed on the sites but at least certain aspects of the morphology and typology of the tools produced (see Chapters 3 and 4).

Studies of raw material sources

The study is still at an early stage. Apart from sporadic references in the earlier literature (e.g. Bordes & de Sonneville-Bordes 1954; Pradel 1954, 1963; Séronie-Vivien 1972; Valensi 1960) it is only during the past 15 years

that it has been taken up as a major study in its own right. The most substantial and detailed body of research has been carried out over the past decade in the Perigord and immediately adjacent areas of southwestern France. As a result of research by workers such as Demars (1982; 1990a,b), Chadelle (1983), Larick (1983, 1986, 1987), Geneste (1985, 1988, 1989a,b, 1990; Geneste & Rigaud 1989), Turq (1988a,b, 1989a, 1990, 1992b) and others, we now have systematic data from a range of sites, extending from the later phases of the penultimate glaciation to the end of the Upper Palaeolithic sequence. At present, this provides the most secure basis for any general assessment of patterns of raw material procurement among Middle Palaeolithic groups and for systematic comparisons with those of the ensuing Upper Palaeolithic.

The success of flint provenancing studies in southwestern France rests on three main factors. The first is the distinctive patterning of the major geological outcrops which make up this region. As seen from Figs 5.1 and 5.2, the various outcrops of flint-bearing rocks are aligned as a series of roughly parallel bands which run broadly from northwest to southeast across the region – commencing with the outcrops of Permotriassic limestone in the east of the region and passing through deposits of successively younger Jurassic and Cretaceous age towards the more recent Tertiary deposits in the west (Demars 1982; Gen-

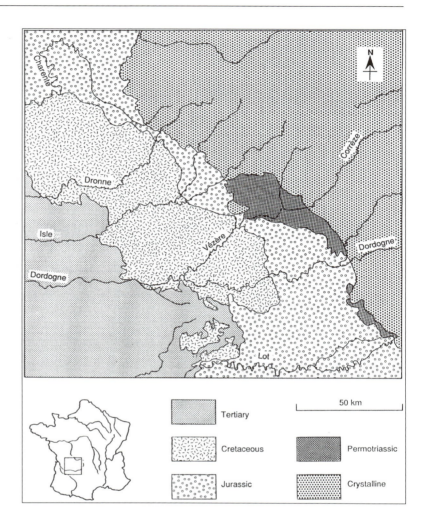

Figure 5.1 *Map of the major geological outcrops in the Perigord and adjacent areas of southwestern France. After Demars 1982.*

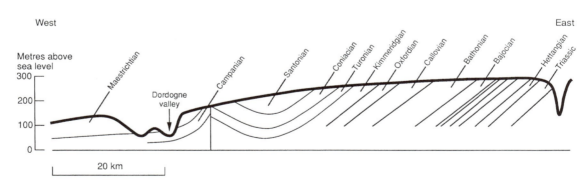

Figure 5.2 *Cross-section of geological outcrops along a west-east transect through the central Perigord area. After Demars 1982.*

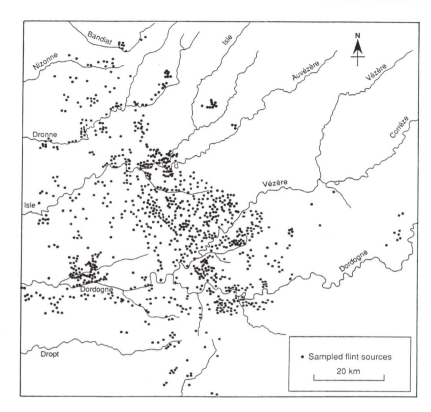

Figure 5.3 *Distribution of geological sources sampled for raw material supplies in the recent raw-material provenancing studies in southwestern France. After Geneste 1985. More recently, many additional sources have been sampled, especially in the areas to the south of the Dordogne (see Turq 1989a, 1992b).*

este 1985, 1989a; Turq 1989a). All these exposures consist of different forms of limestone or related deposits with a variety of either nodular or tabular flint occurring at particular stratigraphic horizons within the deposits. The second factor is the distinctive appearance of many of the individual sources of flint which occur within the different outcrops which makes attribution to particular sources relatively easy. The third is the thoroughness with which most of these outcrops have now been surveyed for potential sources of flint supplies.

The detailed prospecting of these geological formations for potential flint and chert supplies has been carried out partly by geologists (e.g. Séronie-Vivien 1972, 1987; Valensi 1960) and partly by archaeologists working specifically on the problems of Palaeolithic

flint supplies (e.g. Le Tensorer 1981; Morala 1980; Gaussen 1980; Demars 1982; Rigaud 1982; Larick 1983, 1986; Geneste 1985; Turq 1988a, 1989a). Through close collaboration, these workers have now built up an extensive reference collection of samples deriving from well over 1000 different exposures of flint-bearing outcrops, covering most of the 10,000 square kilometres of the Perigord region (Fig. 5.3) – the so-called 'lithothèque du Bassin d'Aquitaine', currently housed in the regional directorate of prehistoric antiquities in Bordeaux (Turq 1989a). All recent progress in flint provenancing studies in the Perigord depends on this extensive reference collection of flint sources and, hopefully, on its representative coverage of most of the raw material resources available to Palaeolithic groups.

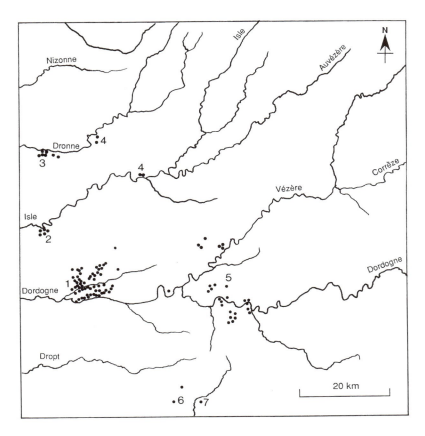

Figure 5.4 *Documented distribution of distinctive varieties of flint in the Perigord and adjacent areas. The sources represented are 1: Bergerac flint; 2 Mussidan flint; 3 Ribérac flint; 4 honey-yellow Santonian flint; 5 beige, jaspoid flint (several sources); 6 Gavaudun flint; 7 Fumel flint. After Geneste 1985.*

As yet, most of the studies of flint sources in the Palaeolithic sites of this region have been based on macroscopic studies of the appearance and texture of the flint – involving such features as colour variations, surface texture, various forms of inclusions, the thickness and character of the adhering cortex, and so on (Demars 1982; Geneste 1985, 1989a; Larick 1986). More detailed analytical studies of the fossil content or trace-element composition of the samples have so far been applied to only a limited amount of material, mainly to corroborate the results of the macroscopic studies (Valensi 1960; Séronie-Vivien 1972, 1987). How far this reliance on simple macroscopic approaches should be regarded as a significant limitation is a mat-

ter of debate. Ultimately, all forms of raw material provenancing studies, whether based on sophisticated analytical techniques or simpler macroscopic approaches, depend on pragmatic criteria. Thus, all these approaches depend not only on a close matching of particular archaeological samples with particular geological sources but also on the ability to demonstrate that these supposedly diagnostic features of different materials are restricted to these sources. Fortunately, many of the varieties of flint documented in the Perigord region do appear to be so visually distinctive that they can be regarded, from all available field evidence, as effectively diagnostic of particular geological outcrops (Fig. 5.4). The best known examples are some of

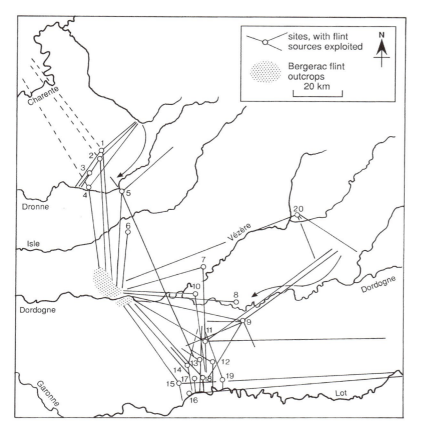

Figure 5.5 *Integrated map of raw material sources exploited from Middle Palaeolithic sites in the Perigord and adjacent areas. The sites represented are: 1 Sandougne; 2 Abri Brouillaud; 3 Le Roc; 4 Fonseigner; 5 Les Festons; 6 Coursac; 7 Le Moustier; 8 Le Dau; 9 Grotte Vaufrey; 10 Roc de Marsal; 11 La Plane; 12 La Lizonne; 13 Ségala; 14 La Burlade; 15 Plateau Cabrol; 16 Moulin du Milieu; 17 Las Pélénos; 18 Les Ardailloux; 19 La Grave; 20 La Chapelle-aux-Saints. After Geneste 1989b.*

the highly coloured and concentrically banded flint nodules from the Maestrichtian deposits of the Bergerac region, the distinctively fine-grained, blue-grey flints from the Angoumian deposits near Fumel (Lot-et-Garonne) and some of the vividly coloured jasper-like flints from the Hettangian formations on the western flanks of the Massif Central (Demars 1982; Geneste 1985; Turq 1989a). By no means all the flint outcrops in the region are so distinctive. In particular, many of the flint sources from the extensive Senonian deposits which occupy a large area of the central Dordogne region appear to be either essentially uniform over large areas or to show localized and erratic variations

within a single source which make exact provenancing impossible. Fortunately, it has now been possible to identify a sufficiently large number of the most distinctive flint sources to document at least the major patterns of movement of these materials over the greater part of the Perigord region.

A final problem which is inherent in all raw material provenancing studies is that of secondary sources – i.e. materials which have travelled (usually by water action) far beyond their geological origin. There is some debate as to how serious this problem may be in the case of the Perigord region. Demars (1982: 72) for example has argued that virtually all flint supplies documented in his studies of Upper Palaeolithic sites in the north-east Dordogne region seem (from the condition of the adhering cortex) to have been derived from essentially primary or near-primary sources and he maintains that river gravel would have been largely avoided for tool manufacture, owing to the effects of battering and frost action on the nodules. Turq on the other hand (e.g. 1988a: 105–7; 1989a) regards stream or river gravels as one of the major sources of flint supplies. This is no doubt the most serious potential ambiguity in current provenancing studies. Against this, however, it should be remembered that most of the river terrace and alluvial deposits in the Perigord region have relatively localized catchment areas from which varied sources of flint could be derived. Clearly, secondary derivation of this kind cannot be invoked for materials which are known to have travelled against the direction of flow of rivers through the region (i.e. for materials which have travelled essentially from west to east) or for materials which have travelled along a north-south transect between the major river catchments (Geneste 1985, 1989a; Turq 1988a) (Fig. 5.5). The strongest argument against widespread use of secondary sources is provided by the highly specialized forms in which the artefacts manufactured from some of the more distant sources are normally represented within archaeological sites – as discussed further below. Hence, while the possibility of secondary sources should always be kept in mind, there are probably relatively few cases in practice where this could be invoked to explain the documented patterns of movement of raw materials in most Palaeolithic sites.

Results of provenancing studies in the Perigord region

Detailed studies of the provenance of raw materials have so far been applied to over 20 different Middle Palaeolithic sites in the Perigord region, incorporating material from

Table 5.1
Principal Middle Palaeolithic sites analysed for raw material sources in southwestern France

1.	Sandougne (Denticulate Mousterian)
2.	Abri Brouillaud (MTA)
3.	Le Roc (Denticulate Mousterian – open site)
4.	Fonseigner (Typical Mousterian, MTA)
5.	Les Festons (Denticulate Mousterian)
6.	Coursac (MTA – open site)
7.	Le Moustier (Typical Mousterian, MTA)
8.	Le Dau (MTA – open site)
9.	Grotte Vaufrey (Typical Mousterian – penultimate glaciation)
10.	Roc de Marsal (Quina Mousterian)
11.	La Plane (MTA – open site)
12.	La Lizonne (MTA – open site)
13.	Ségala (MTA – open site)
14.	Laburlade (MTA – open site)
15.	Plateau Cabrol (Quina Mousterian)
16.	Moulin du Milieu (Quina Mousterian)
17.	Las Pélénos (Quina Mousterian)
18.	Les Ardailloux (MTA– open site)
19.	La Grave (MTA – open site)
20.	Chez-Pourrez (Ferrassie Mousterian)
21.	La Chapelle-aux-Saints (Quina Mousterian)

From Geneste 1989a; Turq 1989a, 1992b; Demars 1982, 1990b. Unless noted otherwise, the sites are cave/rock-shelter localities. For the location of the sites, see Fig. 5.5.

over 40 separate occupation levels (Table 5.1; Fig. 5.5). The majority of these studies derive from the work of Jean-Michel Geneste centred mainly on a series of sites in the northern part of the Dordogne area around the valley of the Dronne and its tributaries (Geneste 1985, 1988, 1989a, 1990; Geneste & Rigaud 1989). Further analyses have been carried out by Geneste for a number of sites further to the south in the valleys of the Vézère and the Dordogne (Le Moustier, Grotte Vaufrey and Le Dau) and by Alain Turq for a range of sites located immediately to the south of the Perigord region between the valleys of the Dordogne and the Lot (Turq 1988a, 1989a, 1992b; Turq & Dolse 1988). To the east of the region, Demars (1982, 1990b) has provided similar data for the sites of Chez-Pourrez and La

Chapelle-aux-Saints in the Department of Corrèze. With one exception, all these sites appear to belong to the classic sequence of Mousterian industries in the region (i.e. dating from some point during the last glacial) and include representatives of all the major industrial variants (i.e. Typical, Denticulate, Mousterian of Acheulian tradition (MTA), Quina and Ferrassie) (see Table 5.1). The one notable exception is the site of Grotte Vaufrey, which includes a long sequence of industries attributed to the later stages of the penultimate glaciation (Geneste 1985, 1988; Geneste & Rigaud 1989). A valuable aspect of these studies is that the data relate to assemblages from both many of the classic cave and rock-shelter locations and also from a variety of open-air sites located on the intervening plateaux.

The data provided in the recent provenancing studies relate in essence to two different aspects of procurement strategies: first, the specific sources of raw materials exploited and the relative frequencies with which these were distributed between different Middle Palaeolithic sites; and second, the precise forms in which these different materials were transported across the landscape and successively reduced or transformed into different technological products. The results of the studies can best be discussed in these terms.

Provenance and distribution of raw materials

The results of the recent provenancing studies reveal three striking features: (1) the strong predominance of materials derived from very local sources in the great majority of Mousterian sites – that is materials derived from distances of at most 4–5 km from the individual site locations; (2) the additional presence in most sites of a surprising range of materials derived from many additional sources – often located along different axes extending in several directions away from the sites (Figs 5.5–5.8); and (3) the occasional presence in almost all sites of more sporadic raw materials derived from much greater distances of up to 80–100 km (Table 5.2). There are localized exceptions to most of these generalizations but as a broad characterization of Middle Palaeolithic procurement patterns in southwestern France, these are the most significant patterns to emerge from recent work (Geneste 1985, 1988, 1989a, b; Turq 1988a, 1989b, 1992b). In more detail, the patterns can be analysed as follows:

Table 5.2
Quantities of raw materials transported over varying distances in Middle Palaeolithic sites in southwestern France, with the degree of utilization of the different raw materials for retouched and utilized tools

Distance over which flint transported (km)	Percentage transported		Percentage utilized
	Perigord	Lot/Quercy	Perigord
Local (<5)	70–98	85–95	1–5
Intermediate (5–20)	2–20*	3–15	10–20
Distant (30–100)	0–5	1–2	75–100

Data for the Perigord sites are from Geneste 1989a: 80-82 and 1989b: 63; data for the Lot and Quercy region are from Turq 1989a: 189; * denotes data from Geneste 1989b: 63.

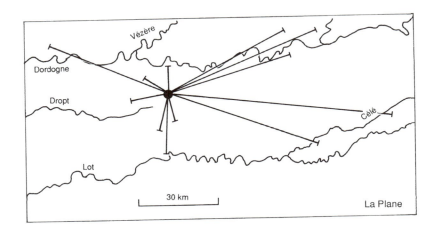

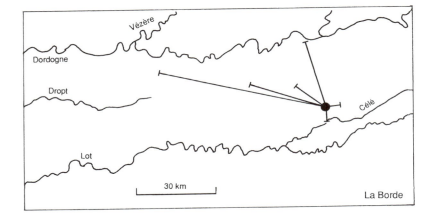

Figure 5.6 *Raw material sources exploited from the Middle Palaeolithic sites of La Plane (upper: MTA) and La Borde (lower: Denticulate Mousterian). After Turq 1989a, 1990.*

1. The tendency to rely predominantly on purely local raw materials has now been documented in effectively all Middle Palaeolithic industries recently analysed in the Perigord region. According to the data provided by Geneste (1985, 1988, 1989a,b) Turq (1988a, 1989a) and Demars (1982, 1990b) the component of raw materials derived from sources of at most 4–5 km from the occupation sites invariably accounts for at least 70 percent of the total lithic assemblages and in several cases rises as high as 95–98 percent (especially in the case of open-air sites, which are often located effectively on the outcrops of raw material) (Table 5.2). As Geneste points out, this could be seen as a simple reflection of the immediate foraging radius from the sites in question – indicating areas which were rapidly and economically exploited within at most one or two hours of travel during a normal working day (Geneste 1985; 1989a). As such the overwhelming predominance of local materials can be seen as a largely predictable aspect of procurement strategies which were 'embedded' in the more general patterns of economic and subsistence activities carried out from particular site locations.

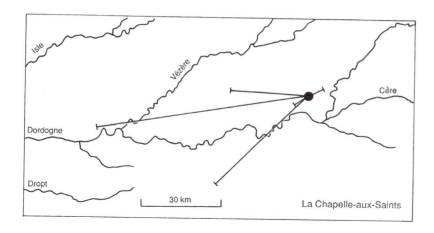

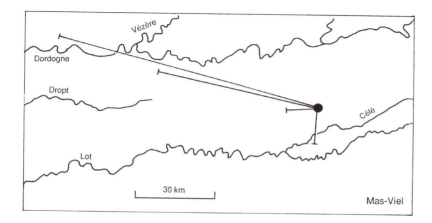

Figure 5.7 *Raw material sources exploited from the Quina-Mousterian cave sites of La Chapelle-aux-Saints (upper) and Mas-Viel (lower). After Demars 1990b, Turq 1989a.*

Closely related is the tendency for many sites to be located virtually on the sources of relatively rich and accessible raw materials. Well documented cases have been recorded, for example, at the site of Plateau Cabrol (Lot-et-Garonne) located on top of local outcrops of Santonian flint (Turq 1977a, 1978, 1988a, 1989a) and at other sites such as Combe-Capelle (Dordogne) where a rich seam of high quality flint nodules occurs immediately adjacent to the site (Peyrony 1943). Similar situations have been documented by Geneste (1985), Duchadeau-Kervazo (1986) and Turq (1989a) at a range of open-air

sites in both the northern and southern Dordogne regions. In these cases there can be little doubt that the immediate accessibility of rich and relatively high quality raw material must have been a primary factor influencing the choice of these particular site locations (see Chapter 8).

2. Far more interesting and unexpected is the presence in the majority of Middle Palaeolithic sites of a surprising variety of raw materials, deriving from distances ranging from 20–30 km to as much as 80–100 km from the site locations. These materials are invari-

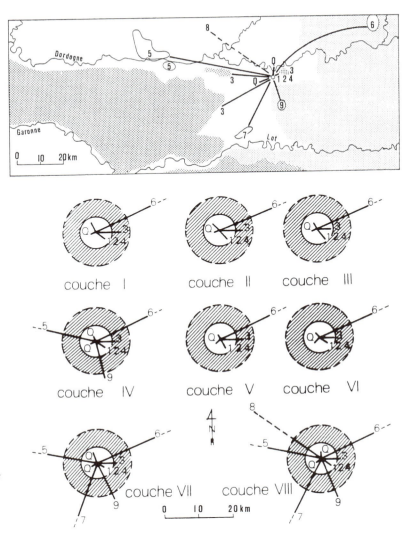

Figure 5.8 *Raw material sources exploited in the different levels of the Grotte Vaufrey. Layers IV–VIII date from the penultimate glacial perod. After Geneste 1988.*

ably present in very small frequencies (Table 5.2) and rarely account for more than ca 5 percent of the total lithic assemblages from particular occupation levels (Geneste 1985, 1988, 1989a,b; Turq 1988a, 1989a). As discussed further below, these materials also tend to be represented in specialized forms, reflecting artefacts brought to sites either as finished tools or as selected flake blanks intended for future tool production. Never-

theless, these materials often derive from widely separated sources, extending along several different trajectories away from the sites (see Figs 5.5–5.8). In the case of La Plane (Fig. 5.6), materials were brought to the site from at least 13 clearly separate sources, from varying distances to the north, south and west of the site location (Turq 1989a: Fig. 6). Similar patterns can be seen at the Grotte Vaufrey (Fig. 5.8) and Fonseigner where most occupation levels incorporate raw materials from at least five or six sources (Geneste 1988, 1989a; Geneste & Rigaud 1989). Exactly how these more distant flint sources were exploited raises several interesting issues to be discussed later. However, many Mousterian groups were able to gain access, if only sporadically, to sources of raw materials located far beyond any normal daily foraging radius from the documented occupation sites and which must have involved either large-scale territorial movements on the part of the individual Mousterian groups (see Fig. 5.17), or possibly some form of exchange relationships with neighbouring groups, extending over virtually all the Perigord and immediately adjacent areas (Geneste 1989a; Geneste & Rigaud 1989; Turq 1989a).

3. As Geneste (1989a) has indicated, when quantities of different raw materials are plotted against the distances over which they were transported, the patterns usually correspond to a roughly exponential decline curve. Well defined patterns of this kind have been documented, for example, by Geneste in several of the levels at Fonseigner and Grotte Vaufrey (Figs 5.9, 5.10) and by Turq in the material from La Plane and Mas-Viel (Geneste 1989a; Geneste & Rigaud 1989; Turq 1989a). None of this is particularly surprising and recalls the kind of fall-off patterns documented in the distribution of obsidian supplies in Neolithic sites in the Aegean and Near Eastern regions (e.g. Renfrew 1969). These patterns could be explained in two possible ways: either as an expression of deliberate energy-economizing behaviour by Middle Palaeolithic groups, which would inevitably place the major emphasis on the most local and easily accessible sources; or alternatively because flint supplies from increasingly distant geological sources were less likely to be encountered in the course of normal daily or seasonal foraging activities. What adds most weight to the deliberate cost-economizing interpretation is that virtually

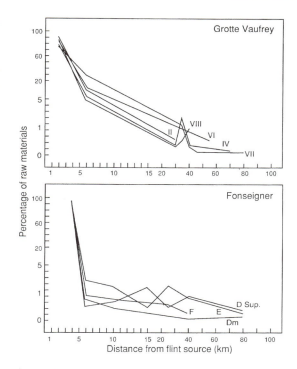

Figure 5.9 *Frequencies of raw materials deriving from varying distances represented in the assemblages from Grotte Vaufrey layers II–VIII (upper) and Fonseigner layers D–F (lower). After Geneste 1989a.*

all the raw materials known to have been transported over substantial distances were relatively high quality varieties of flint presumably preferred and highly valued for their superior flaking qualities (Geneste 1989a). The specific forms in which these materials were transported and used in sites adds further weight to this interpretation, as discussed below.

Some of the most interesting features of the distance-decline curves shown in Figs 5.9 and 5.10, however, relate to exceptions to the exponential patterns of decline. This is particularly clear in the curve for level IV of the Grotte Vaufrey (Fig. 5.9, upper) which shows a sharp secondary peak in the quantity/distance relationship centred on a distance of ca 35 km from the site (Geneste 1988, 1989a).

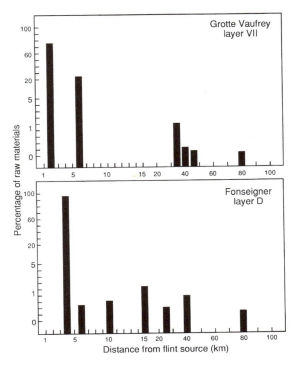

therefore geological outcrops than others. However, it seems significant that all the specific raw materials which do account for secondary peaks in the quantity/distance relationships represent some of the best quality and most highly valued sources of flint. Thus, we may have evidence for deliberate selection of these particular sources and perhaps procurement strategies that were targeted specifically on these locations. If so this could be seen as further evidence for logistical organization in the Middle Palaeolithic with corresponding implications for the planning capacities of the human groups (Geneste 1989a).

4. Finally, these patterns have now been documented for a wide range of Middle Palaeolithic industries in the Perigord region, extending chronologically to the penultimate glaciation (i.e. in the various levels at Grotte Vaufrey) and including virtually all the major industrial variants of the Mousterian (MTA, Typical, Quina/Ferrassie and at least some occurrences of Denticulate Mousterian) (Geneste 1985, 1989a: Fig. 12; Turq 1989a). The only apparent exceptions have been documented in some of the open-air locations in the northern Dordogne region. According to Geneste (1985: Table 16; 1989a), several of these sites (e.g. Le Roc and other sites in the Euche valley: Fig. 5.13) show an almost total reliance on local sources of flint with little evidence for the importation of materials from more distant sources. As he points out, this could indicate either that these particular sites were located effectively on major flint-bearing outcrops, or alternatively could imply very short-term episodes of occupation (Geneste 1989a: 83). Presumably, such short-term occupations would be less likely to involve the exploitation of varied, long-distance resources than would the much longer-term palimpsest occupations represented in most cave and rock-shelter sites. These characteristics of the raw material tend to underscore the specialized nature of the

Figure 5.10 *Frequencies of raw materials deriving from different sources in the assemblages from Grotte Vaufrey layer VII (upper) and Fonseigner layer D (lower). While the distribution at Grotte Vaufrey shows a pattern of progressively decreasing frequencies with increasing distance of transport, the pattern at Fonseigner seems to reflect the preferential exploitation of a number of sources of more distant (and more high-quality) flint. After Geneste 1989a.*

Similar patterns can be seen in the graphs for several of the occupation levels at Fonseigner (Fig. 5.9, lower), which also show secondary peaks at between 15 and 40 km from the site location (Geneste 1989a). One should be cautious about jumping too readily to an economic cost explanation for these patterns, since in some cases secondary resource peaks could arise almost accidentally as a reflection of foraging patterns which, for reasons totally unrelated to the procurement of lithic raw materials, attracted groups more frequently to certain geographical zones and occupations represented in many of the open-air sites in the Perigord region (see Chapter 8). It may also be significant, as Geneste points out, that several of these apparently specialized, short-term open-air occupations (as at Le Roc and other sites in the Euche valley) seem to relate specifically to the Denticulate variant of the Mousterian.

Utilization of raw materials on occupation sites

The varying frequencies with which different materials were transported across the landscape and introduced into sites is only one aspect of raw material procurement strategies. Equally significant are the precise forms in which these materials were transported and their varying patterns of use in the different sites. Geneste argues that this is reflected most clearly in the representation of different technological stages in the overall reduction sequences of the different raw materials (see Chapter 3) and can best be analysed in terms of the three main zones (Geneste 1985, 1988, 1989 a,b):

(a) The most significant zone for the relative abundance and accessibility of raw materials is that located closely adjacent to the sites within a distance of at most 4–5 km. For materials derived from this zone, Geneste argues that virtually all stages of lithic reduction sequences are represented, from initial importation of raw, unmodified blocks of material, to the final production, use and discard of retouched tools (Fig. 5.11). In some cases, as at Grotte Vaufrey, these complete reduction sequences have been documented by refitting flakes to their parent nodules and by the presence in several sites (e.g. Fonseigner and Grotte Vaufrey) of complete nodules which were apparently introduced and stored on the sites for future use. High frequencies of decortication flakes (i.e. those retaining large amounts of cortex from the original nodules) again reflect large-scale

working down of nodules within the occupation sites.

(b) Striking contrasts to the preceding patterns can be seen in the materials transported over much longer distances, ranging between 20–30 km and up to 80–100 km. These far-travelled materials are almost invariably represented on the sites in specialized forms, consisting of essentially 'terminal' products in the lithic reduction sequences – i.e. either fully retouched tools or in some cases primary flake blanks (most frequently Levallois flakes or similar large flakes) probably employed directly as tools without further retouching (Fig. 5.11). Significantly, all forms of cores are virtually lacking in these materials and there is usually little evidence for systematic reduction or primary flaking of materials within the occupation sites. The use of these materials therefore seems to reflect a highly specialized form of exploitation in which only pieces immediately usable as tools were carried over long distances and deliberately introduced into the occupation sites.

(c) The most complex zone in terms of utilization of different materials is that located within the intermediate distances, ranging between ca 5 km and 20 km. As one might expect, materials transported over these distances show a mixture of the strategies documented in the two contrasting zones discussed above (Fig. 5.11). On rare occasions these materials were transported as largely complete nodules. More commonly, they were transported either as partially shaped cores (i.e. the initial stages of cortex removal had been carried out elsewhere) or as selected primary flakes or fully retouched tools. Interestingly, more heavily reduced core forms are generally either lacking or underrepresented in these materials. As Geneste stresses, however, the patterns of use and procurement of these intermediate materials are less well defined than those for the immediate and distant sources and seem to

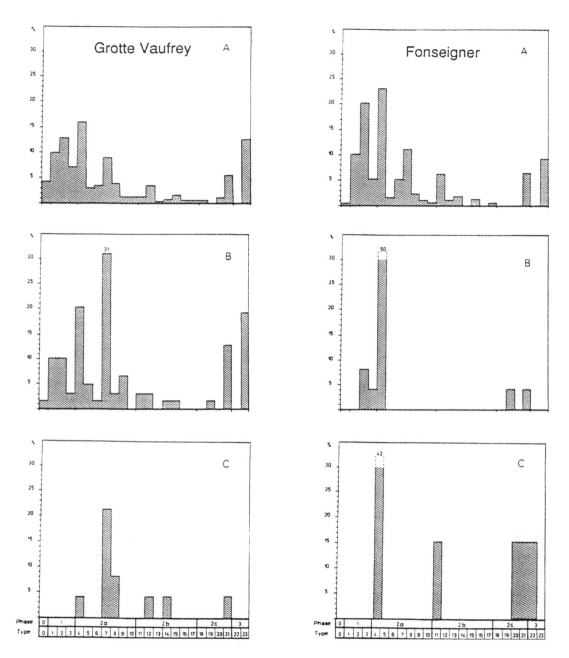

Figure 5.11 *Patterns of utilization of raw materials deriving from varying distances in the assemblages from Grotte Vaufrey layer VIII (left) and Fonseigner (layer D) (right), analysed according to the 'chaîne opératoire' scheme of Geneste 1985 (see Table 3.1 and Fig. 3.1). Both assemblages show that the raw materials imported from the 'intermediate' (graph B) and 'distant' (graph C) sources are represented in a far more selective form than those derived from very local sources. After Geneste 1989a.*

point to different procurement strategies presumably dictated largely by varying patterns of mobility and duration and intensity of occupation in the different sites.

2. Another way of viewing these patterns is in terms of the varying degrees of utilization of the materials on occupation sites, either as systematically retouched tools or as clearly utilized flakes. Again, Geneste argues that some sharply defined patterns can be seen in response to the variable distances over which the materials were transported (Fig. 5.12). In the case of purely local materials (from distances of at most 5 km), the proportion of these utilized products rarely exceeds ca 5 percent and more usually shows values of ca 1–2 percent (Table 5.2). For more distant sources the intensity of utilization increases to around 10–20 percent for materials from the intermediate zone (ca 5–20 km) and between 60 and 100 percent for the most travelled sources derived from distances of between 30 and 80–100 km (Fig. 5.12; Table 5.2). As noted above, this pattern indicates that materials from these distant sources were almost invariably brought to sites either as fully retouched tools or else in forms which could be used or transformed immediately into tools within the sites.

3. Similar patterns can be seen in the relative frequencies in which various core forms were introduced into the occupation sites at increasing distances from raw material sources – with the highest frequencies generally coinciding with the most local materials and the lowest frequencies with the furthest travelled materials. Some informative data have been documented by Geneste (1985: 507–9) in a series of open-air sites located in the small valley of the Euche in the northern Dordogne (Figs 5.13, 5.14). In these sites virtually all raw materials were introduced from one or other of two major flint outcrops, one located close to the village of Remanon near the confluence with the Dronne valley and the

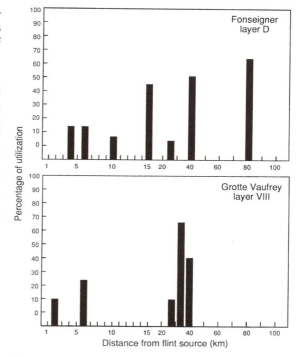

Figure 5.12 *'Intensity of utilization' of raw materials deriving from different sources in the assemblages from Fonseigner layer D (upper) and Grotte Vaufrey layer VIII (lower). Both graphs indicate how the intensity of utilization of the different flint sources (as reflected by the overall frequencies of retouched and utilized pieces in the different raw materials) tends to increase with the overall distances over which the materials have been transported. After Geneste 1989b.*

other located 15 km to the northwest in the vicinity of La Tour Blanche. As shown in Fig. 5.14, when the relative frequencies of cores recorded in the different sites are plotted in relation to the overall distances from these two sources, the results correspond with a pattern of roughly exponential decline, reflecting increasing distances from one or other of these two sources. These results recall some of the broadly similar patterns documented by Munday and Marks in several Middle Palaeolithic sites in the Negev

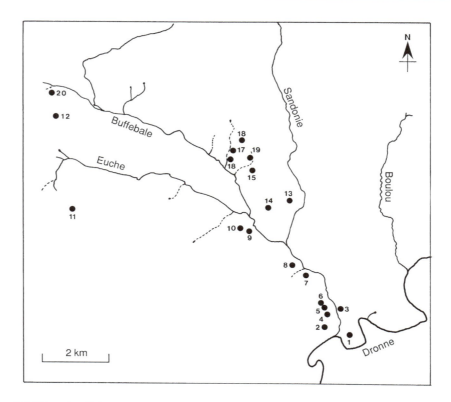

Figure 5.13 *Middle Palaeolithic sites studied by Geneste in the Euche valley, situated at varying distances from two main raw material sources located respectively to the southeast and northwest of the area mapped. From Geneste 1985.*

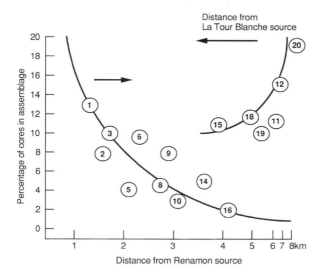

Figure 5.14 *Variable frequencies of cores recorded in the different Middle Palaeolithic sites studied by Geneste in the Euche valley (see Fig. 5.13), located at varying distances from two main raw material sources. The graphs appear to reflect a progressive reduction in the frequencies of cores (expressed as a percentage of the total lithic assemblages) with increasing distances from the two raw material sources. After Geneste 1985.*

area of southern Israel, which again show a progressive decrease in core frequencies with increasing distance from flint supplies (Munday 1976; Marks 1988). These patterns are hardly surprising and presumably reflect simply the natural disinclination of human groups to transport heavy and bulky flint nodules over distances of more than 3–4 km; the obvious reaction would be to produce the maximum number of flakes from each nodule, and therefore to reduce the cores to progressively smaller forms. The data nevertheless provide a further indication of the effects of raw material transportion on the overall composition of lithic assemblages.

4. One final illustration of the way in which increasing distance from raw material supplies can influence patterns of use on occupation sites has been provided recently by Liliane Meignen in her study of the classic Quina-Mousterian industries from the site of Marillac in the Central Charente (Meignen 1988; Meignen & Vandermeersch 1986). In this case raw materials were obtained from

two major sources, one located immediately adjacent to the site in the local outcrops of Jurassic flint and the other from outcrops of much better quality Cretaceous flint from sources 15–20 km to the south-west. As expected from the results described earlier, this led to major contrasts in the patterns of utilization of the two materials, with the more distant and better quality materials accounting for a much higher proportion of the total component of retouched tools on the site and with correspondingly much lower frequencies of cortical flakes and general flaking debris (Fig. 5.15). Meignen's main emphasis, however, is on the evidence for intensive patterns of reworking and resharpening of retouched tools on the site. She points out that this can be documented for many of the typical Quina-type racloirs, both by the very steep, abrupt nature of the retouched edges on many of the tools and the recovery of several examples of typical edge-resharpening flakes. According to Meignen's analysis, this policy of systematic resharpening of the edges of Quina racloirs was

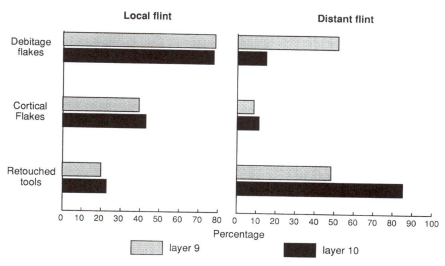

Figure 5.15 *Relative frequencies of cortical flakes, non-cortical flakes and retouched tools manufactured from two different raw materials in levels 9 and 10 of the Quina-Mousterian site of Marillac (Charente). Note the much higher frequencies of retouched tools, and the lower frequencies of cortical and other flakes, in the artefacts from the more distant – and higher quality – flint source. After Meignen 1988.*

restricted almost exclusively to the tools made on the better quality, more distant flint. These results are again hardly surprising and suggest simply that the most distant and better quality flint was more highly valued as a raw material, either because of its inherent scarcity value or, more prosaically, because of its superior flaking qualities. In either case this provides a further illustration of the application of different technological strategies to different raw materials and arguably a further demonstration of the economic curation or maximization of raw materials derived from these more distant and energy-intensive sources.

Evidence for specialized 'extraction' sites in the Middle Palaeolithic

The point has already been made that the majority of documented Middle Palaeolithic sites in the Perigord region appear to reflect a wide range of technological activities, including all stages of the successive production, use and discard of stone tools and usually the use of raw materials from a variety of different geological sources. One invariable component in most sites is the large-scale working down of primary raw materials, carried out within the occupation sites. As the studies by Geneste (1985, 1989a) and Turq (1988a, 1989a) have shown, this can be easily documented from such features as the abundance of primary cortex-removal flakes in the sites and the presence of cores and associated flaking debitage deriving from all major stages of core reduction (Figs 5.11, 5.16). There can be no doubt therefore that this kind of primary reduction of imported nodules of raw material was carried out as a major activity in most of the documented Middle Palaeolithic sites in this region.

One of the central questions raised by these studies is how far one can identify more technologically specialized sites in which the

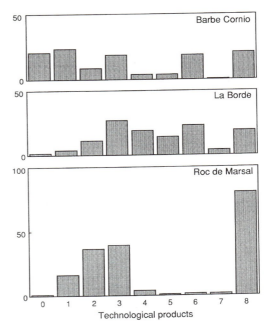

Figure 5.16 *Representation of different categories of flaking debitage and cores in three assemblages from southwestern France (Barbe Cornio, La Borde and Roc de Marsal), as analysed by Turq 1990. The technological products represented are: 0 unworked nodules; 1 cortical flakes; 2 'naturally backed' flakes; 3 'ordinary' flakes; 4 Levallois flakes; 5 core preparation and rejuvenation flakes; 6 cores; 7 retouch flakes; 8 broken and unclassifiable pieces. Note especially the high frequencies of nodules, cortical flakes and cores in the assemblage from Barbe Cornio, which appears to represent a specialized raw-material extraction (or 'quarry') site. The Quina Mousterian assemblage from the cave site of Roc de Marsal, by contrast, includes a much lower proportion of cortical flakes and cores, and appears to represent a 'consumption' location. After Turq 1990.*

large-scale working down of primary raw material supplies was undertaken as a separate, isolated activity within the overall lithic production sequences. In other words, how much evidence is available from the Middle Palaeolithic for what have conventionally been referred to as specialized 'quarry' or 'extraction' sites or primary 'workshop' locations. This point is central to many of the current debates on the strategic or cognitive aspects of Middle Palaeolithic behaviour since, as several workers have pointed out (e.g. Marks 1988; Roebroeks *et al.*1988) recognition of these specialized extraction or quarry sites could be seen as strong evidence for strategic organization in the Middle Palaeolithic and accordingly as evidence for foresight or long-range planning among Neanderthals (Binford 1989).

Any discussion of these issues raises the question of what criteria can be invoked to demonstrate the existence of specialized raw material extraction sites in the archaeological record. Given that primary raw material reduction can be documented in the great majority of Middle Palaeolithic sites and that the majority of known sites are located very close to potential raw material supplies, can we make a meaningful distinction between the conventional notions of 'occupation' sites and specialized 'extraction' sites? To carry conviction, any attempt to identify the latter must presumably be able to demonstrate most of the following features:

1. The location of the site either on or immediately adjacent to a known source of abundant and presumably reasonably high quality raw materials.

2. The strong predominance in the site of artefacts reflecting large-scale working down of these primary raw material sources – including not only evidence of initial extraction, testing and removal of cortex from available nodules, but the systematic reduction of nodules to produce large numbers of either primary flake blanks or specific forms of retouched tools.

3. Evidence for the systematic *removal* from the site of a substantial proportion of the flaked products – presumably as either unretouched primary flakes or various forms of partially shaped or reduced cores intended for further flaking and reduction on the eventual occupation sites.

4. The effective lack of evidence for more general domestic activities on the site – e.g. abundant faunal remains, major hearths or large numbers of heavily utilized and/or reworked or resharpened tools.

5. Presumably the virtual absence of exotic raw materials introduced into the sites from other geological sources.

When defined in these terms, convincing examples of specialized quarry or extraction locations have proved surprisingly difficult to identify in the Middle Palaeolithic sites of the Perigord region. Part of the explanation may lie in the difficulty of recognizing the sites on the ground (Geneste 1985: 515; Turq 1989a: 188). By their nature, the majority of primary extraction sites located actually on the outcrops of raw material are likely to occur in open-air locations and may in many cases lie hidden below deep coverings of later geological deposits. Even when sites have been exposed by erosion or ploughing, the character of the flaking debris may be so poorly characterized (especially in terms of retouched tools) that the sites have either been largely ignored by collectors or have proved impossible to assign to particular periods of the Palaeolithic. Even so it seems surprising that no really well documented examples of obvious quarry sites have been recorded for example on any major outcrops of Bergerac flint – known to have been employed widely as raw material sources in both the Lower and Upper Palaeolithic periods and at least sporadically in the Middle Palaeolithic. According to Geneste (1985: 515–6) the only likely examples of Middle Palaeolithic extraction sites located on these outcrops come from the sites of Campsegret and perhaps Corbiac. As yet, too few details have been published to document their character in any detail.

The most plausible examples of specialized extraction sites have been described by Turq (1988a, 1989a) from a number of localities in the Departments of Lot and Lot-et-Garonne, to the south of the Dordogne. The site of Lascabanes (Lot-et-Garonne) is said to be characterized by a series of nodules of raw material which appears to have been collected from the local flint outcrops on the site and subsequently either 'tested', by scratching the surface or by removal of occasional trial flakes, or partially worked down for the production of simple, non-Levallois flakes (Turq 1988a: 98; 1989a: 188). Few other details are available but the small proportion of retouched tools on the site and the character of the flaking debitage appear to reflect a highly specialized pattern of industrial activity, oriented primarily towards the extraction and initial working of flint supplies.

Broadly similar patterns have been documented at the sites of Barbe Cornio and Lagrave, both in the Lot Department (Turq 1988a: 98–9; 1989a: 188; Jaubert 1984: 92–6). Again the sites are characterized primarily by numerous partially worked or tested blocks of raw material accompanied by large quantities of primary cortex-removal flakes. At Barbe Cornio over 80 percent of flakes recovered from the site retain varying amounts of cortex, as against only 17 percent of non-cortical flakes (Fig. 5.16). Even higher proportions of cortical to non-cortical flakes (approximately 90 percent) are recorded from the site of Lagrave. At both sites retouched tool forms are extremely scarce (accounting for only ca 4 percent of the lithic assemblage at Lagrave) and include many pieces which were apparently unfinished or discarded during manufacture. Rather less consistent with an interpretation of pure extraction sites is the presence of a small number of tools manufactured in clearly non-local materials recovered from Lagrave (Turq 1988a: 99; 1989a: 188). The presence of these pieces could suggest that some other activities, in addition to the extraction and reduction of the local raw material supplies were carried out, at least sporadically, on the site.

How far some of the sites described by Geneste (1985) further north in the Dordogne area fall into the category of specialized extraction sites is more debatable. At the site of Le Dau there is evidence for large-scale working down of nodules derived from local flint sources on the site, and also evidence for selective removal of many of the Levallois flakes evidently produced there (Geneste 1985: 406–10; Rigaud 1969, 1982). As at Lagrave, however, there is evidence for the presence of some other raw materials introduced from more distant sources, including several flakes of Bergerac flint from at least 40 km to the west. It is therefore open to debate how far this could be regarded as a pure extraction site devoted exclusively to the procurement and reduction of local raw material supplies (Geneste 1985: 410). Arguably the site should be seen, in common with most of the other Mousterian sites in the region, as a 'mixed-activity' location combining active extraction and reduction of local raw material supplies with a significant element of more general economic or technological activities (Geneste 1985: 516, 1989a; Turq 1989a: 188).

Extraction of raw materials

There remains the question of how the basic raw materials were located and extracted from the *in situ* geological deposits. All the French workers seem agreed that the deliberate quarrying of flint from the original bedrock formations was a relatively rare occurrence in the Middle Palaeolithic (Turq 1989a: 186; Demars 1982). Certainly, no demonstrable quarry sites of this kind have so far been located in the region and it has been argued that the extremely hard texture of most local limestone formations would have made this a difficult and time-consuming procedure (e.g. Bordes 1984: 169). More significantly, it has been argued that much more easily accessible sources of flint are likely to have been available in most areas, either in the form of talus or solifluction deposits eroding directly from the adjacent flint-bearing outcrops (for example at the foot of many limestone cliffs and escarpments) or in the banks and beds of local rivers or streams (Turq 1988a,1989a). As noted earlier, some workers (e.g. Demars 1982: 72) have argued against the use of river-gravel sources on the grounds that most of the material in these deposits was badly affected either by battering or by frost. Turq (1988a, 1989a), by contrast, has maintained that relatively high frequencies of the flint nodules documented in several Mousterian sites in the southern Perigord region (such as Las Pélénos: Turq 1988a: 107) show clear evidence of derivation from local gravel or stream-bed deposits. As he points out, the material exposed in the beds of streams or rivers could have provided an almost limitless and immediately accessible source of raw material in many contexts. Demars (1990b: 23) concedes that river gravel deposits provided a major source of raw materials (in the form of quartz pebbles) in the Mousterian site of La Chapelle-aux-Saints and points out that river-bed flints were used extensively in some of the Upper Palaeolithic sites in the region – most notably at La Madeleine, located immediately along the banks of the Vézère (Demars 1982: 72). In other contexts it is well known that marine beach deposits were exploited extensively as sources of raw materials in the Middle Palaeolithic – for example in some of the Mousterian coastal sites in Normandy and Brittany (Fosse *et al.* 1986; Fosse 1989) and in the Rissian levels at the Cotte de Saint-Brelade in Jersey (Callow 1986b).

Other issues discussed by Turq and others include the effects of different climatic regimes in exposing various flint outcrops (e.g. the effects of solifluction processes on slope deposits or of vegetation cover on the visibility of flint supplies) and the similar effects resulting from seasonal changes in features such as snow cover, consolidation of deposits by freezing or the varying water-level regimes in rivers and streams (Turq 1988a, 1989a; Jelinek 1988a; Dibble & Rolland 1992; Dibble 1991a). At present, all these issues remain largely speculative and are difficult to evaluate quantitatively. Nevertheless, it is clear that these factors bearing on the relative availability and accessibility of raw material supplies could have had a significant effect on the character of technology practised in many Middle Palaeolithic sites, and must be kept in mind in any assessment of the patterns of technological variability within the Middle Palaeolithic as a whole (see Dibble & Rolland 1992; Dibble 1991a).

Social and mobility implications of raw material distributions

The most valuable aspect of raw material procurement studies is the insight they provide into patterns of mobility of Neanderthal groups and on their possible social or territorial relationships with other groups. Problems in this context should not be minimized, however. We still have extremely limited information on seasonal occupation in different sites and in the absence of this evidence any speculation on annual movement must remain very tentative. However, a number of patterns which emerge from recent raw material studies, must have some direct implications for the associated patterns of seasonal and territorial movement of human groups.

The tendency for the majority of raw materials in Middle Palaeolithic sites to come from strictly local sources (i.e. from at most 5–6 km) is the easiest pattern to account for. As Geneste (1985, 1989a) and others have argued, this can be seen most economically as reflecting the immediate 'foraging radius' of the sites, which could have been exploited easily and efficiently in the course of at most a few hours movement from the individual site locations. Daily foraging territories of this kind are widely documented amongst

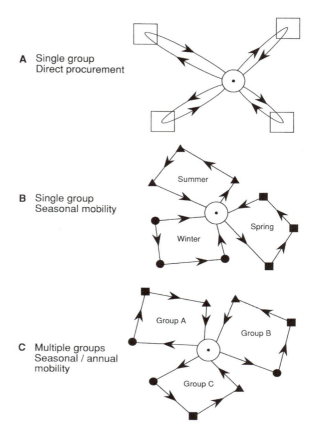

A Single group
Direct procurement

B Single group
Seasonal mobility

Summer

Winter

Spring

C Multiple groups
Seasonal / annual
mobility

Group A

Group B

Group C

Figure 5.17 *Three potential models by which raw materials could be imported into archaeological sites, as a result of different patterns of mobility of the human groups. Model A reflects essentially 'direct' procurement from the available raw material sources by a single group; Model B reflects procurement from a range of more distant sources, in the course of seasonal or annual movements by the same group; Model C reflects the introduction of raw materials into a single site location by a number of different groups, who visited the site at different times.*

modern hunter-gatherers and to this extent the mobility patterns of Neanderthal groups and associated patterns of 'embedded' procurement of local raw materials need be no different from those recorded in modern contexts. There seems no reason to doubt that the raw materials from the intermediate distances of between 6 and 12 km could have been secured on occasions by similar patterns of embedded procurement during more extended daily foraging.

The most significant and intriguing patterns concern raw materials from much more distant sources, between 20–30 km and 80–100 km from the individual site locations – far beyond the scope of any daily foraging

activities. As discussed earlier, four generalizations can be made about the materials from these distant sources. First, they are an almost invariable component of documented lithic assemblages in southwestern France, occurring in at least 80 percent of documented sites; second, the materials invariably occur in very low proportions, rarely exceeding more than 1–2 percent of the total lithic assemblages; third, they tend to occur in very specialized forms, usually as either extensively retouched tools or as large and immediately usable primary flakes; fourth one can usually document their introduction into particular occupation levels from a number of sources, often extending in several directions from the site location (see Figs 5.5–5.8).

As summarized in Fig. 5.17, it is possible to visualize at least three scenarios by which raw materials from these distant sources could have been introduced into Middle Palaeolithic sites. The least likely is that they were introduced into sites as part of a deliberate strategy of direct raw material procurement, involving the movement of either individuals or small groups of Neanderthals over long distances specifically to collect high quality raw material supplies (see Fig. 5.17, Model A). Quite apart from the heavy costs of such long-distance movement in time and energy, the extremely small quantities combined with the technologically specialized forms in which the products were introduced into the sites, would seem to argue against this interpretation. In this context there is a significant contrast with the patterns of procurement of the same raw materials (especially Bergerac flint) by early Upper Palaeolithic groups in the same region, as discussed further below.

Two more plausible scenarios for these patterns of movement are shown in models B and C of Fig. 5.17. Model B represents a situation in which individual Neanderthal groups were involved in a series of far-ranging annual movements, probably in response to shifting seasonal distributions of animal herds or other food resources. Clearly, if these movements extended along several trajectories from a given site location, this could eventually introduce raw materials into the site from a variety of distant and widely dispersed sources. Diagram C, by contrast, models a situation which could lead to almost exactly the same pattern of raw material supplies for any individual occupation level, but which would result from the periodic use of the same site by a number of separate and possibly unrelated human groups, with different patterns of seasonal and territorial mobility. In this case, the variety of raw material sources represented in a

particular occupation horizon would be a simple palimpsest phenomenon, resulting from the intermittent use of the same site location by quite disparate Neanderthal groups. It is difficult to see how these two situations could be clearly differentiated in archaeological terms, at least with the levels of stratigraphic and chronological resolution which can be achieved with current excavation techniques.

The fourth possibility is perhaps the most intriguing in social terms, and raises a prospect which has only rarely been considered in earlier studies of Neanderthal behaviour. This assumes that the complex mix of far-travelled raw materials which can now be documented for the majority of Middle Palaeolithic sites could reflect a complex pattern of social relationships between the individual Neanderthal groups, in which the systematic exchange of raw materials, and perhaps other products, was central to the wider social and demographic relationships maintained between these groups. As Féblot-Augustins (1993) has recently suggested, such exchange relationships could have extended well beyond the purely economic sphere and been tied into wider social relationships based on the exchange of marriage partners between individual, local territorial groups. While it is hardly possible to do more than speculate on this possibility, closely structured exchange relationships of this kind are not merely plausible in Neanderthal contexts but could well have been essential to ensure the viability of relatively small and territorially dispersed populations as long-term demographic units (cf Wobst 1974; Gamble 1983).

How far the striking diversity of raw material sources encountered on Middle Palaeolithic sites can be used to argue for a high degree of territorial mobility on the part of individual Neanderthal groups must therefore remain open. A similar problem emerges from recent studies of the patterns of raw material distribution in Middle Palaeolithic

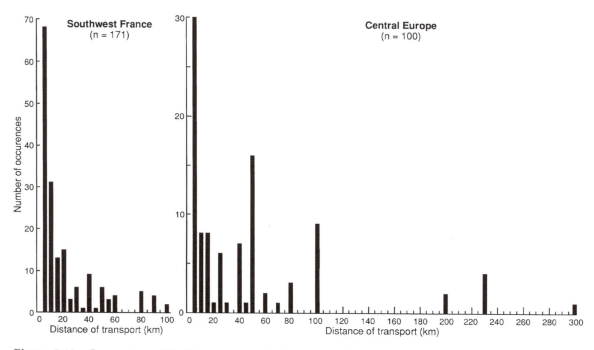

Figure 5.18 *Comparison of the distances over which raw materials were transported in Middle Palaeolithic sites in southwestern France (left) compared with central Europe (right). After Féblot-Augustins 1993. Distances of transport greater than 100 km are only found in the central European sites.*

sites in central and eastern Europe. Here, as both Roebroeks *et al.* (1988) and Féblot-Augustins (1993) have pointed out, it is possible to document even greater distances of movement of high-grade raw materials than those documented in southwestern France, with movements ranging between 100 and 300 km now recorded from at least a dozen different sites (see Fig. 5.18). As in western France, these long-distance movements invariably involved very small quantities of raw materials and the artefacts transported over these distances again occur in highly specialized forms – usually extensively retouched tools. Féblot-Augustin's inclination is to see these movements as evidence for much more wide-ranging seasonal mobility patterns among the Middle Palaeolithic populations of central Europe than in western France, related directly to the more extreme

climatic and ecological contrasts between the winter and summer seasons in the central, continental zones of Europe. While this remains an attractive and plausible explanation, Féblot-Augustins accepts that there could be other explanations for the same patterns, involving either systematic social exchange relationships between different territorial groups or even repeated demographic movements between different regions precipitated by some of the rapid climatic and ecological oscillations which occurred during the early last-glacial period. One factor which has not been sufficiently taken into account in Féblot-Augustin's study, however, is the inherent contrast in the overall geographical distributions of raw materials in the two regions. Whereas in many areas of central Europe high-quality flint supplies are limited to specific and widely separated areas, in southwestern France flint supplies are virtually ubiquitous in the areas with the main concentrations of Mousterian sites. To transport raw materials over distances of 200 km or more into the Perigord region would be tantamount to 'carrying coals to Newcastle' when there were more immediate sources of high quality materials (such as those in the Bergerac region) much closer at hand. While the patterns of raw material distribution recorded in central Europe raise some intriguing possibilities in social and economic terms, therefore, it might be premature to overemphasize the contrasts with the patterns documented in the more western zones of Europe.

Comparisons with Upper Palaeolithic patterns

The final point concerns some of the significant contrasts which can now be documented between raw material procurement and distribution patterns in the Middle Palaeolithic and those in the Upper Palaeolithic. A full discussion of these Upper Palaeolithic patterns is beyond the scope of this study, but from important work carried out by workers such as Demars (1982, 1990b), Chadelle (1983), Larick (1983, 1986, 1987), Bricker (1975) and others, the following generalizations can be made (see Geneste 1989b for a review):

1. The basic sources of raw materials exploited in the Upper Palaeolithic sites of the Perigord region are broadly similar to those documented in the Middle Palaeolithic. Again, most sites tend to show a primary reliance on materials derived from local sources (Fig. 5.20), while the sources of more distant and generally higher quality raw materials correspond essentially with those documented in the Middle Palaeolithic – most notably various sources of Bergerac flint, several varieties of fine-grained jasper and high-quality flints from the Fumel and

Gavaudun areas. The general directions of movement of materials across the region are also similar in the two periods – i.e. predominantly aligned along the major river valleys from east to west but frequently cutting across these river catchments from north to south (Fig. 5.19). On present evidence, therefore, there is no reason to suggest that Upper Palaeolithic groups were securing raw material supplies from more extensive catchment areas than those documented in the majority of Middle Palaeolithic sites in the same region (Geneste 1989b).

2. The major contrast between the Middle and Upper Palaeolithic lies in the relative *quantities* in which raw materials from the more distant and better quality sources were transported by Upper Palaeolithic groups. Whilst the overall quantities of these more distant materials (i.e. materials from distances greater than 30 km) rarely exceed ca 1–2 percent of the total lithic assemblages in the majority of Middle Palaeolithic sites, the corresponding frequencies recorded in Upper Palaeolithic sites may be as high as 20–25 percent (Fig. 5.20). This is especially true of the best quality Bergerac flint, now known to have been used extensively in Upper Palaeolithic industries, from the earliest stages of the Aurignacian onwards (Geneste 1989b; Demars 1982, 1990a).

3. Equally significant contrasts can be seen in the precise forms in which these more distant materials were transported. In the Middle Palaeolithic, materials from relatively distant sources (more than ca 20 km) were almost invariably introduced into the sites in a specific form, normally restricted to fully retouched tools occasionally accompanied by a few carefully selected primary flakes, but hardly ever including cores. In the Upper Palaeolithic, by contrast, distant materials were transported not only in the form of finished tools and specific blank forms but also as either partially or completely shaped

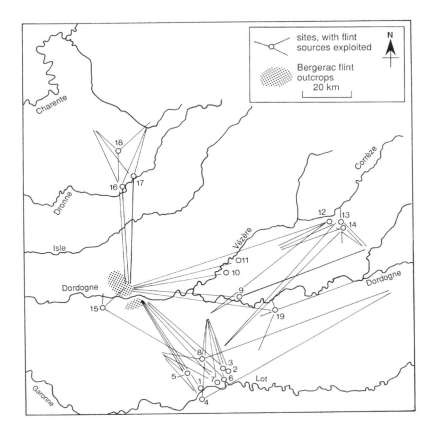

Figure 5.19 *Integrated map of the raw material sources exploited from a range of Upper Palaeolithic sites in the Perigord and adjacent areas, after Geneste 1989b. The overall distribution patterns are generally similar to those recorded for the various Middle Palaeolithic sites shown in Fig. 5.5.*

cores. Thus, raw material supplies seem to have been introduced into sites from these sources not only in more substantial quantities than those in the Middle Palaeolithic but also in forms which would allow more extensive and systematic reduction and working of raw materials within the occupation sites (Geneste 1989b).

4. Many Upper Palaeolithic industries also seem to show a more specialized pattern of use of particular varieties of raw materials for particular forms of retouched tools. In the Aurignacian sites of the Brive region, for example, Demars (1982) has demonstrated that typical *'lamelle Dufour'* forms were manufactured predominantly from certain types

of jasper from specific sources located mainly to the east of the region. Larger and more heavily retouched 'Aurignacian blades', by contrast, were manufactured more commonly of Bergerac flint, while nosed and carinate scrapers show a less selective pattern focused mainly on more local flint supplies (Demars 1982: 140–2). Similar selectivity of particular flint types for particular tool forms has been documented in the production of Noailles burins in the Upper Perigordian industries and in the manufacture of Solutrian bifacial leaf points. As discussed in Chapter 4, there is some limited evidence for raw material selectivity in the production of Middle Palaeolithic tools – notably in the contrast between materials used for rela-

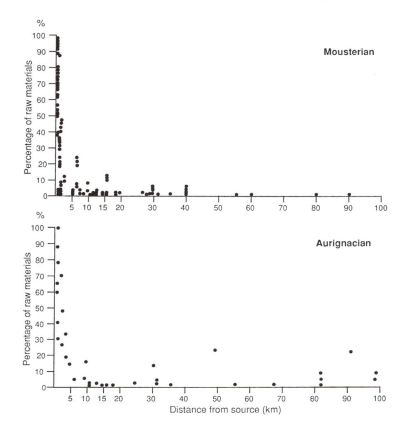

Figure 5.20 *Frequencies of raw materials deriving from varying distances recorded in Middle Palaeolithic sites in southwestern France (upper) compared with those recorded in early Upper Palaeolithic Aurignacian sites in the same region. While both groups of sites show a primary reliance on local flint sources, it can be seen that many of the Aurignacian sites incorporate much higher frequencies of raw materials derived from more distant sources – especially from the high-quality flint sources in the Bergerac region. After Turq 1993.*

tively complex forms such as racloirs, points and bifaces as opposed to simpler, and generally smaller, forms such as notches and denticulates (Tavoso 1984; Geneste 1985). This selection of particular raw materials for particular tool forms, however, seems to be significantly more conspicuous in the Upper than in the Middle Palaeolithic.

5. Finally, for the Upper Palaeolithic we now have much better evidence for the existence of highly specialized 'extraction' or 'quarry' sites or 'workshop locations' (*ateliers de taille*) than we do for the Middle Palaeolithic. The clearest evidence has come from a number of sites located immediately adjacent to high quality flint sources in the Bergerac region

(for example at Champ-Parel and Barbas) which appear to belong to several periods ranging from the early Aurignacian to the later Perigordian (see Geneste 1989b: 64; Chadelle 1989; Bordes & de Sonneville-Bordes 1970: 66). In several sites it has been possible to identify a number of sharply defined, approximately circular concentrations of flaking debitage (probably representing the products of individual flint knappers), indicating intensive, large-scale production of blades and certain specific tool forms located effectively on the sources of high-quality flint supplies. Partially worked and discarded cores and blades are relatively common on the sites but fully retouched tool forms are normally rare. In all respects these locations seem to conform to the definition of highly specialized 'extraction' sites, utilized explicitly, and to all appearances exclusively, for the production of a range of specialized products. As discussed earlier, there is still some doubt as to how far sites of this kind can be identified, with any real confidence, in the Middle Palaeolithic of the same region.

While the patterns discussed above appear to document some real contrasts between patterns of raw material procurement and use between the Middle and Upper Palaeolithic, these should not be exaggerated. As discussed earlier, many basic strategies of raw material exploitation documented in the Upper Palaeolithic sites of the Perigord area do seem to be reflected in an incipient form in many of the Middle Palaeolithic sites of the same region. Nevertheless, there are significant contrasts in the scale on which they were organized (especially for access to the highest quality raw material sources), which could indicate a more systematic and strategically

organized pattern of raw material procurement on the part of Upper Palaeolithic groups (Geneste 1989b: 65). It should also be emphasized that these patterns can now be documented not only in the later stages of the Upper Palaeolithic but also from some of the earliest stages – certainly extending into the earlier phases of the Aurignacian (Fig. 5.20) and apparently (as at La Côte and Roc de Combe) in some of the Châtelperronian sites (Pelegrin 1986; Demars 1982, 1990a). The significance of these contrasts will be discussed further in Chapter 12, in the context of more general comparisons of technological and behavioural changes over the period of the Middle-Upper Palaeolithic transition. That there were significant shifts in raw material procurement strategies over this period must now be regarded as a well documented feature of the archaeological record, with implications not only for changes in lithic technology over this time range, but also for some aspects of the social structure and organization of human groups.

CHAPTER 6

Industrial Taxonomy and Chronology ___

The challenge of documenting, analysing and ultimately explaining the bewildering array of technical and typological variation within the Middle Palaeolithic industries of Western Europe has largely dominated research for the past 40 years. The endless debates which have arisen over the interpretation of this variation will be discussed in Chapter 10. It may be useful to focus here on the more pragmatic issues of the overall scale and character of this patterning and to review some of the attempts which have been made to organize and systematize it in taxonomic terms. The separate but crucial question of the chronological dimension of the variation will be discussed in the later part of the chapter. The primary focus will again be on the classic series of industries from southwestern France, representing the richest and best documented data-base available.

Significant variation in the character and technology of the Middle Palaeolithic industries of this region has been recognized since the last century. In 1864, Lartet and Christy drew attention to the obvious contrasts between the industries recovered from their excavations in the Mousterian levels at Le Moustier and those at Combe Grenal, commenting that the tools from the latter site (presumably deriving from the Denticulate-Mousterian levels in the upper part of the sequence) were generally 'd'un travail gén-

éralement peu soigné' by comparison with those from the classic Mousterian levels at Le Moustier itself. These studies were extended further by Maurice Bourlon at Le Moustier (1905, 1906, 1910, 1911) and by Henri-Martin at La Quina (1907–9), culminating in the systematic attempt by Denis Peyrony to organize the whole of the French Mousterian sequence into two basic groups ('Moustérien Typique' and 'Moustérien de tradition acheuléenne') though with finer divisions within each group (Peyrony 1920, 1930). The distinctive features of what would now be referred to as the 'Denticulate Mousterian' seem first to have been recognized by Bourrinet and Darpeix in their excavations at the Sandougne rock shelter in 1928 (Darpeix 1936).

The definitive study of Middle Palaeolithic variability was provided by François Bordes in a series of publications commencing immediately after the Second World War – initially in association with Maurice Bourgon (Bordes 1948, 1950a, 1953a,b, 1954–55, 1961a,b, 1963, 1972, 1977, 1981, 1984; Bordes & Bourgon 1951; Bourgon 1957). The major features of Bordes' contribution are too well known to require detailed discussion here. The enduring importance of Bordes' contribution was to recognize the need for a rigorous *quantitative* approach to the analysis of industrial variation based on an explicit definition of both the full range of retouched

Table 6.1
Standard type-list of Lower and Middle Palaeolithic flake tools recognized in the taxonomy of François Bordes

1. Levallois flake – typical	32. Burin – typical
2. Levallois flake – atypical	33. Burin – atypical
3. Levallois point	34. Piercer – typical
4. Retouched Levallois point	35. Piercer – atypical
5. Pseudo-Levallois point	36. Backed knife – typical
6. Mousterian point	37. Backed knife – atypical
7. Elongated Mousterian point	38. Natural backed knife
8. Limace	39. Raclette
9. Racloir – single-edged, straight	40. Truncated flake
10. Racloir – single-edged, convex	41. Mousterian tranchet
11. Racloir – single-edged, concave	42. Notch
12. Racloir – double-edged, straight	43. Denticulate
13. Racloir – double-edged, straight/convex	44. 'Bec burinante alterne'
14. Racloir – double-edged, straight/concave	45. Retouch on bulbar surface
15. Racloir – double-edged, biconvex	46–7. Abrupt, thick, alternate retouch
16. Racloir – double-edged, biconcave	48–9. Abrupt, thin, alternate retouch
17. Racloir – double-edged, convex/concave	50. Bifacial retouch
18. Convergent racloir, straight	51. Tayac point
19. Convergent racloir, convex	52. Notched triangle
20. Convergent racloir, concave	53. Pseudo-microburin
21. Déjeté racloir	54. End notch
22. Transverse racloir, straight	55. Hachoir
23. Transverse racloir, convex	56. Plane
24. Transverse racloir, concave	57. Tanged point
25. Racloir on bulbar face	58. Tanged tool
26. Racloir with abrupt retouch	59. Chopper
27. Racloir with thinned back	60. Chopper – inverse
28. Racloir with bifacial retouch	61. Chopping tool
29. Racloir with alternate retouch	62. Diverse
30. End-scraper – typical	63. Bifacial leaf point
31. End-scraper – atypical	

After Bordes 1984: 122. Note that the 'essential' tool percentages used in most of Bordes' analyses are based on the above type list *minus* types 1–3 (i.e. unretouched Levallois flakes and points) and types 45–50 (which are now thought to be accidental forms, caused geological or other damage to the flakes).

tool forms and also the technical aspects of the industries, such as the varying frequencies of Levallois flakes, blades, facetted striking platforms, etc. The core of his methodology was embodied in his comprehensive *'liste typologique'*, consisting initially of 48 distinct tool forms but eventually expanded to a range of 63 types, which was intended to embrace all the major morphological forms of retouched tools encountered in either Middle or Lower Palaeolithic industries (see Table 6.1: Bordes 1961a). As noted in Chapter 4, a large part of this typology was made up of 21 finer subdivisions of his broader *racloir* category, based largely on the position, curvature and number of retouched edges on the tools. The morphologically simpler categories of backed knives, end scrapers, burins and perforators were divided simply into 'typical' and 'atypical' forms. The general category of bifaces was treated separately from the main type list and subdivided, in a separate classification, into 12 major morphological forms. The calculation of various 'technical' features of the industries (reflecting mainly the overall frequencies of Levallois versus non-Leval-

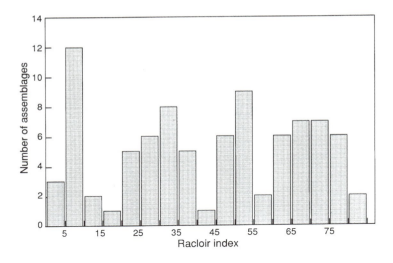

Figure 6.1 *Overall distribution of side-scraper (racloir) frequencies in southwestern French Mousterian assemblages, according to Bordes & de Sonneville-Bordes 1970.*

lois flakes, blades and either strict or broad facetted striking platforms) was based on the total lithic assemblages (i.e. including both retouched and unretouched artefacts) and was expressed as a series of technological indices, indicating the percentage of pieces which fell into individual technological categories.

Bordes' division of the Mousterian industries into five major industrial groupings is equally well known (1953a, 1961b, 1963, 1968a, 1972, 1981, 1984). This taxonomy was based from the outset primarily on the overall frequencies of racloir forms, expressed in relation to the combined frequencies of all other retouched tool types, within the tool assemblages as a whole. In his 1953 paper he pointed out that a frequency distribution of these racloir indices in the documented Mousterian industries from both northern and southwestern France showed a discontinuous, essentially trimodal pattern, which argued for a division of the assemblages into at least three separate groups. A later plot of the same parameter based on a larger group of 88 assemblages revealed a slightly more complex pattern but still suggested a discontinuous distribution of the racloir index

incorporating three or four well separated modes (Fig. 6.1) (Bordes & de Sonneville-Bordes 1970: Fig. 15). The character of this racloir index was therefore taken by Bordes as a reasonable, *a priori* justification for the recognition of three of his major industrial variants – those of the Denticulate, Typical and Charentian (i.e. combined Ferrassie and Quina) Mousterian groupings – though with a possible division of the rather amorphous Typical Mousterian group into two subgroups (Fig. 6.2). The separation from this complex of the Mousterian of Acheulian tradition (MTA) grouping was adopted directly from the earlier taxonomic scheme of Denis Peyrony (1920, 1930) and (as discussed below) was based strictly on the presence in these industries of two distinctive 'type fossils', in the form of cordiform hand axes and typical (i.e. extensively retouched) backed knives. The final component of his taxonomic scheme involved the subdivision of the broader grouping of Charentian industries (i.e. assemblages showing a strong predominance of various racloir forms) into the two subgroupings of Ferrassie and Quina Mousterian based on the dominance of Levallois versus non-Levallois techniques respectively

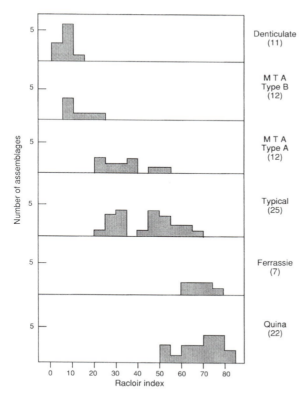

Figure 6.2 *Racloir frequencies recorded in different industrial variants of the southwestern French Mousterian, according to Bordes & de Sonneville-Bordes 1970. The bi-modal distribution of the racloir index in the Typical Mousterian was taken by Bordes to indicate a probable division of these industries into at least two separate entities.*

in the two variants. This separation between the Ferrassie and Quina variants was the only part of Bordes' scheme which depended on the strictly technical as opposed to typological features of the assemblages.

In its most essential form, therefore, Bordes' system of taxonomy, dividing the French Mousterian complex into five major variants is a predominantly quantitative scheme, which relies on the overall, relative frequencies of the major retouched tool forms in the different assemblages (Fig. 6.3). It is this aspect of Bordes' taxonomy which has

been emphasized most frequently in the subsequent literature and which has formed the central focus of recent debates on the behavioural and cultural significance of these industrial divisions, discussed in detail in Chapter 10. What has not been so widely recognized is that whilst these quantitative features were usually employed by Bordes as the primary criteria for distinguishing between the different Mousterian taxa, he frequently made use of several other more idiosyncratic and qualitative rather than quantitative features in support of his taxonomic distinctions, which may be summarized as follows:

1. *'Type-fossil' forms.* The classic use of type-fossil forms in Bordes' taxonomy (as in the earlier taxonomy of Denis Peyrony) was in the definition of the MTA grouping. As Bordes pointed out on numerous occasions, fully typical forms of cordiform and related hand-axe types (e.g. Figs 4.22, 4.25, 4.26) are not only quantitatively rare in all other variants of the Mousterian complex (most notably in the case of the Ferrassie, Quina and Denticulate variants) but apparently totally lacking in all recently well excavated assemblages (Bordes 1953a: 460–3; 1961b: 804–5; 1968a: 101–2; 1981: 78–9; Bordes & de Sonneville-Bordes 1970: 61–3). Especially significant is the total absence of hand axes from all except the uppermost five Mousterian levels in the long and complex Mousterian sequence at Combe Grenal, which comprises 50 levels of all other industrial variants of the Mousterian complex and yielded an aggregate of over 10,000 retouched tools (Bordes 1972). The occurrence of typical cordiform hand axes therefore seems to be in every sense a classic *fossile directeur* of the MTA industries. The same may be true of typical (i.e. extensively retouched) forms of backed knives (Fig. 4.19). As both Bordes (e.g. 1954–55, 1961b, 1981, 1984) and Peyrony (1920, 1930) pointed out, these forms have a close association with typical cordiform hand

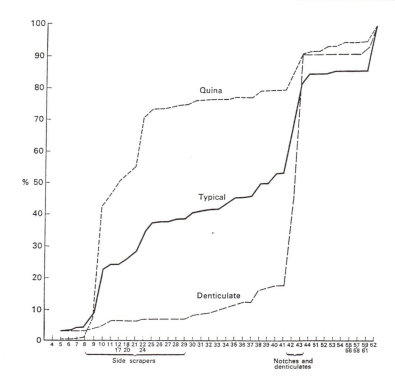

Figure 6.3 *Cumulative frequency curves of tool-type frequencies characteristic of three of the major variants of the southwestern French Mousterian complex (Quina, Typical and Denticulate Mousterian). After Bordes & de Sonneville-Bordes 1970. For a list of the numbered tool types, see Table 6.1.*

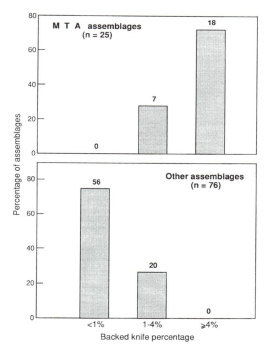

Figure 6.4 *Frequencies of backed knives recorded in southwestern French MTA industries, compared with those recorded in other industrial variants. All industries with backed-knife frequencies higher than 4 percent are attributed by Bordes to the MTA variant.*

axes (in both northern as well as the southwestern French sites) and may rank as much as a type fossil of the MTA variant as the more widely recognized hand-axe types (Fig. 6.4).

How far similar type-fossil forms can be recognized in some other Mousterian variants is more debatable. Bordes argued (1953a: 461; 1961b: 805; 1968a: 101; 1981: 78) that at least two highly distinctive forms are associated closely with industries of the combined Charentian (i.e. Quina + Ferrassie) group – notably, large, bifacially worked *tranchoirs*,

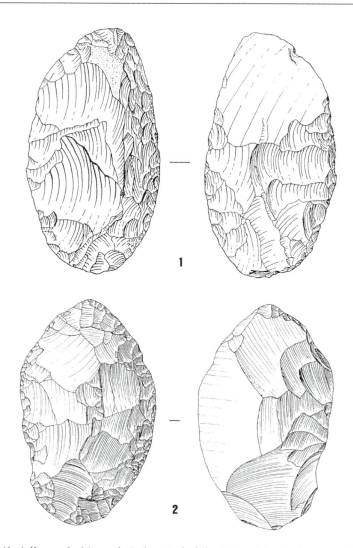

Figure 6.5 *Large, bifacially-worked 'tranchoirs' typical of the Quina-Mousterian assemblages in southwestern France, from the sites of Montgaudier (no. 1) and La Quina (no. 2). After Debénath 1974; Bordes 1961a.*

shaped by a distinctive 'plano-convex' retouch which extends over a large part of the ventral face of the tools (Figs 6.5, 6.6) and thick, symmetrical, double-pointed *limace* forms (Fig. 6.7) (see also Turq *et al.* 1990: 62–3). Outside France similar type-fossil forms are now recognized as characteristic of several other regional variants of the Mous-

terian, including the typical flake cleavers which characterize the 'Vasconian' industries of the Pyrenees and Cantabria (Fig. 4.28), the 'Micoquian' hand axes and related leaf-point forms of the central European industries (Figs 6.8, 6.9) and various tanged and stemmed points of the North African Aterian (Bordes 1968a, 1981, 1984; Klein 1989a; Clark

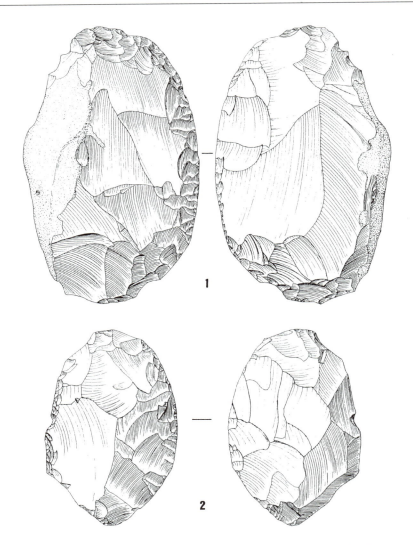

Figure 6.6 *Large bifacial 'tranchoirs' from the Quina-Mousterian assemblages of La Quina (no. 1) and Hauteroche (no. 2). After Bordes 1961a.*

1992). As Bordes pointed out (1977: 39; Bordes & de Sonneville-Bordes 1970: 61 etc.), there is no reason to think that specific type-fossil forms should be any less characteristic of particular industrial variants within the French Mousterian complex than they are in some of the widely recognized regional variants of the Middle Palaeolithic.

2. *Qualitative variations in tool morphology.* The notion that certain more qualitative variations in tool morphology may be characteristic of specific industrial variants of the French Mousterian was frequently emphasized in Bordes' publications. He argued this most emphatically in relation to distinctive 'Quina-type' retouch applied to edges of side

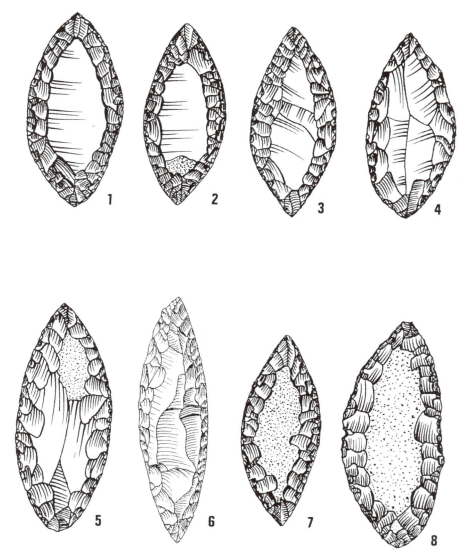

Figure 6.7   *Double-pointed 'limace' forms from Quina and Ferrassie Mousterian assemblages.*

scrapers, points and related forms (Fig. 6.10), which he felt was characteristic (at least in quantitatively significant proportions (Fig. 6.11) of industries belonging to the Quina and Ferrassie groups (e.g. 1961b, 1977, 1981). How far this can be attributed simply to the intensive, repeated resharpening of the tools (along the lines suggested by Dibble, Rolland and others) is still an open question. As discussed in Chapter 4, similar types of retouch are equally well represented in certain assemblages where the abundance of local raw materials provides a powerful argument against intensive tool reduction

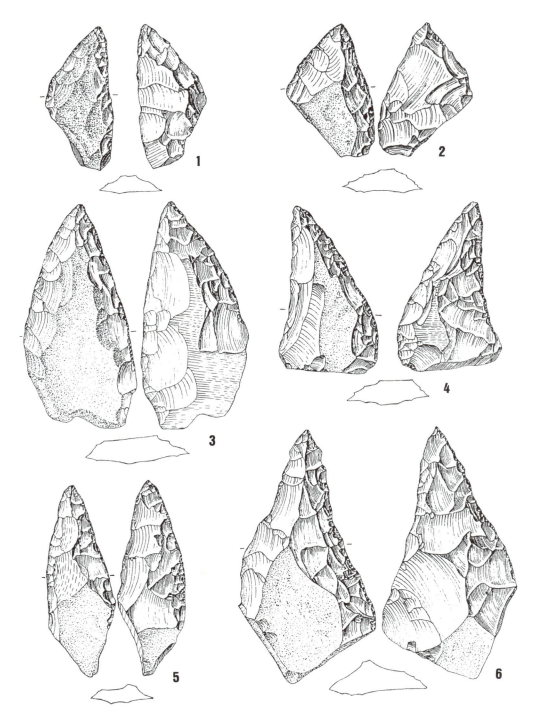

Figure 6.8 *'Micoquian' type bifaces from the site of Klausenische (Germany). After Bosinski 1967.*

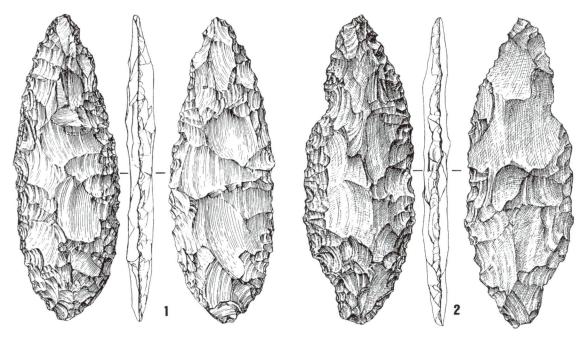

Figure 6.9   *Bifacial leaf points from the late Mousterian levels of the Mauern Cave, southern Germany. After Bohmers 1951.*

Figure 6.10   *Plan and edge views of racloirs showing characteristically stepped, overlapping 'Quina-type' retouch.*

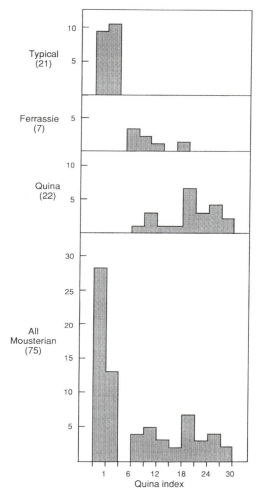

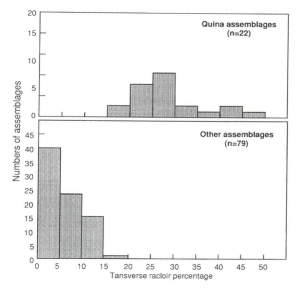

Figure 6.12 *Relative frequencies of transverse racloir forms (expressed as a percentage of all single-edged racloirs) recorded in Quina industries (upper) and other Mousterian industries (lower) in southwestern France. All industries with more than 20 percent of transverse forms are attributed to the Quina variant.*

Figure 6.11 *Percentages of tools shaped by distinctively 'Quina-type' retouch in different industrial variants of the southwest French Mousterian. After Bordes and de Sonneville-Bordes 1970.*

strategies of this kind – most notably perhaps in the case of Combe Capelle Bas, which occurs effectively on top of an outcrop of exceptionally rich and high quality flint (Peyrony 1943; Burgon 1957). Whatever the explanation, Bordes' own figures (1961b: 807, 1977: 38; Bordes & de Sonneville-Bordes 1970: 70) leave no doubt that the relative

frequency of tools shaped by distinctively Quina-type retouch is an effective diagnostic feature of the industries belonging to the Ferrassie and Quina groupings, at least within the specific context of southwestern France (Fig. 6.11).

Similar observations can be made in relation to other aspects of the detailed morphology or techniques of manufacture of particular racloir types. High frequencies of transverse, as opposed to lateral, racloir forms have always been recognized as a diagnostic feature of the Quina Mousterian variant (Fig. 6.12) and in this case clearly cannot be attributed exclusively, or even predominantly, to intensive resharpening techniques (see Chapter 4). The same appears to be true for heavily convex racloir forms, which seem from the limited amount of systematic data available to provide a significant contrast between the combined industries of the Ferrassie and Quina groups and those of the MTA variant (Mellars 1967). Finally, Bordes argued that certain more subtle but to him highly distinctive techniques of retouch applied to the edges of racloirs were specifically characteristic of the MTA variant (Bordes & de Sonneville-Bordes 1970: 72). Bordes had no doubts that these and other details could be seen as 'cultural' or 'stylistic' features of the different Mousterian variants which could in no way be attributed purely to the functional or even technical constraints involved in tool production (e.g. 1959a, 1961b, 1977, 1981; Bordes & de Sonneville-Bordes 1970).

Evaluation of the Bordes taxonomy

There have been a number of attempts over the past 30 years to subject the Bordes taxonomy to more rigorous statistical analysis, employing multivariate analytical techniques. The earliest attempt was published by Doran and Hodson in 1966, based on a small series of 16 assemblages recovered from five different sites in southwestern France (Pech de l'Azé I, Abri Chadourne, Hauteroche,

Ermitage, Mas-Viel). The same type of analysis was later applied to a larger sample of 33 assemblages (principally by incorporating more recently excavated assemblages from the sites of Combe Grenal, Roc de Marsal and Petit-Puymoyen) in the course of my own doctoral research (Mellars 1967). Both analyses depended on the computation of a similarity coefficient between each potential pair of assemblages involved, based on a comparison of the percentage frequencies of the full range of types listed in Bordes' *Liste typologique* but explicitly not incorporating any direct technological information from his various technical indices. This battery of similarity coefficients was then manipulated by means of a computerized multidimensional scaling procedure, to produce a two-dimensional configuration of the assemblages which best approximated to the actual multivariate similarities and contrasts involved.

The result of the analysis of the larger data set is shown in Fig. 6.13. This conforms well in most respects with the basic features of Bordes' taxonomy. It is clear that it is the overall frequency of racloirs in the different assemblages that forms the dominant element, and for this reason the industries characterized by very low percentages of racloirs (those of the Denticulate Mousterian and the

Figure 6.13 *Multi-dimensional-scaling analysis of 33 Mousterian assemblages from southwestern France, from Mellars 1967. The assemblages represented are: nos 1–4 Pech de l'Azé I (layers 4, B, 6, 7); nos 5–6 Hauteroche (lower, upper layers); nos 7–10 Abri Chadourne (layers A, B, C, D); no. 11 L'Ermitage; nos 12–14 Roc de Marsal (layers 6, 10, 11); no. 15 Mas-Viel; nos 16–18 Petit-Puymoyen (cave, C2, breccia); nos 19–33 Combe Grenal (layers 11, 14, 17, 20, 22, 23, 26, 27, 29, 33, 35, 38, 50, 52.*

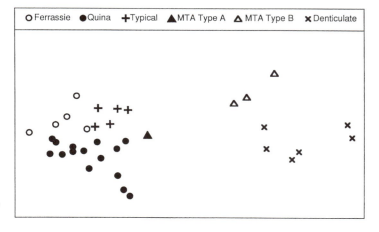

MTA Type B) are separated clearly, on the right-hand side, from those with higher racloir frequencies on the left-hand side (those of the Ferrassie, Quina, Typical Mousterian and MTA Type A). Within these two broad divisions, however, the groupings of individual assemblages correspond closely with Bordes' basic taxonomic divisions. Thus the 15 assemblages attributed to the Quina Mousterian are grouped together closely in the lower, left-hand side of the diagram and in this area no other non-Quina industries occur. In keeping with Bordes' notion of the wider 'Charentian' grouping it is interesting that the group of five Ferrassie-type assemblages are placed immediately adjacent to the Quina group but consistently separate from the latter group and located towards the upper, left-hand zone of the distribution. (It is particularly interesting to see that although Bordes' distinction between the Ferrassie and Quina variants is based on the technical features of the industries – which are not incorporated directly in the present analysis – the separation between the two forms is equally apparent on purely typological grounds). The smaller grouping of five Typical Mousterian assemblages is similarly placed consistently to the upper right-hand-side of the main Quina–Mousterian distribution. The most widely dispersed industries are those of the MTA group. In this case there is a fairly sharp separation of the different assemblages based on the overall frequencies of racloirs, with the single industry of the Type A group on the left of the diagram and those of the Type B on the right. Even so, these industries are separated from those of the other industrial groupings and significantly reveal an overall linear arrangement which corresponds exactly with the documented stratigraphic sequence of the respective assemblages in the different levels at Pech de l'Azé site I.

A second approach to multivariate analysis was published by Callow and Webb (1977, 1981), employing the technique known as Canonical Variates analysis. The aim of this technique is not so much to scan a body of data for its inherent structure but rather to test how far a preconceived perception of clustering in the material can be substantiated in terms of the overall range of the analytical data. In other words, given an *a priori* classification of the material, the technique is designed to assess how far this classification has succeeded in identifying genuine patterns of clustering, or clear discontinuities, in the analytical data. In the same way the technique can assess whether individual units have been ascribed to their taxonomically correct grouping, or whether they would be better assigned to another group.

Callow and Webb's analysis was applied to many more assemblages than that of the multidimensional scaling study discussed above (a total of 96 assemblages in all) but was again restricted entirely to industries from the Perigord and immediately adjacent areas of southwestern France. The analytical data also differ in several respects from those of the earlier study and rely mainly on the principal typological and technological indices employed by Bordes to summarize the major quantitative features of the assemblages. In essence this analysis uses a more restricted range of purely 'typological' features of the assemblages but expands on this data base by incorporating all the major 'technical' features of the industries (i.e. frequencies of Levallois flakes, facetted striking platforms, blades etc.).

The main results of Callow and Webb's analysis are illustrated in Fig. 6.14, which again compresses a multi-dimensional pattern into two main visual dimensions. As they point out, the pattern provides strong support for the major components of Bordes' taxonomy. All the major variants, with the possible exception of the Typical Mousterian, occupy separate zones on the diagram with no significant areas of interchange or overlap between the groups. The Typical Mousterian emerges as the least isolated group and

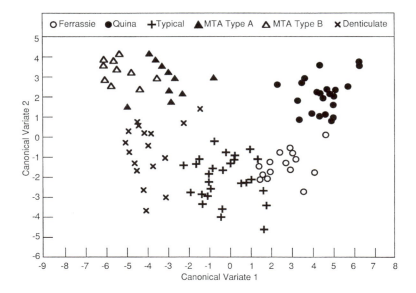

Figure 6.14 *Canonical variates analysis of 96 Mousterian assemblages from southwestern France according to Callow & Webb 1981, with the taxonomic classification of the assemblages according to Bordes.*

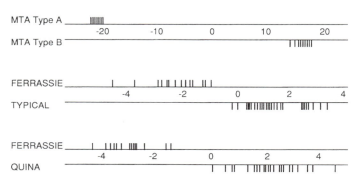

Figure 6.15 *'Discriminant function' analyses of paired groups of Mousterian industries from southwestern France, according to Callow & Webb 1981.*

shows potential areas of overlap with industries of both the Ferrassie and Denticulate groups.

Callow and Webb are at pains to emphasize that although their technique of Canonical Variates analysis is designed to test the validity of Bordes' *a priori* classification of the industries, this technique cannot create patterns of clustering and discontinuities in the data where they do not exist. To explore this further they went on to calculate separate 'discriminant function' measurements for specific pairs of the major groupings, belong-

ing to the most similar and least segregated groups (Callow & Webb 1981). As Figure 6.15 shows, these results provide further support for Bordes' separation of the Quina and Ferrassie industries and the two major variants of the MTA industries (Type A and Type B). They also support a clearer pattern of separation of the Ferrassie and Typical industries than that suggested by the more general two-dimensional plot – though this distinction remains the most difficult to substantiate in objective quantitative terms. Their results were summed up by Callow and Webb:

It should be apparent ... that there is considerable justification for the partition of the southwest French Mousterian assemblages proposed by Bordes. Not only are the typological and technological data multimodal in character, but it is possible to identify several discrete clusters of assemblages corresponding to his variants. The primary goal of this investigation, to confirm or deny the existence of such clusters, would, therefore, seem to have been achieved.

(Callow & Webb 1981: 137)

The chronological dimension

How far the various approaches to multivariate statistical analysis discussed above provide a convincing vindication of Bordes' industrial taxonomy of southwestern French industries is no doubt debatable. It is sometimes argued that most multivariate clustering procedures are designed to produce such patterns and may sometimes (especially perhaps in the case of the Canonical Variates and related techniques) tend to exaggerate patterns of similarity and contrast within and between the major groups identified in the analyses. Nevertheless, the results of the studies described above are fully consistent with the major features of the Bordes taxonomy even if some of the finer distinctions in his scheme (most notably those between some of the Ferrassie and Typical Mousterian industries or between some of the later (Type B) MTA industries and those of the Denticulate group) remain less clear.

The observation which gives by far the strongest support to the major features of the Bordes taxonomy is the clear evidence for the stratigraphic and chronological distribution of the principal industrial variants in the southwestern French sites. As I have discussed in detail elsewhere (1965, 1969, 1970, 1986a,b, 1988, 1989, 1992), the evidence for this is strong for three of the most typologically distinctive of Bordes' variants, notably those of the Quina, Ferrassie and MTA

variants. Since the relevant evidence has been discussed so fully in earlier publications, the most critical observations can be summarized here.

1. The most explicit evidence for a chronological separation of the Ferrassie, Quina and MTA industries in southwestern French sites is provided by the stratigraphic distribution of these three variants in the long and detailed sequence recorded in Bordes' excavations at the site of Combe Grenal (Fig. 6.16). As discussed in Chapter 2, this site contains 55 separate levels of Mousterian occupation which appear to span the chronological range from the opening stages of the last glaciation (ca 110,000 BP) to almost the end of the Mousterian succession (ca 45–50,000 BP). Within this sequence, the stratigraphic distribution of the industries attributed by Bordes to the Ferrassie, Quina and MTA variants shows a simple pattern (Fig. 6.16): all industries attributed to each variant are confined to a narrow span of the total sequence, and the stratigraphic ranges occupied by the three variants are entirely separate, i.e. six levels of Ferrassie Mousterian (in layers 35–27), overlain by nine layers of Quina Mousterian (in layers 26–17), overlain (in the uppermost part of the sequence) by five levels of MTA (layers 1–5) (Bordes 1955a, 1961b, 1972). Clearly, this sequence provides no evidence whatever for any significant overlapping, extensive synchronism or direct interstratification of industries belonging to the Ferrassie, Quina and MTA groupings, as claimed in many of the earlier discussions of the southwestern French Mousterian sequence. It should also be emphasized that the attribution of the different industries in the Combe Grenal sequence to these particular industrial variants was established by Bordes (1955a, 1961b), long before any of the subsequent debates over the taxonomic or chronological significance of these variants surfaced in the literature.

COMBE GRENAL

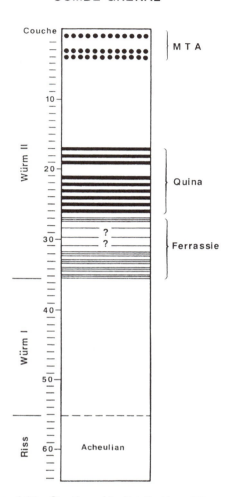

Figure 6.16 *Stratigraphic distribution of Ferrassie, Quina and MTA industries within the Mousterian succession at Combe Grenal, according to Bordes 1961b; 1972 etc. The industries from layers 28–31 were classified in Bordes' original publication of the site (1955a) as representing an 'attenuated Ferrassie' variant, but were later reclassified as 'Typical Mousterian enriched in racloirs'.*

2. Further support for the stratigraphic patterns documented at Combe Grenal is provided by all other sites in the southwestern French region which contain directly superimposed levels of either Quina, Ferrassie or

Table 6.2
Sites with levels of MTA (represented by typical cordiform hand axes) overlying levels of either Quina or Ferrassie Mousterian in western France

1. Combe-Grenal	Dordogne
2. Combe-Saunière	Dordogne
3. Combe-Capelle Bas	Dordogne
4. La Gane	Dordogne
5. Grotte XVI	Dordogne
6. Les Merveilles	Dordogne
7. Le Moustier upper shelter	Dordogne
8. Pech de l'Azé I/II	Dordogne
9. Pech de Bourre	Dordogne
10. Roc de Marsal	Dordogne
11. La Rochette	Dordogne
12. Abri du Chasseur	Charente
13. Hauteroche	Charente
14. La Quina	Charente
15. Chez-Pourrez	Vienne
16. Roc-en-Pail	Maine-et-Loire

For details of sequences, see Mellars 1969: 164-5; 1988: 108. Information on the recently-excavated sequences at Combe Saunière, Grotte XVI and La Quina were provided by J.-M. Geneste, J.-P. Rigaud, A. Debénath and A. Jelinek.

MTA industries. So far, no less than 16 different cave and rock-shelter sites have been recorded in this region in which levels of MTA (i.e. levels characterized by the presence of typical cordiform hand axes) are superimposed directly above levels of either Quina or Ferrassie type Mousterian, with as yet no well documented instance of a reversal of this stratigraphic sequence (Table 6.2: Mellars 1969, 1988, 1989c). As I have pointed out elsewhere, the probability of this stratigraphic situation arising entirely by chance, i.e. on the null hypothesis that the MTA, Ferrassie and Quina industries were distributed over the same time spans within this region, is almost nil – of the order of 1 in 30,000. Sites which contain superimposed levels of both Ferrassie and Quina Mousterian industries, although less frequent, show an equally consistent stratigraphic succession. All recorded instances (notably at

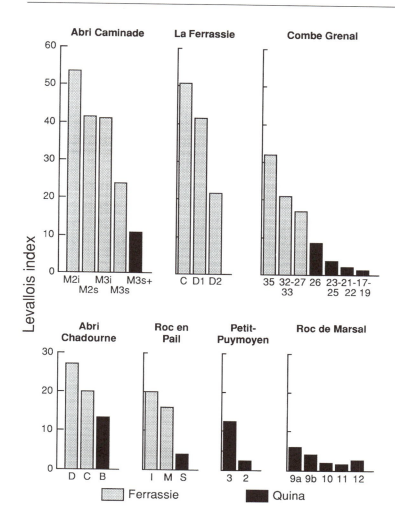

Figure 6.17 *Levallois indices recorded in stratified sequences of Ferrassie and Quina Mousterian assemblages at Abri Caminade, La Ferrassie, Combe Grenal, Abri Chadourne, Roc-en-Pail, Petit-Puymoyen and Roc de Marsal. After Mellars 1988: Fig. 2.3, with some additions and corrections. Data for the Roc de Marsal are taken from Turq 1988b.*

Abri Chadourne, Abri Caminade-est and Roc-en-Pail and most probably also at Roc de Marsal, Chez-Pourrez and Pech de l'Azé IV) conform to the pattern documented at Combe Grenal in showing a clear superpositioning of the Quina Mousterian levels above those of the Ferrassie variant (Mellars 1988, 1989c).

3. The strongest argument for a chronological separation of the Ferrassie and Quina Mousterian assemblages, however, is provided by the evidence for a relatively simple

pattern of technological evolution which can be documented consistently in all sites which contain stratified sequences of one or other of these variants (Fig. 6.17). Sequences of this kind have now been recorded from at least seven different sites in the Perigord and immediately adjacent areas – notably at Combe Grenal (with 15 separate Ferrassie/ Quina levels), La Ferrassie (three levels), Abri Chadourne (three levels), Abri Caminade-est (three levels), Petit-Puymoyen (two major levels), Roc de Marsal (five main lev-

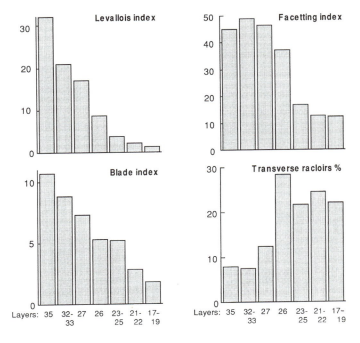

Figure 6.18 *Technological and typological features of the stratified sequence of Ferrassie and Quina Mousterian assemblages in layers 35–17 at Combe Grenal – apparently reflecting a gradual technological 'evolution' from the Ferrassie to the Quina variant.*

els) and Roc-en-Pail (three main levels) (Mellars 1969, 1988; Rolland 1988b). As shown in Fig. 6.17, in all these sites it is possible to observe a simple pattern of gradual, step-by-step decrease in the Levallois index of the assemblages between the lower and upper levels in the sequences. Significantly, these patterns can be seen not only between the main blocks of Ferrassie and Quina assemblages (at Combe Grenal, Abri Chadourne, Abri Camindade and Roc-en-Pail) but also clearly *within* the various stratified levels of Ferrassie Mousterian (at La Ferrassie, Combe Grenal, Abri Chadourne, Abri Caminade and Roc-en-Pail), and *within* the stratified levels of Quina Mousterian at Combe Grenal, Roc de Marsal and Petit-Puymoyen. Thus, these sites reveal a progressive decrease in the Levallois component of the assemblages which cuts clearly across the Ferrassie-Quina interface and suggests a gradual pattern of linear technological 'evolution' from one var-

iant to the other (Fig. 6.18). Again, the probability of these stratigraphic trends arising entirely by chance – through so many different levels in so many different sites – is extremely small – less than 1 in 10,000.

4. The implications of a relatively late chronological position for the majority of the MTA industries in the Perigord sites are supported further by the frequency with which MTA industries have been found stratified directly beneath levels containing early Upper Palaeolithic industries within the cave and rock-shelter sites. As shown in Table 6.3, at least 13 well documented occurrences of this kind have now been recorded in the Perigord region (Mellars 1969, 1988). Of course, direct stratigraphic superpositioning of this kind cannot be taken as an automatic reflection of close proximity in time, since major stratigraphic hiatuses can and do occur in cave and rock-shelter sequences. Never-

Table 6.3
Sites with levels of MTA directly underlying levels with Upper Palaeolithic industries in southwestern France

1. Abri Audi	Dordogne
2. La Cavaille	Dordogne
3. La Combe	Dordogne
4. Grotte XVI	Dordogne
5. Laussel	Dordogne
6. Les Merveilles	Dordogne
7. Le Moustier upper shelter	Dordogne
8. La Rochette	Dordogne
9. Le Chasseur	Charente
10. Grotte Marcel-Clouet	Charente
11. La Gane	Lot
12. Roc de Combe	Lot
13. Quinçay	Vienne

For details of these sequences see Mellars (1969), Debénath (1971), Bordes & Labrot (1967), Lévêque (1987).

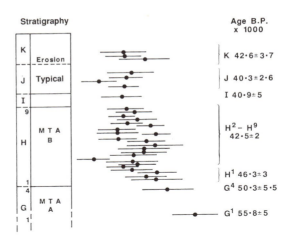

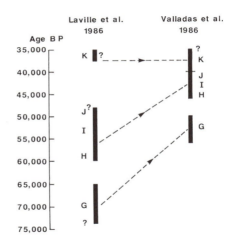

Figure 6.19 *Thermoluminescence (TL) dates for the Mousterian succession in the lower shelter at Le Moustier, after Valladas et al. 1986. In the lower diagram the results of the TL dating are compared with the original chronology of these levels proposed by Laville (1975, 1988; Laville et al. 1986 etc.). After Mellars 1988.*

theless, the frequency with which sequences of this kind have been recorded in southwestern France provides strong additional support for the hypothesis of a relatively late position of the MTA industries in the Mousterian succession as a whole.

5. Finally, direct evidence for the absolute ages of the major industrial variants in southwestern French sites is still limited but is nevertheless fully consistent with the stratigraphic evidence discussed above. The most significant dates in this context are those secured by Hélène Valladas (based on the TL dating of burnt flint samples) for the classic sequence of MTA levels in the lower shelter at Le Moustier (Valladas *et al.* 1986; see also Mellars & Grün 1991). As shown in Fig. 6.19, these confirm that the MTA levels at Le Moustier do indeed belong to the final stages of the Mousterian sequence, and therefore contradict earlier suggestions by Laville and others that these levels are broadly contemporaneous with the long succession of Ferrassie and Quina-Mousterian levels at Combe

Grenal (Fig. 6.20: see Mellars 1986a,b, 1988; Laville 1973, 1975, 1988). Closely similar dates for typical MTA industries have been obtained for the sites of Fonseigner and La Quina, further confirming that these levels

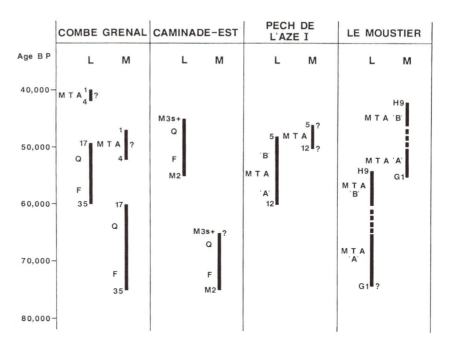

Figure 6.20 *Comparison of the relative and absolute chronologies proposed by Laville (L) and Mellars (M) for various Mousterian successions in southwestern France. From Mellars 1988.*

date from the final stages of the Mousterian sequence, broadly within the time range of ca 55–40,000 BP (Valladas *et al.* 1987 and personal communication). As noted above, these results are consistent with the general stratigraphic position of the MTA industries which almost invariably occur immediately below levels with early Upper Palaeolithic industries in southwestern French sites.

Unfortunately, direct absolute age determinations for levels of Ferrassie and Quina Mousterian are still lacking from sites in southwestern France. However, dates obtained for these industries in a number of areas to the east of the Perigord (for example at Baume-Vallée in the Haute-Loire and La Roquette and Brugas in the Gard) all conform closely with the dates predicted above – i.e. consistently earlier than ca 55–60,000 BP (see Valladas *et al.* 1987; Raynal & Huxtable 1989).

For many of the Perigord sites, however, there is now increasing evidence from geological and other palaeoclimatic indicators that these industries date largely from the period of rigorous, full-glacial climate which characterized stage 4 of the oxygen-isotope sequence – i.e. broadly within the time range of ca 60–75,000 BP. As discussed in Chapter 2, this is clearly the position occupied by the entire sequence of Ferrassie and Quina industries within the overall climatic succession at Combe Grenal (Fig. 2.23) and is equally strongly suggested by the faunal and other associations of similar industries at other sites (Mellars 1969, 1988, 1992a). This point has recently been argued explicitly by Delpech (1990), Guadelli (1990) and others for the Quina-Mousterian levels at Le Regourdou, Pech de l'Azé II, Grotte Vaufrey, La Chapelle-aux-Saints, Roc de Marsal etc.,

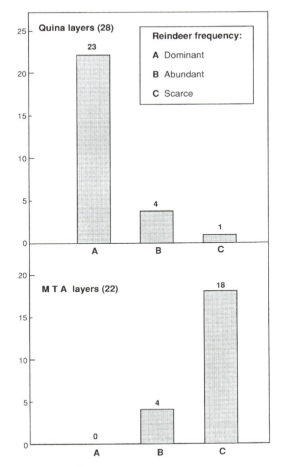

Figure 6.21 *Frequencies of reindeer remains found in association with Quina and MTA industries in southwestern France. From Mellars 1969 and other sources; where no precise quantitative data are available, the attributions follow the more qualitative descriptions given in the original site reports.*

all of which are characterized by high frequencies of reindeer remains and a virtual lack of faunal indicators of more temperate climatic conditions (see Fig. 6.21). In this respect the faunal assemblages contrast strikingly with those documented from the great majority of MTA levels in the same region, which are almost invariably dominated by remains of either aurochs, bison or red deer –

suggesting much milder climatic conditions than those which prevailed during the occupation of the Quina-Mousterian horizons (Mellars 1967; 1969: 145, 1992a). This provides further support for the conclusion that the MTA industries date largely from the predominantly interstadial conditions of isotope stage 3, rather than from the much colder conditions of isotope stage 4.

6. The last point to be emphasized concerns the strong evidence for a pattern of typological and technological development within the overall sequence of MTA industries. This was clearly recognized by Bordes himself (1954–55, 1959a, 1961b, 1972, 1981, 1984) and formed the basis for his formal division of the MTA variant into two major subgroups, designated respectively Type A and Type B. The distinction between the two variants rests first on a much higher frequency of typical hand axes and various racloirs in the earlier (Type A) than in the later (Type B) industries; and second, on a general increase in the frequencies of typical backed knives and in most cases notched and denticulated tools in the later (Type B) industries. Bordes had no doubt that these two variants should be regarded as parts of a single technological and chronological development and pointed out that in at least five separate sites within the Perigord region one could demonstrate a direct superpositioning of levels of Type B assemblages over those of Type A (notably at Le Moustier, Pech de l'Azé sites I and IV, La Rochette, and the Abri Blanchard: see also Bourgon 1957; Delporte 1962; Delporte & David 1966). One of the most significant features of these sequences is that in none of the long and multilayered successions of MTA industries is there any trace of the intercalation of any other (i.e. non-MTA) variants of the Mousterian. This adds further weight to the argument against the hypothesis of an extensive synchronism or overlapping of the time ranges occupied by the different industrial variants of

the Mousterian complex in southwestern France.

Without stressing the implications of stratigraphic and chronological data further, it is clear that all available data are consistent with a relatively simple, successive pattern in the occurrences of the Ferrassie, Quina and MTA variants in southwestern French sites and categorically opposed to a hypothesis of an extensive synchronism, overlapping or direct interstratification of these three variants in this region. The sharp localization and complete separation of these industries in the long and complex industrial succession at Combe Grenal remains the most explicit and impressive single observation in this regard (Fig. 6.16). How far this pattern can be extrapolated to other regions of France is of course a separate question to be answered on the basis of evidence from each individual region (see Mellars 1969: 147–50). The evidence from the southwestern French sites remains nevertheless clear, internally consistent and largely self explicit. As a further independent endorsement of the general integrity and coherence of the Ferrassie, Quina and MTA variants as significant technological entities within the southwestern French sites, their observed chronological patterning stands as a critical observation.

The Denticulate and Typical Mousterian variants

In contrast to the sharply defined patterning of the Ferrassie, Quina and MTA industries, the various occurrences of so-called Typical and Denticulate Mousterian industries in southwestern French sites are of a very different character. From the Combe Grenal sequence alone it is evident that both variants can occur in at least two separate points in the stratigraphic sequence (Fig. 6.22) and the same pattern is confirmed by evidence from a number of other sites in the same region, notably Pech de l'Azé sites II and IV, Roc de Marsal, Abri Chadourne, Moulin du Milieu

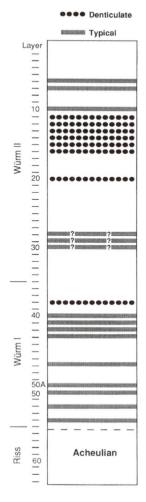

Figure 6.22 *Stratigraphic distribution of Denticulate and Typical Mousterian industries throughout the Mousterian succession at Combe Grenal, according to Bordes 1961b, 1972 etc. The industries from layers 28–31 were originally classified by Bordes (1955a) as representing an 'attenuated Ferrassie' variant, but were later reclassified as 'Typical Mousterian enriched in racloirs.*

etc. (Mellars 1969; Bordes 1961b, 1963, 1975a, 1981, 1984 etc.).

As I have argued in more detail elsewhere (1969: 158–61) the more dispersed stratigraphic and chronological patterning of Denticulate and Typical Mousterian assemblages becomes less surprising when the specific typological and technological features which supposedly define these two variants are examined in detail. In terms of the classic Bordes taxonomy, the definition of each variant rests almost exclusively on a single basic typological parameter, namely the overall frequency of racloirs (in relation to the combined frequencies of all other tools) in the tool assemblages as a whole. Thus the Denticulate variant has always been defined primarily by an exceptionally low percentage of racloirs (below 20 percent and usually less than 10 percent) while the Typical Mousterian is defined by a 'moderate' percentage of racloirs ranging between ca 20 and 55 percent (e.g. Bordes 1953a, 1961b, 1963, 1981, 1984). The other features used by Bordes to define these two variants are almost entirely negative – i.e. the absence of features specifically characteristic of the other three better defined variants (i.e. typical hand axes and backed knives, significant frequencies of tools shaped by Quina retouch, the scarcity or absence of bifacially retouched racloirs, typical limaces etc). The only possibly more positive feature employed in Bordes' definitions of these two variants is the relatively high frequency of notched and denticulated tools (usually between ca 40 and 60 percent) which is taken as an essential feature of the Denticulate variant (Bordes 1963). Since this is determined mainly by the scarcity in these assemblages of more extensively retouched forms, and is also characteristic of some later [Type B] MTA assemblages, its status as a significant positive feature of the Denticulate assemblages is very tenuous.

The implications of these observations are evident. Defined primarily in negative typological and technological terms the status of the Denticulate and Typical variants as significant or meaningful industrial or taxonomic entities must be open to grave doubt. There are indications that Bordes himself was well aware of these problems. In discussing the Denticulate assemblage from the upper levels of the Abri Chadourne, for example, he commented 'il reste possible que le Moustérien à denticulés represente un phenomène de convergence, un cul de sac d'ou aboutiraient, par dégénérescence, d'autres types d'industries' (Bordes et al. 1954: 249). He was even more explicit in discussing the various industries attributed to the Typical Mousterian group. In his 1977 paper he commented 'Typical Mousterian is the most variable, and may represent, in the present state of our knowledge, a kind of "rag bag" in which are put all the assemblages which are neither MTA nor Quina-Ferrassie or Denticulate' (Bordes 1977: 38). On several other occasions he suggested that the industries previously attributed to the Typical grouping certainly required subdivision into two or more separate groups (e.g. Bordes & Sonneville-Bordes 1970: 63, 64, 68; Bordes 1977: 38, 1981: 79, 1984: 151). When viewed in these terms the failure of the Typical and Denticulate industries to occupy a neat, consistent and well defined stratigraphic and chronological position within the total sequence of Mousterian industries in the southwestern French sites is hardly surprising.

A closer examination of the stratigraphic positions of the Denticulate and Typical industries in the Perigord sites, however, reveals a stronger element of chronological patterning than has generally been recognized in the past (Mellars 1969). As several workers have pointed out (e.g. Rolland 1988b; Dibble & Rolland 1992) there are now strong indications that occurrences of taxonomically Typical Mousterian industries are particularly common during the earliest stages of the last-glacial sequence (i.e. during the various stages of isotope stage 5) and indeed with two or three possible exceptions may well represent the only industrial variants recorded from this part of the last glacial sequence within the Perigord sites – reflected for example in all the well documented successions of Würm I assemblages recorded at

Combe Grenal (Fig. 6.22), Pech de l'Azé sites II and IV, Roc de Marsal, Le Regourdou and La Chaise. The only other well defined occurrences of Typical Mousterian industries date from the middle or later stages of the ensuing Würm II phase and are best represented by the block of assemblages stratified between the uppermost levels of Denticulate Mousterian and the overlying levels of MTA in layers 6–10 at Combe Grenal (Bordes 1972) (Fig. 6.22). As I have discussed elsewhere (1969: 159; 1988), the block of three Typical Mousterian industries stratified within the main sequence of Ferrassie Mousterian levels at Combe Grenal (in layers 28–30) were described in Bordes' (1955a) initial publication of the site as representing an 'attenuated Ferrassie' variant and differ from the immediately adjacent industries only by a slight reduction in the overall percentage of racloirs in the assemblages. In most respects therefore these industries could be seen as an integral part of the long and continuous succession of Charentian industries on the site.

As regards the Denticulate Mousterian, it now seems clear that these industries are similarly restricted largely to two separate time spans in the Perigord sites. The earlier is represented by single levels of Denticulate Mousterian found stratified within the longer sequences of Typical Mousterian assemblages in the later Würm I levels at Combe Grenal (layer 38), Pech de l'Azé II (layer 4B), the Roc de Marsal (layers 2–3) and Moulin du Milieu (layer X). The later and better documented horizon is represented by occurrences of characteristically Denticulate assemblages stratified immediately above the rich and typical sequences of Quina-Mousterian industries at Combe Grenal (layers 16–11), Abri Chadourne (layers A and A-B), Hauteroche (layer 3), and (further to the north) Roc-en-Pail (Maine-et-Loire). The single, isolated level of Denticulate Mousterian recorded by Bordes within the uppermost levels of Quina Mousterian at Combe Grenal (in layer 20 – separating the main block of Quina industries in layers 26–21 from the three final Quina levels in layers 19–17) appears to be closely related to this main grouping of post-Quina Denticulate assemblages and stands as the only well documented example of industrial interstratification so far recorded in the numerous long and multilayered successions of Mousterian industries in southwestern French sites. In particular it should be recalled that despite the existence of many other long and complex sequences of Charentian industries at several other sites in this region (e.g. Abri Chadourne, Roc de Marsal, Chez-Pourrez, Abri du Chasseur, Abri Caminade, La Ferrassie etc.) no other occurrences of this kind have so far come to light.

CHAPTER 7

Middle Palaeolithic Subsistence _____

Few topics are more important in prehistory than studies of subsistence patterns, yet few raise more potential problems. The importance of the topic is self-evident. The first requirement for any biological species is to secure an adequate food supply, which is both reliable and predictable in the long term and takes account of more short-term, month-to-month or even day-to-day variations in supplies. The success of these food-procurement strategies will dictate – or at least set strict limits on – the density of population which can be supported in particular habitats, and will ultimately control the extent to which certain regions can be occupied permanently (Gamble 1983, 1986; Whallon 1989). Similarly, the character and distribution of food resources influence the patterns of settlement and movement of the human groups, the overall range and extent of foraging territories and the technology needed to exploit the resources effectively. These factors in turn impinge on the social and residential structures of the groups, and the kinds of organization and planning essential to their survival in particular environments. It is hardly necessary to adopt an ecological determinist viewpoint to recognize that these issues are not merely relevant but ultimately critical to a proper understanding of all aspects of human behaviour and organization in the past.

The difficulties of studying subsistence patterns are equally obvious. The overriding problem throughout the whole of prehistory is evaluating the importance of plant-food resources, given the poor prospects of survival of most plant residues in the great majority of archaeological sites. Even with comparatively durable food residues such as animal bones, however, analysis and interpretation are by no means straightforward (Klein & Cruz-Uribe 1984; Binford 1978, 1981). In many contexts animal bones may have disappeared entirely from the archaeological record (for example in the majority of Palaeolithic open-air sites) or been subjected to varying degrees of selective destruction which can distort the surviving faunal assemblages in complex and significant ways. In certain contexts there are also problems of distinguishing between human food residues and those introduced into the sites by carnivores or other predators. Superimposed, are the limitations of the excavation and recording methods employed in the original collection of faunal remains: few sites have been excavated over the total extent of the original occupied areas and for even fewer sites do we have detailed information on precise spatial distribution of different categories of faunal material over the occupation surfaces. Finally, almost all available samples of archaeological faunal remains are likely to represent some kind of occupational palimpsest combining the compound products of multiple episodes of activity on the sites. Thus the residues of brief (e.g. seasonal) episodes of occupation may be impossible to separate from more complex patterns possibly spanning decades or even centuries of repeated occupation on the same site (e.g. Bordes 1975b; Villa 1983; Binford 1982b; Chase 1986a, 1988).

With such complexities in the nature of the surviving data it is hardly surprising that interpretation of subsistence strategies in the Middle Palaeolithic has become a minefield of debate (Binford 1984, 1985, 1991; Chase 1986a, 1988, 1989; Klein 1979, 1989b; Stiner 1990, 1991a,b, 1992, 1993a etc.). The main issues which have emerged in the recent literature may be summarized as follows:

1. What were the respective roles of hunting versus scavenging in the exploitation of animal resources by Middle Palaeolithic groups?

2. Regardless of the exact mode of exploitation, can we identify clear patterns of 'specialization' in animal exploitation strategies – i.e. strategies focused selectively on certain species of game?

3. If deliberate hunting in the Middle Palaeolithic can be documented, what techniques or strategies were employed? To what extent were different patterns of technology, planning or social cooperation involved?

4. How were animal carcasses processed? For example, were they butchered at the location of kills or in adjacent occupation or home-base sites? How intensively were they utilized? To what extent was systematic processing or storage of meat carried out in different site locations?

5. Finally, can we identify patterns of seasonality in the exploitation of particular species of game in different site locations?

It is these questions, focused primarily on the evidence from southwestern France, which will be addressed in this chapter.

Faunal exploitation patterns in southwestern France

The topic on which we have most data and in which problems of differential survival and other sources of bias in the faunal data are least severe is the total range of animal resources exploited in southwestern France and how far this exploitation was focused on particular species of game. At least for larger species of herbivores, which evidently contributed the bulk of the meat supply, survival of bone remains in archaeological deposits is likely to be roughly comparable for different species, so that estimates of their relative frequencies can usually be calculated with a fair degree of confidence (Binford 1978, 1981; Chase 1986a; Klein & Cruz-Uribe 1984). Similarly we can be fairly confident in most sites that we are dealing with the products of human activity in the archaeological faunal assemblages, rather than the mixed products of human activities and those brought into the sites by carnivores. The majority of available faunal assemblages derive from open rock-shelter sites rather than enclosed cave sites, and in most cases the sparse representation of carnivore remains makes it unlikely that they contributed more than a minor part of the total faunal remains. Notable exceptions are one or two cave sites (e.g. the Grotte Vaufrey, Le Regourdou and Pech de l'Azé II) where hyenas or other carnivores evidently did contribute significantly to the total bone assemblages (cf. Binford 1988; Laquay 1981). Arguably the major limitation of the present data is the scarcity of information on the detailed spatial distribution of different species in the sites. From the limited data available there is no reason to think that different species were treated very differently in this regard and no obvious evidence for highly localized patterns of distribution of partic-

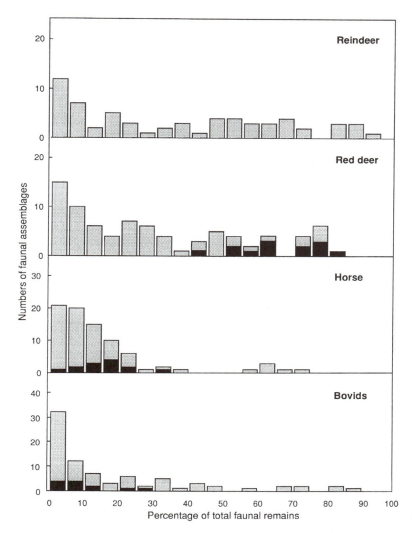

Figure 7.1 *Frequencies of the four principal faunal taxa recorded in last-glacial Mousterian faunal assemblages from cave and rock-shelter sites in southwestern France. The figures are based on the total numbers of identified specimens (NISP) in the faunal assemblage as a whole and include only assemblages with more than 50 identified remains. Data from Laquay 1981; Guadelli 1987, 1990; Delpech 1988; Madelaine 1990; Jaubert 1984; Niederlender et al. 1956. The areas of the graphs shaded in black indicate data from the Würm I levels (= isotope stage 5) at Combe Grenal. 'Bovids' include remains of both* Bos primigenius *and* Bison.

ular taxa over the occupation surfaces (see Chapter 9). Nevertheless, the possibility of some patterning of this kind should always be kept in mind, especially for material recovered from some of the smaller excavations.

Leaving aside these issues, the composition of Middle Palaeolithic faunal assemblages from southwestern France shows several clear patterns. The first point to emphasize is the diversity of the species represented in the majority of the sites. In the

case of the large and medium-sized herbivores the Mousterian groups had access to many species, all of which were exploited to varying degrees at different sites (Bordes & Prat 1965; Chase 1986a; Laquay 1981; Guadelli 1987; Madelaine 1990). The bulk of the faunal remains in all the sites derives from one or more of five major species: reindeer (*Rangifer tarandus*), red deer (*Cervus elaphus*), horse (*Equus caballus*), aurochs (*Bos primigenius*) and the steppe bison (*Bison priscus*). While remains of any of these may be domi-

nant in the faunal material from particular
occupation levels (see below) there is usually
evidence that several different species were
exploited on a relatively substantial scale
(Figs 7.1, 7.3). The same pattern can be seen in
the exploitation of several rarer species. In
addition to the five principal taxa listed
above, most sites have yielded more sporadic
remains of several other herbivorous species,
including wild boar (*Sus scrofa*), roe deer
(*Capreolus capreolus*), ibex (*Capra ibex*), the
steppe ass (*Asinus hydruntinus*), chamois
(*Rupicapra rupicapra*) and (more rarely) giant
deer (*Megaceros giganteus*), rhinoceros (*Rhinoceros tichorhinus*) and mammoth (*Mammuthus primigenius*). Regardless of whether
these species were exploited by hunting or
scavenging, the overall impression is of
groups who were fairly eclectic in their
exploitation of different animal species, and
who were able by one means or another to
make use of many different resources within
the respective catchment areas of the major
occupied sites.

Specialization in animal exploitation patterns

The question of 'specialization' in the exploitation of animal resources has been debated
at length in the recent literature on Middle
Palaeolithic subsistence patterns, and raises
complex and interrelated issues (see Chase
1986a, 1987a, 1989; Mellars 1973, 1982, 1989a;
Orquera 1984; Klein 1989b; Stiner 1992).
Clearly, much in these debates depends on
the precise meaning of specialization in the
context of animal exploitation. If defined in
purely quantitative terms, then any faunal
assemblage which shows a predominance of
one particular species is in this sense relatively specialized. But to have any meaning
in cultural or behavioural terms, specialization must presumably imply deliberate *selection* of particular species from the total
resources potentially available for exploitation. It is this demonstration of conscious

selection of particular species that must form
the core of the argument for deliberate specialization in the subsistence strategies of
Middle Palaeolithic groups.

Inevitably, this issue is closely linked with
the ecology and behaviour of different species. Unless we have a clear idea of what
animal resources were available in particular
locations – including seasonal variation in
distribution and abundance – it is impossible
to disentangle the effects of any human selection from purely natural variation in animal
population numbers. There is an obvious
danger of circularity of reasoning here. In
most contexts our only information on the
distribution and relative abundance of different animal species is provided by the composition of the archaeological faunas. How in
this situation one can differentiate objectively between the effects of human as
opposed to purely natural ecological specialization in the composition of animal communities in particular locations has not been
resolved (cf. Mellars 1973; Chase 1987a).

A classic illustration of this dilemma is
provided by the long and rich succession of
faunal assemblages recovered from the site of
Combe Grenal in the Dordogne valley. As
discussed earlier, the total Mousterian
sequence at this site spans over 12 metres of
deposits apparently covering the period from
ca 115,000 to around 50,000 BP – i.e. embracing the greater part of stage 5, stage 4 and at
least the earlier part of stage 3 of the oxygen–
isotope succession (Bordes 1972; Guadelli &
Laville 1990; see Chapter 2, Figs 2.20–2.23).
Abundant faunal assemblages have now
been recovered and fully documented from
virtually the whole of this sequence, incorporating material from 55 distinct occupation
levels (Bordes & Prat 1965; Chase 1986a;
Laquay 1981; Guadelli 1987). As an illustration of changing frequencies in the exploitation of different animal species, this sequence
is at present unparalleled in the Middle Palaeolithic of western Europe.

The patterns shown in Figs 7.2 and 2.22

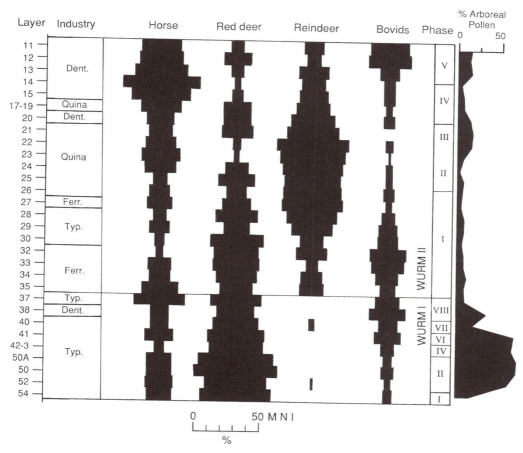

Figure 7.2 *Frequencies of the four principal faunal taxa recorded throughout the Mousterian succession at Combe Grenal, calculated in terms of Minimum Numbers of Individuals (MNI). After Chase 1986a. Only levels with sufficient faunal remains to allow reliable quantitative estimates are included. The arboreal pollen frequencies shown on the right relate specifically to the levels from which faunal data are presented.*

illustrate the changing frequencies of the major species of herbivores documented throughout this sequence, based respectively on the minimum numbers of individuals (MNI) of the different species recorded in each occupation level (Fig. 7.2) and on the total numbers of identified specimens (NISP) of each species (Fig. 2.22). The most relevant figures here are those based on the MNI since it is these figures which reflect most accurately the actual numbers of carcasses of different species which contributed to the overall faunal assemblages brought into the

site. It is now generally recognized that the figures for the NISP can in many cases be heavily influenced by the patterns of dismemberment and butchery of the carcasses which, however interesting and significant in other contexts, are clearly not directly relevant to the original numbers of animals exploited in the different levels. It is true that MNI estimates may in certain cases inflate the apparent frequencies of some rarer species, but this source of distortion is likely to be much less significant than that resulting from major differences in the patterns of butchery of different species (Grayson 1978; Klein & Cruz-Uribe 1984; Binford 1981).

From the viewpoint of specialization, the data from Combe Grenal can be viewed in two ways. Arguably the most striking feature in Fig. 7.2 is the way in which the changing frequencies of most of the major species of herbivores can be seen to correspond with major changes in climatic and ecological conditions during the earlier stages of the last glaciation. As discussed in Chapter 2, the predominance of red deer throughout all the Würm I levels in the site (layers 55–37) clearly reflects the comparatively temperate climatic conditions during the greater part of this phase, corresponding to the various phases of oxygen–isotope stage 5. By contrast, the shift to reindeer-dominated faunas in levels 36–17 indicates a shift to far more rigorous climatic conditions during isotope stage 4 (Fig. 2.23) (Guadelli 1987; Guadelli & Laville 1990). The faunal percentages documented in the uppermost part of the sequence (above layer 17) are more variable but can be seen in general as a response to the predominantly milder, if oscillating, climatic conditions during the earlier part of isotope stage 3.

Another striking feature of Fig. 7.2 is the relatively gradual nature of most of these faunal transitions. In most levels the frequencies of the major species change in a fairly gradual, highly patterned way (Mellars 1970; Chase 1986a, b) which do not seem to relate to changes in either the associated lithic industries or any other major behavioural variables (such as changing butchery patterns) in the overall patterns of human occupation of the site. The whole of this faunal sequence could therefore be seen as reflecting a massive degree of climatic and ecological control over the composition of the different faunal assemblages, with little evidence for any deliberate specialization or intentional selection of different species on the part of the various human groups who successively occupied the site.

A rather different interpretation of the Combe Grenal sequence has been presented by Chase (1986a: 17–24) in the context of his detailed reanalysis and reappraisal of the faunal assemblages from the site. Chase argues in essence that many of the faunal changes documented throughout the sequence are *too* gradual and *too* simply structured to be explained purely in terms of changing climatic and ecological conditions. He points out for example that the relative uniformity of the faunal assemblages recovered from the various Würm I levels (all heavily dominated by red deer remains) can be seen to override much documented variation in the various climatic indicators for these levels (e.g. sedimentological parameters) and also the character of local vegetational conditions as reflected in the associated pollen spectra. Similarly, he points out that a number of brief but apparently significant climatic oscillations recorded in the upper part of the sequence (notably in levels 22–20 and 13–11) do not seem to be reflected in the composition of the associated faunas (Fig. 7.2). To him this provides a strong argument that the varying composition of the faunal assemblages *must* indicate some component of deliberate preference and selection by the human groups for different species of game which to a large extent overruled the purely ecological variations in the local faunal communities.

Chase's points are well taken and may provide an argument for an element of specialization in the exploitation of certain economically preferred species of game. The fundamental problem in this context lies in our ignorance of how the natural population dynamics of particular species respond to short-term episodes of climatic and vegetational change. It has often been pointed out that several species of herbivores (most notably perhaps red deer) have relatively wide environmental tolerances and may not respond immediately to minor episodes of ecological change. Animal communities in general seem to have a degree of inertia to short-term ecological shifts and it was possibly only under conditions of relatively prolonged and sustained climatic and vegetational changes that major shifts in the overall composition of the local faunal communities would have occurred. At present we know far too little about the precise ecological adaptations or population dynamics of various species of last-glacial herbivores to assess how far Chase's arguments may be valid.

Leaving aside these arguments, we are left with a number of rather impressionistic suggestions that certain species of herbivores may be associated with particular segments of the Middle Palaeolithic sequence in a way which may not be explicable in simple environmental terms. Bordes frequently maintained that the majority of Denticulate Mousterian levels showed an apparent specialization in the exploitation of horse resources, as reflected in several levels at both Combe Grenal and Pech de l'Azé II (Bordes 1961b: 809, 1972: 70, 112, 1978: 192; Bordes & de Sonneville-Bordes 1970: 71). Laquay (1981) has made the same suggestion for the association between high frequencies of roe deer and the levels of so-called 'Asinipodian' Mousterian in the Würm I sequence at Pech de l'Azé IV (see also Bordes 1978: 191). While these associations may well have some validity, it is debatable whether either of them will stand up to close scrutiny. Chase (1986a,b) has pointed out that the association between horse-dominated faunas and Denticulate Mousterian levels at Combe Grenal is by no means as clear-cut as Bordes implied and could well represent a largely accidental correlation between two simultaneous but unrelated patterns of climatic and industrial change (Fig. 7.2). The same could be true of the high frequencies of roe deer remains recorded in the Asinipodian levels at Pech de l'Azé IV, which again is by no means a one-to-one correlation (Laquay 1981) and could be a simple reflection of local environmental conditions in this particular location at this point in the early Wurmian succession.

The whole issue of specialization therefore remains problematic, at least within the context of the sites discussed above. On purely *a priori* grounds it could be argued that no pattern of food acquisition either by human or non-human predators is likely to be entirely random and opportunistic, in the sense of exploiting food resources exactly in relation to their relative abundance in the local environment. It could also be argued that even if most of the documented changes in animal exploitation patterns throughout the Middle Palaeolithic sequence can be seen to correlate with simultaneous changes in climatic and ecological conditions, this need not exclude the possibility that an *additional* element of selection was imposed by the human groups over and above these natural fluctuations in animal population frequencies. In other words it could be argued that the relative frequencies of different species observed in the archaeological record are a reflection of both ecological shifts in the natural frequencies of these species combined with a strong element of selection for particular species by the human groups. For the present all of this remains largely speculative. All the arguments so far advanced for clear patterns of faunal specialization at Combe Grenal and other related sites are at best rather tenuous and have still to be demonstrated in unambiguous terms.

It is now becoming clear that the most convincing evidence of economic specialization in the Middle Palaeolithic is likely to come not from the well known cave and rock-shelter sites in the major river valleys of southwestern France but from some open-air locations now being found on more exposed plateaux of the region, discussed in the final part of this chapter. As yet, these sites remain poorly documented and few have produced substantial and well preserved faunal assemblages (see Rigaud 1982; Geneste 1985; Duchadeau-Kervazo 1986). The available

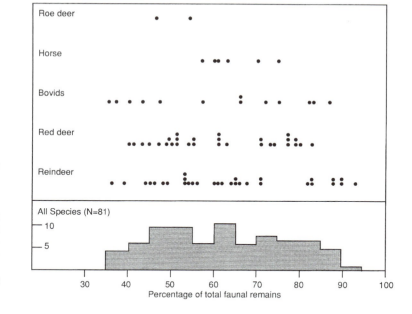

Figure 7.3 *Frequencies of the numerically dominant faunal taxa recorded in last-glacial Mousterian faunal assemblages from cave and rock-shelter sites in southwestern France, expressed in terms of numbers of identified specimens. In only one of the 81 assemblages does the overall frequency of any species exceed 90 percent. For data sources see Fig 7.1; only assemblages with more than 50 identified remains are included.*

data, however, suggest that some of these sites may provide a very different picture of the economic activities of Middle Palaeolithic groups. The recently discovered site of Le Roc in the Euche valley (Dordogne) is one such locality. Few details are yet available but from the provisional data provided by Geneste (1985: 94–5) it seems that effectively the whole of the faunal material from the site consists of remains of either *Bos* or *Bison*, associated with a broadly 'Denticulate' form of lithic industry. More fully documented examples of the same pattern have been recorded at a number of open air sites to the south of the Perigord region – notably at La Borde and Coudoulous in the Department of Lot, and further to the south at the site of Mauran in Haute-Garonne (Jaubert *et al.* 1990; Jaubert 1984; Girard *et al.* 1975; Girard & David 1982; Farizy & David 1992). All these sites have yielded rich and well documented faunal assemblages in which remains of either one of the two main species of large bovids (either bison or aurochs) account for between 93 and 98 percent of the total faunal assemblages. Unless we are to envisage an

extraordinarily specialized and exceptional combination of ecological conditions in the immediate vicinity of these particular sites, the argument for a real element of economic specialization in the subsistence activities at these particular locations would seem impossible to deny.

The final comparison relates to the much more widespread evidence for economic specialization which can now be documented from many of the Upper Palaeolithic sites in southwestern France (Mellars 1973, 1982, 1989a; Chase 1987a). In this case comparisons can be made on a strictly equivalent basis, since the great majority of the Upper Palaeolithic settlements occupy precisely the valley habitats, and often the same site locations, as those of the Middle Palaeolithic cave and rock shelter sites. In the case of the Upper Palaeolithic sites, the evidence for clear economic specialization has never been seriously questioned, and is marked by overall frequencies of reindeer remains in many sites of up to 95–99 percent of the total faunal assemblages (see Fig. 7.4: Mellars 1973; Delpech 1983; Chase 1987a; Boyle 1990). One

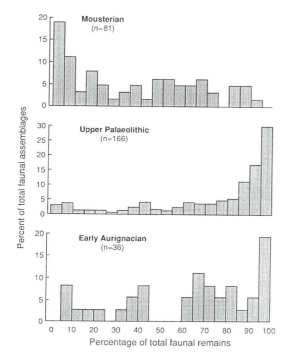

Figure 7.4 *Percentages of reindeer remains recorded in last-glacial Mousterian levels in southwestern France, compared with those recorded in Aurignacian (lower) and all Upper Palaeolithic levels (middle) in the same region, expressed in terms of numbers of identified specimens. While 9 out of 36 of the Aurignacian faunas show frequencies of reindeer above 90%, only one of the Mousterian assemblages shows a comparable frequency. Data on the Aurignacian faunas are taken from Boyle 1990: Appendix 1; for sources of data on Mousterian faunas, see Fig. 7.1.*

ferent sites – Abri Pataud, Roc de Combe, La Gravette and Le Piage (Mellars 1989a: 357; Boyle 1990). The contrast between these patterns and those documented in the various Middle Palaeolithic cave and rock-shelter sites in the same region is striking. As noted earlier, overall percentages recorded for any single species within the various Mousterian cave and rock-shelter sites never rise higher than ca 85–90 percent of the total faunal assemblages (whether calculated in terms of MNI estimates or total numbers of identified remains) and usually show much lower values of ca 60–70 percent (Fig. 7.3). Regardless of whether certain Mousterian groups practised a significant element of deliberate economic specialization in the exploitation of particular animal resources, it is clear that these patterns became much more sharply defined, and more widespread, during the earliest stages of the Upper Palaeolithic sequence (Mellars 1989a).

The processing and utilization of animal carcases

Studies of how the carcases of exploited animals were systematically introduced and processed in archaeological sites are critical to the current debates on the general behaviour and organization of Middle Palaeolithic communities on several different levels. Most obviously they can shed light on exactly how different species were utilized by human groups and how far the different stages of processing and use of particular species were carried out in different site locations. These studies are central to all of the recent discussion on the specific functional orientation of Middle Palaeolithic sites and how such activities may be reflected in the character and composition of the associated stone-tool assemblages (see Chapter 10). Similarly, patterns of bone-element frequencies are equally relevant to the current thorny debates on the relative roles of hunting and scavenging in

point which has not been so generally recognized is that these exceptionally specialized reindeer faunas can be documented not only from the later stages of the Upper Palaeolithic sequence (i.e. from the Upper Perigordian, Solutrian and Magdalenian phases) but also from some of the earliest levels, dating from the initial stages of the Aurignacian, around 32–34,000 BP (Fig. 7.4). Thus, overall reindeer frequencies of 95–98 percent have now been documented from early Aurignacian levels in at least four dif-

Middle Palaeolithic adaptations which will be discussed later.

As yet, this research is in its infancy. Most of the earlier studies of Palaeolithic faunal assemblages have focused largely if not entirely on the relative frequencies of different species in particular occupation levels – usually seen more as evidence of climatic variations than in terms of patterns of human exploitation of these species (Bordes & Prat 1965; Delpech 1983; Guadelli & Laville 1990 etc.). Despite some precocious and astute observations made by Henri-Martin (1907, 1909), Bouyssonie *et al.* (1913) and others early this century, it is only during the past two decades that these issues have been taken up systematically by faunal analysts. For the southwestern French sites we have access at present to only one really systematic body of data, that published by Philip Chase (1986a) on the long sequence of faunal assemblages recovered during the excavations of François Bordes at Combe Grenal. As a result of Chase's study, we now have reasonably full information on the frequencies of at least the principal skeletal elements of the major faunal species throughout the majority of the 55 levels of Mousterian occupation in the site (Figs 7.5, 7.6). For other sites in the region we have only limited and uneven information: some data on the assemblages recovered from Bordes' excavations at Pech de l'Azé sites II and IV, provided in the doctoral dissertation of Guy Laquay (1981); preliminary analyses of faunal assemblages recovered during much earlier excavations of Denis Peyrony at a number of sites in the Vézère valley (Le Moustier, Les Merveilles, La Ferrassie and Gare de Couze) provided by Stephane Madelaine (1990); and more systematic analyses of the assemblages from layer VIII at the Grotte Vaufrey and material from Bonifay's unpublished excavations at Le Regourdou provided respectively by Binford (1988) and Delpech (in press). While none of these assemblages has been published in the same detail as that

provided by Chase for the Combe Grenal assemblages, they provide useful points of comparison for some of the more general and apparently significant patterns documented by Chase in the Combe Grenal data (Fig. 7.7).

Any studies of the varying frequencies different skeletal elements in archaeological faunal assemblages are beset by taphonomic and related problems which have been discussed at length in the literature (Binford 1978, 1981, 1984; Binford & Bertram 1977; Klein & Cruz-Uribe 1984; Chase 1986a). The most obvious problem is that of variable patterns of survival of different bone elements within the archaeological deposits, either because of chemical destruction of the bones or as a result of various mechanical processes such as human trampling on the deposits or the effects of frost action. If these processes acted differentially on different bone elements, this could lead to major distortions in the relative frequencies of survival of particular segments of the animal carcases in different occupation levels. Closely related are the effects of carnivore action on the bone residues, as well as the effects of human processing of particular bones related, for example, to intensive smashing of bones for marrow extraction or even deliberate burning of bones as fuel. Superimposed on these problems are the effects of variable patterns of discard of bones within the archaeological deposits. Clearly, if different segments of the animal carcases were systematically discarded in different areas of the occupation sites, this could lead to significant variations in the relative frequencies in which different bones were recovered during archaeological excavations, depending entirely on the particular areas of the site excavated (see Chapter 9). Evidently therefore, any study of the variable frequencies of different bones in archaeological faunal assemblages must be handled with considerable caution and with these sources of bias or distortion in mind.

Whether these potential sources of distortion are as serious in practice as in theory is more debatable. Thus, Binford, Chase and others have pointed out that the great majority of faunal remains recovered from the Combe Grenal deposits appear to be remarkably well preserved and show few signs of heavy chemical decomposition or erosion of the bones. The clearest illustration of this is the excellent preservation of many smaller and more delicate bones including those of rodents and other small game such as hares and rabbits (Laquay 1981; Guadelli 1987). If these relatively fragile bones have survived, it seems unlikely that many of the bigger and more robust bones from the larger herbivorous species (reindeer, red deer, horse, bovids etc.) have been totally destroyed. Similarly, both Chase (1986a) and Laquay (1981) have pointed out that whilst a few bones from Combe Grenal show signs of carnivore chewing, this is rare in the faunal assemblages as a whole. Finally, it should be noted that even where bones have been systematically and intensively fractured for the extraction of marrow, this rarely leads to the total destruction of the articular ends of the bones, which are the parts most commonly used for analysis and quantification of different carcase elements.

According to Chase, the most serious effects of differential destruction or survival of different bones within the Combe Grenal deposits are likely to be confined to some of the most delicate and thin-walled bones, such as the dorsal and lumbar vertebrae, ribs and some of the more fragile mandible and skull bones. Chase (1986a) quotes data to suggest that in certain levels up to 70 percent of the original ribs and vertebrae may have been destroyed – either through natural *in situ* decay or the difficulties of retrieving bones intact during the excavations. He suggests similar rates of destruction for some of the more delicate parts of the upper and lower jaws, based on comparisons of frequencies of fragments of maxillary and mandibular bones in relation to the numbers of associated

teeth. Whether the poor representation of phalanges in certain occupation levels can be attributed entirely to heavy fragmentation of these bones for marrow extraction, as Chase suggests, is more questionable. As noted above it seems unlikely that this fragmentation would lead to the destruction of the articular ends of the bones, although heavy fragmentation could possibly reduce their prospects of recovery in archaeological excavations.

The main point to emphasize here, however, is that the most significant patterns documented by Chase in the Combe Grenal data relate not so much to variations in bone element frequencies between different stratigraphic levels of the sequence but rather to the striking contrasts which can be documented between *different species* of exploited animals, frequently within the same occupation levels – as for example between remains of red deer and reindeer, or horses and large bovids. These comparisons are between equivalent parts of the carcases of different species which must have had similar prospects for survival and recovery in the same occupation levels. It is this kind of patterning which forms the central core of Chase's analysis and which quite clearly cannot be dismissed as a simple artefact of differential survival of the different skeletal elements within the Combe Grenal sequence.

Skeletal patterns at Combe Grenal

The detailed patterns of skeletal frequencies documented by Chase at Combe Grenal can be examined from two different perspectives: first in terms of contrasts between different species of exploited animals in particular levels of the occupation sequence; and second in terms of variations in these bone-element frequencies between different stratigraphic and chronological segments of the archaeological sequence. This section of the chapter will examine some of the patterns documented by Chase primarily as an exercise in pattern recognition, largely detached from potential interpretations of the data in economic or behavioural terms (Figs 7.5, 7.6). The more controversial issues of behavioural interpretation will be discussed in the following section.

Red deer
Some of the most consistent patterns documented in Chase's analysis relate to the assemblages of red deer remains (Fig. 7.5). As noted earlier, these form the dominant element in the faunal assemblages recovered from all the Würm I levels in the site (coinciding with the relatively temperate climatic conditions of isotope stages 5d-5a) and continue in reduced frequencies throughout all the ensuing Würm II levels (Fig. 7.2) (Bordes & Prat 1965; Guadelli & Laville 1990). The most abundant and fully documented assemblages of red deer remains derive from the long succession of Typical Mousterian levels between layers 54 and 36. Chase points out that while remains of red deer in these levels come from all parts of the carcase (with the notable exception of vertebrae and ribs), the bone assemblages as a whole are characterized by three major features (Fig. 7.5): first, by a strong representation of bones and teeth from the heads of the animals (i.e. both upper and lower jaws); second by a clear over-representation of bones from the lower (distal) portions of front and rear limbs (i.e. ulnae, radii, tibiae and metapodials); and third by a clear under-representation of bones from the upper limbs (i.e. humeri, femora and scapulae). These seem to be constant in almost all the red deer bone assemblages throughout the Combe Grenal sequence and appear to reflect a general tendency for bones from the meatier parts of the skeleton to be strongly under-represented on the site, and those from the more marginal, less meaty parts to be over-represented. The possible explanations for this will be discussed further in the following section, but the consistency of this pattern, repeated in many

different levels, leaves no doubt that it is a general feature of the treatment and utilization of red deer carcases throughout the greater part of the Combe Grenal sequence. One more curious feature of the red deer bone assemblages from the Würm I levels is the strong representation of innominate (i.e. pelvic) bones. The significance of this is more enigmatic but it seems to be a particularly characteristic feature of these levels in the Combe Grenal sequence.

The pattern of red deer remains in the overlying sequence of Würm II levels (coinciding essentially with the main sequence of Ferrassie and Quina Mousterian levels between layers 35 and 20) repeats in most respects that documented in the Würm I levels. The main contrasts are a sharp increase in the frequency of head remains in the lowest levels of the Würm II sequence (layers 35–26) combined with an apparent absence in these levels of pelvic bones (Fig. 7.5). Conversely, frequencies of scapulae (from the upper front limbs) increase in the final layers of the sequence (levels 25–20). Most of these assemblages, however, contain fairly small numbers of postcranial bones, so it is perhaps debatable how much significance should be attached to these features.

Reindeer
The patterns of reindeer remains throughout the Combe Grenal sequence show some similarities to those of red deer, combined with some striking contrasts (Fig. 7.5). Reindeer remains are effectively lacking throughout all the Würm I levels and only appear in strength in the Würm II levels where they effectively dominate the faunal assemblages between layers 35 and 17 (Fig. 7.2) (Bordes & Prat 1965; Guadelli & Laville 1990). Interestingly, in the earlier part of the Würm II sequence (layers 35–26, coinciding essentially with the main sequence of Ferrassie Mousterian levels) patterns of reindeer remains are broadly similar to those of red deer in the same levels. Again, there is a

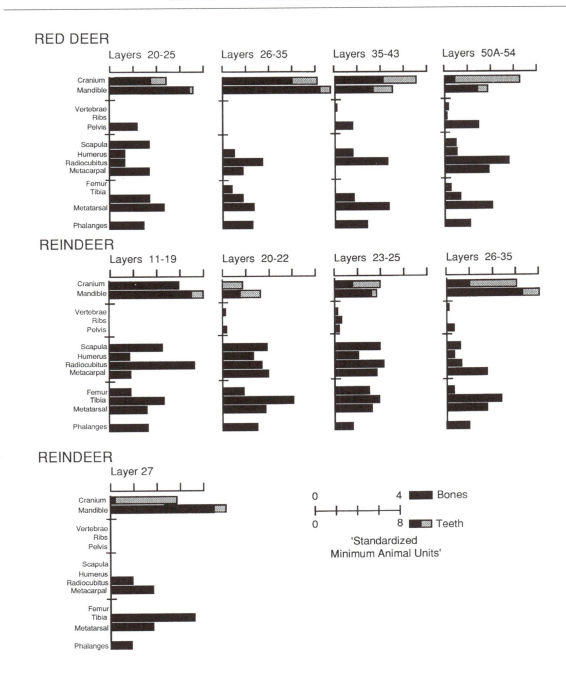

Figure 7.5 *Relative frequencies of different skeletal elements of red deer and reindeer in different stratigraphic levels at Combe Grenal. The figures compare the actual numbers of each skeletal element with those which would be expected if complete carcases of the animals had been introduced into the site. Note that the frequencies of teeth are shown on a different scale from the that of the 'bony' parts of the skeleton. After Chase 1986a: Fig. 8 and Tables A32–47.*

strong over-representation of head remains (especially mandibles and lower teeth in layer 27) together with lower limbs, and a marked under-representation of remains of upper limbs (Fig. 7.5). Again, the emphasis seems to be on the more marginal, meat-poor parts of carcases. The only clear contrast with the red deer pattern is the relatively weak representation of pelvic bones – though it may be significant that these bones are also lacking in the small assemblages of red deer bones recovered from this specific segment of the Würm II succession (i.e. levels 35–26).

One of the most striking contrasts documented in the whole of the faunal succession at Combe Grenal can be seen in the pattern of reindeer remains in the middle part of the Würm II sequence, between levels 25 and 20. There are three conspicuous changes in this part of the sequence. First, the frequencies of teeth, jaws and other cranial remains show a sudden decline compared with those documented in the earliest Würm II levels – especially marked in the material from levels 22–20. Second, there is an equally marked shift in the frequencies of bones from the upper parts of the limbs (i.e. humeri, femora and scapulae), all of which increase sharply compared with those in the underlying levels. Third, most of the levels in this part of the sequence show a fairly consistent, if infrequent, representation of both ribs and vertebrae of reindeer; as discussed earlier, the overall representation of both the latter bone elements may have been influenced by poor survival and recovery of remains, but both are significantly more frequent in this part of the sequence (especially in levels 25–23) than in any other levels on the site. As Chase points out, these features seem to reflect a major change in the patterns of bone representation of cervids at Combe Grenal which contrasts sharply with the patterns documented for both red deer and reindeer in the underlying levels of the sequence. The clear implication is that in these levels much more emphasis was placed on the more economic-

ally useful, main meat-bearing parts of the carcase. Interestingly, these distinctive patterns of reindeer carcases coincide with the major sequence of Quina-Mousterian industries between levels 26 and 20 of the succession.

A third and final shift in the patterns of reindeer representation at Combe Grenal can be seen in the later stages of the Würm II sequence between levels 19 and 11 (i.e. corresponding with the final Quina-Mousterian levels and the overlying levels of Denticulate Mousterian). The assemblages of reindeer bones from these levels are admittedly small (with a combined total of only 44 postcranial bones, together with 150 teeth) but nevertheless two major changes can be documented (Fig. 7.5): a renewed increase in the frequencies of head remains, which now heavily dominate the faunal assemblages – in contrast to the scarcity of heads and teeth in the immediately underlying levels; and an increase in the frequencies of lower segments of both the front and rear limbs, seen especially in the frequencies of radii and tibiae. Despite the limited size of the faunal assemblages, both changes appear to be significant and to indicate a reversion to broadly similar patterns in the treatment and utilization of reindeer carcases to those documented in the lower levels of the Würm II sequence (levels 35–26).

Horse
The patterns of horse remains can only be studied in detail for the upper part of the Combe Grenal sequence (layers 25–11), since it is only in these layers that they occur in sufficient quantities for a reliable analysis – and even here samples of postcranial bones, as opposed to teeth, are relatively small. However, there would seem to be evidence for two different patterns in the representation and treatment of horse remains (Fig. 7.6). The first is represented in levels 25–20 corresponding with the main block of Quina-Mousterian levels. Whilst the broad pattern

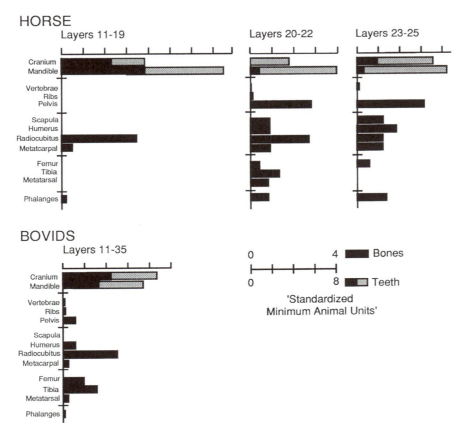

Figure 7.6 *Relative frequencies of the main skeletal elements of horses and bovids in different stratigraphic levels of Combe Grenal, calculated as in Fig. 7.5. Bovids include remains of both* Bos *and* Bison; *owing to the scarcity of remains of bovids throughout the sequence, all of the remains of these taxa from the Würm II levels (layers 11–35) have been combined. From Chase 1986a: Fig. 8, Tables A48–51.*

of horse remains in these levels shows some similarities to that of reindeer in the same levels, there are three important differences: first, the ratio of head and tooth remains to those of postcranial bones is much higher for the horse than for reindeer; second, horse remains show an unusually high frequency of pelvic bones, substantially higher than those for any other species in any other level of the site; and third, there is a curious over-representation of bones from front limbs (especially in levels 25–23) over those of rear limbs. The significance of the latter features remain enigmatic, although it may be recal-led that a similar over-representation of pel-vic remains was also documented in many of the assemblages of red deer bones in the Würm I levels of the site. Whatever the sig-nificance of this feature in economic or car-case-processing terms (cf. Chase 1986a: 54) there is a marked contrast between relative frequencies of pelvic remains of horses and reindeer recorded in these particular levels.

The second distinctive pattern of horse remains is seen in the later part of the Würm II levels (layers 19–11) corresponding with the final Quina-Mousterian levels and the overlying levels of Denticulate Mousterian.

The total assemblage of horse bones from these levels is small (including only 19 post-cranial remains, together with 528 teeth) but nevertheless shows a number of idiosyncratic features suggesting a major change in carcase utilization patterns (Fig. 7.6). In the first place, these levels show an even higher proportion of head and tooth remains than those documented in the immediately underlying levels – with remains of heads and jaws now massively dominating the faunal assemblage. Second, the postcranial bones consist overwhelmingly of a single skeletal element – i.e. ulnae from the lower front limbs. With the exception of isolated specimens of radii, carpals, metapodials and phalanges, all other parts of the skeleton are entirely lacking. Even allowing for the exceptionally small sample size, this suggests a curiously different pattern of skeletal representation from that documented in any of the underlying levels of the Combe Grenal sequence.

Bovids

Finally, Chase presents some interesting if limited data on the representation of large bovid remains – i.e incorporating the remains of both aurochs (*Bos primigenius*) and steppe bison (*Bison priscus*) which are effectively impossible to separate for the majority of bones. The remains of these species are too sparse in most parts of the sequence to allow a detailed, layer by layer analysis, and to obtain a sample of adequate size Chase had to combine material from all parts of the Würm II levels into a single composite sample. Because of this and the impossibility of differentiating between the remains of *Bos* and *Bison*, any detailed analysis of these remains would, as Chase points out, be rather premature. Nevertheless, the general pattern which emerges from the combined Würm II sample appears to show several features which are in many ways similar to those recorded for red deer in the Würm I levels and to some extent those of horse in levels 25–20 (Fig. 7.6). In general, remains from the

lower limbs (particularly the radiocubitus, ulna and tibia) dominate those of the upper limbs (femur and humerus) and there is also an over-representation of remains from the head and teeth. These patterns again emphasize the more marginal parts of the carcase in preference to the main meat-bearing bones – especially in view of the total absence of scapulae and virtual absence of ribs and vertebrae, which in the case of these very large animals are extremely large and robust bones, unlikely to be easily destroyed or overlooked in excavations. Perhaps the most curious feature of the bovid remains is the virtual absence of extreme distal portions of limbs, represented by the metapodials and phalanges – again, large and easily identified bones which are unlikely to be destroyed or missed in excavations. This contrasts with the patterns documented for almost all other species in the site and must surely indicate either that the feet of these large bovids were rarely brought into the Combe Grenal site or that they were systematically removed or discarded elsewhere.

Skeletal patterns at other sites

One of the central questions in this context is how far the particular patterns of bone element frequencies documented by Chase at Combe Grenal can be regarded as typical of Middle Palaeolithic faunal assemblages from southwestern France and how far they may be in some way specific to this particular location. Detailed information from other sites in the region is still limited. The best data come from the studies by Guy Laquay (1981) on several of the faunal assemblages recovered from Bordes' excavations in the Würm I levels at Pech de l'Azé sites II and IV (Bordes 1972, 1975a) and from the studies of Stephane Madelaine (1990) on the assemblages excavated by Denis Peyrony from the sites of Le Moustier, La Ferrassie, Les Merveilles and Gare de Couze and now stored in the National Museum of Prehistory at Les

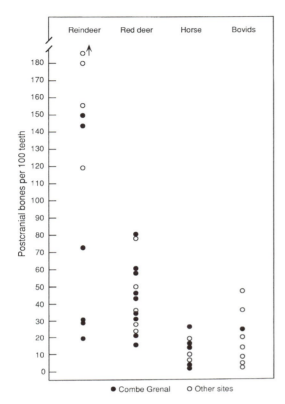

Figure 7.7 *Relative proportions of postcranial remains to teeth recorded for different species in Mousterian faunal assemblages from southwestern France. The data show clearly that remains of horses and bovids are represented in general by much lower overall frequencies of postcranial remains to teeth than those of both red deer and (especially) reindeer. Infilled circles indicate data from Combe Grenal. From Chase 1986a; Laquay 1981; Madelaine 1990 and other sources.*

Eyzies. Further analyses have been provided by Binford (1988) for the material from the major occupation horizon (layer VIII) at the Grotte Vaufrey and by Delpech (in press) for the material from levels 2–5 at Le Regourdou. In common with Combe Grenal all these sites are either cave or rock-shelter locations situated within or closely adjacent to major river valleys (Fig. 8.1).

Few of these sites have provided sufficiently large samples of faunal remains to permit the kind of detailed analyses provided by Chase for the Combe Grenal assemblages and for the majority of assemblages we are reduced to some rather gross comparisons based mainly on overall frequencies of cranial (mainly teeth) as opposed to postcranial parts of the skeleton (Fig. 7.7). For the assemblages recovered from earlier, pre-war excavations (Le Moustier, La Ferrassie, Les Merveilles and Gare de Couze) it should also be kept in mind that the precise standards of faunal recovery employed during the excavations are unknown – although from the composition of the surviving collections there is no reason to think that the material was collected in a significantly more selective or biased way than in more recent excavations. For the two cave sites (Le Regourdou and Grotte Vaufrey) there is the added complication that a part of the faunal material almost certainly derives from intermittent use of the caves as carnivore dens, rather than from residues of human occupations (Binford 1988; Delpech, in press). Even so, there are some interesting and significant patterns which emerge from these studies, which provide some useful comparisons with the better documented material from Combe Grenal (Fig. 7.7).

1. In at least five of the assemblages documented here remains of reindeer are represented by relatively high ratios of postcranial bones to teeth, strongly reminiscent of those documented throughout most of the occupation levels at Combe Grenal, especially from the main Quina-Mousterian levels in layers 25–20 (cf. Figs 7.5, 7.7). This can be seen, for example, in the assemblages from the Quina or Ferrassie-Mousterian levels at Le Regourdou (layer 2), Les Merveilles (lower layer) and Pech de l'Azé IV (layer I1), as well as in the MTA levels at both Gare de Couze and Les Merveilles (upper layer). The consistency of this pattern, now documented in so many sites suggests strongly that reindeer carcasses in general were treated differently from those

of most other species, and were more often introduced into the sites as relatively complete carcases rather than as more selective parts. Some of the possible implications of this are discussed in the following section.

2. A second, sharply contrasting pattern can be seen in remains of the much larger species of large bovids (either *Bos* or *Bison*) and horse. Detailed analyses of these taxa are frequently hampered by small sample sizes but for almost all the faunal assemblages discussed here, remains of these species are consistently represented by much lower ratios of postcranial bones to teeth than for reindeer (Fig. 7.7) – for example in the assemblages from the Typical and Charentian Mousterian levels at both Pech de l'Azé II (layers 4B and 4C2) and Pech de l'Azé IV (layer I2), and in the MTA levels at Gare de Couze and Le Moustier (layers G and H). The clear implication is that these large species were usually introduced into the sites with a much heavier emphasis on heads and jaws, rather than on the postcranial carcase – once again recalling patterns recorded for the same species at Combe Grenal. That this pattern cannot be dismissed simply as an artefact of selective preservation (i.e. as indicating better survival prospects for teeth than bones in occupation deposits) is clearly demonstrated by the fact that remains of reindeer recovered from precisely the same occupation levels (at Gare de Couze, Les Merveilles and Pech de l'Azé IV) show effectively the reverse of this pattern. The only exception is the small assemblage of horse remains recovered from the Rissian level at Grotte Vaufrey, which shows a predominance of postcranial bones and an exceptionally low incidence of head and teeth remains. As Binford (1988) has pointed out, however, this could conceivably be either an artefact of small sample sizes or, more likely, an indication that the horse carcases were introduced into the site mainly by carnivores, rather than by humans.

3. Remains of red deer show a pattern which appears to be generally closer to that of the larger herbivore species (i.e. bovids and horse) than to reindeer. In five of the documented assemblages (layers J3a and X at Pech de l'Azé IV, layer 4B at Pech de l'Azé II, layer VIII at Grotte Vaufrey and layer J at Le Moustier) there is a much stronger emphasis on heads and teeth than on the postcranial skeleton (Fig. 7.7) – again recalling Chase's results for the same species in the majority of the occupation levels at Combe Grenal (Fig. 7.5). One notable exception is the assemblages from levels 2–5 at Le Regourdou, where remains of red deer are represented overwhelmingly by various postcranial bones and with only sparse representation of heads and teeth (Delpech, in press). This raises the possibility of a significantly different pattern of utilization of red deer carcases at this site which could possibly be related to the peculiar structural or even ritualistic features reported by Bonifay (1964) from these levels. However, Delpech warns that many of the red deer remains at Le Regourdou could just as easily result from carnivore activity in the cave, rather than from the relatively ephemeral episodes of human occupation. To regard the data from Le Regourdou as an exception to the normal pattern of red deer representation in southwestern French sites would almost certainly be premature from the evidence to hand.

4. Remains of species other than those discussed above are too poorly represented in the majority of southwestern French sites to allow any reliable assessment of patterns of bone element frequencies. Laquay (1981), however, has presented some interesting if limited data on remains of roe deer recovered from level 52 at Combe Grenal and layer J4 at Pech de l'Azé IV. According to Laquay each assemblage is characterized by a ratio of postcranial to cranial remains which is much higher than those recorded for either red deer or horse in the same levels, and which is

comparable with the patterns of reindeer remains in the various sites discussed earlier. Similar patterns have been documented, on rather small samples, for the remains of wild boar from two sites – Le Regourdou (layer 2) and Pech de l'Azé IV (layer J3a) (Laquay 1981). It may be significant that both of these are relatively small species, with body weights not too different from those of reindeer (Van den Brink 1967). While it would be premature to draw too many conclusions from such limited samples, it may well be that the carcases of these smaller animals were exploited and processed by Middle Palaeolithic groups in a broadly similar way to those of reindeer.

Carcase utilization patterns

To demonstrate clear patterns in bone element frequencies of the kind documented by Chase at Combe Grenal and at some of the other sites discussed above, is one thing. To formulate clear-cut explanations for such patterns in human behavioural terms is a far more difficult challenge, especially with the very limited data at present available on such questions as the seasonal use of different species, the age and sex distribution of animals, the overall spatial patterning of the faunal remains in the sites and so on. These issues have generated much discussion in the literature leading to a range of divergent interpretations (Binford 1978, 1981, 1984, 1988, 1991; Speth 1983; Chase 1986a, 1988, 1989; Stiner 1991a, b; Klein 1986, 1989b, c; Turner 1989; O'Connell et al. 1988a, b). All that will be attempted here is to focus on some of the most striking and well documented patterns which have emerged from recent studies at Combe Grenal and elsewhere, and to evaluate some of the alternative behavioural models which could possibly account for them. More in-depth analysis of these questions will be possible when some of the current studies of the faunal assemblages from Combe Grenal and

elsewhere have been completed and fully published (see Binford 1992).

Red-deer processing patterns

One of the most consistent patterns documented by Chase (1986a) in the Combe Grenal data relates to the assemblages of red deer remains recorded in both the Würm I and the earlier part of the Würm II levels on the site. These assemblages have three characteristics (Fig. 7.5): (1) a strong representation of heads and teeth; (2) an equally strong representation of lower limb remains; and (3) a general under-representation of remains from the major meat-bearing parts of the main axial skeleton and upper limbs. Interestingly, this pattern has been documented not only in the remains of red deer from both the Würm I and Würm II levels at Combe Grenal but also in the remains of reindeer for the earliest levels of the Würm II sequence (levels 35–26). Similar patterns have also been recorded in red deer remains from several other sites in the same region (Pech de l'Azé II and IV, Le Moustier, Grotte Vaufrey).

In assessing the behavioural significance of these patterns, there are at least three quite separate models which could possibly be advanced to account for the bone element data:

Scavenging
Chase accepts that in many ways the most obvious interpretation for these patterns would be in terms of systematic scavenging of abandoned carcases from either natural animal deaths or carnivore kills. All the recent research on carnivore predation has emphasized that carnivores tend to consume initially the meatiest parts of the carcase (i.e. mainly the axial region and upper parts of main limbs) and to leave behind the least meaty portions represented by the head, lower limbs and feet (Binford 1978, 1981; Blumenschine 1986, 1987; Stiner 1991a, b). It is those parts of the skeleton left over at

carnivore kills which are therefore most frequently available to human scavengers (Fig.
7.14) and it is precisely these elements which
are best represented in the various assemblages of red deer bones from Combe Grenal.

Whilst acknowledging the general
strengths of this argument, Chase has rejected virtually the whole of the scavenging
model for the faunal material from both these
and other occupation levels at Combe Grenal
(Chase 1986a, 1988, 1989). The detailed arguments will be discussed more fully later but
in essence Chase feels that the combination of
other arguments in favour of systematic
hunting throughout the Combe Grenal
sequence can be used to dismiss any simple
scavenging theory to explain these bone frequencies. He argues for example that patterns of cut marks and breakage patterns on
many bones are not consistent with large-
scale utilization of remains from partially
desiccated or frozen carcasses (Fig. 7.11). He
also argues that several other lines of evidence appear to document competent and
systematic hunting of relatively large animals well before the time range of the Combe
Grenal sequence. Although not discussed
specifically by Chase in this context, he could
also have drawn attention to other aspects of
the faunal assemblages to argue against a
simple scavenging hypothesis, including for
example the relatively strong representation
of pelvic bones of red deer (a relatively meaty
bone which one might expect to have been
removed by carnivores well before any scavenging of carcasses by human groups) and the
weak representation of phalanges and other
foot bones (which one would normally
expect to be brought into the site attached to
lower limbs) (Fig. 7.5). Significantly, similar
patterns of bone-element frequency can be
documented for red deer remains in several
of the later Würm II levels at Combe Grenal
(levels 20–25), where there is almost conclusive evidence for systematic hunting in
the exploitation of reindeer (see below). To
account for these particular patterns it would

be necessary to suggest quite different strategies (i.e. scavenging for red deer and hunting
for reindeer) were used for these two species
of deer during precisely the same periods of
occupation.

*On-site butchery and processing of animal
carcases*
The interpretation advanced by Chase himself (1986a, 1988) to account for the patterns
discussed above is related directly to the
processing and utilization of red deer carcases on site. Chase argues in essence that the
Combe Grenal site served principally as a
'primary' hunting and butchering location in
which largely complete carcases of hunted
animals were taken for systematic processing, fairly close to the location of the kills. His
hypothesis is that the majority of the red deer
carcases were dismembered and butchered
on site and most of the prime, meaty parts
removed from the site – perhaps after drying
to aid storage – for use and consumption in
another complementary type of activity or
settlement location. According to this
hypothesis, the animal parts that were abandoned at Combe Grenal consisted of the less
economically useful parts of the carcases (i.e.
mainly heads and lower limbs) which were
processed on site to extract certain delicacies
such as the brains and tongues, together with
marrow from the main marrow-bearing
bones (metapodials, ulnae, tibiae and phalanges). To use his own words:

> In either case it would appear that the meat-
> iest portions of the carcass were being trans-
> ported elsewhere for consumption, while
> the less meaty portions were being con-
> sumed on the spot. This would imply that
> the site served as a hunting camp for red
> deer, and that there existed another site or
> sites whose occupation depended to some
> extent upon the hunting carried out in the
> vicinity of Combe Grenal
>
> (Chase 1986a: 50).

On the face of it, this model is no less
coherent or internally consistent than that of

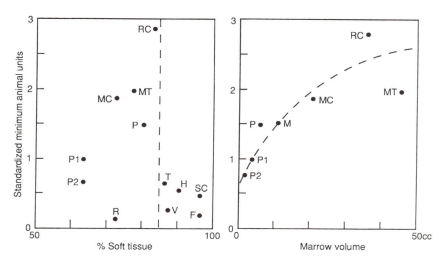

Figure 7.8 *Frequencies of different skeletal elements of red deer recorded in levels 50A-54 at Combe Grenal, compared with the relative amounts of 'soft tissue' (i.e. meat) (left) and volumes of marrow (right) associated with the different bones. After Chase 1986a. The symbols for the different bone elements are: CR cranium; F femur; H humerus; M mandible; MC metacarpal; MT metatarsal; P pelvis; P1 first phalange; P2 second phalange; R radiocubitus; SC scapula; T tibia; V vertebrae.*

the scavenging hypothesis discussed above. When examined in detail, however, it reveals a number of obvious problems. First, it leaves open the question of the precise location and character of the supposedly complementary parts of the settlement and exploitation system of which Combe Grenal, by implication, formed only one specialized part. Where exactly are these other sites which hypothetically should contain all the more meaty, economically useful parts of red deer carcases lacking at Combe Grenal? Chase of course could well dismiss this objection on the grounds that other economically specialized sites could well exist, for example, at open-air locations on the plateaux of the region or at other as yet undiscovered or unexcavated sites. While one can hardly dismiss these arguments, the model does rely heavily on missing data and on the existence of specialized consumption or home-base sites for which direct evidence is, as yet, lacking.

There are other problems with Chase's hypothesis. As he points out (1986a: 47) the theory cannot easily account for the relative abundance of pelvic bones of red deer at Combe Grenal, nor for the apparent under-representation of phalanges and other foot bones – unless this can be attributed entirely to heavy fragmentation and accordingly poor recovery of phalanges. More significantly it fails to explain the striking contrasts between the patterns of red deer bones compared with those of reindeer in the later Würm II levels at Combe Grenal (layers 20–25) (Fig. 7.5). Again, one would need to explain why very different patterns of carcase utilization were applied to two similar species in the same occupation levels. In other words, why should the same groups practise a systematic policy of transporting most prime parts of red deer carcases away from the site, whilst apparently processing and consuming most of the same components of reindeer carcases on the site itself?

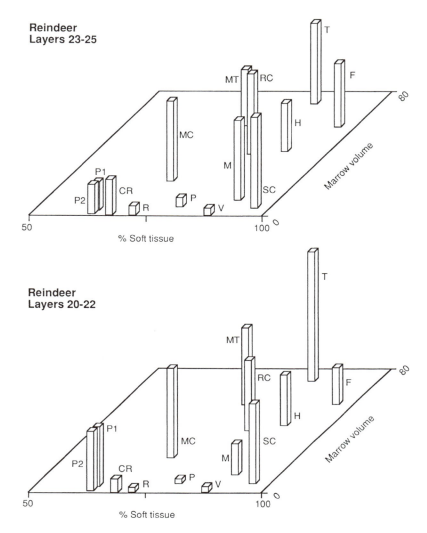

Figure 7.9 *Frequencies of different skeletal elements of reindeer in layers 23–25 (upper) and layers 20–22 (lower) at Combe Grenal, compared with the amounts of soft tissue and marrow volumes associated with the different bones. Abbreviations for the different bone elements follow those in Fig. 7.8. After Chase 1986a.*

Transport-distance effects

The third model could be described by what Perkins and Daly (1968) referred to as the 'schlepp effect'. This asserts that the extent to which particular parts of animal carcases are brought back to occupation or home-base sites depends partly on the distances over which the remains have to be carried, partly on the overall costs (in terms of work and effort) involved in transporting them, and partly on the perceived economic value of different parts of carcases. Where remains of relatively large animals are involved, such as those of large ungulates, one would normally

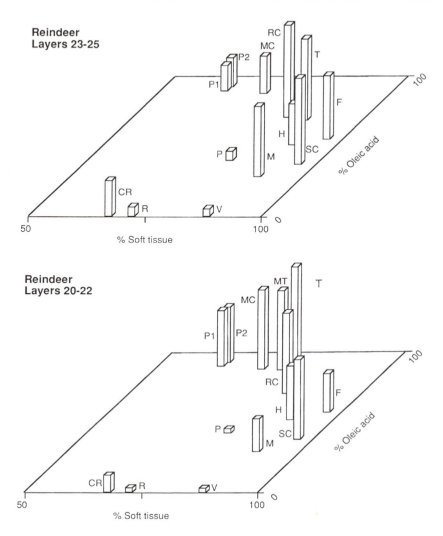

Figure 7.10 *Frequencies of different skeletal elements of reindeer in layers 23–25 and 20–22 at Combe Grenal, compared with the amounts of soft tissue and percentages of oleic acid (as a reflection of fat content) associated with the different bones. Abbreviations for the bone elements follow those in Fig. 7.8. After Chase 1986a.*

expect them to be dismembered and partially butchered at or very close to the point of killing, and only selected, especially valued parts of carcases transported over substantial distances to the main base camps or occupation sites.

This hypothesis could be invoked to account for several aspects of the Combe Grenal assemblages. It assumes that most red deer were killed at some distance from Combe Grenal – perhaps on the various plateaux areas adjacent to the Dordogne valley – and that the particular parts of red deer carcases most frequently transported back to the site were perceived to be of some special value. As Chase (1986a: 46–50) has suggested, this might well be applicable to both head remains and many lower limb bones. The value of heads presumably depended on the use of the brains and tongue – both relatively tasty and fat-rich foods which frequently have been reported in the ethnographic literature as special delicacies among hunting and gathering groups (Speth & Spielmann 1983; Speth 1987, 1990; O'Connell et al. 1988a, b; Stiner 1991a). The same observation could apply to the major marrow-yielding bones, which provide an equally rich and nutritious source of food. In addition, Chase points out that the particular distal limb segments brought back to Combe Grenal seem to be those with the highest concentration of fatty acids and which also retain substantial amounts of marrow throughout the annual cycle of deer growth and development (Figs 7.9, 7.10) (Chase 1986a: 51; Speth and Spielmann 1983; Speth 1987, 1990). Viewed in these terms Combe Grenal would have served primarily as a special food-processing location where these selected parts of the red deer carcasses would be introduced for intensive processing and extraction of the food delicacies – procedures which apparently involved time-consuming breakage of the main marrow bones, similar processing of the crania and mandibles to extract brains and tongue and probably (in the case of heads) cooking or roasting over a substantial hearth (Binford 1984: 160–1; 1992). All these activities might well have involved the use of special processing locations, possibly located some distance away from the actual procurement sites of the game.

This hypothesis could of course be accommodated equally well within either the deliberate hunting or opportunistic scavenging models. In the case of hunting, the implication would be that most of the immediately accessible and edible parts of the carcasses were consumed either on the site of the kills or perhaps at other closely associated butchery or carcase processing locations. If so, this would present an obvious challenge to future archaeological researchers to identify examples of these primary kill and butchery locations. For scavenging, the pattern could be much simpler. In this case, the kind of intensive head and marrow-processing sites documented in these early Würm I levels at Combe Grenal might well represent the *only* substantial occupation or special processing sites potentially visible within the archaeological records of the Middle Palaeolithic.

Reindeer processing patterns

The second distinctive pattern documented by Chase at Combe Grenal relates to the assemblages of reindeer remains in some of the later Würm II levels – notably those from the main sequence of Quina-Mousterian levels in layers 25–21 (Fig. 7.5). Bone-element frequencies of reindeer recorded in these levels contrast in several major respects with those of red deer remains from both the Würm I and Würm II levels of the site. In the reindeer remains from levels 25–20, heads (including both upper and lower jaws) are poorly represented, whilst the main upper limbs (humerus, femur and scapula) are more strongly represented than those in any of the earlier levels in the site. In most respects this pattern is almost the reverse of the red deer bones in showing an increase of principal meat-bearing bones and a reduction of more marginal parts, especially those of the head and lower jaws. The only parts of the main meat-bearing skeleton which are conspicuously under-represented are the ribs and vertebrae. As discussed earlier, however, this may have more to do with poor survival of these particular bones than their original absence from the site.

In this case there seems to be a general concensus shared by both Chase (1986a, 1988) and Binford (1984, 1985, 1991) that the exploitation of reindeer in these later Würm II levels must have involved some component of systematic hunting. Binford has suggested, from a preliminary analysis of the age distribution of the remains, that most of the bones were from young individuals, probably killed shortly after birth or as yearling calves, and probably within their late spring and early summer calving grounds, which could have been located directly in front of the Combe Grenal site. According to this hypothesis the heads would probably have been removed at the site of the kills (to reduce the weight of the carcases) while the rest of the carcase was brought into the site, apparently for consumption on the spot. The overall distribution of the bone elements is regarded by both Chase and Binford as providing effectively conclusive evidence for the deliberate hunting of reindeer in this particular segment of the Combe Grenal sequence.

Bovids and horse processing

Interpretation of the available data on the exploitation of large bovids and horses at Combe Grenal presents a number of problems. Remains of these species are much less abundant in the Combe Grenal sequence than either red deer or reindeer so that we are confronted with relatively small sample sizes, especially for the postcranial parts of the skeleton (Chase 1986a: Tables A48–A51). For both horses and bovids there are additional difficulties in the precise taxonomic identification of the remains. In the case of horse it is difficult for most of the bone elements to distinguish between the remains of the steppe ass (*Asinus hydruntinus*) and the caballine horse (*Equus caballus*), although where this distinction has been made, the data seem invariably to indicate a heavy predominance of the latter in the southwestern French sites (Delpech 1983; Laquay 1981; Guadelli 1987). Similarly with bovids it is difficult to distinguish between remains of the steppe bison (*Bison priscus*) and aurochs (*Bos primigenius*) for the majority of the bone elements (Delpech 1983). For all of Chase's analyses at Combe Grenal, therefore, the remains of these species have been grouped

simply into the broad divisions of 'equids' and 'large bovids'. Inevitably, this raises the question of potential ecological and behavioural distinctions between these species. What is clear, nevertheless, is that these are all relatively large animals (with the exception of the rare steppe ass) with body weights substantially greater than those of either reindeer or red deer (Van den Brink 1967; Delpech 1983). This presents an interesting opportunity to compare how the carcases of these larger species were exploited and processed in comparison with the generally smaller deer forms.

Even allowing for the taxonomic and sampling problems noted above, one feature is immediately clear from the data summarized in Figs 7.6 and 7.7. Remains of horses and bovids almost invariably show a strong predominance of heads and jaws over the rest of the postcranial skeleton. This is clear not only in the data from Combe Grenal, but also in the material from several other sites in the same region – Le Moustier, Gare de Couze, Les Merveilles and Pech de l'Azé II and IV. In all these assemblages, ratios of postcranial bones to combined totals of upper and lower teeth are consistently below 50 percent and frequently as low as 5–10 percent (Fig. 7.7). That this pattern cannot be attributed merely to differential patterns of survival of teeth as opposed to bony remains in the archaeological deposits is clearly demonstrated by the fact that in several cases (as in Combe Grenal levels 25–20 and in the material from Les Merveilles and Gare de Couze) remains of reindeer recovered from precisely the same occupation levels show much higher overall ratios of bones to teeth. The data summarized in Fig. 7.7 leave no doubt that a strong predominance of heads over postcranial remains is a constant feature of large bovid and horse remains in at least the majority of documented Mousterian occupation levels in southwestern France

In broad terms, therefore, the patterns of horse and bovid remains are similar to those

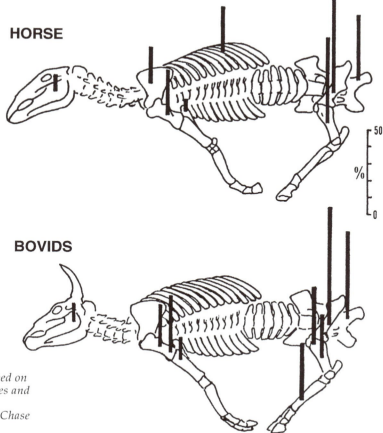

Figure 7.11 *Location and relative frequencies of butchery marks observed on different parts of the skeleton of horses and bovids in the Mousterian faunal assemblages at Combe Grenal. After Chase 1989.*

of red deer discussed above, through with an even stronger emphasis on heads and jaws and a weaker representation (at least in the case of bovids) of extreme lower limbs. The possible explanations for these patterns would also seem to be similar. Binford sees this pattern as strongly indicative of scavenging activities, with particular emphasis on the parts of carcases (especially heads) most frequently available in the leftovers of carnivore kills. Against this, Chase (1986a, 1988, 1989) has argued that the main emphasis was on upper rather than lower parts of limbs of bovids and horse (although this does not emerge clearly from his own data – with the possible exception of levels 23–25 at Combe

Grenal) and that there is no evidence for heavy chopping or hacking through the bones that might be expected to result from the dismemberment of partially desiccated or frozen carcases (Fig. 7.11). Chase also draws attention to Levine's (1983) studies of the age distribution of horse teeth for three different levels at Combe Grenal (Fig. 7.12), which appear to indicate an essentially 'catastrophic' profile in the age distribution of the animals killed (see below) and which is arguably much more consistent with deliberate hunting than with any kind of scavenging activities. These objections have been countered by Binford (1991) who argues that even in the case of scavenged animals, some

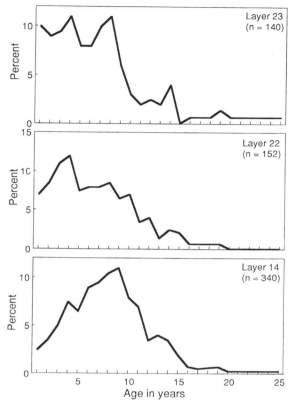

Figure 7.12 *Estimated age distribution of remains of horses in three different stratigraphic levels at Combe Grenal, based on crown-height measurements of the molar teeth. Allowing for the underrepresentation of very fragile teeth from the youngest age classes, the distributions seem to reflect a 'catastrophic' age profile, comparable with that to be expected in a living herd (see Fig. 7.13). After Levine 1983.*

the same levels? And why is there such a conspicuous scarcity of the feet and lower limbs of large bovids which one would normally expect to be abandoned at the primary kill or butchery location as the least useful parts of the carcase? As Chase points out, perhaps the most economical interpretation for the remains of large bovids, at least, would be in terms of the 'schlepp effect' model discussed earlier. In view of the massive differences in the body weights of large bovids and reindeer this may be the most plausible way of accounting for the striking differences in the patterns of bone-representation of these two species – as well as those of horse – in the southwestern French sites. But any such interpretation would leave entirely open the question of whether the basic exploitation of these large species involved systematic hunting or simply opportunistic scavenging of remains from either natural animal deaths or carnivore kill sites.

Head processing patterns

Finally, some interesting details which have emerged from the studies of both Chase and Binford relate to the variable representation of upper and lower jaw remains of different species. From the data assembled by Chase (1986a) it can be seen that whether calculated in terms of numbers of teeth from the upper and lower jaws or from the surviving fragments of maxillary and mandibular bones, there is much variation in these ratios both between different species and in different levels of the Combe Grenal sequence (Figs 7.5, 7.6). In most levels (both at Combe Grenal and elsewhere) there seems to be a strong tendency for lower jaw remains to be represented more frequently than upper jaw remains – sometimes by a ratio of three or four to one (for example in the reindeer remains from levels 11–21 at Combe Grenal and red deer remains from levels 31–35). But in other contexts these ratios may be reversed; thus a predominance of upper over

degree of selection of the more useful parts of the carcase would probably be involved. Similar objections could be raised against the interpretation of the horse and bovid remains in terms of primary butchery locations from which selected parts of carcases were systematically removed for further processing or consumption elsewhere. Why in this case should the pattern of horse remains (notably in levels 25–20 at Combe Grenal) be significantly different from that of reindeer in lower jaws and teeth is apparent in the red deer remains from levels 54–43 at Combe Grenal (Fig. 7.5) and from the contemporaneous levels at Pech de l'Azé II (layer 4B) and at Pech de l'Azé IV (layer J3a). The same can be seen in large bovid remains from layer C at La Ferrassie and levels G and H at Le Moustier. Since this cannot be attributed simply to poor survival or incomplete recovery of the remains during excavation, the pattern must have some significance in economic or behavioural terms.

Recently, Binford (1984, 1991) has argued that these patterns probably reflect deliberate economic strategies by Palaeolithic groups, related to different approaches to processing animal brains and tongues as especially nutritious and favoured food resources. From a study of modern ethnographic sources, he points out that while some groups deliberately separate the mandible from the remainder of the skull at the site of the kills, other groups tend to transport complete heads back to occupation sites for systematic processing there (Binford 1984: 222–3). In some cases complete skulls may be cooked over substantial hearths, leaving distinctive patterns of burning on the lower parts of the mandibles. In other cases the crania may be cooked separately from lower jaws (i.e. with mandibles detached) producing evidence of burning selectively on the maxillary teeth (Binford 1984: 160–1).

Most of these data remain to be systematically analysed for the Mousterian faunas under discussion here. It may be significant, however, that all the samples from Combe Grenal which show a clear pattern of over-representation of upper teeth relate to the assemblages of red deer bones in the Würm I levels of the site. Remains of reindeer, red deer, horses and bovids from all of the documented Würm II levels, by contrast, show an emphasis on teeth from the lower jaws, in some cases with three or four times as many lower as upper teeth. As noted above, the only other samples which show a predom-

inance of teeth from the upper jaws are those of large bovid remains from the principal Ferrassie-Mousterian level at La Ferrassie and from the two MTA levels (layers G and H) at Le Moustier. Thus there could well be a significant element of behavioural patterning in this data, relating either to different industrial variants of the Mousterian or to different chronological stages in the last glacial sequence.

Binford has provided a few interesting additional details. From his own analyses of the Combe Grenal data he has suggested that skulls and mandibles may have been treated differently according to the age of the animals. Thus in the Denticulate Mousterian levels of layers 16–14 the majority of lower jaws of horses seem to come from relatively old (or at least adult) animals, while the upper jaws are mainly from young individuals (Binford 1984: 222–3). He suggests by analogy with the behaviour of recent Nunamuit eskimo that this may indicate a deliberate preference for processing the brains of young animals and the tongues of older and larger animals. More detailed information on these patterns will no doubt be presented in the forthcoming monograph on the Combe Grenal material (see Binford 1992). From the data provided by both Chase and Binford it is clear that there is some fascinating behavioural information still to be derived from some of the more detailed features of the faunal assemblages from Combe Grenal and other sites.

Hunting versus scavenging

The respective roles of hunting versus scavenging – and the criteria by which these can be identified – have largely dominated discussions of Middle Palaeolithic subsistence patterns over the past decade (Binford 1982b, 1984, 1985, 1991; Chase 1986a, 1988, 1989; Klein 1982, 1986, 1989b,c; Stiner 1990, 1991a,b, 1993a). The topic is linked with wider issues of Neanderthal behavioural patterns, involving the general organization and planning involved in subsistence activities and many related issues of the mobility and settlement patterns of the human groups, the levels of coordination and integration between different individuals or social segments of Neanderthal communities and the kinds of technology involved in the procurement and subsequent processing of animal carcases. The issues of hunting and scavenging are equally central to an understanding of the specific ecological niche which Neanderthal groups occupied within the local ecosystems (Gamble 1986; Stiner 1990), and therefore to the levels of population density which could be maintained, permanently and securely, in different environments. These issues are of course by no means confined to the Middle Palaeolithic and continue to dominate discussions on much earlier periods of the Palaeolithic, reaching back to the earliest stages of the Pleistocene (Binford 1981, 1985; Isaac 1984; Potts 1988).

Much of the discussion has centred on the specific criteria by which these strategies can be reliably identified from the surviving faunal assemblages (see for example Binford 1984, 1988, 1991; Klein 1986, 1989b,c; Blumenschine 1986, 1987; Chase 1986a, 1988; Turner 1989; Stiner 1990, 1991a,b). Potentially, several different aspects of faunal evidence can be brought to bear on the issue – including data on the age distribution of exploited animals (Stiner 1990, 1991b), selection of parts of carcasses brought back to sites (Stiner 1991a), evidence for specific patterns of butchery or cut marks on bones (Binford 1984, Blumenschine 1986, 1987) and evidence for varying degrees of specialization in the species exploited. Many of the relevant aspects of the evidence from the southwestern French sites have already been touched on in the earlier discussions. Since the whole question of hunting versus scavenging remains in many ways central to an understanding of behavioural patterns in the Middle Palaeolithic, the most critical data are worth reviewing more closely.

As regards the evidence from southwestern France, it seems universally agreed that systematic hunting must be accepted at least for the remains of reindeer recovered from the majority of the Würm II levels at Combe Grenal. This point is now accepted by both Chase (1986a, 1988) and Binford (1991), and seems to be clearly reflected in patterns of bone-element frequencies for this species – with a strong emphasis on the main meat-bearing bones and an underrepresentation of heads and terminal limbs. Binford has suggested that the hunting was targeted principally on young and probably female reindeer, probably within their late spring and early summer calving grounds, closely adjacent to the Combe Grenal site. As he points out, the hunting technology involved need not have been impressive, and may have amounted to little more than that involved in the clubbing of young seals.

Binford's views on the exploitation of red deer have yet to be spelled out in detail, although from a few scattered comments in the literature he would appear to opt for predominantly scavenging for the majority of the red deer bone assemblages at Combe Grenal. Chase (1986a, 1988), by contrast, has argued specifically for the hunting of this species. The major debate at present hinges on patterns of exploitation of the largest herbivore species – the large bovids (either *Bos primigenius*, or *Bison priscus*) and horse (*Equus caballus*). In this case we are confronted by two sharply opposed alternatives: whereas Binford (1982b, 1984, 1991) has argued strongly for purely scavenging of these species, Chase (1986a, 1988, 1989) has argued equally forcefully for deliberate, systematic hunting. The respective arguments may be summarized as follows:

1. In at least some samples of horse remains at Combe Grenal, upper meat-bearing limbs are more strongly represented than lower limbs. Chase sees this as a powerful argument against simple scavenging of residues

from left-overs of carnivore kills. In fact, the samples on which these generalizations are based are all rather poor and the patterns described by Chase perhaps not as clear cut as he has implied (Figs 7.6, 7.27). Nevertheless, the general validity of this observation has apparently been endorsed by Binford (1991) based on his own analyses of the Combe Grenal material and receives some support from the composition of faunal assemblages recovered from other sites in the southwestern French region (for example, large bovids in layers G and H at Le Moustier and horses from layer I2 at Pech de l'Azé IV and the Gare de Couze: Laquay 1990).

2. The second argument relates to the clear patterns of butchery marks which can be documented on many of the major limb bones of horses and large bovids at Combe Grenal. Two points in particular have been emphasized by Chase (1986a, 1988, 1989): first, the lack of evidence of heavy chopping or hacking through the limb bones, which he argues would be predicted for the scavenging of residual meat supplies from remains of partially desiccated or frozen carcases; and second, the frequent occurrence of cut marks and incisions concentrated around either the main joints or along the shafts of long bones, which point to the dismemberment and removal of meat from carcases while the bodies were still in a relatively fresh condition (Fig. 7.11). In support of these arguments, Chase quotes partly from recent studies of modern carnivore predation patterns reported by Blumenschine (1986, 1987) and partly from similar arguments advanced by Binford himself (1984) in his analysis of the faunal remains from the Middle Stone Age site of Klasies River Mouth in South Africa. How far these arguments can be seen as providing a categorical case for hunting is perhaps more debatable. Presumably if human groups were able to gain access to animal carcases shortly after death, there is no reason why the butchery strategies

employed in dismembering them and removing any remaining meat should differ significantly from those employed in the processing of deliberately hunted animals. The data cited by Chase are therefore suggestive, but hardly conclusive.

3. Possibly a more forceful argument in favour of the hunting hypothesis is provided by recent studies of age-mortality profiles of horses and large bovid remains from French Middle Palaeolithic sites. From a study of the crown-height profiles of horse teeth recovered from three separate levels at Combe Grenal (layers 14, 22 and 23), Levine (1983) has argued that these appear to reflect an essentially catastrophic age profile, analogous to that expected in the natural age composition of a living herd (Figs 7.12, 7.13). Jaubert and Brugal (1990; also Slott-Moller 1990) have recently advanced the same interpretation for aurochs remains recovered from the open-air site of La Borde, in the adjacent Department of the Lot, and for bison remains from the site of Mauran in the Haute Garonne (Farizy et al. 1994) (Fig. 7.24). As both Levine (1983) and Chase (1988, 1989) have pointed out, none of these age profiles conforms to the patterns which would be predicted for simple scavenging of material either from natural death carcases or the remains of carnivore kills. The latter would normally be expected to show a preponderance of either the youngest or oldest age classes (i.e. the most vulnerable components in the herds) or alternatively a dominance of larger and more mature animals, whose remains would be most likely to survive the effects of predation by carnivores and therefore remain available for the longest periods to human scavengers (Chase 1988). Both Chase and Levine argue that these age profiles would be more consistent with some kind of large-scale, unselective hunting strategies involving, for example, driving or stampeding animals over cliff faces or similar natural obstacles.

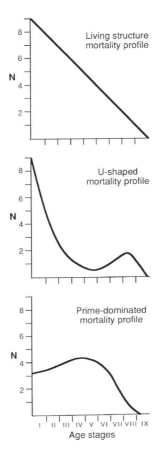

Figure 7.13 *Contrasting age-mortality profiles for exploited animal populations. The 'catastrophic' profile (upper) reflects the age distribution to be expected in a natural, living population. The 'prime-dominated' (lower) and 'attritional' (i.e. young + old dominated) profiles by contrast indicate some highly selective patterns in the in the hunting or other death processes of the animals. After Stiner 1990.*

4. Probably the strongest single argument for the deliberate hunting hypothesis, however, is provided by the composition of the faunal assemblages recovered from several recently explored open-air sites in southwestern France, discussed more fully in the final section of this chapter. From four separate sites (Mauran, La Borde, Coudoulous and Le Roc) we now have evidence of rich faunal assemb-

lages in which remains of either one or other of the two main species of large bovids (i.e. aurochs or bison) account for virtually all the recovered faunal remains – ranging from ca 93 percent at La Borde, to over 98 percent at Le Roc and Mauran (Girard & David 1982; Farizy & David 1992; Jaubert 1984; Jaubert & Brugal 1990; Geneste 1985). Exactly how these extremely specialized faunal assemblages can be accounted for in terms of opportunistic scavenging has yet to be explained. Almost by definition, opportunistic scavenging is likely to sample the full range of dead animals present in particular habitats, presumably in roughly direct proportion to their availability in the local environment. Unless we are to envisage a situation in which virtually all the local animal communities consisted of a single species, or alternatively some extraordinarily selective form of scavenging, arguments in favour of systematic hunting of large bovids in these particular contexts would seem impossible to avoid.

Binford has not yet responded in detail to the various arguments discussed above, beyond brief comments in a recent review (Binford 1991). While apparently accepting Chase's first argument on the over-representation of upper limb bones of bovids and horses at Combe Grenal, Binford argues that no form of scavenging is likely to be entirely random and that it could well involve some selection of the most productive and meaty parts of any carcases available at carnivore kills. He has advanced precisely this argument for the remains of red deer in the late Rissian levels at the Grotte Vaufrey, which show a similar emphasis on the more meat-rich upper limb bones (Binford 1988). He has further suggested that at Grotte Vaufrey there is evidence for differential treatment of young and older red deer – the older animals being represented mainly by remains of heads while the younger, immature deer appear to be represented mainly by selectively scavenged bones from upper limbs and

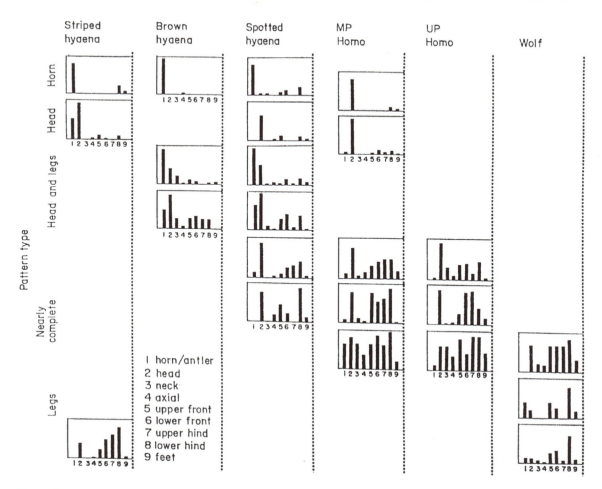

Figure 7.14 *Relative frequencies of different skeletal elements of medium-sized ungulates represented in faunal assemblages accumulated by different types of carnivores, after Stiner 1991a. Stiner argues that 'obligate scavengers' (such as the striped and brown hyena) accumulate mainly heads, or heads and legs, whereas hunters (such as the wolf) accumulate fewer heads and more limb segments. Faunal assemblages from Italian Middle Palaeolithic sites (shown in the fourth column) appear to exhibit both patterns, while those from Upper Palaeolithic sites (in the fifth column) exhibit the typically hunting pattern.*

trunk. He has also contested the cut-mark evidence, arguing that the remains of scavenging from carcasses of recently killed animals need not show significantly different patterns of butchery from those of deliberately hunted animals. Binford's main argument, however, rests on the dramatic and apparently consistent over-representation of

remains from heads and jaws of horses and large bovids which can be documented throughout the whole of the sequence at Combe Grenal, as well as at several other sites in the same region (Fig. 7.7). To him, this overwhelming head-dominated pattern is totally inconsistent with the hypothesis of large-scale, systematic hunting (see Stiner

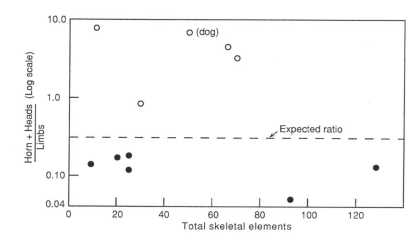

Figure 7.15 *Relative frequencies of horn and head parts to limb bones (expressed by the ratio H+H/L) for medium-sized ungulates, in faunal assemblages accumulated by scavengers (striped and brown hyena – marked by open circles) and hunters (wolves – marked by closed circles). After Stiner 1991a.*

1991a,b, for a systematic analysis of the modern ethnographic and zoological data on this point: see Figures 7.14, 7.15):

> The crucial point however is that the vast majority of horse and bovid remains at Combe Grenal are head parts. For instance, an estimated 132 horses are represented at Combe Grenal (Würm I and Würm II only) by head parts alone, whereas for appendicular skeletal parts the best one can do is 8.5 horses for the same levels ... I know of no hunting people who abandon muscle-bearing parts in favour of heads with such astonishing regularity!
>
> (Binford 1991: 113).

Against this one could cite the arguments for the Schlepp-effect factor discussed earlier, which for these very large animals is likely to have had a major effect on the butchering and transportation of carcasses. The remaining arguments advanced by Chase on the age structure of horse remains at Combe Grenal and the remarkable specialization in the exploitation of large bovids documented at Mauran, La Borde, Le Roc and other sites, have not yet been answered directly by Binford. No doubt these issues will be addressed fully and forcefully in the forthcoming monograph on the Combe Grenal material.

How one can reconcile the various arguments for and against hunting or scavenging is still not clear. Chase accepts (1986a) that the strongly head-and-lower-limb dominated pattern documented for red deer remains and several other species in the southwestern French sites could provide a reasonable *a priori* argument for scavenging. At the same time the character and composition of the highly specialized faunal assemblages from Mauran, Le Roc, La Borde and other sites (reflecting an almost exclusive exploitation of large bovids) seems to provide powerful if not conclusive evidence for deliberate hunting of these species in these particular sites. The same could be said, with slightly more reservations, for Levine's (1983) studies of the age profile of horse remains from Combe Grenal (Fig. 7.12), and for the similar studies of bovid remains from La Borde and Mauran (Fig. 7.24) (Slott-Moller 1990; Jaubert & Brugal 1990; Farizy *et al.* 1994). What is common ground to both Chase and Binford is that by at least the middle stages of the Würm II succession, the deliberate killing of local reindeer herds must be regarded as a well documented feature of the faunal record. In other words, it is now effectively agreed that in at least certain contexts

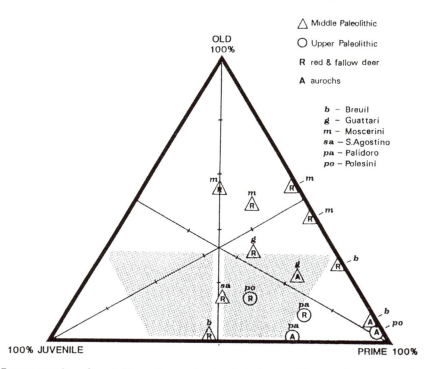

Figure 7.16 *Representation of mortality patterns of animals in terms of three main age groupings (juveniles, prime-aged and old) expressed as triangular scattergram plots, after Stiner 1990. Stiner argues that typical hunters (as represented by north American Indian groups) exploit mainly prime-aged animals, whereas scavengers tend to accumulate higher frequencies of old animals. The distributions show that whereas Middle Palaeolithic faunal assemblages from four sites in western Italy (Grotta Breuil, Guattari, Moscerini and San Agostino) seem to reflect both kinds of exploitation pattern, the assemblages from two Upper Palaeolithic sites (Palidoro and Polesini) reflect characteristic prime-dominated hunting patterns.*

Middle Palaeolithic groups were competent in both their behaviour and technology to secure certain species of game by means of deliberate hunting strategies. The same conclusion has been drawn recently by Mary Stiner (1990, 1991a,b) from her studies of a series of Mousterian faunal assemblages from west-central Italy, where she believes one can demonstrate a variety of exploitation strategies ranging from almost pure scavenging (at the sites of Grotta Guattari and Grotta dei Moscerini) to specialized, focused hunting adaptations, seemingly identical to those

documented in the Upper Palaeolithic sites in the same region (at San Agostino and Grotta Breuil) (Fig. 7.16). As Stiner (1991b) emphasizes, the main issue hinges on the relative scale on which these hunting strategies were practised – i.e. whether as a regular and dominant pattern of animal exploitation, or as a subsidiary employed at certain times and locations to supplement subsistence strategies based mainly on scavenging of animal carcasses. More systematic data on both the age structure of different species in the faunal assemblages and the detailed taphonomic patterns of remains in different sites, is needed to resolve these issues more conclusively.

Hunting strategies

Finally, what evidence do we have for the actual techniques or strategies of hunting in Middle Palaeolithic contexts? The evidence is admittedly limited, but may provide some critical insights into behavioural patterns.

1. First, there is now strong evidence for the use of some forms of heavy-duty thrusting or penetrating spears in certain Middle Palaeolithic contexts, and apparently even from the later stages of the Lower Palaeolithic. The best documented examples are the yew-wood spear with a carefully shaped and fire-hardened tip recovered from the earlier Hoxnian deposits (? ca 350,000 BP) at Clacton on Sea (Essex) (Oakley *et al.* 1977), and the similar specimen recovered from deposits of probably last interglacial age from Lehringen in Germany (Movius 1950; Adam 1951) (Fig. 7.17). The interpretation of these artefacts as hunting weapons has occasionally been questioned but the evidence seems generally persuasive. Clive Gamble's (1988) suggestion that they could represent snow probes for locating the carcases of animals buried beneath deep snow drifts is intriguing and ingenious but perhaps not entirely convincing. As Chase (1988) has pointed out, the fact that the Lehringen spear was found apparently lying beneath the bones of an adult elephant would seem to pose obvious difficulties for this interpretation.

The arguments for and against interpreting various forms of Mousterian points as hafted tips of spears or other hunting weapons have been discussed in Chapter 4. Clearly, not all these pointed forms can be interpreted with any confidence as weapon tips, and many probably served simply as hand-held or hafted butchery or skinning knives. Nevertheless, the evidence for characteristic impact

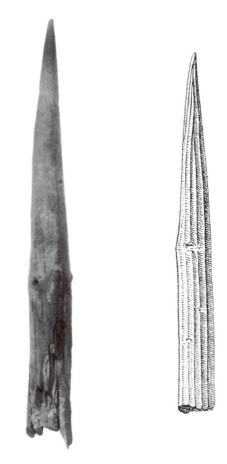

Figure 7.17 *Points of wooden spears recovered from the Hoxnian interglacial deposits (?ca 350,000 BP) at Clacton on Sea, England (left), and last interglacial deposits (ca 110–130,000 BP) at Lehringen, northern Germany (right). The former spear is of yew, while the latter is of unknown wood. The Lehringen spear is said to have had an original length of 2.4 m, and was found associated with the skeleton of a straight-tusked elephant. After Oakley* et al. *1977 and Jacob-Friesen 1959.*

damage apparent on the tips of several typical Mousterian points from the penultimate glacial age occupation levels at the Cotte de St Brelade (see Fig. 4.17) appears to support the weapon-tip interpretation (Callow 1986a). Similar arguments have been ad-

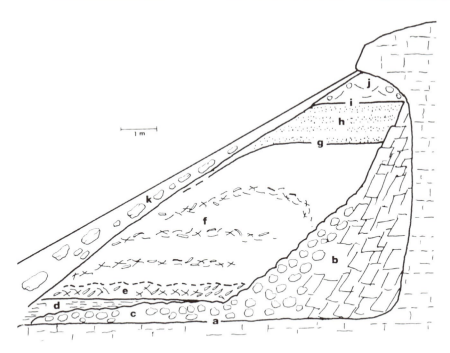

Figure 7.18 *Schematic section of the Mousterian deposits at La Quina (Charente), showing the position of the bone accumulations (in layers f and e) at the foot of a limestone cliff, which may have served as a cliff-fall hunting location. After Jelinek* et al. *1988.*

vanced on the basis of micro-wear studies for many of the classic Levallois point and related forms from Middle Palaeolithic sites in the Middle East (Shea 1989; but see Holdoway 1989 for a different view). It is also worth recalling in this context the early discovery of an apparently typical Levallois point lying amongst the bones of a mammoth skeleton at Ealing in the Thames Valley (Wymer 1968: 261).

Even if the existence of these spears provides clear evidence for the deliberate killing of game in certain Middle Palaeolithic and even earlier contexts, this in itself tells us little about the actual techniques or strategies of hunting. Binford has argued that thrusting spears are only viable in very close-range encounters with animals and could well have been used for the final despatching of ani-

mals that were already in some way disadvantaged – such as sick or injured animals, or those just born or partially disabled by carnivore attacks. Use of short-range thrusting spears need not imply hunting on any very enterprising scale, and as yet there is no convincing evidence for use of throwing spears or other long-range weaponry in Middle Palaeolithic contexts (despite the occasional, speculative claims for the use of bolas slings or similar devices: Henri-Martin 1923). On balance it is more likely that these long-range projectiles developed much later in the Palaeolithic sequence – perhaps not until the appearance of spear throwers and associated barbed bone and antler points in the later stages of the Upper Palaeolithic sequence (Sonneville-Bordes 1960; Peterkin 1993).

2. Some of the most persuasive evidence for the actual strategies of hunting have come from a number of sites which appear to reflect some form of collective hunting techniques, involving the use of cliff-fall locations. At the site of La Quina, for example, excavations by Henri Martin (recently resumed by Jelinek and Debénath) revealed a massive accumulation of bones of horses, large bovids and reindeer extending over a distance of at least 100 metres along the base of a steep cliff, which represents the only such topographic feature within several kilometres of the site (Fig. 7.18). Jelinek *et al.* (1988) have argued that this could have acted as a typical 'jump' or cliff-fall hunting location where individual animals or small herds were driven over the cliff face and subsequently dismembered and butchered effectively on the kill site (see also Chase 1989). At the site of Mauran in the Pyrenean foothills a massive accumulation of bison bones again occurs in close proximity to a steep riverside escarpment (Fig. 7.20) (Girard *et al.* 1975; Girard & David 1982; Farizy & David 1992; see below), and at Puycelsi in the Tarn a large accumulation of remains of horse, bison and other species occurs in a similar cliff-side location (Tavoso 1987). More reliable interpretation of all these sites will be possible when the current studies of the composition and taphonomy of the bone assemblages have been completed and published.

Probably the most convincing case for deliberate cliff-fall hunting techniques has been advanced by Kate Scott (1980, 1986, 1989) from material recovered from two of the later 'Rissian' (isotope stage 6) levels at Cotte de St Brelade (Jersey). Here, Scott was able to document remains of at least twenty individual mammoths and five woolly rhinos, all concentrated in two separate levels in the cave filling (Fig. 7.19) lying immediately at the foot of the steep rock face which drops sharply into the cave from the headland above. The clear patterns of cut marks on the bones and the organized way in which the

remains had been piled into heaps (Fig. 7.19) leaves no doubt that the remains reached this position by deliberate human action. As Scott points out, the presence of at least seven complete mammoth skulls (each weighing 300–400 kg) makes it highly unlikely that the remains were deliberately carried into the cave from kill and butchery locations elsewhere. The strongest argument for the deliberate driving of the game into the cavern rather than accidental falls over the cliff face is that the remains were restricted entirely to these two specific points in the stratigraphic sequence and were totally lacking at other points in the succession (Scott 1986, 1989). From this and other evidence, arguments for deliberate fall-hunting strategies in this particular site would seem to be highly persuasive, if not virtually conclusive.

3. Finally, some further insights into the character and organization of hunting activities may be provided by studies of the age-structure of particular species in Middle Palaeolithic sites. As noted earlier, Levine's studies (1983) of the age profiles of horse remains from three separate levels at Combe Grenal appear to indicate an essentially 'catastrophic' pattern similar to that in a living herd (Figs 7.12, 7.13). As she points out, these age profiles would conform best with some kind of unselective mass-killing strategies, such as the deliberate driving of family groups or small herds over cliff faces or similar obstacles, or unselective ambush hunting of herds during seasonal migrations. Similar patterns are seen in Slott Moller's (1990) study of the age distribution of aurochs remains from the site of La Borde and David's analysis of the bison remains from Mauran (Fig. 7.24) (Jaubert & Brugal 1990; David & Farizy 1994). Similar arguments have recently been advanced by Mary Stiner (1990, 1991b) for the age profiles of red deer and aurochs remains from a number of later Mousterian sites (notably San Agostino and Grotta Breuil) in west-central Italy (Fig. 7.16).

Layer 3

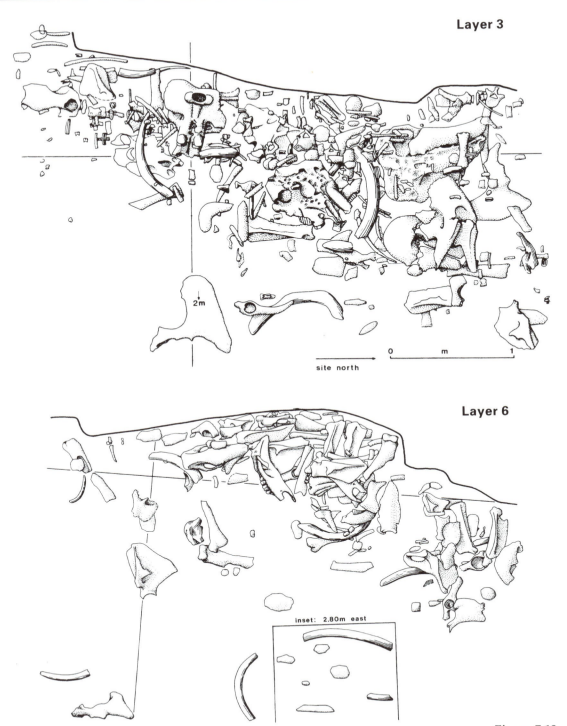

site north

Layer 6

inset: 2.80m east

Figure 7.19

None of these patterns can provide more than a hint of the actual hunting strategies involved but nevertheless suggest that in some contexts the hunting of game by Middle Palaeolithic groups may have been carried out to some degree on an organized and socially cooperative basis.

A very different pattern of age structure has been postulated by Binford for reindeer remains in several of the later Würm II levels at Combe Grenal – notably those from the main sequence of Quina Mousterian levels in layers 21–25. As noted above, initial studies by Binford of the cranial and postcranial remains suggest that the majority of these animals seem to have been killed either as very young calves, shortly after birth, or as yearlings at the end of the first year of life. This is taken by Binford to suggest systematic culling of the young deer, together with the associated females, probably within their late spring and early summer calving grounds, very close to the Combe Grenal site. As he points out, the killing of young deer under these conditions could have been a remarkably easy activity. To Binford this is precisely the kind of strategy that one might anticipate in the earliest stages of the development of deliberate hunting and would have involved very little monitoring or prediction of herd movement, or any other systematic strategies in the pursuit and killing of game.

Specialized bovid-hunting sites

All the sites so far discussed in this chapter have certain features in common. Almost all are located in the central or northern parts of the Perigord region, mostly within or closely adjacent to the valleys of the Dordogne or the Vézère. All the sites are typical cave or rock shelter locations, situated within easy access of major river valleys (Fig. 8.1). And the faunal assemblages recovered from the great majority show a relatively generalized pattern of species representation, comprising substantial frequencies of at least two or more herbivorous species (Figs 7.1, 7.3). Granted these similarities in the location and character of the sites it is perhaps not surprising that we can detect some general regularities in the patterns of skeletal element frequencies at the different sites.

A sharply contrasting pattern has been documented over the past few years in a number of sites which lie, with one exception, some way to the south of the Perigord region, in the areas between the Lot valley and the northern foothills of the Pyrenees – notably the sites of La Borde and Coudoulous in the Department of Lot, Mauran in the Haute-Garonne Department of the Petites-Pyrenées and, within the Dordogne itself, Le Roc in the Euche valley (Jaubert *et al.* 1990; Jaubert 1984; Girard *et al.* 1975; Farizy & Leclerc 1981; Farizy & David 1992; Farizy *et al.* 1994; Geneste 1985). The remarkable feature which links all of these sites is the highly specialized character of the associated faunal assemblage, focused almost exclusively on the exploitation of one of the two main species of large bovids – i.e. either steppe bison (*Bison priscus*) or aurochs (*Bos primigenius*) – with frequencies between ca 93 and 100 percent of the total faunal assemblages in the different sites. Interestingly, the sites share other features: all are either fully open-air locations (as at Mauran and Le Roc) or essentially 'aven' or collapsed cavern localities (as at La Borde and Coudoulous) which can probably be regarded as effectively open sites at the time of the human occupations; all have yielded unusually high densities of faunal

Figure 7.19 *(facing page)   Accumulations of mammoth and rhinoceros bones encountered in levels 3 and 6 of the penultimate-glacial age deposits at the Cotte de St Brelade (Jersey). The positioning of the bones, and the highly selective skeletal elements represented in the two bone heaps, indicate a deliberate human accumulation, rather than a natural faunal accumulation in the cave. After Scott 1986.*

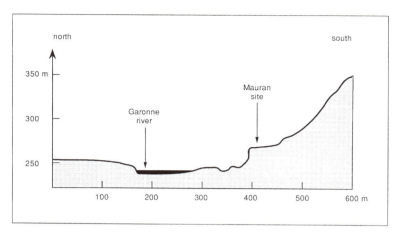

Figure 7.20 *Location of the bison-butchery site of Mauran (Haute-Garonne), showing its relationship to the Garonne river, and the adjacent steep slope immediately to the south. After Farizy & Leclerq 1981.*

remains in relation to associated lithic artefacts; and the character of the lithic industries recovered from the sites is similar, with a dominance of notched and denticulated tools, and variable amounts of more heavy-duty choppers or pebble-tool forms (Figs 7.28–7.31). The most enigmatic aspect of the sites hinges on their chronology. As discussed further below, the sites could span various points of the Middle Palaeolithic sequence from the penultimate glaciation to the final stages of the Mousterian. Before turning to the detailed character of the faunal assemblages recovered from the sites, it may be useful to sketch in the locations and archaeological associations in more detail.

The site of Mauran lies almost 200 km to the south of the Perigord region, adjacent to the left bank of the Garonne river in the northernmost foothills of the Pyrenees. Excavations carried out by Catherine Farizy between 1974 and 1981 revealed an enormous concentration of bison remains (cf. *Bison priscus*: see Fig. 7.21) contained in a deposit approximately 30–40 cm thick which extends over a total area of at least 800–1000 square metres (Girard *et al.* 1975; Girard & David 1982; Farizy & Leclerc 1981; Farizy & David

1992; Farizy *et al.* 1994). The location is particularly interesting (Fig. 7.20). The site stands on the edge of a level escarpment or terrace which dominates the adjacent valley of the Garonne (at an absolute altitude of ca 270 metres) and commands sweeping views to the north, west and south. The flat, terrace-like feature contains a major depression which probably contained a small lake or pond at the time of the human occupation. The lithic industry from the site is much less abundant than the faunal assemblage but consists predominantly of simple notched and denticulated forms, together with a few typical racloirs, accompanied by much larger pebble or chopper tools, manufactured predominantly from local pebbles of quartzite derived from the adjacent gravels of the Garonne (Figs 7.28, 7.30) (Farizy *et al.* 1994). The age of the site remains controversial: while geological indications suggest a date during the earlier phases of the last glaciation (probably during isotope stage 5) recently secured ESR dates obtained directly on samples of bison teeth from the site point to an age of around 35–45,000 BP (Grün *et al.* 1994).

The site of La Borde lies much closer to the Perigord region, approximately 15 km to

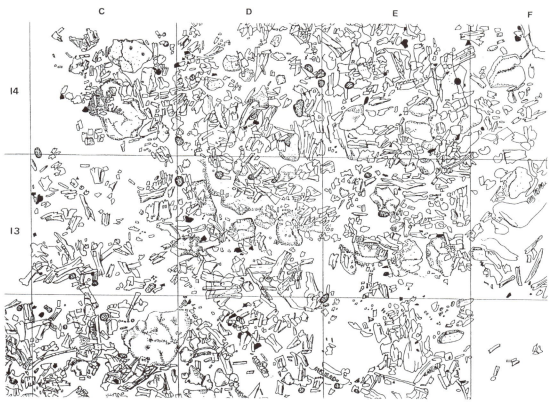

Figure 7.21 *Detail of the bison-bone accumulation at Mauran, showing the highly fragmented nature of the remains and the high frequency of teeth. Lithic artefacts are shown in black. After Farizy & David 1992.*

the north of the Lot valley and only 40 km to the south of the Dordogne (Jaubert *et al.* 1990). The site appears to be a typical aven or collapsed cavern feature situated on a high (ca 320 metres) plateau with no immediately adjacent rivers or other major valley features. The closest water course is that of the Célé stream, 3 km to the south (Fig. 7.22). The excavations on the site were restricted to a small area and were carried out essentially as a salvage operation following the destruction of the major part of the site by building operations in 1971. By systematically sieving large quantities of the disturbed deposits, however, it was possible to recover a rich

assemblage of both fauna and lithic artefacts, all apparently derived from a single stratigraphic level ca 55 cm thick at a depth of approximately 2.5 metres below the original surface of the karstic depression (Fig. 7.23). The area of the deposits is impossible to estimate precisely but is likely to have been at least 100 square metres. The faunal assemblage in this case consists almost entirely of aurochs remains, which account for ca 93 percent of the total faunal material (Slott-Moller 1990; Jaubert & Brugal 1990). The associated lithic assemblage is manufactured overwhelmingly from local quartz pebbles with a small component of imported flint and

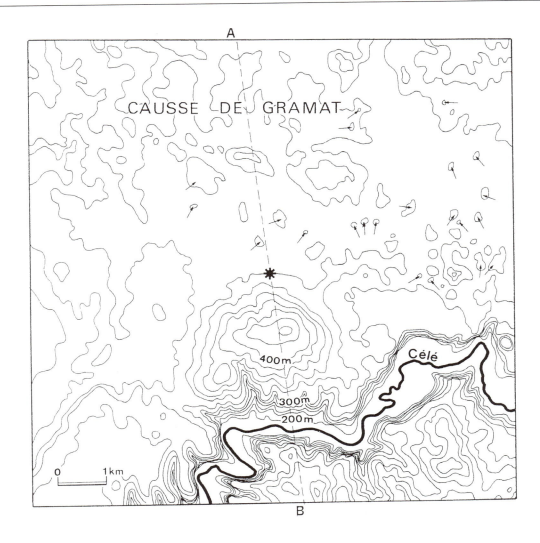

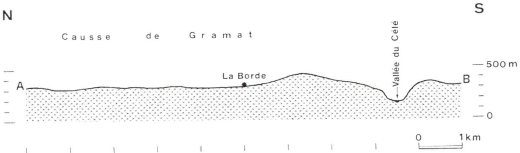

Figure 7.22 *Location of the La Borde site (Lot) occupying a relatively high location in the 'Causses' region of southwestern France, with no immediately adjacent water courses. After Jaubert et al. 1990.*

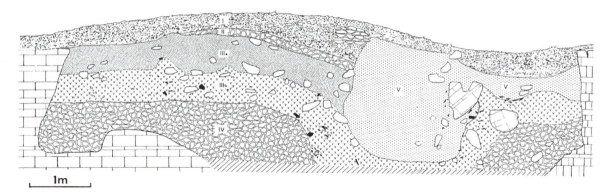

Figure 7.23   *Cross-section through the collapsed cavern ('aven') deposits at La Borde. Level IIIb contained the main concentration of aurochs (*Bos primigenius*) remains, and associated stone industry. After Jaubert et al. 1990.*

consists of the same range of forms as those documented at Mauran – i.e. a clear predominance of notched and denticulated tools, combined with large numbers of heavily flaked pebble and chopping tools (Jaubert 1990; Jaubert & Turq 1990) (Figs 7.29, 7.31). The age of the site is again controversial but on climatological grounds is thought to coincide with either the earliest stages of the last glaciation (isotope stage 5a or 5c?) or perhaps some point during the preceding interglacial (Laville 1990). All the geological indications suggest that the site dates from a temperate period – as indeed does the character of the fauna.

The site of Coudoulous (near Tour de Faure, Lot) parallels in most respects La Borde (Jaubert 1984; Jaubert & Turq 1990). The site again appears to be a collapsed cavern located only 15 km to the southwest of La Borde, close to the confluence between the valleys of the Célé and the Lot. It occupies the upper part of a steep limestone escarpment and commands wide-ranging views over the adjacent valleys. The major contrast with La Borde is that the overwhelmingly dominant faunal species (as at Mauran) is bison rather than aurochs – which accounts for over 95

percent of the total faunal remains (Jaubert & Brugal 1990). According to Jaubert (1984; Jaubert & Turq 1990) the composition of the associated artefacts is almost identical to that of La Borde and Mauran: the industry is manufactured once again overwhelmingly from local quartz pebbles and is made up from the same combination of crudely worked notches, denticulates and heavy-duty chopping tools as that at Mauran and La Borde. The age of the site remains unknown but has been attributed by Bonifay (on the basis of rather tenuous geological evidence) to one of the later interstadial episodes of the penultimate glaciation.

The final site of Le Roc is the only site which lies within the Perigord region. This is a fully open-air location situated on a small terrace along the southern flank of the Euche valley (see Fig. 5.13) approximately 4 km above its confluence with the Dronne (Geneste 1985: 94–8, 385–92). Few details of the site have as yet been published and the material was collected mainly in the form of surface exposures supplemented by a few small test trenches. The salient feature, once again, is that virtually all the recovered faunal material consists of large bovid remains, though their precise taxonomic identification (as *Bison* or *Bos*) remains to be established. The data assembled by Geneste shows that these remains were distributed over a relatively large area (at least 400 square metres) and were associated with flint artefacts consisting almost entirely of notched and denticulated forms – though in this case apparently without more heavy-duty chopper or pebble tools. The age of the site is at present unknown.

Faunal patterns and behavioural interpretations

The remarkable character of the four sites discussed above combined with the detail in which the faunal material from two of the sites (La Borde and Mauran) has now been published, provides us with some of the most explicit evidence at present available for a specific behavioural interpretation of Middle Palaeolithic sites in western France. A perceptive analysis of these issues has been provided recently by Jaubert and Brugal (1990) in their general review of the results of the La Borde excavations – although as will be seen my own interpretation of the data differs in some respects from theirs. The salient points are as follows:

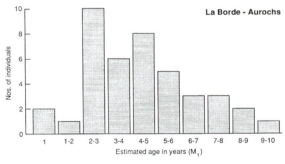

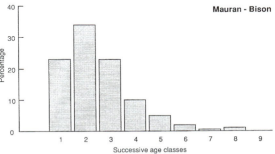

Figure 7.24 *Estimated age distribution of the remains of aurochs from La Borde (upper) and remains of Bison from Mauran (lower), based on crown-height measurements of molar teeth. Allowing for the selective destruction of the youngest and most fragile teeth, both patterns seem to reflect a 'catastophic' age profile, similar to that to be expected in a living herd. After Slott-Moller (1990) and David & Farizy 1994.*

1. There can be no doubt that the composition of the faunal assemblages recovered from the four sites provides a powerful if not conclusive argument for the deliberate hunting of large bovids in these particular contexts, as opposed to opportunistic scavenging. As noted earlier, the extraordinarily specialized character of the faunal assemblages (comprising between 93 and 100 percent of one or other of the two main species of bovid) would seem virtually irreconcilable with a hypothesis of purely scavenging activities. To account for these patterns in terms of scavenging, one would have to assume either that the natural animal populations in these particular locations consisted almost entirely of a single herbivorous species or alternatively that the human groups practised an extraordinarily selective pattern of scavenging the remains of these particular species from the total range of local carnivore kills. Both scenarios seem unlikely. The same conclusion is reinforced by the overall age profiles of the animals documented at two of the sites (La Borde and Mauran), which appear to reflect an unselective, 'catastrophic' profile, similar to that of a living herd (Fig. 7.24) (Slott-Moller 1990: 47; David & Farizy 1994).

Again, this conflicts with the kind of 'attritional' age profile (focused mainly on the youngest and oldest age classes) that would normally be expected from the scavenging of either natural death carcases, or the residues of carnivore kills (Fig. 7.13). Overall, the arguments against a purely scavenging hypothesis to account for the faunal assemblages from these four sites would seem virtually conclusive.

2. Jaubert and Brugal go on to offer some interesting speculations as to the specific methods of hunting large bovids in these particular locations. Their suggestion is that the kind of aven or collapsed-cavern localities documented at La Borde and Coudoulous could have served as natural traps or pit-fall locations into which the animals could have been deliberately driven as part of a systematic cliff-fall hunting strategy (Jaubert & Brugal 1990: 137–8). One of the direct implications of this suggestion of course is that most of the faunal remains from these sites are derived from animals that were killed and butchered on the spot. As discussed below, my own impression is that certain aspects of the composition of the bone assemblages may be rather difficult to reconcile with this hypothesis. Nevertheless the possibility remains that at least some of the carcases represented at La Borde and Coudoulous were derived from this kind of hunting strategy. As discussed earlier, broadly similar cliff-fall hunting strategies could well be envisaged for many other Middle Palaeolithic sites in southwestern France, including both Mauran and La Quina, as well as many of the other cliff-side locations within the various river valleys of the Perigord and adjacent areas.

3. The questions of the size and character of the human groups who occupied the sites and the duration of individual episodes of activity on sites are inevitably much more difficult issues to approach from the perspective of the archaeological data. As Jau-

bert, Brugal, Farizy and others have pointed out, there can be no doubt whatever that all the sites must represent the products of numerous, repeated visits to the same location, probably extending over periods of several decades if not several centuries (Jaubert & Brugal 1990; Girard & Leclerc 1981; Farizy & David 1992). This can be seen in both the stratigraphic depth of the occupation levels recorded in the different sites (ranging from ca 30 to 60 cm of apparently continuous deposition) and the enormous quantities of bone remains and lithic artefacts introduced into the sites. As Jaubert and Brugal (1990: 139) have pointed out, it seems unlikely that remains of over 40 individual aurochs and almost 100 kilograms of lithic remains were introduced into the La Borde deposits during a single episode of occupation! Once this point is accepted, then it becomes almost impossible to make any reliable estimate of the sizes of the human groups present during individual episodes of occupation. Even at Mauran – estimated to extend over at least 100 square metres – it is impossible to assess whether the site represents the activities of relatively large human groups or simply the compounded products of numerous, widely dispersed occupations by very small groups. Based on the evidence for the highly intensive processing of the carcases at Mauran, and the large amounts of food that even a single bison carcase would provide, Farizy has recently suggested that the site may have been visited by groups of perhaps 30 or more individuals (Farizy et al. 1994: 241).

One point which is clear is that many of the individual episodes of occupation must have covered a sufficiently long period to allow a relatively intensive and varied pattern of economic and technological activities to be carried out. Occasional fragments of heavily burned bone, for example, have been recorded from at least three of the sites, leaving no doubt that fires were frequently lit on the sites. And at Mauran and Coudoulous there are reports of at least one clearly defined hearth within the occupation levels (Jaubert 1984; Farizy *et al.* 1994: 235). In addition, there is evidence from all the sites for intensive butchery and fragmentation of animal bones, for the extraction of marrow, and for the *in situ* flaking of cores and other nodules for the production of stone artefacts (Jaubert & Brugal 1990; Farizy & David 1992; Farizy *et al.* 1994). However the data are interpreted, they suggest that the sites represent something more than ephemeral stop-over locations devoted purely to rapid extraction of principal meat-bearing bones from carcases of butchered animals. Assessed in these terms, Jaubert and Brugal suggest that the sites are likely to represent at least several hours of intensive industrial and food-processing activities during individual episodes of occupation. Farizy has envisaged individual episodes of occupation at Mauran of up to a month or more (Farizy *et al.* 1994: 241).

4. Seasonality of exploitation is equally crucial to any assessment of the human utilization of animal populations but the available data for the present sites remain rather sparse. In discussing the age structure of aurochs remains from La Borde, Slott-Moller (1990) suggested that these appear to indicate an essentially continuous seasonal pattern, indicating sporadic killing of animals throughout most seasons of the year. Jaubert and Brugal (1990: 135) have contested this, suggesting that if one focuses on the youngest age classes, there may be evidence for a concentration of mortality patterns during two or three specific seasons – around March–June, September and November–December. There are similar hints of seasonality in the bison remains from Mauran. Here, David & Farizy (1994: 180) have suggested that remains of the youngest teeth may point to the existence of two major age classes, concentrated at around 3–5 and 16–18 months. If this pattern is substantiated by further research it may point to a concentration of human activity on this site principally during the late summer and autumn months.

The seasonal patterns therefore remain rather tenuous. Uncertainties are increased by a lack of information on the general seasonal movements of bison and aurochs populations under the varying glacial and interglacial conditions of the Pleistocene and the exact age and palaeoenvironmental context of the different sites. Nevertheless, as Jaubert and Brugal (1990: 135) point out, it is likely that both these species were to some degree migratory in their habits and it seems fair to assume that their intensive exploitation in these locations was related to major seasonal aggregations or migratory movements of the animal herds (Farizy & David 1994: 180).

5. One of the most critical questions concerns the precise character of the processing applied to the animal carcases within the sites. As noted earlier Jaubert and Brugal and Farizy are inclined to regard the sites essentially as primary kill and butchery locations, devoted mainly to the processing of carcases of animals that were killed effectively on the spot (Jaubert & Brugal 1990; Farizy & David 1992). My own impression is that some of the specific features of the bone assemblages recovered from the sites are difficult to reconcile with this interpretation (see Figs 7.25–7.27). At both La Borde and Mauran (the only sites for which detailed anatomical analyses are available) there is a very strong representation of head and teeth remains combined with a striking under-representation of main axial skeleton remains (i.e. ribs and vertebrae) and – perhaps most significantly – from the pelvic region (Jaubert & Brugal 1990: 138; Farizy & David 1992: 95). Even if the poor representation of ribs and vertebrae could be due in part to the reduced survival prospects of these more delicate bones, the virtual absence of pelvic remains can hardly be explained in these terms. One possibility of course is that some of the missing parts of carcases were deliberately removed from the sites by the human groups, in the form of prime meat-bearing bones for

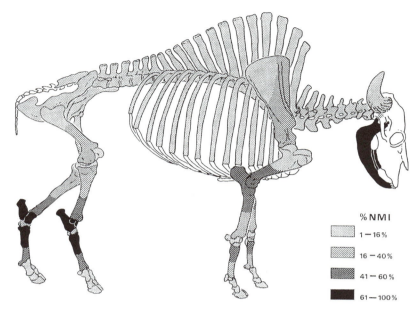

Figure 7.25 *Relative frequencies of the different skeletal elements of bison in the faunal assemblage from Mauran. The percentages express the observed frequencies of the different bone elements compared with those which would be expected if complete carcases had been introduced into the site. After David & Farizy 1994.*

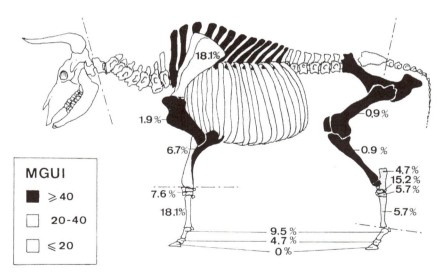

Figure 7.26 *Relative frequencies of different skeletal elements of aurochs (Bos primigenius) in the faunal assemblage from La Borde. The percentages shown against each bone express the actual numbers of the different skeletal elements recovered from the site compared with those which would be expected if complete carcases had been introduced into the site. The shading of the different bone elements indicates the relative economic utility of the different parts of the carcase, expressed in terms of Binford's 'Modified General Utility Index'. After Jaubert & Brugal 1990.*

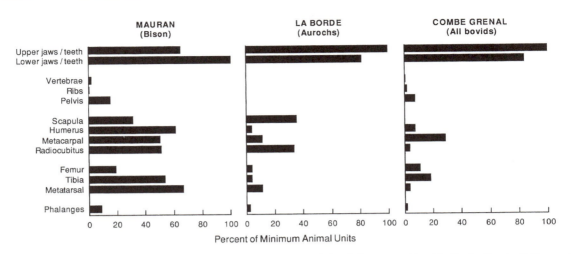

Figure 7.27 *Comparison of relative frequencies of different skeletal elements of bovids (including both* Bos *and* Bison*) represented in the faunal assemblages from Mauran, La Borde, and Combe Grenal (layers 11–35) After David & Farizy 1994; Slott-Moller 1990; Chase 1986a.*

consumption in another location. While this could possibly account for the relative under-representation of some of the upper limb bones (e.g. femur, pelvis, humerus etc.), it is difficult to see how it could explain the dramatic under-representation of other elements such as ribs, vertebrae and certain other parts of the main axial skeleton, which in the case of very large animals such as bison and aurochs one would normally expect to be stripped of any usable meat at the site of the kill and subsequently discarded on site (Fig. 7.27). Conversely, Farizy herself has drawn attention to the fact that some of the much richer, meat-bearing bones, such as the humerus, do appear to be well represented in the Mauran assemblage (Farizy & David 1992: 88). Certain other specific features of the bone assemblages (such as the scarcity of all phalanges at Mauran and the total absence of terminal phalanges at La Borde) would seem equally difficult to reconcile with the notion of these sites as primary kill-site locations, where all the more marginal and less useful parts of carcases would be detached and discarded actually on the sites.

Interestingly, many of the bone-element patterns documented at both La Borde and Mauran show at least broad similarities with those documented for the remains of either *Bos* or *Bison* in the various occupation levels at Combe Grenal (see Fig. 7.27). At all three sites we can see the same strong emphasis on remains of teeth and jaws combined with a variable but generally strong representation of the main lower limbs (i. e. tibia and/or ulna). Remains of ribs and vertebrae are conspicuously scarce in all the sites. The main contrasts with Combe Grenal lie in the much stronger representation of all the main bones (with the exception of the femur) at Mauran, and the relatively high frequencies of the scapula and radiocubitus at La Borde (Jaubert & Brugal 1990: 138; Farizy & David 1992: 95; David & Farizy 1994). However, the overall similarities in the composition of the bone assemblages from Combe Grenal and La Borde in particular are sufficient to suggest that much the same patterns were involved in the introduction and processing of large bovid remains within these sites.

Unless we are to suggest that Combe Grenal was also the location of large-scale, primary butchery of bovid carcases (which seems highly unlikely on several grounds) the implication is that the sites of La Borde and possibly also Mauran were devoted primarily to the intensive processing of certain selected carcase parts of animals that were killed at some distance from the immediate area of the archaeological sites. As at Combe Grenal, the main bones introduced into the sites would appear to be those which yielded either the largest amounts of bone marrow from the main lower limb segments, or rich and highly prized supplies of brains and tongues from heads. As at Combe Grenal, we could see the sites primarily as specialized locations for the extraction and processing of these particular elements, which were deliberately collected from the actual kill sites and transported into the archaeological locations specifically for these processing and extraction activities. This need not imply that the kill locations were at any great distance from the processing sites. On the contrary, the massive accumulations of bone remains recorded at all four sites suggests that the animals were procured fairly close to these locations. But to regard the archaeological sites as lying effectively on the kill sites would seem to be in conflict with several specific aspects of the faunal assemblages.

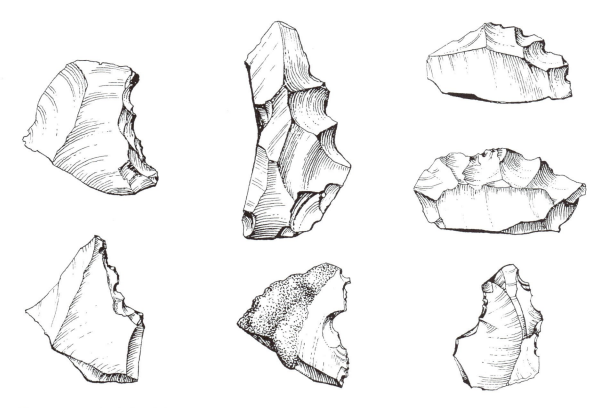

Figure 7.28 *Notched and denticulated tools of flint associated with the bison-bone assemblage from Mauran. After Girard et al 1975.*

Figure 7.29 *Flake tools of flint and quartz associated with the aurochs-bone assemblage from La Borde. After Jaubert* et al. *1990.*

6. The final question relates to the functional orientation of the lithic industries recovered from these sites. As noted above, these were characterized in all the sites by a similar range of artefacts, heavily dominated by simple notched and denticulated forms and (in three of the sites) associated with large numbers of heavy-duty choppers and associated pebble tools (Jaubert & Turq 1990) (Figs 7.28–7.31). We are still poorly equipped with reliable information on the specific, intended functions of any of these tools. The probability remains, nevertheless, that all these forms were associated in these particular sites primarily with the intensive processing of large animal carcasses, involving either the removal of meat from the bones or the more heavy-duty separation of the different bone and limb segments. It is tempting to suggest that some of the larger choppers and pebble tools

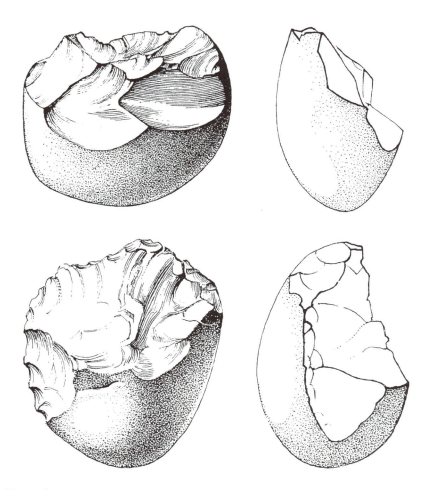

Figure 7.30 *Heavy-duty 'chopping tools' of quartzite associated with the bison-bone assemblage at Mauran. After Girard et al. 1975.*

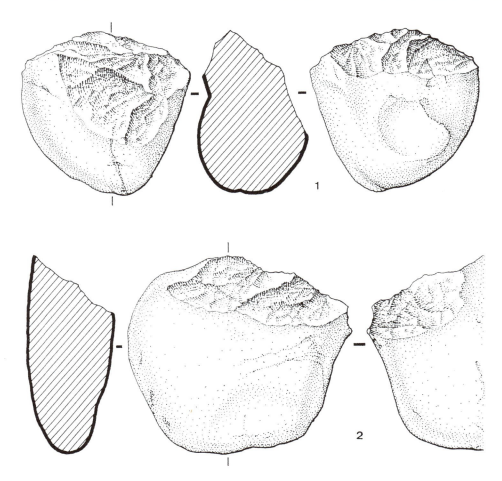

Figure 7.31   *Chopping tools of quartz associated with the aurochs-bone assemblage at La Borde. After Jaubert* et al. *1990.*

would have been particularly well suited to these tasks – especially in the case of massive animals such as bison and aurochs. For the moment this remains largely as speculation. There is the possibility, however, that the curious lithic industries recovered from La Borde, Mauran, Coudoulous and Le Roc provide one of the rare but more convincing cases of clear functional specialization in the composition of Middle Palaeolithic stone-tool assemblages.

CHAPTER 8

Sites in the Landscape _____

The aim of this chapter is to examine the spatial patterning of Middle Palaeolithic sites on a regional basis – i.e. in terms of the overall distribution of sites across the landscape, and in relation to the various environmental, topographic or other factors which seem to have influenced both the specific location of sites and their distribution in different ecological or environmental zones. What patterns if any can we discern in the overall spatial distribution of Middle Palaeolithic sites? What factors seem to have been important in influencing the choice of particular locations? And how were these different locations related to the particular economic, social or technological activities carried out in the sites? Clearly, all of these issues are central to any understanding of both the overall settlement and mobility patterns of the human groups, and the ways in which these patterns may be reflected in the character of the associated lithic assemblages from the different sites (see Chapter 10).

This chapter will again take the evidence from the Perigord and adjacent areas as the central focus of study. For a number of reasons it is convenient to discuss the sites under the broad headings of 'cave and rock-shelter' and 'open-air' sites. While in some cases this can be a slightly arbitrary distinction (notably in the case of some of the rock face sites such as the Abri Brouillaud, Pech de l'Azé IV and Combe Capelle Bas), it forms a convenient framework for studying some of the more

obvious and basic features of the documented site distributions.

Cave and rock-shelter sites

The classic cave and rock-shelter sites have largely dominated research on the Middle Palaeolithic of southwestern France throughout this century. It is only during the past two decades that a systematic attempt has been made to extend this research to open-air sites in the region (Rigaud 1969, 1982; Le Tensorer 1981; Turq 1977a,b, 1978, 1988a, 1989a, 1992b; Geneste 1985). How far this may have distorted our understanding of Middle Palaeolithic settlement and activity patterns will be discussed in the following section. Nevertheless, the various river valleys of the Perigord and immediately adjacent areas have produced a remarkable concentration of cave and rock-shelter sites (Fig. 8.1), many of which contain exceptionally detailed records of human activity spanning long periods of the Middle Palaeolithic succession. As a reflection of highly intensive patterns of occupation of cave and rock-shelter locations, this concentration of sites is unique within Europe.

It is surprising that no systematic study has ever been devoted to the distribution, location and topographic aspects of all Middle Palaeolithic cave and rock-shelter sites in this region – though several partial studies focused on specific areas have been pub-

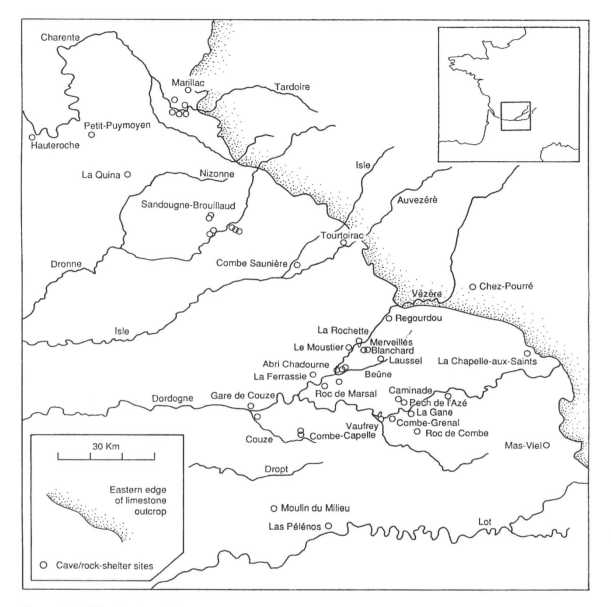

Figure 8.1 *Distribution of the principal Middle Palaeolithic cave and rock-shelter sites in the Perigord and adjacent areas of southwestern France.*

lished (Rigaud 1982; Duchadeau-Kervazo 1982, 1984, 1986; Le Tensorer 1981; Geneste 1985; Jaubert 1984). Any study of these issues encounters widely recognized problems – the varying intensity of site prospecting in each area, the possibility of selective erosion or destruction of occupation deposits in different geological contexts and the inevitability that previous research will have tended to concentrate on the richer and more deeply stratified sites in preference to the smaller and archaeologically poorer ones (cf. White 1985; Rigaud 1982; Geneste 1985). Even allowing for these potential sources of bias in the recorded distributions, there are several fairly clear and well documented patterns which have emerged from these studies which are worth reviewing here.

1. The first and most obvious point is the strong concentration of sites on the various limestone regions which outcrop in a broad band approximately 40–60 km wide around the western and southern foothills of the Massif Central (see Fig. 8.1). Obviously the dominating factor here is the natural distribution of caves and major rock shelters, which are largely confined to these geological outcrops. However, the pattern is not quite as simple as it might at first appear. As shown in Fig. 8.1, sporadic cave or rock-shelter occupations can be traced in certain areas some way to the east of the main limestone outcrops, as for example in some sandstone regions of the eastern Perigord (e.g. Chez-Pourrez in the Corrèze) and extending into the granitic and crystalline outcrops of the Massif Centrale (e.g. Baume Vallée and Rond du Barry in the Haute-Loire). To the west of this zone, only a single well documented site is known – the cave of Pair-non-Pair, located on an isolated outcrop of limestone in the Gironde department, close to the Atlantic coast.

As Turq and others have pointed out, however, some of the critical factors in these distribution patterns may have been related not only to the inherent distribution of cave and rock-shelter sites but also to the availability of lithic raw material supplies. There seems to be a tendency for both cave/rock-shelter and open-air sites to be concentrated mainly on outcrops which provided the most abundant, accessible and high quality flint, such as the extensive Senonian outcrops in

the southern Perigord and some of the outcrops of Maestrichtian and related deposits further to the north-west. Other ecological and environmental factors which may have influenced these distributions have been discussed in Chapter 2 – notably, the potential importance of sheltered, valley habitats as refuges for several species of game during the colder, full-glacial episodes of the Pleistocene and the probable role of these valleys as primary migration routes for species such as reindeer and horse between the higher elevations of the Central Massif and more low-lying areas of the Atlantic Plain (see also Mellars 1985). For many reasons these sheltered, limestone valley regions of southwestern France are likely to have provided an almost ideal combination of environmental and ecological resources of critical importance to human groups throughout the whole of the last glacial succession.

2. When the distribution of cave and rock-shelter sites is examined in relation to the main river valley drainage, some other clear patterns emerge. As seen in Fig. 8.1, the highest densities occur in three major river catchments – those of the Dordogne and its immediate tributaries (c. 15 sites), the Vézère and related tributary valleys (c. 20 sites) and the various valley systems of the Dronne and Charente in the northern Dordogne region and southern Charente (c. 15 sites). Beyond this core zone, sites are distributed more sporadically, for example, in the catchment zone of the Lot river immediately to the south of the Dordogne and in some more northern valleys in the adjacent Departments of the Vienne, Corrèze, Maine-et-Loire and Deux-Sèvres (Fig. 8.1). As discussed further in Chapter 12, these distributions reveal some subtle but interesting contrasts when compared with the distribution of Upper Palaeolithic sites in the same regions (Fig. 8.2), which may indicate significantly different patterns of mobility and economic strategies during the two periods.

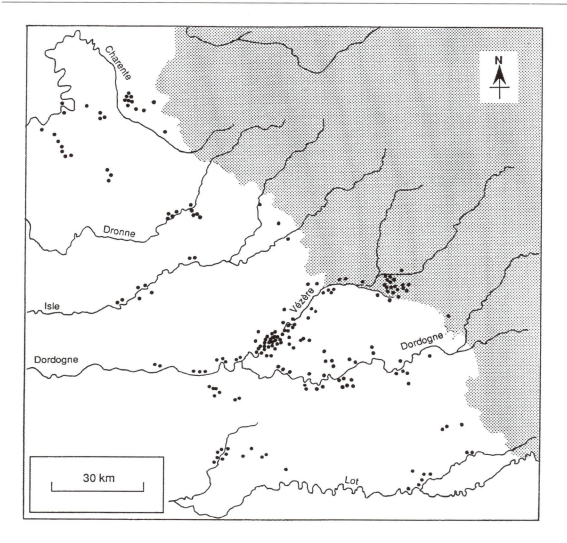

Figure 8.2 *Distribution of Upper Palaeolithic sites in the Perigord and adjacent areas – including both cave/ rock-shelter and open-air sites. After Demars 1982.*

3. If the site distribution is examined in relation to more localized topographic features, other interesting patterns are seen. Several workers have noted that many of the most intensively occupied sites (e.g. Combe Grenal, Pech de l'Azé, Laussel, Combe Capelle) tend to occur not within the main course of the major river valleys (of the Dordogne, Vézère, Lot etc.) but in some of the minor tributary valleys which feed into them. Local concentrations of sites in these locations have been recorded for example in the small valley of the Céou to the south of the Dordogne (Rigaud 1982, 1988), in the valley of the Beune on the north bank of the Vézère, and in the valleys of the Boulou and Rébières, to the

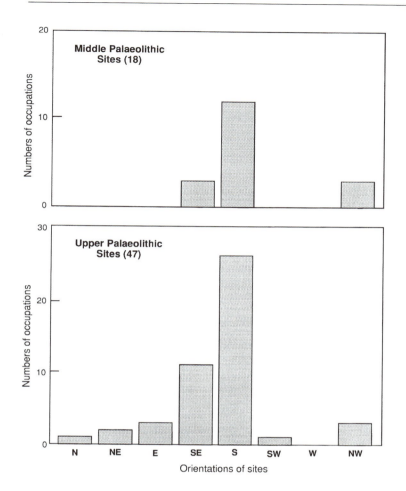

Figure 8.3 *Solar orientation of Middle and Upper Palaeolithic cave and rock-shelter sites in the northern Perigord region, as documented by Duchadeau-Kervazo (1982). In both cases the sites show a strong tendency to be oriented towards the south – presumably to obtain the maximum solar radiation.*

south of the Dronne (Duchadeau-Kervazo 1982, 1986; Geneste 1985) (Figs 8.1, 8.5). It would seem that in many contexts these smaller tributary valleys offered the most attractive habitats for Middle Palaeolithic groups – perhaps because of the better degree of climatic protection provided by these sheltered narrow valleys, or perhaps because of local vegetational or other micro-habitats which would have attracted concentrations of animals or other economic resources to these locations (see Chapter 2 and White 1985). It may be significant, however, that most of these sites are usually located only a few kilometres from the confluence with larger valleys and would therefore have

allowed easy access to the economic resources available within the wider and more exposed floodplains of the major rivers.

A more universal pattern is the tendency for sites to occur predominantly along the south or south-east facing flanks of valleys, in preference to the north or north-west-facing slopes (Fig. 8.3) (Duchadeau-Kervazo 1982, 1986; White 1985). The explanation in this case is almost certainly related to simple climatic factors. South-facing locations inevitably benefit from the maximum exposure to sunlight, and enjoy correspondingly higher temperatures in all seasons of the year. Protection from winds was no doubt an equally important factor, in an area where the prevailing wind direction is mainly from the west, and where the coldest and harshest winds come mainly from the north (White 1985). This would no doubt have been an especially crucial factor during the winter months, when local temperatures during the colder periods of the Upper Pleistocene could well have fallen below −20°C (see Chapter 2). Even today local temperature differences of up to 25°C can occur between the north and south facing slopes of certain river valleys in the Perigord (Duchadeau-Kervazo 1986: 57). Thus it is not surprising that almost all the known cave and rock-shelter sites, from both the Middle and Upper Palaeolithic periods, are located specifically on northern flanks of principal river valleys (see Fig. 8.3; Duchadeau-Kervazo 1986; White 1985). The only clear shift in this settlement pattern occurs during the final stages of the Upper Palaeolithic sequence (late Magdalenian and Azilian), when year-round temperatures improved rapidly during the closing stages of the last glaciation (Duchadeau-Kervazo 1986: 57).

A third, more variable pattern can be seen in the altitudes at which cave and rock-shelter sites occur above the floor of adjacent valleys. Some sites were located very close to the floodplain of valleys – in some cases in locations which were evidently subject to flooding during periods of high river level (for example in the lower shelter at Le Moustier and at Trou de la Chèvre, Sandougne, Les Ourteix and La Quina). More frequently, however, the sites were located well above the river floodplains, beyond the reach of seasonal flooding. Several factors could have been involved in the choice of these higher locations. As Duchadeau-Kervazo pointed out (1982, 1986; see also White 1985) most of these sites would have benefited from longer exposure to direct sunlight than those in the valley-bottom locations (an especially significant factor during the late autumn and winter months) and would have commanded

more extensive views over local valley habitats. As vantage points for observing the distribution and movement of game or the location of other economic resources, these high-level locations would have offered obvious advantages (Fig. 8.4).

There are sporadic sites at even higher elevations, in positions which would have required quite a strenuous climb from the adjacent valley floors. Some of the best documented examples have been recorded in the Céou valley, immediately to the south of the Dordogne – most notably the sites of Grotte Vaufrey and Grotte XVI, which stand at heights of around 100 metres above the adjacent valley floor (Rigaud 1982, 1988). These and similar sites (such as the Roc de Marsal and La Ferrassie, close to the Vézère valley) would probably have been more easily accessible from the neighbouring plateaux than from the valley habitats immediately below (Turq 1988b; Delporte 1984). As Rigaud (1988) pointed out, some of these (most notably the Grotte Vaufrey and Grotte XVI) seem to indicate much more sporadic occupation than sites in the more accessible valley locations.

4. Finally, there can be little doubt that the immediate availability of good-quality flint supplies was a major factor in the location of many if not the majority of documented cave and rock-shelter sites in the Perigord area. As noted earlier, both Turq (1988a, 1989a) and Duchadeau-Kervazo (1982, 1984, 1986) have pointed out that the distribution of Middle Palaeolithic sites seems to mirror in many ways the distribution of good quality flint supplies, while Geneste (1985, 1989a; Geneste & Rigaud 1989) has maintained that the presence of raw material supplies within a distance of at most 2–4 km is an almost invariable feature of the sites studied by him in the central and northern Perigord. In several cases there are examples of cave or rock-shelter sites which are located effectively on the source of high quality flint outcrops, for

Figure 8.4   *View of the Dordogne valley, close to the site of Combe Grenal (facing north).*

example at Combe Capelle in the Couze valley (Peyrony 1943), and at Moulin du Milieu in the Lot (Turq 1989a: 196). Perhaps more significantly, studies by Duchadeau-Kervazo (1982, 1986) and Turq (1988a, 1989a) have suggested that in areas where flint supplies are either scarce or of poor flaking quality, Mousterian sites tend to be more sparsely distributed, for example on the outcrops of Jurassic limestone in the northern Dordogne and on similar outcrops between the valleys of the Dordogne and the Lot (Figs 8.8–8.10). The clarity of these patterns leaves little doubt that the immediate accessibility of suitable raw material was a critical factor influencing the location of the majority of cave and rock-shelter locations in the southwestern French region.

Summary

Reviewing the patterns of site distributions discussed above, two general features are apparent. First, most cave and rock-shelter sites in southwestern France share several basic environmental and topographic features. Most of the sites are located in well sheltered locations, usually in positions offering extensive and wide ranging views over adjacent valley habitats, and almost invariably with easy access to abundant and high quality raw materials. While the immediate economic catchment zones accessible from the sites would have been mainly within the local valley areas, the human groups would have had relatively quick and easy access to a range of very different habitats – either on the local, exposed plateaux bordering the principal river valleys or within the extensive flood-plain zones of the major valleys themselves (notably those of the Dordogne, Vézère, Isle, Dronne and Lot). As Turq (1989a: 196) has emphasized, the dominating feature of most of these site locations would seem to have been the diversity of the ecological and environmental zones which could be exploited, easily and efficiently, from the different locations.

The second conclusion which follows directly from the above is that most if not all the cave and rock-shelter sites would have been ideally located to serve in some sense as 'central places' from which diverse economic and technological activities could be carried out (cf. Stiner 1991a). The sites could be seen in other words as pivotal locations providing not only well sheltered living space but also the maximum range of economic resources which could be exploited within a short radius of movement. The same point has been made by Duchadeau-Kervazo:

> La densité des sites occupés, le plus souvent a plusieurs reprises, et la quantité de vestiges recoltés (silex, faune ...) semblent attester l'existence d'habitats recouvrant de longues périodes. Ils constituent de veritables lieux de ralliements, points fixés autour desquels il est possible de rayonner, contribuant ainsi a une meilleure exploitation du territoire.
>
> (Duchadeau-Kervazo 1984: 48).

As discussed further below, these features may be critical in assessing the precise role played by these sites in the overall mobility and economic strategies of Mousterian groups. To find evidence for more varied and contrasting types of site locations we must turn to the open-air sites in the same region, which may provide evidence for a very different and potentially much more specialized pattern of activities.

Open-air sites

The systematic study of the abundant Middle Palaeolithic open-air sites in southwestern France is still in its infancy. Despite over a century of avid collecting from these sites by amateur and professional archaeologists, it is only during the last twenty years that any serious and controlled attempt has been made to survey and document the sites in any detail (Rigaud 1969, 1982; Le Tensorer 1973, 1981; Turq 1977a,b, 1978, 1988a, 1989a, 1992b; Geneste 1985; Duchadeau-Kervazo 1982, 1986). Despite the recent advances it is clear that our understanding of open-air occupations remains in many ways not only limited but possibly subject to distorting factors. As Rigaud (1982), Geneste (1985) and others have pointed out, the major limitations in the available data are as follows:

1. First, the number of systematic excavations carried out on open-air sites is still very small. There are important exceptions but for the great majority we are dependent on surface collections of material, or at best material recovered from very small scale, sondage excavations designed mainly to document the stratigraphic and geological context of the finds, rather than to study the detailed character and content of the sites (Rigaud 1982; Le Tensorer 1981; Turq 1978; Duchadeau-Kervazo 1982, 1986). Inevitably, such collections are likely to contain material from several periods of activity on the sites, potentially spanning many different stages of the Palaeolithic and even post-Palaeolithic periods. Material recovered from several of the best-known sites, e.g. Le Dau, Plateau Baillard, Plateau Cabrol, La Plane, Metayer etc., demonstrates that many open-air locations were major centres of activity in periods ranging from the Lower Palaeolithic through to the Upper Palaeolithic and even Neolithic periods (Fig. 8.14) (Rigaud 1969, 1982; Le Tensorer 1973, 1981; Turq 1977a, 1978, 1988a; Geneste 1985). Studies based on different states of patination have generally proved totally

inadequate to separate material from these different periods of occupation (Rigaud 1982).

Equally problematical is the highly selective nature of the collecting known to have taken place on these sites in the past. Inevitably, earlier collectors have tended to concentrate on the more visually impressive finds and seem to have had a strong preference for collecting hand axes or more heavily retouched points and side scrapers rather than simpler notched or denticulated forms, let alone cores or unretouched debitage flakes (Rigaud 1982; Duchadeau-Kervazo 1982; Jaubert 1984). Only for the sites which have been surveyed and collected under closely controlled conditions can we be reasonably confident that all material has been recovered in an unbiased and representative way. Needless to say, studies of the quantitative aspects of assemblages can only be made with any confidence for these recently collected series – and even then problems of occupational mixtures and palimpsests can never be ruled out in the absence of extensive, controlled excavations.

2. Studies of the overall distribution of open-air sites are subject to similar limitations. The problems in this case derive partly from the uneven nature of the surveys and prospecting carried out in different regions and partly from the effects of modern vegetation and agriculture in either obscuring or revealing sites (Rigaud 1982; Geneste 1985; Jaubert 1984; Duchadeau-Kervazo 1986; Turq 1989a: 182). Inevitably, areas which are known to contain rich open-air sites will tend to attract the attention of collectors and therefore boost the numbers of known sites in particular regions (for example in the Bergerac region, or on some of the plateaux between the Vézère, Dordogne and Lot valleys). The large areas of the Perigord covered by uncultivated woodland (e.g. Geneste 1985: Fig. 18) makes prospecting effectively impossible in many localities, while the zones of cultivated land tend to be concentrated mainly on the more

level plateaux or river floodplain areas or on the deeper, richer soils (White 1985). In the case of river valleys and slopes there is the additional problem of burial of sites under deep layers of alluvium or slope-wash deposits (Turq 1989a).

3. A third major limitation to the study of open-air sites in functional or economic terms derives from the total lack of organic remains from all but a handful of sites. There are a number of rare and highly important exceptions, discussed in the preceding chapter (Mauran, Le Roc, Puycelsi etc.), but for the great majority of sites any speculations about the activities carried out on the sites, or the seasonality of occupation, must rely almost entirely on the character and composition of the lithic assemblages, assessed in the context of the topographic location and character of the sites.

4. Finally, there is the problem of dating the sites. In the absence of preserved organic material, the only clue to the chronology of the great majority of open-air sites is provided by the character of the associated geological deposits. Despite the excellent work carried out for example by Laville (1975), Le Tensorer (1981), Texier (1982), Debénath (1974), Kervazo (1989) and others, the problems of chronostratigraphic interpretations and correlations based on sedimentological and related evidence are now widely recognized and can rarely provide more than a tentative and frequently highly controversial indication of either the relative or absolute ages of different sites (cf. Mellars 1986a, 1988). At present the majority of open-air sites must be regarded as undated and could belong to almost any period of the Middle Palaeolithic ranging from the penultimate glaciation through to the end of the Mousterian sequence (Turq 1988a: 98).

When allowance is made for these factors it is clear that any generalizations on the character, chronology, or even overall distribution of Middle Palaeolithic open-air sites must be treated with considerable caution. There are nevertheless some apparently clear and well documented patterns which can be observed in the available data, which are crucial to an assessment of patterns of Middle Palaeolithic activity in southwestern France. The following discussion will focus on two major aspects: first, the general distribution of sites; and second, the specific character of these occupations, insofar as this can be assessed from the available lithic assemblages and some specific features of site locations.

distribution of open-air sites

1. Arguably the most striking feature of Middle Palaeolithic open-air sites in southwestern France is their abundance. If a site is defined by any occurrence of one or more artefacts which are recognizably Middle Palaeolithic in character, then the total of recorded open-air sites within the Perigord and immediately adjacent regions must run into several hundred and possibly more than a thousand. Within the northern Dordogne region alone (coinciding essentially with the Dronne valley drainage) Duchadeau-Kervazo (1982, 1986) has recorded over 120 find-spots of apparently Middle Palaeolithic material (Figs 8.5, 8.6), while Rigaud (1982), Le Tensorer (1981), Geneste (1985), Turq (1988a, 1989a, 1992b), Jaubert (1984, 1985) and others have documented similar numbers from areas further south between the valleys of the Isle, Vézère, Dordogne and Lot (Figs 8.10, 8.11). This is an impressive total, which clearly dwarfs the number of Middle Palaeolithic cave and rock-shelter locations in the same areas and greatly exceeds the numbers of open-air finds of Upper Palaeolithic material from the same region (cf. Rigaud 1982; Duchadeau-Kervazo 1982, 1986; White 1985) (e.g. Fig. 8.6).

If attention is focussed on the quantities of material recovered from individual sites, then the patterns become rather less dramatic. Thus Duchadeau-Kervazo (1982:

Tables 88–96) has reported that of a total of 129 separate find-spots of apparently Middle Palaeolithic artefacts recorded in her survey of the Dronne valley drainage, 37 are represented by finds of only single artefacts (principally isolated cordiform hand axes or other distinctive tool forms), while only 40 sites are represented by collections of 10 or more artefacts (Fig. 8.7). Only seven sites (less than 6 percent of the total) yielded more than 100 artefacts. Clearly, these are minimal figures, since it is unlikely that the available surface collections from each location represent more than a fraction of the total artefacts originally present on the sites and of course the numbers of sites waiting to be discovered within the same areas (potentially obscured by woodland or later geological deposits) are unknown. Nevertheless, these figures provide a rather more realistic impression of the relative intensity of utilization of open-air versus cave and rock-shelter locations than might be gleaned from a rapid inspection of either total site numbers or the impressive scatters recorded on some distribution maps (see Figs 8.5, 8.6, 8.11). Although no precise figures are available, it is unlikely that the total number of artefacts (or at least systematically retouched tool forms) recovered from the various Middle Palaeolithic open-air sites within the northern Perigord region runs into more than a few thousand, which is much less than that recorded from the aggregated total of cave and rock-shelter sites in the same region (see Geneste 1985; Duchadeau-Kervazo 1982). Above all, it should be remembered that while the majority of open-air sites probably represent only a single, or at most a few repeated, visits to the same site, the normal pattern in cave and rock-shelter sites is to find a massive palimpsest of repeated occupational episodes superimposed within the same location. Clearly, we shall never be able to estimate the relative intensity of utilization of open-air versus cave and rock-shelter sites with any precision – granted the limitations on potential field survey tech-

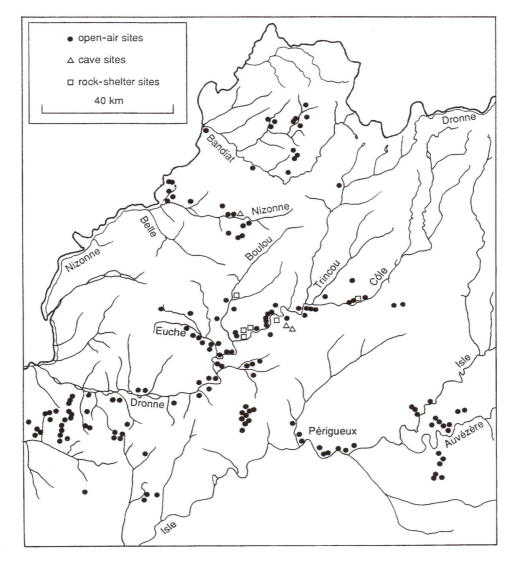

Figure 8.5 *Distribution of Middle Palaeolithic open-air and cave/rock-shelter sites in the northern Perigord region, as documented by Geneste (1985).*

niques, differential patterns of site destruction and the other factors discussed above. But to assume that open-air sites represent the dominant component of Middle Palaeolithic activity patterns in areas such as the Perigord region – equipped with large numbers of naturally sheltered and intensively

utilized cave and rock-shelter sites – would almost certainly be premature with the evidence at present in hand.

2. The overall distribution of Middle Palaeolithic open-air sites shows some obvious similarities to that of cave and rock-shelter sites,

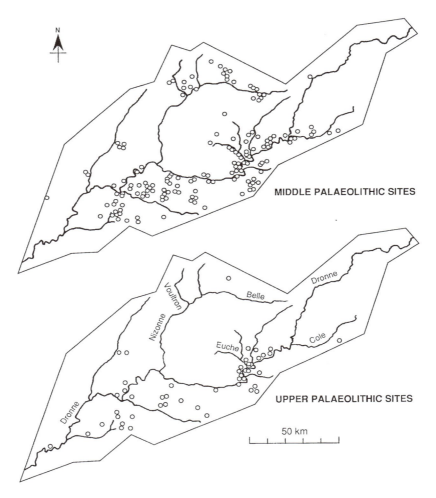

Figure 8.6 *Comparison of the overall distributions of Middle Palaeolithic and Upper Palaeolithic sites in the Dronne valley drainage of the northern Perigord region, as documented by Duchadeau-Kervazo (1982, 1986). As shown in Fig. 8.5, the Middle Palaeolithic distribution is characterized by a much higher frequency of open-air sites.*

together with some significant differences (Duchadeau-Kervazo 1982, 1984, 1986; Turq 1989a). From the areas surveyed in detail it is clear that many of the most densely occupied and exploited areas coincide with the major limestone plateaux which separate the major river valleys of the Perigord – for example between the valleys of the Lot and the Dordogne, the Dordogne and the Vézère, and the various river and tributary systems of the Isle

and Dronne catchments in the northern Dordogne and southern Charente (Rigaud 1982; Geneste 1985; Turq 1988a, 1989a; Le Tensorer 1981; Debénath 1974; Duchadeau-Kervazo 1982, 1986) (see Figs 8.5, 8.6, 8.10). All of these are essentially interfluvial locations which lie between areas of dense concentrations of cave and rock-shelter sites within major river valleys. But clearly the distribution of open-air sites is much less tightly controlled by the

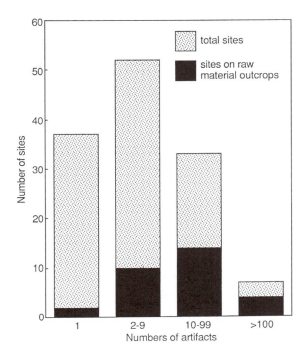

Figure 8.7 *Occurrence of Middle Palaeolithic sites in relation to sources of lithic raw materials, as documented by Duchadeau-Kervazo (1982) in the northern Perigord region. It will be seen that the richest sites (in terms of numbers of artefacts) show a strong tendency to be associated with raw material outcrops.*

the Garonne and their tributaries (Fig. 8.11) (Le Tensorer 1976,1981; Jaubert 1984, 1985; Turq 1988a, 1989a). To the north, similar distributions can be traced into the Departments of Vienne, Deux-Sèvres and Maine-et-Loire (Mazière & Raynal 1976; Gruet 1976; Pradel 1954 etc.). Finally, there are some less well documented extensions of open-air sites well to the east of the limestone regions, on areas of sandstone or crystalline formations around the western margins of the Massif Central (Mazière & Raynal 1976; Delporte 1976; Jaubert 1985) (Fig. 8.11). Thus, the distribution of Middle Palaeolithic open-air sites covers a broad region of southwestern France, running from the central foothills of the Massif Central westwards to the Atlantic coast and southwards to the Pyrenees. The most conspicuous concentration of sites nevertheless corresponds – as for cave and rock-shelters – predominantly with the major belt of limestone and associated flint-bearing outcrops running from north-west to south-east through the central Perigord region.

3. A third series of interrelated patterns can be seen in the more specific location of open-air sites in the Perigord region. All of the earlier studies have emphasized that open-air sites tend to occur predominantly in relatively high level exposed locations, usually on the highest parts of the interfluvial plateaux between the major river valleys. Concentrations of this kind have been recorded, for example, by Rigaud (1982) on the high plateaux between the Vézère and Dordogne; by Turq (1977a, 1978, 1988a, 1989a, 1992b) and Le Tensorer (1973, 1981) in the regions between the Dordogne and the Lot; and by Duchadeau-Kervazo (1982, 1986) in the northern Dordogne and southern Charente. What has become clear recently, however, is that this pattern applies predominantly to the larger, richer and more intensively occupied sites and is probably less true of smaller and more ephemerally occupied sites. Duchadeau-Kervazo (1986: 61) has pointed out that

associated bedrock formations than are cave and rock-shelter sites, and the overall distribution of open-air sites extends well beyond the major limestone belt which contains almost all the known occurrences of cave and rock-shelter locations. Thus, dense scatters of finds have been recorded, for example, immediately to the west of the major limestone formations in the vicinity of Bergerac (apparently related to the distribution of rich flint outcrops in this area: Guichard 1976) and also sporadically further west around the Garonne estuary and along the Atlantic coast (Lenoir 1983). To the south, further scatters of finds extend into the limestone regions of the Causses du Quercy and onto many of the river terraces of the Tarn,

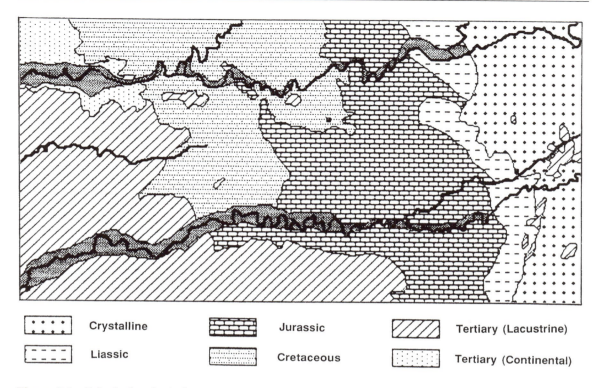

| Crystalline | Jurassic | Tertiary (Lacustrine) |
| Liassic | Cretaceous | Tertiary (Continental) |

Figure 8.8   *Principal geological outcrops in the areas between the Dordogne and Lot valleys. After Turq 1989a.*

in the northern Perigord area virtually all sites containing more than 100 artefacts occur in these higher plateaux locations, whereas smaller and poorer sites (represented by between 1 and 100 artefacts) are more widely distributed over both high level and more low-lying areas, either on slopes of river valleys or in some cases within the floodplain zone. Similar patterns have been documented in detailed surveys by Geneste (1985) on some smaller tributary valleys of the Dronne drainage (notably those of the Euche and Buffebale: see Fig. 5.13) and by Turq (1988a, 1989a) in the areas between the Lot and Dordogne. Some of the possible implications of these patterns for the character of the activities carried out in the different locations will be discussed further below.

4. Finally, there would seem to be strong correlations once again between the distribution of open-air sites and the immediate availability of lithic raw material supplies. Thus Turq (1989a: 182–5, 194–6) has demonstrated that within the southern Perigord region, the densest concentrations of open-air sites seem to coincide with the limestone outcrops offering the most abundant and high quality materials – notably those of the Senonian formation, and other outcrops close to the junction between the Bathonian and Bajocian deposits (Figs 8.8–8.10). Duchadeau-Kervazo (1984, 1986) has emphasized similar patterns in the northern Perigord – i.e. with a major concentration of sites on the various Senonian deposits and much lower densities on the adjacent Jurassic limestone

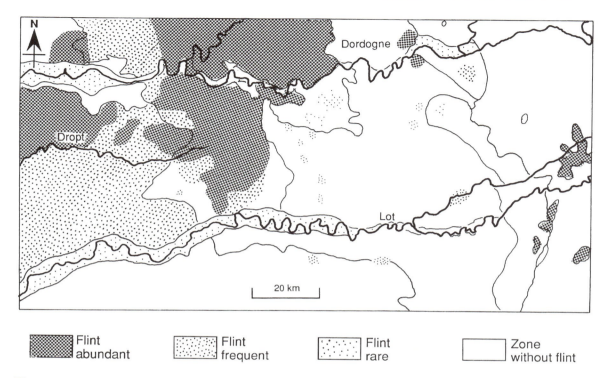

Figure 8.9 *Geological distribution of flint sources in the areas between the Dordogne and Lot valleys, after Turq 1989a. The richest flint sources correspond with the various Cretaceous and Tertiary limestone outcrops, as shown in Figure 8.8. The alluvial deposits of the major river valleys also provide a potential source of secondary flint supplies.*

and crystalline formations. One of the most conspicuous examples of this effect is apparent on the various Maestrichtian deposits in the Bergerac region, where there is a major concentration of sites clearly related to the high quality outcrops of zoned and banded flint in these formations (Guichard 1976).

A further illustration of the same pattern is provided by the frequency with which open-air sites (especially the larger and more intensively occupied sites) occur either on or in immediate proximity to rich and high-quality raw material sources – for example at sites on the Plateaux Cabrol (Lot-et-Garonne), at La Plane (Lot), Le Dau and Coursac (Dordogne) and further to the north at the well-known site of Fontmaure, located directly on exposures of high-quality jasper in the Vienne (Turq 1977a, 1978, 1989a; Rigaud 1969, 1982; Geneste 1985; Pradel 1954). Indeed, Turq (1988a: 99) has estimated that approximately 80 percent of the larger and more intensively occupied sites in the southern Perigord area were located either directly on or very close to major raw material supplies. Overall, the coincidence between raw material distributions and the occurrence of the richer and more intensively occupied open-air sites is one of the most striking characteristics of Middle Palaeolithic site distributions in southwestern France.

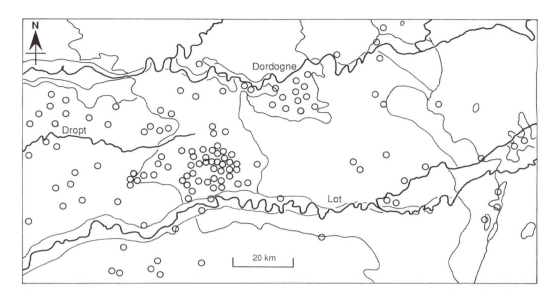

Figure 8.10 *Overall distribution of Middle Palaeolithic sites (including both cave/rock-shelter and open-air sites) in the areas adjacent to the Dordogne and Lot valleys. After Turq 1989a. The major concentrations of sites can be seen to correspond fairly closely with the areas of richest flint supplies, as documented in Fig. 8.9. The outlines of the major geological outcrops correspond with those shown in Fig. 8.8.*

Character of open-air sites

Any attempt to identify the activities carried out in open-air sites is inevitably beset by the various problems outlined earlier – especially by the lack of organic remains from all except a small handful of sites and the lack of information from controlled excavations on the overall extent, size and internal spatial organization of sites. Some of the clearer and apparently well documented patterns which have emerged from recent work can be summarized as follows:

1. The feature which has been commented on most frequently in the earlier literature is the high proportion of open-air sites in south-western France which appear to belong to the Mousterian of Acheulian tradition (MTA) variant. This diagnosis of course rests on the presence of typical cordiform or triangular hand axes which are to all appearances iden-

tical to those found in the classic MTA industries in local cave and rock-shelter sites. Whether this can be taken to imply a strict chronological synchronism with the analogous industries from the rock-shelter sites is of course a separate question which, in the absence of clear dating evidence for most of the sites, remains difficult to assess (cf. Mellars 1969: 147–9). Leaving this aside, characteristic cordiform or triangular hand axes occur, frequently in large numbers, on a high proportion of documented open-air sites within the Perigord region and also in many adjacent areas to the north, south and west. Precise quantitative data are hard to establish but in the northern Perigord region, for example, over two-thirds of the recorded open-air Middle Palaeolithic sites are said to have produced typical hand axes (Ducha-deau-Kervazo 1982, 1986), while similar patterns are suggested by the studies of Rigaud

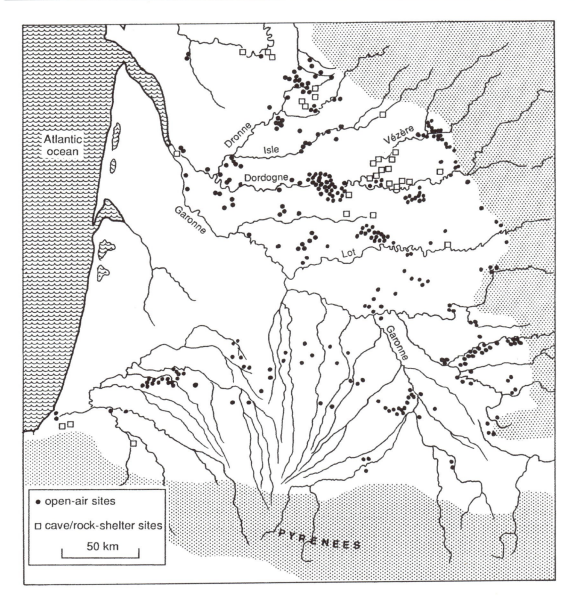

Figure 8.11 *Distribution of finds of cordiform and related hand-axe forms within the southwestern French region, as documented by Jaubert & Rouzaud (1985). Although all these finds are conventionally attributed to the MTA it should be noted that almost all the finds from open-air sites are at present undated.*

(1982) in the areas between the Vézère and the Dordogne and by Turq (1988a, 1989a, 1992b), Le Tensorer (1981) and others further south between the Dordogne and the Lot

(Fig. 8.12). The map compiled by Jaubert (Fig. 8.11) shows over 300 separate find-spots of apparently Mousterian hand axes from open-air localities, distributed over a broad zone

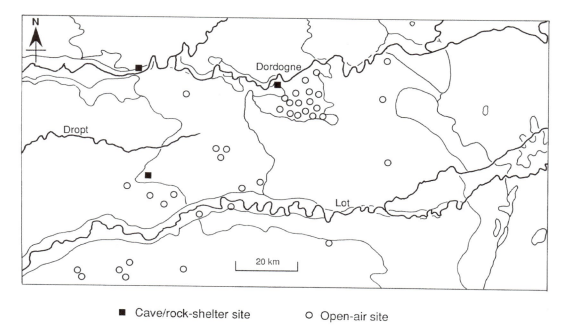

■ Cave/rock-shelter site   ○ Open-air site

Figure 8.12 *Distribution of MTA industries in cave/rock-shelter and open-air sites in the areas between the Dordogne and Lot valleys, as documented by Turq 1992b.*

extending from the northern foothills of the Pyrenees through to the Loire valley and the western Atlantic Plain.

While these figures are impressive and leave no doubt as to the widespread distribution of technologically MTA industries within the open-air sites of southwestern France, any interpretation of these statistics should be treated with caution. As noted above, the great majority of recorded open-air sites are represented by finds of either single artefacts or by small groups of associated finds (see Fig. 8.7, and Duchadeau-Kervazo 1982, 1986). Many of the finds of supposedly MTA industries at open sites are therefore likely to represent either tools which were casually lost or discarded in the course of foraging activities or at best scatters from very brief episodes of activity on the sites. Equally significant is the relative visibility of hand axes in the course of surface

collecting (due to their size and distinctive appearance) and the obvious attractions which these tools have had for generations of amateur flint collectors over the past century (Rigaud 1982). As Jaubert (1984) has emphasized, the apparent abundance and widespread distribution of hand axes in surface collections have almost certainly been greatly inflated by the combination of these two factors in the course of previous surface collecting on the sites.

Any impression that other industrial variants of the Mousterian are effectively lacking at open-air sites, however, has been clearly contradicted by discoveries over the past 20–30 years. Apparently typical occurrences of Quina-Mousterian assemblages, for example, have been recorded from the sites of Chinchon (Gironde) and Puycelsi (Tarn), as well as from at least five or six sites on the higher plateaux between the valleys of the

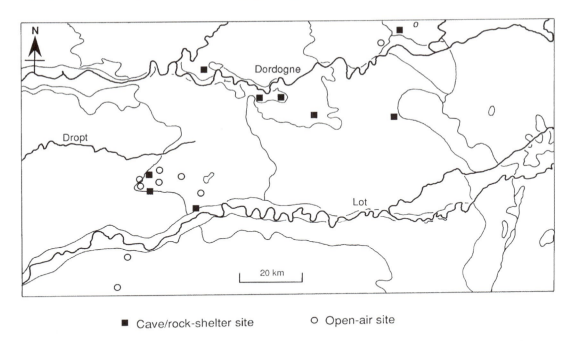

Figure 8.13 *Distribution of Quina Mousterian industries in cave/rock-shelter and open-air sites in the areas between the Dordogne and Lot valleys. After Le Tensorer (1981) and Turq (1992b).*

Lot and the Dordogne (Fig. 8.13) (Sireix & Bordes 1972; Tavoso 1987a; Le Tensorer 1973, 1981). Similar occurrences of Ferrassie type industries have been reported from open-air sites at Pons (Charente-Maritime) and Rescoundudou (Aveyron) (Lassarade *et al.* 1969; Jaubert 1983, 1989), while occurrences of seemingly characteristic Denticulate assemblages have been reported from a number of sites in the Euche valley in the northern Dordogne (Geneste 1985: 76–105) and further south from the sites of La Borde in the Department of Lot (Jaubert *et al.* 1990) and Mauran in the Haute Garonne (Girard *et al.* 1975; Jaubert & Brugal 1990). Occurrences of taxonomically Typical Mousterian assemblages are more difficult to document from open-air sites (largely owing to the poorly characterized nature of this variant: see Chapter 6), but have been reported provisionally from the sites of Corbiac and Fon-

seigner (Dordogne) and Tour-de-Faure (Lot) (Guichard 1976; Bordes 1984; Geneste 1985; Turq & Dolse 1988). Even if the majority of documented open-air sites appear to relate to the MTA variant, it is now clear that occurrences of most if not all the other recognized industrial variants of the Mousterian are well represented in these locations.

2. Information on total areas of occupation at open-air sites and the intensity or duration of occupation episodes is inevitably difficult to derive from the available field data. Obviously, this information can only be used with any confidence from extensively excavated sites where the overall spatial and stratigraphic extent of the occupation zones have been accurately defined. For surface collections we face not only the inevitable problem of occupational palimpsests but also the likelihood that sites which may have originated

as relatively small, discrete concentrations will have been extensively disturbed and redistributed in the course of repeated ploughing or other agricultural activities on the sites.

The tendency in the recent literature has been to make a general distinction between the categories of relatively rich, extensive sites and poorer or more restricted sites (Geneste 1985; Turq 1988a, 1989a). Some of the better documented examples of the latter category have been reported by Geneste (1985: 76–105) at a number of sites in the Euche valley in the northern Dordogne (Fig. 5.13) and by Turq (1988a, 1989a) for sites in the Lot and Lot-et-Garonne regions. Two general features of these smaller and poorer sites have been reported: first, that they tend to occur predominantly in lower locations, either along the slopes of valleys or close to the floodplain zones of valley bottoms (Turq 1988a: 104; 1989a; Duchadeau-Kervazo 1986); and second that the lithic assemblages recovered from the sites are not only relatively restricted in quantative terms, but often seem to reflect some clear specialization in the particular forms of retouched tools or different categories of flaking debitage represented (Turq 1988a, 1989a). Some of the possible implications of these patterns are discussed further below.

As noted earlier, one clear consensus which has emerged from recent work is that the majority of the much richer and more extensive open-air sites occur predominantly on higher and more exposed locations of the major plateaux of the Perigord (Rigaud 1982; Duchadeau-Kervazo 1986; Turq 1988a). Examples of such sites have been reported by Rigaud (1969, 1982) at a number of localities on the Meyrals plateaux between the valleys of the Vézère and the Dordogne, and by Turq on some of the similar high plateaux areas between the Dordogne and the Lot. To the north of the Perigord similar examples of extremely rich, extensive open-air sites have been reported at Fontmaure in the Vienne

(Pradel 1954, 1963) and at La Croix-Guémard in Deux-Sèvres (Ricard 1980).

The central problem in all these rich and extensive open-air sites is to know whether they genuinely reflect substantial episodes of occupation by relatively large human groups or simply compounded palimpsests of frequently repeated visits by much smaller groups to the same location. As noted above, one of the most conspicuous features of these sites is that they can often be shown to have been occupied at many different periods of the Palaeolithic, ranging from the Acheulian through to the Aurignacian or Upper Perigordian (for example at Le Dau, Coursac, Plateaux Baillard, Plateau Cabrol, La Plane, Metayer etc.: Rigaud 1969, 1982; Le Tensorer 1973, 1981; Turq 1977a, 1978, 1988a; Duchadeau-Kervazo 1986; Geneste 1985). Clearly, many of these sites were highly favourable locations, which attracted human occupation repeatedly throughout the Palaeolithic succession. Equally significant is the fact that when systematically surveyed, they can usually be seen to break down into smaller and discrete concentrations. This has been recorded, for example, for several sites on the heavily occupied Meyrals plateau in the Sarlat region, on the Plateau Cabrol and Plateau Baillard areas in Lot-et-Garonne and at the large open-air site of La Croix-Guémard (Fig. 8.14) in Deux-Sèvres (Rigaud 1982; Turq 1978; Ricard 1980). Such indications reinforce the impression of frequent, repeated visits to the same location by relatively small human groups. The only indication that these sites might represent rather different activities from those represented at the archaeologically poorer sites is provided by the composition of some lithic assemblages recovered from the sites, as discussed further below.

3. The most thorough attempt to analyse these patterns of the varying character and composition of lithic assemblages from open-air sites has been published recently by Turq (1988a, 1989a, 1992b) based on his surveys in

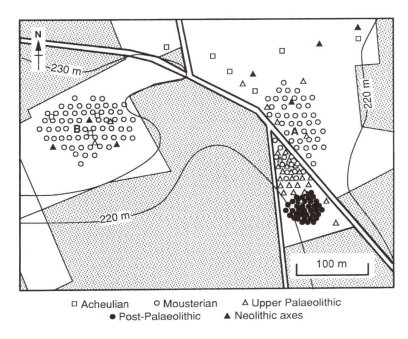

□ Acheulian   o Mousterian   △ Upper Palaeolithic
      ● Post-Palaeolithic   ▲ Neolithic axes

Figure 8.14 *Distribution of surface finds of Acheulian, Mousterian, Upper Palaeolithic and Neolithic artefacts at the site of La Croix-Guémard (Deux-Sèvres), after Ricard 1980. This site illustrates the common tendency for open-air sites to show occupation at many different periods – especially when (as in this case) located in proximity to a source of abundant flint. The Mousterian finds are represented mainly by large quantities of cordiform hand-axes.*

the Lot and Lot-et-Garonne regions between the Dordogne and the Lot. Turq suggests that the majority of lithic assemblages recovered from open-air locations in this region can be categorized into four basic facies reflecting, predominantly, the varying representation of lithic extraction and production activities, as opposed to more general use and discard of tools on the different sites. Briefly, the main elements of his scheme can be summarized as follows (Turq 1988a: 98–9; 1989a; see also Geneste 1985: 515–7):

'Extraction and exploitation' sites
As discussed in Chapter 5, these are the sites which in other contexts have usually been described as 'quarry' sites, devoted primarily to the extraction and initial working of flint resources from local, usually high quality,

raw material outcrops. Turq (1988a: 98; 1989a: 188) describes one such major site from this region – that of Lascabannes, located immediately adjacent to a rich flint source on the Senonian outcrop in Lot-et-Garonne. The assemblage is characterized predominantly by a concentration of flint nodules, most of which exhibit signs of deliberate scratching or scoring of the surface (to assess the quality of the flint) followed by preliminary testing of the nodules by occasional flake removals, or more systematic removal of outer (cortical) flakes to reduce the weight of the nodules for transportation from the site. An apparent example of a similar extraction site has been reported briefly by Geneste (1985: 515) from the site of Campsegret in the Bergerac region, but has not yet been described in any detail.

'Extraction and production' sites

Turq (1988a: 98) categorizes these sites as essentially 'workshop' (*atelier*) locations, devoted partly to the extraction of raw materials from local flint outcrops, and partly to the systematic working down of nodules into fully prepared primary flakes (such as Levallois or related flakes) or to certain specific forms of retouched tools. He quotes two major sites of this type, those of Barbecornio and Lagrave, both in the Lot Department and both located immediately adjacent to rich and high-quality flint supplies (on the Santonian and Bajocian/Bathonian outcrops respectively) (Turq 1988a: 98; 1989a: 188). The former site seems to have been devoted primarily to the production of simple, non-Levallois flakes while the latter was focused on the production of hand axes manufactured mainly from large flake blanks. The distinguishing features of these assemblages, in Turq's terms, consist of high frequencies of debitage flakes, in which flakes retaining large amounts of cortex (from initial trimming of nodules) usually represent over 50 percent of the total flake component (Fig. 5.16). The majority of finished flakes and tools are assumed to have been removed from the sites for use in other locations, so that frequencies of major primary flakes and finished tool forms are low. Significantly, most of the hand-axe forms from the Lagrave site are said to be unfinished or broken specimens, apparently abandoned in various stages of manufacture (Turq 1989a: 188).

'Mixed strategy' sites

These sites account for around 20 percent of those documented in Turq's surveys and are assumed to represent the widest and most varied economic and technological activities (Turq 1988a: 99). He emphasizes that all the sites in this category include some substantial component of basic flaking and reduction of raw flint nodules carried out within the sites themselves, and that most are located very close to good-quality flint outcrops.

Their main distinguishing feature is the occurrence not only of these primary stages of flake and tool production but also high frequencies of retouched tools and, in most cases, large numbers of heavily worked and reduced cores. As examples, Turq quotes the sites of La Plane, La Burlade, Les Ardailloux and Plateau Cabrol, all represented by large and well documented flint assemblages and all apparently belonging to the MTA variant (Turq 1988a: 104; 1989a: 189).

Turq points out that in terms of overall frequencies of different lithic forms (nodules, cores, primary flakes, retouched tools etc.) it is these assemblages which appear to correspond most closely to those documented from the great majority of cave and rock-shelter sites within the region (Turq 1988a: 104; 1989a; see also Geneste 1985: 516). They seem, in other words, to represent what would normally be thought of in the context of cave or rock-shelter sites as essentially 'domestic' or 'residential' sites, reflecting a relatively wide and diverse range of activities carried out at the same location. Significantly, Turq (1988a: 104) points out that it is these 'mixed strategy' sites which occur most commonly on the higher plateaux of the region and which usually command extensive views over the surrounding habitats and ecological zones. They also generally correspond with the most archaeologically rich sites, suggesting either single periods of prolonged activity on the sites or at least frequent and repeated visits to the same location.

'Episodic' occupations

Finally, Turq (1988a: 99) describes a more heterogeneous group of sites which he refers to collectively as 'episodic' occupations or temporary stopover (*halte*) sites. This grouping overlaps in some respects with the other site types discussed above, but is characterized in his terms by evidence for much more ephemeral occupations suggesting very brief periods of activity on the sites. Regrettably, few examples of specific sites are quoted, but he claims that all the sites appear to reflect some specialized, short-term activities, related either to the localized working of a few flint nodules, or to the use of a restricted range of retouched tool forms or unretouched flakes within small, spatially restricted areas (Turq 1988a: 99). His comments on the general character and distribution of the sites are similar to those of Duchadeau-Kervazo (1982, 1986) and Geneste (1985) for sites further north in the northern Dordogne area. These small, short-term occupations are said to account for most of the documented occurrences at open-air sites (estimated at approximately 80 percent of his sample: Turq 1988a: 99) and the sites are distributed fairly widely over plateau-top and river-valley locations as opposed to selectively on the higher plateaux (Turq 1988a: 104). It is unfortunate that so few of these sites have as yet been published. Potentially they could provide evidence for the kinds of discrete individual activities or occupation episodes which are notoriously difficult to identify at more intensively occupied sites.

How far Turq's analysis can be extrapolated to other regions of the Perigord remains to be seen. There seem to be some general similarities with the distribution patterns reported by Duchadeau-Kervazo (1982, 1986) for the northern Perigord, and also with those documented by Geneste (1985) for sites in the Isle and Dronne valleys. In common with Turq, Geneste (1985: 515–6) has postulated the existence of specialized extraction and production (*atelier*) sites, though admitting that available information on these sites in the northern and western Perigord remains rather limited. Geneste has, however, documented a series of small-scale sites located at various points in the Euche valley (Fig. 5.13), which may correspond broadly with Turq's category of episodic or stopover sites (1985: 76–105). None of the sites has so far been investigated in great detail and their

overall extent and total content of lithic material remains uncertain. Nevertheless Geneste points out that most of these sites seem to be characterized by relatively high proportions of retouched tools and usually high frequencies of heavily reduced cores (1985: Table 16) (Fig. 5.14). One of the most intriguing sites was recorded by Geneste (1985: 84–8) in the Carrière Thomasson, where a small group of artefacts (comprising four racloirs, two naturally backed knives, a single denticulate, two flint cores, and a group of 26 flakes) was found in close association with the bones of a single mammoth. In this case we would seem to have a rare glimpse of a localized butchery site, apparently focused on a single carcase of either a hunted or possibly scavenged animal (Geneste 1985: 517). This remarkable and as yet unique discovery may provide a clue to the kinds of ephemeral activities represented at some of the smaller Middle Palaeolithic open-air sites in southwestern France.

4. Finally both Geneste (1985, 1989a) and Turq (1989a) have reported some interesting patterns in the types and frequencies of raw material supplies employed on open-air sites. Geneste in particular emphasizes that open-air sites in general tend to show a stronger emphasis on the exploitation of local flint resources than those documented in the majority of cave and rock-shelter sites. Thus he points out (1985: 504; 1989a: 79) that in the case of open-air sites the frequencies of purely local raw materials (derived from distances of at most 4–5 km) range between 82 and 98 percent (with a mean value of 94 percent), whereas comparable percentages recorded in various cave and rock-shelter sites show generally lower values ranging from 66 to 89 percent (with an average of 78 percent). To some extent this may reflect simply that most open-air sites are located directly on or very close to major sources of raw materials. He also points out, however, that this pattern is equally consistent with the idea that most occupations at open-air sites

were restricted either to very brief periods of time or to the exploitation of food and other resources within a short distance from the sites. These patterns tend to reinforce the notion that a high proportion of open-air sites were relatively short-term, special-activity sites, reflecting more ephemeral activities than those documented in the majority of the cave and rock-shelter occupations.

This is not, however, an invariable pattern at open-air sites. At the site of La Plane, for example, Turq (1989a: 191) has documented that whilst approximately 89 percent of the raw material was derived from local sources (up to 2 km from the site) there is evidence for the utilization of at least ten other flint sources, located at distances ranging between 10 and 80 km to the north, east and west of the site location (Fig. 5.6). Similar patterns have been documented at the site of Le Dau (with at least six separate flint sources up to 80 km from the site) and at Tour-de-Faure (with at least four sources, up to 40 km away) (Geneste 1985: 406–7; Turq & Dolse 1988). The evidence seems to reinforce other indications discussed above that some of the larger open-air sites may well represent the focus of relatively intensive and wide-ranging economic activities, extending over substantial areas of the surrounding habitats.

CHAPTER 9

The Spatial Organization of Middle Palaeolithic Sites

Studies of the detailed spatial distribution of occupation residues on well defined living surfaces can be one of the most productive lines of research into the organization of Palaeolithic groups. Potentially, these studies can shed light on several features of crucial importance to understanding the behaviour and organization of the communities – the size of the social groups who occupied the sites, the duration of the individual episodes of occupation, the nature of the economic and technological activities carried out on the sites, and the ways in which all these activities were organized in relation to the total living space available within the sites. The kinds of patterns documented at such well known Upper Palaeolithic settlements as Pincevent, La Verberie and Etiolles in France, and Meer in Belgium, provide impressive illustrations of the behavioural insights which can be secured from detailed spatial studies of this kind (Leroi-Gourhan & Brézillon 1972, Leroi-Gourhan 1976, Audouze 1987, 1988, Cahen *et al.* 1979).

Interpretations from these studies must of course be treated with caution. As discussed earlier, it is rarely possible to be certain that the spatial distributions recorded in any occupation level reflect the residues of only a single, relatively brief episode of activity on the sites, and to this extent most of the recorded patterns must be seen as potential palimpsests of several separate episodes of occupation on the same location. There are also questions of variable patterns of survival of different kinds of occupation residues on the sites (especially in the case of faunal remains, as discussed in Chapter 7) and the possibility of various forms of post-depositional disturbance of the original occupation patterns by either human or natural means. Whether these problems are as serious in practice as they may seem in theory is perhaps more debatable. As Meignen (1993: 161–3) and others have pointed out, there are reasons to think that in many sites (especially cave and rock shelters) the basic spatial constraints imposed by the size and character of the available occupation space within the sites led to at least broadly similar patterns of use during separate episodes of occupation, and accordingly to some relatively clear patterns in the resulting archaeological residues. Similarly, the relatively rapid rates of sedimentation and generally favourable conditions of preservation in many cave and rock-shelter sites seem to have led to remarkably complete survival of the original occupation residues.

The number of Middle Palaeolithic sites which have been subjected to this kind of detailed recording and analysis of occupation surfaces is unfortunately still fairly limited. The aim of the present chapter is to focus on a number of French sites where the most detailed recording of this kind has been carried out, and to examine some of the patterns which emerge. The more general questions of the economic and social inferences which can be drawn from these studies will be discussed in the later sections of the chapter.

Grotte Vaufrey

The Grotte Vaufrey stands approximately 100 metres above the small valley of the Céou, only 2 km to the south of its confluence with the Dordogne valley (Rigaud 1988). The cave is cut into the local Coniacien limestone formation and faces almost due west, receiving full sunlight during the middle and later parts of the day. It has a high and impressive entrance but extends as a major chamber for a distance of only 15–20 metres from the cliff face (Fig. 9.1). The total area of the sheltered part of the cave is around 170 square metres.

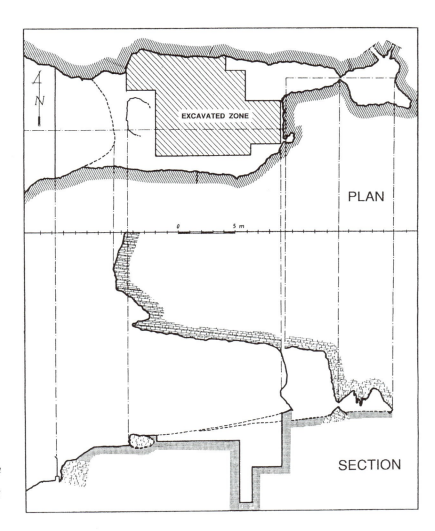

Figure 9.1 *Plan and section of the excavated zone in the Grotte Vaufrey. After Rigaud 1988.*

In front of the cave the hillside drops away steeply towards the valley, requiring a strenuous climb to gain access from the valley floor. Access from the adjacent plateau above the cave is rather easier but still requires navigation of a steep and difficult slope. In view of these difficulties of access it is perhaps not surprising that the site seems to have been occupied by Palaeolithic groups on a relatively ephemeral, sporadic basis throughout the whole of the occupation sequence.

Excavations carried out by Jean-Philippe Rigaud between 1969 and 1982 explored 5 metres of deposits, apparently spanning the period from isotope stage 10 or 11 to the earlier part of the last glacial (Rigaud 1988). Most of the archaeological levels proved to be rather poor in artefacts and only level VIII provided a sufficient concentration of material within a single, well defined occupation horizon to allow the finds to be analysed in detailed spatial terms. Taxonomically, the industries from this and the immediately adjacent levels have been attributed by Rigaud to an early form of Typical Mousterian, characterized by a moderate percentage of racloirs and notches/denticulates, and showing highly developed use of Levallois techniques. On various geological and faunal grounds the occupation of this level is attributed to isotope stage 7 or the earlier part of stage 6 and is dated on the basis of a uranium series of measurements of adjacent flowstone deposits to ca 170–200,000 BP.

The total area occupied by the deposits of layer VIII is estimated by Rigaud to be in the region of 160 square metres, of which some 90 square metres were explored and fully documented in the recent excavations (Fig. 9.1). It is not certain that this represents the whole of the occupied area in this level but the overall distribution of occupation material suggests strongly that the excavated area incorporates at least the major zone of human activity in the site. To summarize analyses provided by Rigaud & Geneste (1988), Simek (1988) and

Binford (1988), the most significant features of the distribution patterns are as follows:

1. Perhaps the most striking feature is the highly localized distribution of the archaeological material, concentrated mainly towards the southern wall of the cave. As shown in Fig. 9.2, over 90 percent of the artefacts were found here, occupying an area of only 30–40 square metres and mainly distributed between a scatter of large limestone blocks which were evidently present on the cave floor before the human occupation. As Rigaud and Geneste (1988) point out, there can be little doubt that the human groups who occupied the site were very small – possibly no more than three or four individuals during any episode of occupation.

2. The distribution of lithic flaking debitage in the site (particularly that represented by very small flakes, less than 1.5 cm in length) reveals three main concentrations, each measuring only 3–4 square metres (Fig. 9.2). The principal concentration is located along the south wall of the cave (between the scatter of large stone blocks noted above), while the others lie approximately 3–4 metres to the north and east. Refitting studies demonstrate that flaking was carried out *in situ* in these locations (Fig. 9.3) but have not so far produced clear links between the material in the different clusters. Whether or not the different episodes of flaking represented in the clusters were carried out during precisely the same phase of occupation on the site or during a succession of separate occupations must remain open.

3. The distribution of other categories of lithic artefacts (Levallois flakes, complete nodules of raw material and various forms of retouched tools) conforms broadly to the same distributional pattern and shows the same general concentration mainly towards the southern wall of the cave. Nevertheless, it is clear that these categories of material are rather less tightly concentrated in this partic-

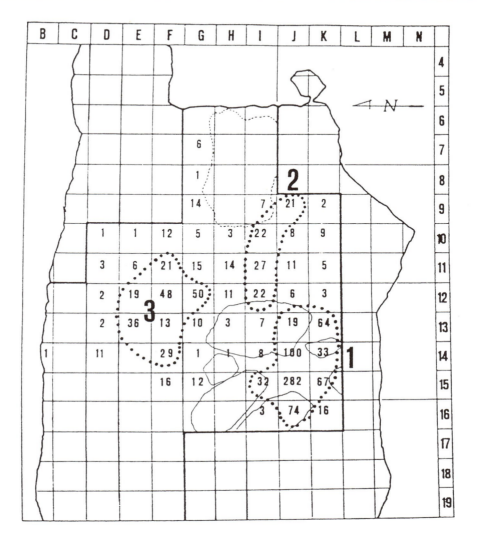

Figure 9.2 *Overall distribution of lithic artefacts in layer VIII of the Grotte Vaufrey. The zones of highest density are enclosed by dotted lines. The outlines of the large stone blocks on the floor of the cave are also indicated. After Rigaud & Geneste 1988.*

ular area and more widely distributed over the whole of the cave interior. Thus separate concentrations of complete Levallois flakes were detected towards the eastern end of the cave (Fig. 9.4), while the distribution of unworked or partially flaked flint nodules was similarly dispersed and patchy in several different areas of the site (Fig. 9.5). Unfortu-

nately, no detailed information is provided in the published report on the distribution of flint cores.

4. A detailed study of the distribution of different categories of retouched tools is unfortunately hampered by the small numbers of pieces recovered for most of the tool

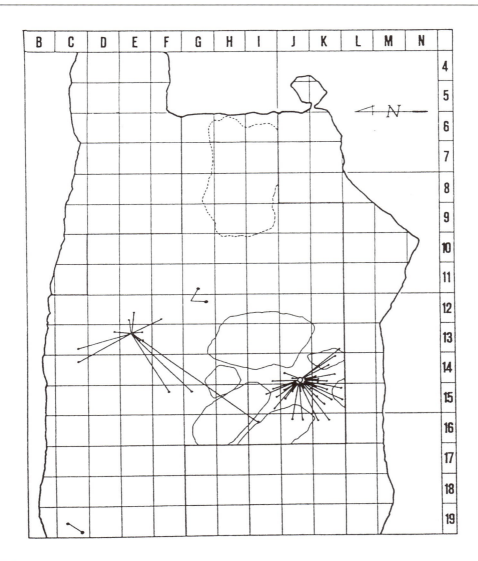

Figure 9.3 *Spatial distribution of two major groups of refitted artefacts in layer VIII of the Grotte Vaufrey (see also Fig. 3.2). After Rigaud & Geneste 1988.*

types. However, there is an indication that retouched tools are generally more widely distributed on the site than are the products of primary flaking debitage, and there are some hints of variable patterns in the distribution of the two major retouched tool categories – racloirs and notches/denticulates. From the data provided by Rigaud and Geneste (1988: 608–9) it appears that notches and denticulates are more concentrated adjacent to the main debitage clusters (especially in the main area of debitage along the south

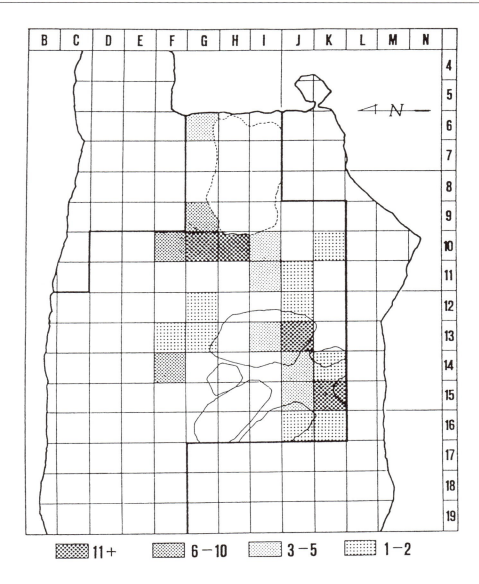

Figure 9.4 *Distribution of Levallois flakes in layer VIII of the Grotte Vaufrey. After Rigaud & Geneste 1988.*

wall) while racloirs seem to be more widely distributed in the peripheral parts of the occupation zone (Figs 9.6, 9.7). In view of the small sample sizes of these tool types, however, it might be premature to attach too much weight to this pattern.

5. The most difficult and controversial patterns to interpret in the Vaufrey deposits are those relating to the distribution of animal remains (Fig. 9.8). Whilst there is general agreement that some component of the animal remains almost certainly derives from

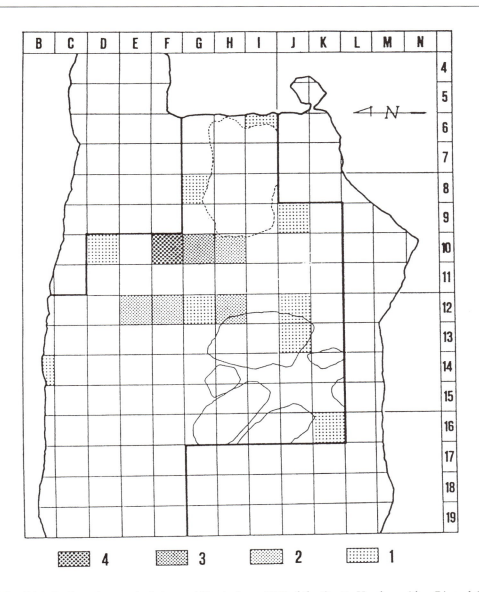

Figure 9.5 *Distribution of unworked river cobbles in layer VIII of the Grotte Vaufrey. After Rigaud &*
Geneste 1988.

occasional use of the cave as a carnivore den, there is considerable debate about how much this has contributed to the faunal assemblage. From a study of various taphonomic aspects of the bones, Binford (1988) has suggested that most of the faunal remains recovered from the main area of industrial activity along the southern wall of the cave probably derive from carnivore activity, apparently preceding the main period of human occupation in this level. This has been disputed by Simek (1988) from the results of his own

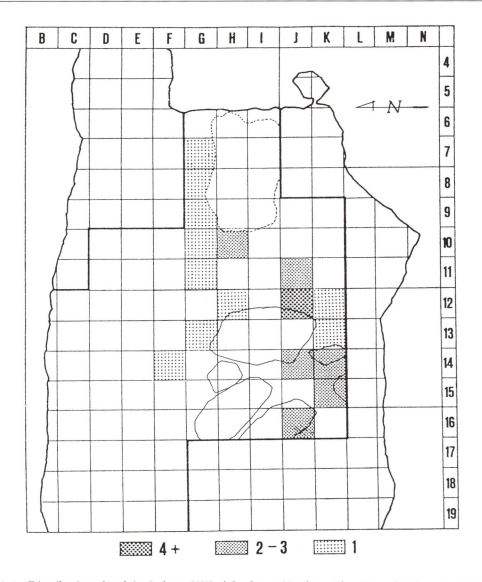

Figure 9.6 *Distribution of racloirs in layer VIII of the Grotte Vaufrey. After Rigaud & Geneste 1988.*

statistical analyses of distributions of faunal remains, who suggests that most of the faunal remains from all parts of the site are probably related directly to human occupation. In either case, it is apparent from the published plans that faunal remains in general are widely distributed in the site, with much less concentration along the southern wall of the cave than for the lithic remains (Fig. 9.8). An interesting feature of the faunal material is a clear concentration of heavily fragmented bone splinters recorded within a small area

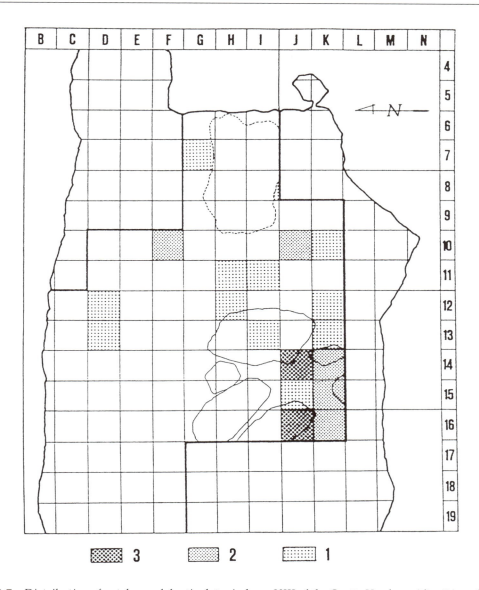

Figure 9.7 *Distribution of notches and denticulates in layer VIII of the Grotte Vaufrey. After Rigaud & Geneste 1988.*

towards the rear, eastern, end of the cave (Fig. 9.9). This apperently indicates a special location of intensive smashing of red deer and horse bones, presumably for marrow extraction and could well relate to one particular episode of human activity on the site.

6. Finally, Rigaud & Geneste (1988) note that while no well defined hearths were detected in this occupation level, nor indeed in any of the occupations at Grotte Vaufrey, there appeared to be recognizable concentrations of burned bones or charcoal at two separate

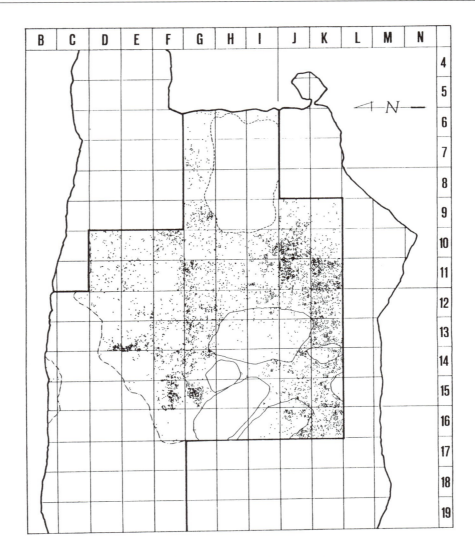

Figure 9.8 *Overall distribution of faunal remains in layer VIII of the Grotte Vaufrey. As discussed in the text, many of these remains are likely to derive from contemporaneous use of the site as a carnivore den. After Rigaud & Geneste 1988.*

points in the deposit (Fig. 9.10), each fairly close to one of the main flint flaking areas discussed above. In all probability these are the remains of brief fires lit for heat and perhaps cooking or food preparation within the main zone of industrial activity on the site. Again, whether different fires were lit during the same phase of occupation or during a sequence of separate occupations can hardly be decided from the available data.

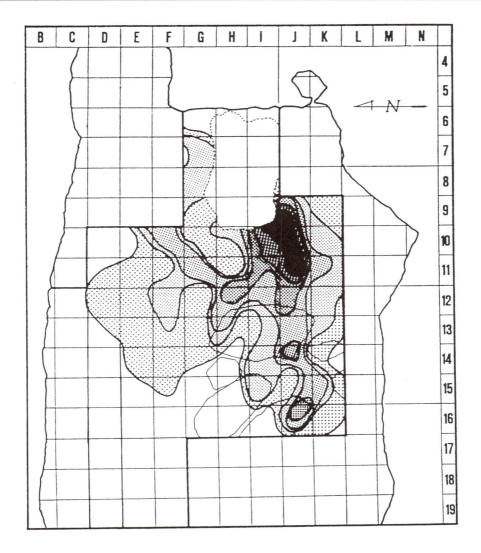

Figure 9.9   *Distribution of small long-bone splinters in layer VIII of the Grotte Vaufrey, plotted as an 'iso-density' distribution diagram. After Binford 1988.*

The spatial distributions documented at the Grotte Vaufrey, therefore, provide some useful, if limited, data on the patterns of human use of the site. Rigaud & Geneste (1988) argue that the distributions almost certainly represent a palimpsest of repeated phases of site occupation, possibly spanning a long period. Nevertheless, the overall distribution of the archaeological material leaves no doubt that the human groups who occupied the site were very small and can hardly have used the site for more than fairly short-term visits. Despite this, there is evidence for a wide range of activities in the site

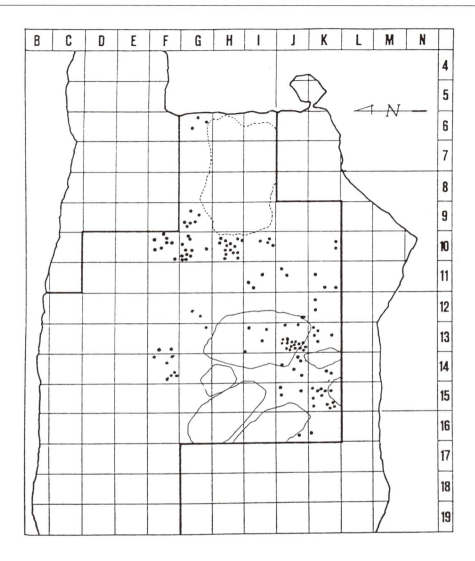

Figure 9.10 *Distribution of traces of burning in layer VIII of the Grotte Vaufrey, aparently showing two main concentrations (in squares F-I/9–10 and J-K/13–15), possibly representing the location of two hearths. After Rigaud & Geneste 1988.*

– intensive flaking of imported nodules of raw material, the production, use and discard of several varieties of tools, occasional use of fires and apparently the introduction and processing of several different species of animals (red deer, roe deer, horse, bovids, chamois). Rigaud & Geneste (1988) speculate

that the area of concentrated lithic and flaking debitage along the south wall of the cave – which receives the longest exposure to sunlight – may have served as the main location for various daytime activities, while the area along the northern wall, apparently less affected by draughts, may have served primarily for overnight sleeping. Perhaps the main point to keep in mind is the relative inaccessibility of this site and the sparse distribution of archaeological material throughout the occupation sequence as a whole. In this respect the site may reflect a very different pattern of use to that represented in many of the archaeologically richer and more intensively occupied sites located in more low-lying and accessible positions in the major river valleys of the region.

Grotte du Lazaret

The spatial patterns documented by Henry de Lumley (1969a) in the later Acheulian levels of the Lazaret cave, in the Department of Alpes Maritimes, close to Nice, provide interesting comparisons with those documented in the Grotte Vaufrey excavations discussed above. The site consists of a large cave directly overlooking the Mediterranean coast and located, at the time of occupation, about 500 metres from the contemporary coastline. Vertically, the cave stands approximately 100 metres above the estimated sea level at the time of occupation and, as in the case of the Vaufrey site, would have required a strenuous climb to gain access from the main adjacent areas of economic and subsistence activities. The complete sequence of deposits in the cave occupies over 7 metres, apparently spanning the period from the 'Mindel-Riss' interglacial to the earlier part of the last glacial. The excavations carried out by Octobon and de Lumley explored all parts of this sequence but a detailed account has been published only for the final Rissian levels, which include the much-discussed 'living floor' and the controversial traces of a supposed hut structure (de Lumley 1969a; de Lumley et al. 1969). The excavations in this level (layer 5) were confined to the interior of the cave and extended over an area of 55 square metres (Fig. 9.11). From various geological and faunal evidence, the age of the occupation level has been correlated with

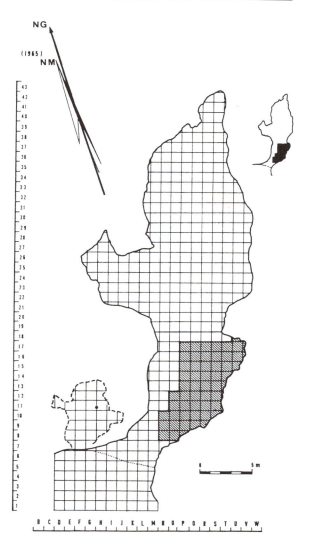

Figure 9.11 *Plan of the Grotte du Lazaret, with the excavated zone shaded. After de Lumley 1969a.*

isotope stage 6, probably in the region of 130–150,000 BP (i.e. slightly younger than the estimated age of the Grotte Vaufrey occupation discussed above). The character of the associated lithic industry shows broad similarities with that from the Grotte Vaufrey and consists of moderate percentages of racloirs, pointed forms and notched or denti-

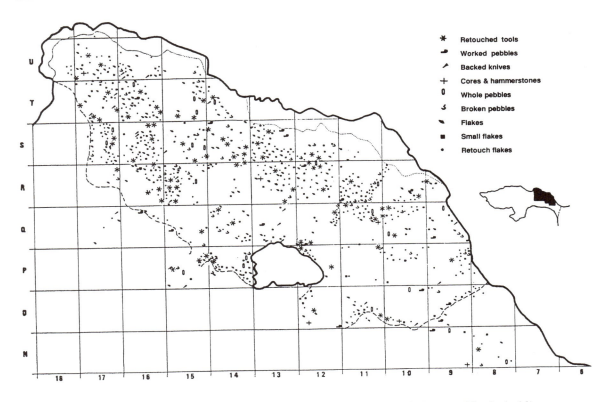

Figure 9.12   *Distribution of various forms of lithic artefacts in the Grotte du Lazaret. The dashed line indicates the inferred limits of the main occupation zone. After de Lumley* et al. *1969.*

culated tools – though with much less use of typical Levallois techniques than at Vaufrey. Despite de Lumley's tendency to describe the industry as 'Late Acheulian', the main occupation level documented in layer 5 did not produce any trace of characteristic hand axes. The associated faunal assemblage is dominated by remains of red deer, ibex and fallow deer with more sporadic remains of chamois, horse and bovids. From studies of the age distribution of ibex teeth, de Lumley estimates that occupation in this level was concentrated primarily during the full winter months, between mid-November and mid-April.

De Lumley's very full publication of the

material from level 5 provides information on the distribution of most of the major categories of both lithic and faunal material, with a series of detailed distribution plans (de Lumley *et al.* 1969). Despite the inevitable problem of occupational palimpsests, the distributions reveal a number of sharply defined patterns, in certain respects strongly reminiscent of those documented in the Grotte Vaufrey:

1. The most striking feature of the distributions (as at the Grotte Vaufrey) is the concentration of archaeological material within a relatively small and sharply defined area of the cave. As at Vaufrey, the material is con-

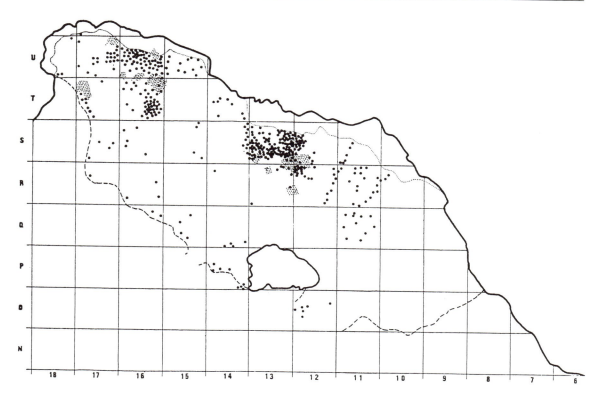

Figure 9.13   *Distribution of burning traces in the Grotte du Lazaret. Charcoal fragments are indicated by black dots; ashy zones are indicated by stippling. The distributions point to the existence of two separate hearths, located adjacent to the east wall of the cave. After de Lumley et al. 1969.*

centrated mainly along the south wall and occupies about 35 square metres – similar to that documented in the Vaufrey excavations (Figs 9.12–9.15). The conclusion seems inescapable that this particular activity area must have been the product of a very small social group, probably no more than five to ten individuals – possibly smaller if repeated, superimposed episodes of activity on the same spot are taken into account.

2. Evidence is present for two separate hearth areas, both located close to the south wall of the cave, approximately 3 metres apart (Fig. 9.13). As at Vaufrey, no evidence for deliberate construction of the hearths was

detected, but the sharp localization of both charcoal fragments and general ashy zones can be used to pinpoint their location. From the lack of evidence for heavy burning immediately below the hearths, de Lumley suggests that the fires lit there were probably short-lived and rarely reached high temperatures.

3. Two features of the archaeological material seem to show a particularly close association with the areas immediately adjacent to the north and west of the hearths: the distribution of flint tools and flaking debris (Fig. 9.12) and small splinters of heavily fragmented bones (Fig. 9.14). Both would appear

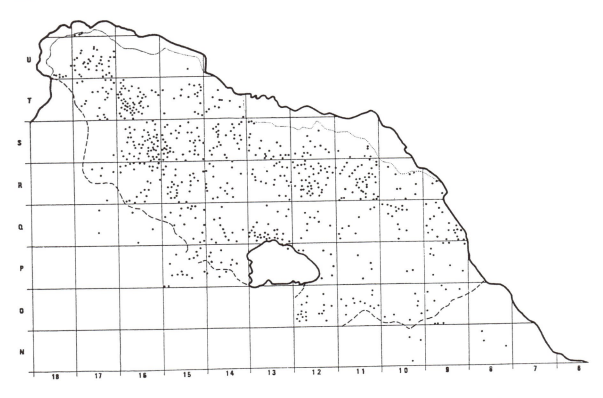

Figure 9.14   *Distribution of small long-bone splinters in the Grotte du Lazaret. After de Lumley* et al. *1969.*

to indicate that the areas immediately adjacent to the hearths served as major centres of industrial and processing activities on the site, involving both the *in situ* flaking of flint nodules and the deliberate smashing of bones for extraction of marrow. Interestingly, a similar concentration of flint debitage and bone splinters was recorded around the edges of a large stone block at the western edge of the central activity area, suggesting that this served as a seat or anvil stone for similar activities at some point during the occupation sequence. The distribution of larger fragments of bones, represented mainly by articular ends of long bones, metapodials etc., shows some similarity to that of the more fragmented bones, but is generally wider over the more western parts, further

from the two major hearths. This suggests that these zones could have been used for more general butchery or processing of animal carcases, or perhaps as marginal areas for the discard of larger and more complete bone fragments.

4. The total number of retouched stone tools and discarded cores recovered from the excavations is unfortunately too small to allow a detailed analysis of the distribution patterns. From the data provided by de Lumley there is no clear indication of very different patterns of distribution of any of these forms from those of the general flaking debris, and no reason to infer a significantly contrasting pattern for different types of retouched tools (racloirs, points, denticulates, etc.). In view of

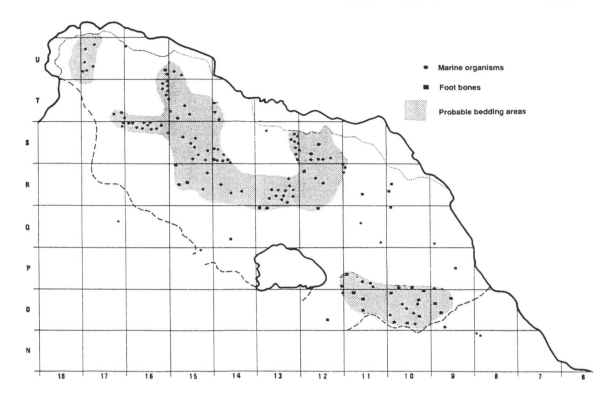

Figure 9.15 *Distribution of small sea shells and foot bones of fur-bearing animals in the Grotte du Lazaret. De Lumley suggests that these may indicate the location of bedding areas, consisting of piles of seaweed covered with animal skins. After de Lumley et al. 1969.*

the small sample sizes, however, it is hardly possible to comment in any detail on these aspects of the distributions.

5. Finally, de Lumley recorded some interesting if rather enigmatic data on the distribution of small marine molluscs in the deposits – principally those of *Littorina neritoides* (Fig. 9.15). He argues that these must have been deliberately introduced into the site and suggests that they may have been attached to masses of seaweed possibly used as bedding for the site occupants. As he points out, the distribution of these molluscs seems to be tightly concentrated around the edges of the two main hearth areas and he

suggests that these would be obvious places for sleeping, immediately adjacent to the hearths. He goes on to argue that the distribution of foot bones of various fur-bearing animals (mainly wolf and fox) show a broadly similar distribution (Fig. 9.15), and suggests that these could derive from the remains of animal skins used as coverings for the bedding areas. Clearly, there are other possible explanations for these patterns (such as the introduction of seaweed as a source of food) and it would be premature to accept de Lumley's interpretations without caution. Nevertheless, the distribution of these components shows an obvious pattern and must have specific implications for the character of the activities carried out in the cave.

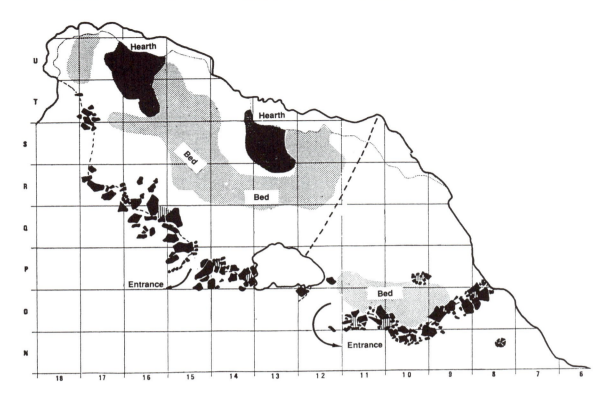

Figure 9.16 *De Lumley's reconstruction of the probable organization of space within the Lazaret cave. The line of stones (marked in black) is thought to indicate the base of a tent covering, supported by a series of vertical stakes (marked by vertical shading) surrounded by rings of stones. After de Lumley* et al. *1969.*

6. The question of a deliberately constructed 'hut' or 'tent' in the Lazaret cave has generated a good deal of controversy. The evidence cited by de Lumley included an apparently linear concentration of small stone blocks immediately surrounding the main zone of occupation, the sharp lateral limits of the lithic and bone distributions, coinciding fairly closely with the stone distribution and the presence of eight roughly circular clusters of small stones which he argued could have served as the basal supports of wooden stakes or poles (although there was no evidence of post-holes penetrating below the level of the stones) (Fig. 9.16).

These features have been questioned by other workers on the grounds that the stone 'arrangements' could possibly reflect natural areas of rock falls from the cave roof or at best simply an attempt to clear stones away from the main occupied area to its edges. On the basis of the published evidence it is hardly possible to resolve these questions. One can only note that the whole issue of a deliberately constructed living structure in the Lazaret cave remains controversial, and can hardly be accepted in isolation as a demonstration of the architectural abilities of Neanderthal groups (Fig. 9.17).

Leaving aside these reservations over the

Figure 9.17 *De Lumley's reconstruction of the hypothetical hut structure in the Lazaret cave. After de Lumley et al. 1969.*

claimed living structure, there can be no doubt as to the importance of the distributions documented in de Lumley's meticulous excavations of the Lazaret cave. As noted above, they show striking similarities in several respects to those recorded in the Vaufrey cave and may indicate a broadly similar pattern of human use of the site – in both economic and social terms. The very small and sharply defined area of occupation leaves little doubt that the human groups were also small and the overall distributional pattern would seem to indicate a succession of relatively brief episodes of occupation in the cave apparently associated, as at Vaufrey, with a spectrum of different subsistence and tech-

nological activities. To find these patterns at broadly the same date and in two very different environmental and ecological contexts is possibly one of the most interesting features of the current archaeological record of the Middle Palaeolithic.

Les Canalettes

The excavations of Liliane Meignen in the rock shelter of Les Canalettes (Aveyron) have recently been published as a detailed monograph (Meignen 1993). The rock shelter occupies the western side of a small dry valley, located at an altitude of almost 700 metres in the central limestone region of the southern

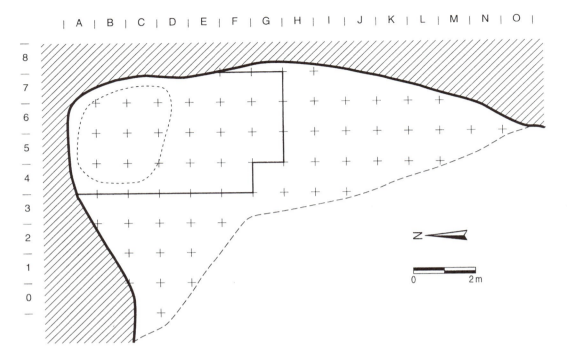

Figure 9.18 *Plan of the excavated area in layer 2 of the Les Canalettes rock shelter. The position of the present rock overhang is indicated by the outer dashed line. The distributions plotted in Figs 9.19–9.21 relate to squares D–G/5–7. After Meignen 1993.*

Languedoc area, approximately 60 km to the north-east of Montpellier. The site faces almost due south and at the time of occupation would probably have provided an occupation area of up to 60 square metres under the directly sheltered part of the rock overhang. The industry recovered from the site appears to represent a broadly Typical Mousterian facies, comprising roughly equal proportions of racloirs and notched or denticulated tools – not dissimilar in overall composition to the assemblages from Grotte Vaufrey and Lazaret. Here, however, the site is much later and is dated by TL measurements on burnt flints in the region of 73,000 BP – i.e. close to the major climatic transition from stage 5 to stage 4 of the last glacial sequence (Valladas *et al*. 1987; Meignen 1993). The associated faunal remains reflect a pri-

mary emphasis on the exploitation of red deer and horse – probably, to judge by the high altitude of the site, concentrated mainly during the summer months.

Meignen recorded separate spatial plans for three different levels in the sequence – layer 2 (close to the top of the sequence) and the upper and lower parts of the underlying layer 3. While the distributions in all three levels show a broadly similar pattern (Meignen 1993, Figs 59–77) she suggests that the distribution plots recorded for layer 2 show the clearest patterns, probably least affected by occupational palimpsests. She stresses that the areas excavated and recorded in detail cover only part of the total potential zone of occupation in the rock shelter, and must therefore be treated with some caution (Fig. 9.18). Nevertheless, the dis-

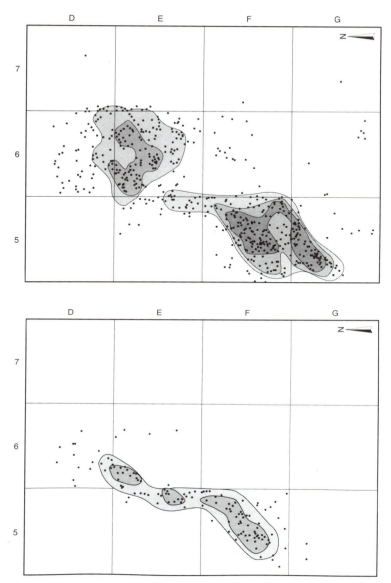

Figure 9.19 *Distribution of flint flakes (upper) and quartz artefacts (lower) in layer 2 of Les Canalettes. After Meignen 1993.*

tribution plots reveal a number of very clear patterns, which in certain respects recall those recorded at the Grotte Vaufrey and Grotte du Lazaret (see Figs 9.19–9.21).

1. A striking aspect of the distribution once again is the concentration of the occupation residues within a relatively small area of the total living space available within the site. Even if the material towards the front of the rock shelter may have been partially truncated by erosion, the abrupt limits recorded towards the inner part of the shelter evidently cannot be explained in these terms.

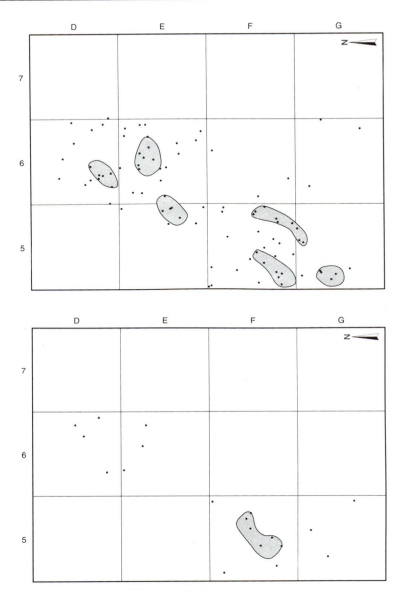

Figure 9.20 *Distribution of
retouched tools (upper) and cores
(lower) in layer 2 of Les Canalettes.
After Meignen 1993.*

Meignen suggest that these parts of the
potential occupation surface may perhaps
have been deliberately avoided by the occupants because of the very stony deposits in
this part of the site. The alternative would be
to see this as a more general feature of the
occupation pattern, possibly reflecting the
location of sleeping areas in this heavily sheltered part of the site (broadly along the lines
suggested by Rigaud & Geneste (1988) for the
similarly sheltered area along the northern
wall of the Vaufrey cave). Whatever the
explanation, it is clear that the distribution of
both lithic and faunal material in the site is

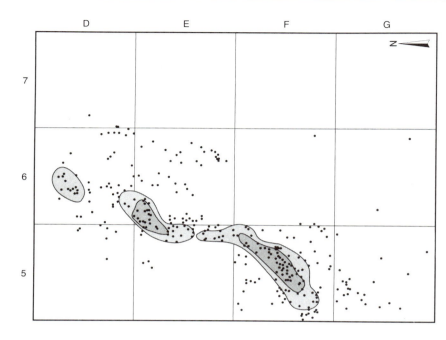

Figure 9.21 *Distribution of total faunal remains in layer 2 of Les Canalettes. After Meignen 1993.*

highly localized and again points strongly to the activities of a relatively small human group.

2. The distribution of lithic material in the site breaks down into two major concentrations, located towards the northeast and southwest ends of the excavated area, lying approximately 2 metres apart (Fig. 9.19). Meignen emphasizes that these concentrations contain exceptionally high densities of very small flakes and spalls, clearly indicating the flaking of cores and nodules on the spot. Relatively few cores were recovered but their overall distribution coincides closely with that of flaking debris, as does that of the retouched tools (Fig. 9.20). A more significant contrast can be seen in the distribution of the two major raw materials worked on the site – flint and poorer quality quartz – both apparently derived from sources relatively close to the site. While both raw materials occur to some degree in both the principal concentrations, the ratio of quartz to flint is very much

higher in the southern concentration (Fig. 9.19). While we cannot be certain that these two materials were employed during precisely the same episodes of occupation, there was apparently a significant separation in the working of these two materials on the site.

3. The distribution of faunal remains (consisting overwhelmingly of small, heavily fragmented splinters of bone and isolated teeth) coincides almost exactly with the main zones of lithic debitage (Fig. 9.21). In this case, there was evidently a very close correspondence in the areas used to flake the lithic raw materials and to process many of the introduced carcase parts – mainly, apparently, for the extraction of bone marrow from the principal long bones. Meignen suggests that the detailed distribution of fragmented faunal remains seems to coincide especially closely with the distribution of flaked quartz fragments, and wonders whether this may reflect the habitual use of quartz nodules or pounders to crack the marrow bones. Per-

haps the most interesting feature of the faunal distributions is the clear tendency for most of the larger and more complete bones to occur towards the northern edge of the distribution, closely adjacent to the north wall of the rock shelter. Meignen reports that this pattern seems to be equally evident in the underlying levels of the occupation sequence and can apparently be regarded as a general pattern in the overall distribution of faunal material on the site – again recalling the patterns documented in the Grotte Vaufrey, and apparently also in the Mousterian levels of the Grotte de l'Hyène at Arcy-sur-Cure (Farizy 1990a).

4. Finally, a potential contrast with the Grotte Vaufrey and Grotte du Lazaret is the apparent absence of clear traces of hearths in this particular level of the Les Canalettes deposits. In view of the ephemeral nature of hearths documented in many Mousterian sites (for example at Grotte Vaufrey) this may or may not be significant and could indicate simply that associated fires left no visible traces in the occupation levels or indeed that hearths were located just beyond the limits of the excavation. However, Meignen records that in the immediately underlying level (layer 3) clear traces of a major hearth were apparent centred on a square F5, immediately to the west of the main concentrations of lithic and faunal remains. It is probable, therefore, that here (as at Vaufrey and Lazaret) most of the industrial and food processing activities were carried out in close proximity to a hearth, in this case situated towards the outer part of the sheltered zone under the rock overhang.

Combe Grenal

The classic excavations of François Bordes at the site of Combe Grenal have already been referred to at several points in earlier chapters. The site at present consists of a small cave, preceded by a rock overhang, located on the south-facing slope of a small dry

valley only one kilometre to the south of its confluence with the main Dordogne valley. Excavations carried out by François Bordes between 1953 and 1965 revealed a sequence of almost 13 metres of rich archaeological deposits, apparently spanning the period from the final stages of the penultimate glaciation (isotope stage 6) through to almost the end of the Mousterian succession, around 45–50,000 BP (see Figs 2.20, 2.23). As explained in Chapter 2, the deposits occupy a succession of major erosional steps in the underlying bedrock formation, which become progressively younger in age and retreat progressively towards the north (i.e. away from the adjacent valley) during the different stages of the occupation sequence (Fig. 2.20). The sediments, pollen, fauna and (to a lesser extent) lithic industries in this sequence have now been documented in some detail (see Bordes 1955, 1971a, 1972; Bordes & Prat 1965; Bordes et al. 1966; Laville 1975; Laville et al. 1980). In all, 55 distinct levels of Mousterian (i.e. last glacial) occupation have been identified in the sequence, underlain by at least nine levels with late Acheulian industries.

Bordes' untimely death in 1981 prevented the full publication of the extremely rich archaeological assemblages from Combe Grenal and as yet only partial and preliminary accounts have been published – the most detailed being that provided in *A Tale of Two Caves* (Bordes 1972; see also Bordes 1971a). During the early 1960s Bordes established a close working relationship with Lewis Binford (initially in collaboration with Sally Binford) and committed the study of detailed distributions of archaeological and faunal material in the different occupation levels of the site to Binford's own research programme. To date, only brief accounts of this work have appeared in print (see Binford 1992). A major monograph, however, is now in preparation, setting out the full results of the spatial analyses. In advance of publication, Binford has generously provided information on the main results of these studies, and the provisional interpretations he draws from them.

The most fully documented parts of the Combe Grenal sequence are those belonging to the earlier Würm I levels, between layers 50 and 55. In these levels, Bordes excavated about 60 square metres of the deposits, estimated by Binford to comprise at least 70–80 percent of the available occupation space within the rock-shelter during this phase of the sequence. Binford's studies of these levels were based on two main categories of data: the distribution of the main forms of lithic artefacts (retouched tools, cores, larger primary flakes etc.) and the various categories of faunal remains – principally the larger and more identifiable fragments of bones and teeth. Less immediately identifiable material, such as smaller fragments of lithic debitage and animal bone splinters, were recorded mainly according to the relevant metre squares and could therefore only be studied in much broader spatial terms.

The detailed analysis of the distribution patterns in the different occupation levels must of course await the forthcoming monograph. In general terms, however, Binford claims to have recognized a consistent and repeated pattern in the major occupation levels in the Würm I sequence (i.e. levels 50, 50A, 52 and 54 – all containing industries of essentially Typical Mousterian form) and which he believes is repeated in a similar form in many later levels in the sequence (i.e. between levels 35 and 10) which contain a succession of taxonomically very different industries ranging from the Ferrassie and Quina to the Denticulate variants. In most of these upper levels, however, the clarity of the spatial patterns is seriously disrupted by the earlier excavations of Peyrony and others on the site (as well as the construction of a medieval stable) and cannot be analysed in the same level of detail as applied to the Würm I levels. The account given below is therefore focused mainly on the character of the patterns which

Binford believes he can identify with some confidence in the well documented sequence of levels 50–54. The specific social and functional interpretations advanced by Binford for these patterns, and some of the difficulties with these interpretations, will be discussed further in Chapter 11.

1. One of the clearest patterns is claimed to be discernible in the distribution and character of different forms of hearths in the Combe Grenal deposits. In essence, he believes that two quite separate forms of hearths can be identified. First there are those represented by large, irregular distributions of ashy material, measuring in some cases up to several square metres and concentrated mainly in the more central parts of the sites, well inside the area protected by the contemporaneous rock overhang. These larger, central hearths are claimed to be situated mainly on areas of relatively 'soft' sandy or silty sediments and are frequently marked by slight depressions in the underlying deposits, probably resulting from repeated clearing of burnt material from the hearths during successive episodes of occupation. The second type of hearth is represented by much smaller and more localized concentrations of burnt material, often including heavily burnt bones, which are distributed mainly in the outer or more marginal parts of the occupation zone, usually close to the estimated drip-line of the contemporary rock shelter and frequently associated, apparently, with either isolated blocks of limestone or scatters of large blocks. As discussed further below, Binford suggests that these two types of hearths served as the foci for very different economic and technological activities in the site and are associated with significantly different patterns of lithic and faunal residues.

2. A second pattern, closely associated with the two major hearth forms, is said to be discernible in the distribution of the two major categories of retouched tool forms – i.e.

various forms of racloirs and point forms versus simpler notched and denticulated forms. Binford claims that in most of the occupation levels the distribution of notches and denticulates seems to be associated predominantly with the larger, more central hearths, while the distribution of various side-scraper forms is associated mainly with the smaller and more peripheral hearths. There is no suggestion of course that these are entirely separate distributions, and the individual distribution plots compiled by Binford for the different occupation levels reveal a massive degree of overlap of the different types. Nevertheless, he believes that some statistically significant separation can be recognized in the distribution of these two major tool categories in at least some of the occupation levels at Combe Grenal.

3. Binford claims that a similar pattern is evident in the distribution of particular categories of faunal remains. Thus the larger and more central hearths seem to be associated predominantly with parts of the crania and upper jaws, together in many cases with concentrations of heavily fragmented bone which evidently derive from the intensive smashing of long bones, metapodials etc. for the extraction of marrow. The smaller, more outlying hearths, by contrast, are said to be associated predominantly with either larger and more complete bones, or with the detached, articular ends of the major long bones – particularly those of the lower limbs and feet. In some cases remains of mandibles also seem to be associated with the more peripheral hearths. Binford sees these peripheral areas therefore as reflecting more 'primary' butchery activities, where particular segments of the animal carcases would be processed as a prelude to intensive cooking and extraction of brains and marrow from the remains in the more central parts of the site.

4. Clear patterns are apparently less easy to identify in the distribution of other forms of

lithic and faunal remains. For example, it would appear that localized concentrations of flaking debitage can occur at almost any point on the occupation surfaces – though perhaps generally more frequently towards the front of the occupied areas. Discarded cores are also widely distributed but are perhaps more frequently associated (especially in the case of irregular, non-Levallois cores) with the main concentrations of notched and denticulated tools in the more central parts of the site. In other cases clusters of cores seem to be associated with larger, more bulky animal bones (such as pelvic bones) perhaps indicating localized dumps of material deliberately collected and discarded away from the centres of 'domestic' or processing activities on the site.

5. Finally, Binford claims to have documented some significant patterns in the distribution of different types of lithic raw materials. As discussed in Chapter 4, there seems to be a general tendency in Mousterian sites for some of the smaller and morphologically simpler tool forms (particularly notches and denticulates) to be manufactured preferentially from relatively poor quality raw materials – usually obtained very close to the site – whereas larger and more complex forms such as racloirs, points and bifaces are more frequently manufactured from more distant and generally better quality flint supplies. Binford maintains that this pattern is reflected to some degree in the overall distribution of different varieties of flint in several occupation levels at Combe Grenal. Thus the distribution of more local and poorer quality flints tends to be associated mainly with the notched and denticulated tools in the more central parts of the site, and the better quality, further travelled flints with the main distributions of racloirs, points etc. in the more peripheral areas. However, it seems that this is not an invariable pattern; in some levels the situation may be reversed, with the denticulates being manufactured predomi-

nantly from better quality and more distant flint supplies than those used for the racloirs and pointed forms. Binford's provisional explanation is that this could be related to the variable mobility patterns of the human groups, and the extent to which different animal species were exploited either very close to the site (as in the case of reindeer in the main sequence of Würm II levels) as opposed to more distant locations (as in the case of horses and bovids which were possibly exploited mainly on the various plateaux at some distance from Combe Grenal). He argues that in some cases this could lead to the use of predominantly local flint supplies for manufacturing 'primary' butchery tools (i.e. racloirs and points) whilst more distant flint supplies could have been used to process further travelled carcases (principally horses and bovids) which were introduced into the site in the form of complete skulls and detached lower limb segments, mainly for the extraction of brains and bone marrow.

Clearly, the various arguments in this context become rather complex. Nevertheless, the central notion is that the procurement and use of different flint sources could have been related fairly directly to the particular economic resources exploited in different occupation levels, and the consequently varying territories – and therefore flint-bearing deposits – covered by the human groups in the course of these day-to-day activities.

The various caveats which must be expressed with regard to all of these claimed spatial patterns in the Combe Grenal data are obvious. Until the full supporting data and analyses have been published, it will be impossible to assess the precise character and clarity of any of the spatial patterns and correlations, or how much significance can be attached to the patterns in statistical terms. As discussed further in Chapter 11, there are reasons to suspect that many of the distributions may be less clear and statistically robust than Binford's preliminary accounts of the

data (e.g. 1992) suggest. Clearly, if these patterns are borne out by the full battery of analytical and statistical data, then this will stand as a critical contribution to our understanding of the overall character of the internal structure and organization of Middle Palaeolithic sites. Binford's interpretations of the possible sociological significance of these patterns will be discussed in Chapter 11.

Structural features

Directly related to the character and spatial organization of Middle Palaeolithic sites is the nature and distribution of any associated structural features. Before considering some of the more general implications of the spatial patterns discussed above, therefore, it is worth reviewing the available evidence for various forms of structures in Middle Palaeolithic contexts, under the conventional headings of hearths, paving, walls, pits and postholes.

Hearths

Traces of hearths of one form or another have been recorded from many Middle Palaeolithic sites and can probably be regarded as an almost ubiquitous component of sites of this period (de Lumley & Boone 1976a,b; Perlès 1977; Bordes 1972). The controlled use of fire in living sites can be documented as far back as the 'Mindel' or 'Elster' glaciations (for example at Terra Amata, Vertesszölös, Lunel-Viel etc.), in both caves and open-air locations (de Lumley & Boone 1976a). Presumably, the fires served a variety of functions ranging from a simple source of heat for social or sleeping areas, to defence from carnivores and cooking food. As Binford (1988) and others have pointed out, the occurrence and location of hearths are potentially of considerable interest in pinpointing a focal point for the organization of social and economic or technological activities on the occupation surfaces. This section will examine briefly the variety of hearth forms recorded in Middle Palaeolithic contexts and assess how far any clear patterns can be identified in either their construction or specific location in occupation levels.

Open hearths

The most common forms of hearth reported from Middle Palaeolithic sites consist simply of localized areas of burning at one or more points on the occupation surface. These would appear to represent places where fires were lit on the existing ground surface without any attempt to prepare the surface or restrict the scatter of fuel or ash. Numerous examples have been reported in the literature. At the Grotte Vaufrey for example there would appear to be evidence for two such separate hearth areas, each represented by concentrations of burned bones and charcoal, localized within the southern half of the cave (Rigaud & Geneste 1988). A similar pattern was reported in the Grotte du Lazaret, where two zones of concentrated charcoal, burnt bones and burnt flints were located immediately adjacent to the south wall of the cave (de Lumley *et al.* 1969) (Figs 9.10, 9.13). Other apparently similar features were found in the deep fissure at the Hortus cave (de Lumley 1972). Most such reported hearths are fairly small, rarely measuring more than 40–50 cm in diameter. The slight traces of burning documented in the underlying sediments are usually taken to indicate that the fires were short lived, and rarely attained very high temperatures. There are occasional reports of much larger hearths of this essentially open type. Binford (1992) has reported that many of the more centrally located hearths in the Würm I levels at Combe Grenal seem to cover large, irregular areas measuring up to a metre in diameter, apparently reflecting either the location of more extensive fires or repeated use of the same general area for successive fires. Similarly, Bordes (1971a, 1972) has reported that similar 'non-constructed' hearths in the Rissian levels of the

Pech de l'Azé II cave can attain a metre or more in size.

Constructed hearths

Claims for the deliberate construction of hearths in Middle Palaeolithic contexts are by no means rare, but the evidence advanced for these interpretations is frequently controversial. One sometimes suspects that excavators expect to find deliberate stone arrangements in association with hearth areas and to interpret available evidence in these terms. Granted the virtual ubiquity of small to medium-sized stones in most cave and rock-shelter deposits, claims for such deliberate stone settings must be viewed with caution. Some of the better documented examples are as follows.

Paved hearths

In discussing the range of hearth types recorded in the Rissian levels at Pech de l'Azé II, Bordes (1971a, 1972) described a number of what he regarded as clearly paved hearths. Few details are given, but he claims that several of these could be seen to consist of deliberate arrangements of flattened limestone blocks placed at the base of the hearth deposits and usually showing clear evidence of burning on their upper surfaces (Fig. 9.22). He claims that some of these hearths reached a metre or so in diameter, and seemed to be located preferentially deep inside the cave rather than at the entrance. He suggests that the hearths were intended explicitly for cooking – i.e. hearths in which fires would be lit to heat up the stones, followed by deliberate removal of ashes so that the heated stones could be used for cooking meat or other foods. Without further evidence it is difficult to evaluate this suggestion – although there are of course numerous ethnographic examples of the use of similar stone-lined hearths for precisely this kind of food preparation. Other examples of apparently paved hearths have been reported from the earlier Acheulian levels at Terra Amata (de Lumley 1969b;

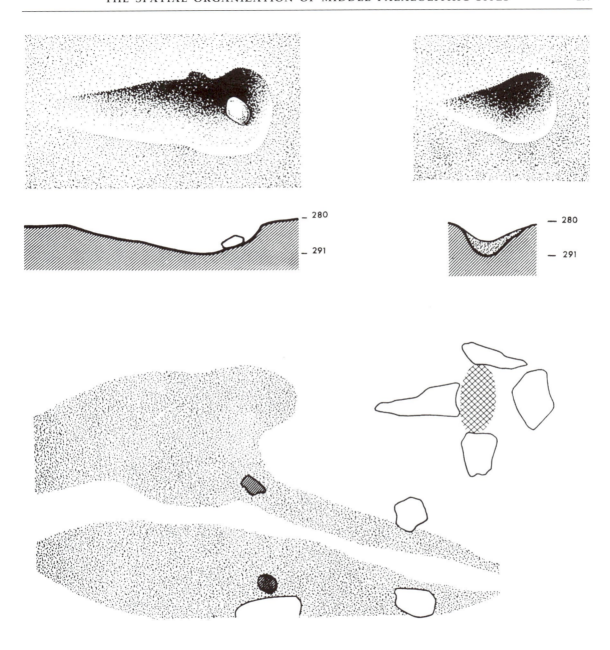

Figure 9.22 *Three forms of hearths recorded by Bordes in the Acheulian (penultimate glacial) levels of the Pech de l' Azé II cave (Dordogne). While some of the hearths are marked simply by a diffuse scatter of ashy material and burned stones (lower), others are marked either by clearly defined scooped depressions (upper) or by burnt areas surrounded by limestone blocks (middle). After Bordes 1971a.*

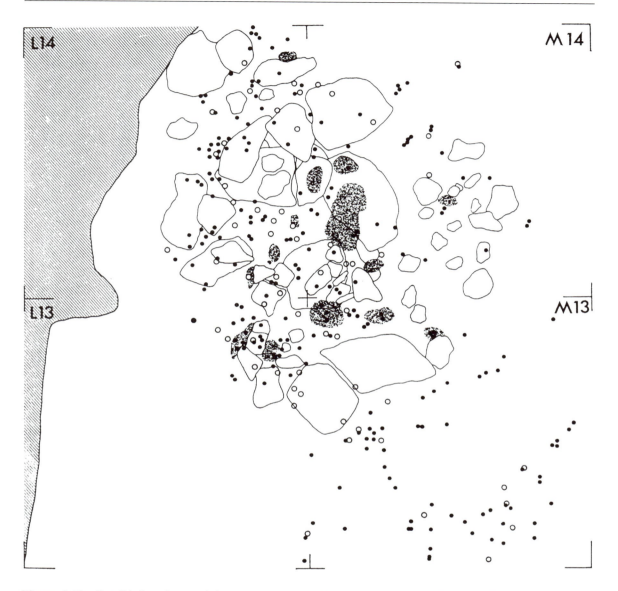

Figure 9.23 *Possible hearth recorded in the Denticulate Mousterian level (layer 3) of the Hauteroche rock shelter, marked by concentrations of burned-bone fragments (indicated by stippling) associated with a scatter of apparently unburned limestone fragments, occupying a scooped depression in the underlying deposits. Stone tools and flakes are marked by open circles and black dots respectively. Grid numbers indicate square metres. After Debénath 1973.*

de Lumley & Boone 1976a) and from the Mousterian levels at Fonseigner (Geneste 1985) and a number of rather poorly dated sites (La Roche Gélétan, Port Pignot etc.) on the Normandy coast (Fosse *et al.* 1986; Michel 1990).

Examples of deliberate stone arrangements placed around the edges of hearths have again been reported from a wide range of Middle and Lower Palaeolithic sites. In the supposedly 'Mindel' levels at Terra Amata, de Lumley reported that many of the hearths on the dune areas of the site appeared to have been deliberately protected by stone settings placed selectively around their north-west edges to shield the fires from the prevailing winds (de Lumley 1969b; de Lumley & Boone 1976a). Similar claims have been made for 'encircled' hearths in the Acheulian levels at Pech de l'Azé II (Fig. 9.22), Lunel-Viel and Orgnac, in the Mousterian levels at Baume les Peyrards (Vaucluse), and in some of the open-air sites on the Normandy coast (de Lumley & Boone 1976a,b; Fosse et al. 1986; Michel 1990).

Two more fully documented examples of deliberate stone settings in association with hearths have been reported from late Mousterian levels in the Grotte-à-Melon (near Hauteroche, Charente) and the Grotte du Bison cave at Arcy-sur-Cure (Yonne) (Figs 9.23, 9.24). Both of these features seem remarkably similar and illustrate some of the difficulties inherent in the interpretation of these stone 'constructions'. At the Grotte-à-Melon, Debénath (1973) described two levels of stones apparently associated with a hearth deposit (Fig. 9.23): a lower level of large flat stones (20–40 cm in diameter) which he believed had formed a basal paving for the hearth and an upper level of smaller stones (less than 20 cm in diameter) which were dispersed more generally around the central hearth area. As will be seen from Fig. 9.23, the actual distribution of the stones is rather diffuse and can hardly be reconstructed into an organized pattern without some element of faith. The pattern documented at the Grotte du Bison is rather clearer (Farizy 1990a). Here there was a well defined central ashy zone (about 25 cm in diameter) which appears to be broadly encircled by a scatter of small to medium-sized stones (Fig. 9.24). Approximately half of these stones are said to show evidence of burning and were pre-

sumably associated in some way with the original construction of the hearth or with some activities immediately around it. Of course it could be that in both cases the original arrangement of stones around the hearths was more sharply defined, and has since been disturbed and obscured by later activities on the sites; even so, to describe either hearth as deliberately 'constructed' may perhaps be stretching the available evidence to its limits.

Excavated hearths

Claims that hearth areas were excavated or scooped into the underlying deposits have again been made frequently for Lower and Middle Palaeolithic sites – for example for several of the hearths recorded at Terra Amata and Lunel-Viel, for at least one of the hearths at Champlost (Burgundy) and in several of the recently excavated sites on the Normandy coast (de Lumley 1969a; de Lumley & Boone 1976a; Villeneuve & Farizy 1990; Michel 1990; Fig. 9.25). Even where deliberate deepening of deposits below hearths can be clearly documented, there is always the question of whether this was a deliberate feature (presumably to provide shelter from draughts) or whether it resulted simply from repeated clearing of fuel and ashes during successive periods of use (cf. Michel 1990).

The most explicit evidence for this form of deliberately excavated hearth has been reported by Bordes from the late Rissian levels at Pech de l'Azé II (Bordes 1971a; 1972: 60–3). The hearths described by Bordes are curious; they are said to be extremely small (usually only 15–25 cm in diameter), to extend for depths of around 10 cm into the underlying deposits, and usually show indications of some kind of vent or channel running from the centre of the hearth towards one side (see Fig. 9.22). Bordes claims that the material surrounding these hearths was clearly reddened by heat and the depressions were filled with blackened, ashy deposits. There seems little doubt therefore that the

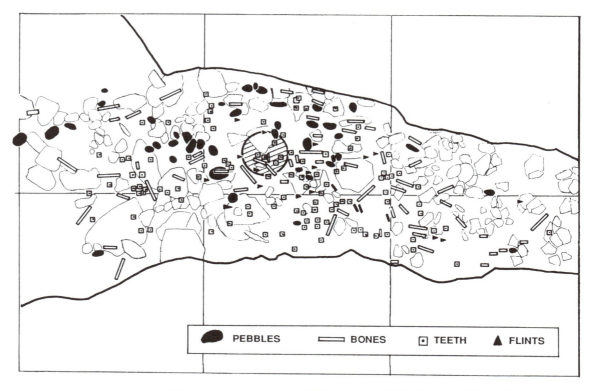

PEBBLES   BONES   ⊡ TEETH   ▲ FLINTS

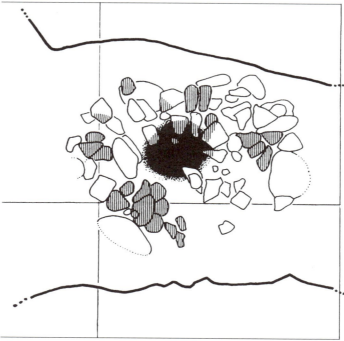

Figure 9.24

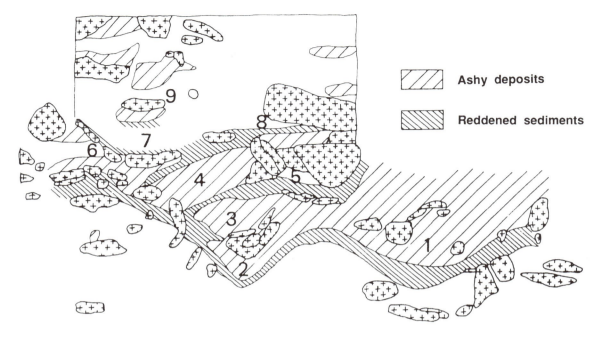

Ashy deposits

Reddened sediments

Figure 9.25 *Sequence of stratified hearths recorded at the early last glacial site of Saint-Vaast-la-Hougue on the Normandy coast. Stones are indicated by crosses. After Fosse 1989.*

depressions were associated in some way with *in situ* burning, but interpretation of their function presents obvious difficulties. From the extremely small size of the hearths it is difficult to see how they could have functioned either for cooking or as a source of heat in any of the usually envisaged ways. Beyond noting that they tend to occur mainly in the deeper, interior areas of the cave, Bordes gives no further hint as to their function. He does, however, claim that similar features have been reported from several Upper Palaeolithic sites – notably in the Russian site of Kostienki XIX, in the Solutrian

levels at the Cueva del Ambrosia (southern Spain) and in the Aurignacian or early Upper Perigordian levels at the Roc de Combe and Corbiac (France) (Bordes 1972: 61–2).

Paving

The existence of clearly defined areas of pebble or stone paving has been documented from many Upper Palaeolithic sites in France, ranging from the Upper Perigordian to the late Magdalenian. The usual presumption is that the paving was intended to regularize and consolidate the occupation surfaces or,

Figure 9.24 *(Opposite page) Clearly defined circular hearth, associated with burned stones (marked by vertical shading), recorded in the later Mousterian levels of the Grotte du Bison at Arcy-sur-Cure (Yonne). The upper diagram shows the location of the hearth in relation to the cave interior and distribution of artefacts and faunal remains in the same level. The grid indicates square metres. After Girard 1976.*

Figure 9.26   *Area of cobble paving recorded in the 'Rissian' (?isotope stage 6) levels of the Baume-Bonne cave in Provence (Basses-Alpes). The paved area covered approximately 6 square metres and contained up to 185 cobbles per square metre, apparently obtained from the alluvial deposits immediately below the site. After de Lumley & Boone 1976a.*

perhaps more likely, to provide a drier and more stable living surface in areas prone to waterlogging (e.g. Leroi-Gourhan 1976; Gaussen 1980; Combier 1982; Sackett 1988).

The existence of analogous features in earlier Palaeolithic contexts is therefore by no means implausible, and seems to be supported by two or three reasonably well documented occurrences from French sites. The best evidence comes from the later Rissian levels in the Baume-Bonne cave, near Quinson in the Basses-Alpes. Here, de Lumley & Boone (1976a) reported several sharply prescribed areas, each covering about 10 square metres, containing continuous and densely packed concentrations of quartz and quartzite pebbles derived from the adjacent gravels of the Verdon river (Fig. 9.26). De Lumley reported densities in places of up to 185 of these cobbles per square metre. The fact that all the pebbles must have been carried deliberately into the site, combined with the relatively sharply defined limits of the paved areas and the fact that most of the fractured pebbles were apparently laid with the convex (i.e. smoother) faces uppermost, would appear to provide strong if not conclusive evidence for the deliberate construction of the paving. De Lumley's explanation is that the pebbles were placed selectively in certain naturally damp areas of the cave floor, where muddy pools would have formed during periods of heavy rains. From the available evidence there seems no reason to doubt either the deliberate nature of the paving or de Lumley's plausible explanation for its origin.

Other claims for deliberate paving in Middle Palaeolithic sites are either less well documented in the literature, or more open to doubt. Localized areas of pebble paving apparently similar to those at Baume-Bonne have been reported from the Mindel-Riss levels of the Aldène cave (Herault), covering about 6 square metres, and from certain levels of the Mas des Caves cave at Lunel-Viel (Herault) (de Lumley & Boone 1976a). According to Tuffreau, a concentrated zone of flint nodules, covering approximately 3 square metres, recorded in level D1 at Biache-Saint-Vaast (Pas-de-Calais) could conceivably represent the remnants of similar paving – although he points out that this could be interpreted alternatively as a localized concentration of stored flint supplies, intended for later tool manufacture on the site (Tuffreau & Marcy 1988). Possible traces of even earlier pebble paving were described by de Lumley (1969b; de Lumley & Boone 1976a) from some of the occupation levels on the dune areas at Terra Amata.

Possibly similar areas of stone paving have been reported so far from only two sites in southwestern France. The most widely reported example was recorded by Denis Peyrony (1934) in his excavations in the large shelter at La Ferrassie. Intercalated between the two principal Ferrassie-Mousterian levels in layers C and D1 he reported a roughly rectangular area of approximately 15 square metres containing what he interpreted as a deliber-

ately laid floor of flattened limestone slabs. Peyrony's account of this feature is brief and from the information provided it is hardly possible to evaluate his interpretation in any detail. Since natural geological accumulations of limestone slabs are known to occur in many rock-shelter deposits, any interpretation of this as a deliberately constructed feature needs to be viewed with caution. An apparently similar layer of limestone slabs was recorded by Bordes in the lower levels of Pech de l'Azé site I and interpreted by him as most probably a natural feature (Bordes 1954–55: 406).

A possibly more convincing feature was recorded by Geneste (1985) in his excavations in the open-air site of Fonseigner in the Dronne Valley. In one of the lower occupation levels (layer I) he recorded a dense concentration of pebbles from the adjacent river, which he describes as having a an apparently organized pattern strongly reminiscent of the undisputed pebble pavements recorded in many of the well known Upper Palaeolithic open-air sites in the same region (Solvieux, Le Cérisier, Guillassou, Le Breuil etc.) (Geneste 1985: 43, Fig. 10; cf. Sackett 1988). This discovery is reminiscent of those reported from Baume-Bonne and other cave sites in the Provence region discussed above and may rank as a further, well documented example of the use of deliberate pebble-paving to stabilize occupation surfaces in Middle Palaeolithic sites.

Stone walls

Whether or not traces of deliberate stone walling can be identified in Middle Palaeolithic sites depends not only on the character of the archaeological evidence but also on the definition of 'wall'. De Lumley (1969b) has claimed that as far back as the Mindel glaciation several occupation areas at Terra Amata were partially surrounded by stone blocks or boulders, which he believes were used either to delimit the occupation zones or hold down

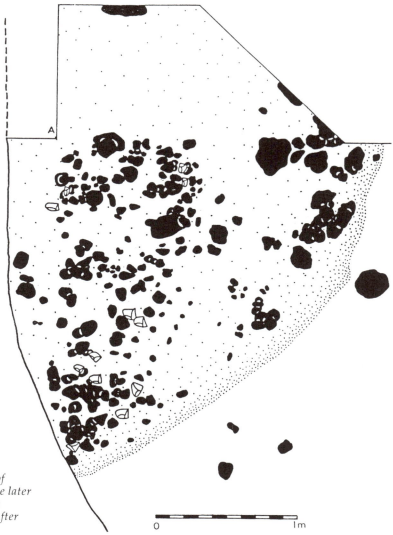

Figure 9.27 *Claimed remnants of stone walling recorded in one of the later Mousterian levels (layer 17) of the Cueva Morín cave in Cantabria. After Freeman 1988.*

0   1m

the skin coverings of tents or similar structures. Similar claims have been made for the hypothetical hut structure at the Lazaret cave and for supposedly similar living structures in the early last-glacial levels of the Baume des Peyrards rock-shelter (Vaucluse) (de Lumley 1969a; de Lumley & Boone 1976b). Further north, a possible structural align-

ment of granite blocks has been reported from the site of Saint-Vaast-la-Hougue, located on top of last-interglacial beach deposits on the Normandy coast (Fosse 1989).

As already discussed in the context of the Lazaret site, these finds are open to problems of interpretation. How far do the stone arrangements represent deliberate features and how far do they reflect purely natural configurations of stones on the sites? Even if the stone arrangements cannot be dismissed as entirely natural, could they just indicate *ad hoc* attempts to improve living areas by clearing away some of the more cumbersome stone blocks from the central occupation areas towards the peripheries? In either case, the stones could hardly be described as 'walls' in any meaningful sense of the word.

Potentially more significant are the occasional claims for segments of 'dry-stone walling' – implying the deliberate piling up of stones in two or more courses. A widely reported case was described by Bordes (1954–55; 1972: 16) in the lower MTA levels (layer 4) at Pech de l'Azé shelter I. Here, he believed he could identify a length of approximately 2 metres of deliberate, wall-like construction, which apparently delimited the main area of human occupation and associated hearths on the site and appeared to extend the natural line of the rock wall of the adjacent cave. No plans, drawings or photographs of this wall have been published and since it is said to have stood only about 25 cm high, there must inevitably be some doubt as to whether it could represent, at least in part, a natural, geological accumulation. Other traces of possible walling are equally poorly documented in the literature. At the cave of Le Rigabe (Var) de Lumley and Boone (1976b) reported a short length of dry-stone walling, said to be almost 40 cm high and approximately one metre in length, which appeared to have been used to protect the entrance to a small area of human occupation in one of the side chambers of the cave. At the Cueva Morín (near Santander, in Cantabria) the evidence reported by Freeman (1988) consisted of a series of localized piles of limestone fragments (said to contain in some cases up to 27 stones, standing at heights of up to 25 cm) distributed mainly around the edges of the most intensively occupied part of the cave (Fig. 9.27). Freeman suggests that

these piles may represent 'either the remnants of a fallen dry stone wall or vestiges of stone heaps used to support the base of a curtain wall of some sort ... demarcating the zone of intense human occupation from the rest of the cave' (Freeman 1988: 22–3). Since similar, smaller, piles of stones were found at several other points in the occupation area (Fig. 9.27), this interpretation is perhaps not without its problems.

Pits

Evidence for deliberate pits in Middle Palaeolithic sites is unquestionably better than that for most of the other 'structural' features discussed above. The best evidence comes from three of the classic Mousterian sites in the Périgord or immediately adjacent areas – Combe Grenal, Le Moustier (lower shelter) and La Quina (Figs. 9.28, 9.29). The most fully documented was recorded by Bordes in one of the earlier Würm I levels at Combe Grenal (layer 50). Here Bordes (1972: 134–6) recorded a roughly circular pit measuring approximately 90 cm in diameter and excavated for a depth of approximately 40 cm into the underlying deposits, with an essentially U-shaped cross-section (Fig. 9.28). The pit was filled with red earth (apparently derived from the deposits of layer 50A) and contained a concentration of small limestone slabs approximately mid-way in the filling. A series of similar but shallower pits were said to be present in the same occupation level but have yet to be described in detail.

Bordes compared this find with a similar pit-like feature documented by Denis Peyrony (1930) in the much later Mousterian levels in the lower shelter at Le Moustier (Fig. 9.29). The pit documented by Peyrony was roughly circular, approximately 70–80 cm in diameter and extended for a depth of approximately 60 cm into the underlying deposits, extending from the surface of layer I into the underlying deposits of layer H. As at Combe

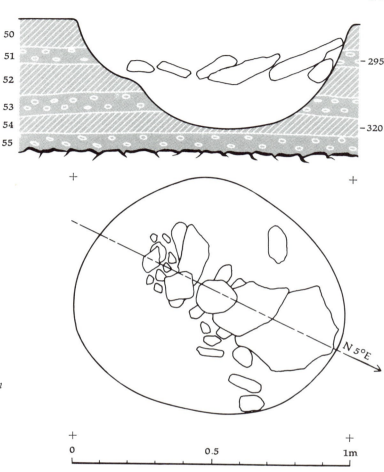

Figure 9.28 *Cross-section and plan of the pit recorded by Bordes in the earlier Mousterian levels (layer 50) at Combe Grenal. Bordes suggested that the pit might originally have contained a human burial. After Bordes 1972.*

Grenal, the upper levels of the pit contained a series of limestone slabs, apparently placed deliberately at the surface of the filling (Bordes 1972: 136–7).

A third well documented pit has been recorded more recently in the current excavations of Jelinek and Debénath in the uppermost Mousterian levels at La Quina (Charente). The dimensions of the pit are similar to those recorded at Combe Grenal and Le Moustier (approximately 70–80 cm in diameter and approximately 50 cm deep) and the pit is said (as at Le Moustier) to show a

flattened base. In this case, the filling of the pit consisted almost entirely of variably sized limestone fragments, up to 30 cm across, which are assumed to have been inserted deliberately to infill and perhaps protect the outlines of the pit. Stratigraphically, the pit appears to relate to one of the levels of the MTA occupation (layer 6A1) close to the top of the Mousterian sequence.

Perhaps one of the most interesting aspects of these pits is the technology implied in their construction. The excavation of pits up to 60 cm deep into consolidated stony deposits

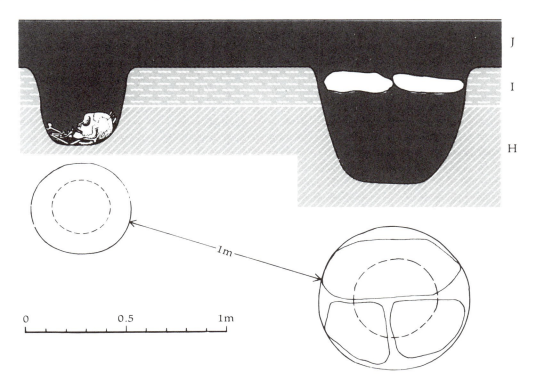

Figure 9.29 *Two adjacent pit features recorded by Peyrony in the uppermost Mousterian level (layer J) in the lower shelter at Le Moustier. The smaller pit contained the burial of a young child. After Bordes 1972; see also Peyrony 1930.*

must have involved a good deal of effort and could presumably only have been achieved with the use of wooden or conceivably bone or antler digging sticks. The intended functions of the pits remains much more enigmatic. Bordes (1972: 134–7) suggested that those recorded at Combe Grenal and Le Moustier may have been intended as human burial pits – pointing out that a second, smaller pit recorded by Peyrony (1930) in the same stratigraphic level at Le Moustier was found to contain the skeleton of a young child (Fig. 9.29); the absence of similar burials in the larger pit at Le Moustier and at Combe Grenal could, he suggested, reflect either poorer conditions for bone survival in these features or the burial of even younger infants whose bones could hardly be expected to

survive. A very different interpretation has been advanced by Jelinek and Debénath for the pit at La Quina. They argue that this and, by implication, other similar pits might have been intended for the long-term storage of meat or other food supplies – particularly over the winter months when permanently frozen soil conditions could well have allowed storage of meat supplies for several weeks or even months. Inevitably, both interpretations remain highly speculative from current evidence. Nevertheless, the very clearly defined nature of these pits in well documented Middle Palaeolithic contexts is highly significant and provides some of the best evidence at present available for the deliberate construction of occupation features in Middle Palaeolithic sites.

Post-holes

Finally, there is some sparse but interesting evidence for the occasional occurrence of post-holes on Lower and Middle Palaeolithic sites. Several apparent post-holes were reported by de Lumley (1969b) in association with the various occupation structures on the beach and dune areas at Terra Amata. Whilst the interpretation of these features has generated much controversy in the literature (cf. Villa 1977, 1983; Gamble 1986), there is far less ambiguity surrounding the single, very clearly defined post-hole recorded by Bordes in the upper part of the Mousterian succession (layers 14/15) at Combe Grenal (Fig. 9.30). Bordes describes this post-hole (which was carefully excavated and subsequently cast in plaster) as extending for a depth of at least 20 cm into the underlying deposits and showing a roughly circular cross-section approximately 4 cm in diameter. The tip of the post appears to have been pointed and apparently mushroomed slightly as it was driven against a hard stone in layer 21. Even if the precise form and function of the wooden stake remains hypothetical, the reality of this feature is difficult to deny. Bordes reported no further post-holes in this part of the site, but points out that other related features could have existed in adjacent parts of the deposits which were removed during earlier excavations by Peyrony and others. Any speculation as to whether the post-hole may have formed part of a larger structure on the site is therefore impossible to resolve. The main interest of this find is that it indicates not only the occasional use of wooden posts on Middle Palaeolithic sites but also, by implication, the associated wood-working technology required to produce them.

Summary

The problems inherent in any analysis of the spatial organization of Middle Palaeolithic

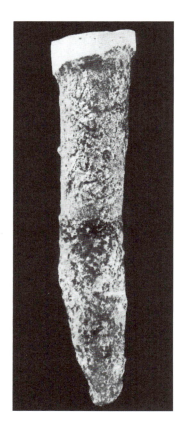

Figure 9.30 *Cast of the post-hole recorded by Bordes in one of the later Mousterian levels (layer 14, containing a Denticulate Mousterian industry) at Combe Grenal. The stake was approximately 4 cm in diameter, and penetrated for a depth of at least 20 cm into the underlying deposits. After Bordes 1972.*

sites, or indeed those of most periods in prehistory, have been emphasized in the introduction to this chapter – the virtual inevitability of occupational palimpsests, incomplete or biased patterns of survival of different kinds of occupation material in different areas of the living surfaces, the effects of natural and human disturbance of the occupation residues etc. – all combined with the limitations of the particular excavation and recording techniques employed in the original excavations. Making allowance for all these potential sources of distortion of the original spatial patterns, it might seem unrealistic to expect to discern any coherent patterns in the distributions documented for specific sites. In fact, the data which has emerged from the studies described above provide much more information on several of these issues than might have been anticipated. The available sample of sites is of course extremely limited, and one should be cautious of offering generalizations from such limited case studies. Certain patterns, nevertheless, seem reasonably clear.

1. One of the clearest and possibly most significant patterns is the way in which the distributions of occupation residues can often be seen to occupy relatively restricted and sharply defined areas of the available occupation surfaces. This seems to be clearly reflected in the distributions recorded in at least three of the sites discussed above – Grotte Vaufrey, Grotte du Lazaret and Les Canalettes. The obvious implication is that the human groups who utilized these sites must have been relatively small, and can hardly have comprised more than five or ten individuals. As discussed in Chapter 11, it would be premature to assume that all Middle Palaeolithic social groups were of similar size and it is possible that some of the larger and more intensively occupied sites reflect the activities of much larger social aggregations. Nevertheless, the clear documentation of these small, restricted occupation areas in several Middle Palaeolithic sites is one of the most positive contributions to have emerged from recent spatial studies of Middle Palaeolithic sites.

2. The distribution and location of hearths on well-defined occupation surfaces reveals some equally interesting patterns. As noted above, various forms of hearths are relatively common features in many Middle Palaeolithic sites and are often distributed widely over the occupation areas. At least two well defined hearths were identified at the Grotte du Lazaret, apparently two similar features at the Grotte Vaufrey and, according to the studies of Binford, equally if not more frequent hearths in several of the Würm I levels at Combe Grenal. Equally frequent hearths have been reported, though not yet fully described, in the excavations of Bordes in sites I and II at Pech de l'Azé (Bordes 1954–55, 1972). The positions of the hearths, however, appear to be extremely variable. In some cases (as at Combe Grenal and, apparently, at the Grotte du Bison at Arcy-sur-Cure and the Grotte-à-Melon) they may be located in a fairly central position, with respect to the total space available in the sites (Figs 9.23, 9.24: Binford 1992; Farizy 1990a; Debénath 1973). More commonly, however, hearths seem to be distributed in more marginal areas of the sites – for example against the rock wall of shelters, close to the estimated drip line, or (as at the Grotte du Lazaret [Fig. 9.13] and, apparently, at Pech de l'Azé II) relatively deep inside the cave interiors (de Lumley et al. 1969; Bordes 1972). What seems to be lacking in most of the sites is a tendency to place hearths in a consistent location during different, successive episodes of occupation. Some of the patterns reported by Binford (1992) in the Würm I levels at Combe Grenal may be an exception to this, but the lack of a repeated, regular location of hearths may be a general feature of Middle Palaeolithic sites. Interestingly, a similar pattern was reported in the much discussed mammoth-bone 'structure' in the Ukranian site of Molodova V, where traces of hearths are reported from at least eight or nine separate locations in the most densely occupied area of the site (Păunescu 1989; Soffer 1989a).

3. Despite the variable location of hearths, there may be some regularities in the specific kinds of occupation and industrial residues found in association with them. At both the Grotte Vaufrey and Grotte du Lazaret it can

be seen that the major concentrations of lithic flaking debitage (apparently resulting from *in situ* flaking of nodules) are associated closely with the areas immediately adjacent to the hearths (Figs 9.2, 9.12), indicating that these areas served as major locations for both certain stages of primary flaking and the subsequent stages of tool production. It seems that here and also at the Grotte-à-Melon and in several of the more centrally located hearths at Combe Grenal the same areas tend to contain the highest concentrations of small bone splinters (Figs 9.9, 9.14), suggesting that they were also used extensively for the extraction of marrow from the long bones of exploited animals (Debénath 1973; Binford 1992). Similar concentrations of animal teeth in close association with hearths at Combe Grenal, Grotte Vaufrey, the Grotte du Bison (Fig. 9.24) and Grotte-à-Melon may suggest that the same areas frequently served for the systematic breakage and processing of animal skulls – presumably to extract the brains and tongues. These patterns suggest that the places immediately adjacent to hearths habitually served as major centres both for various industrial activities and for the intensive processing of at least certain kinds of food.

However, these patterns are not invariable. In certain instances there are indications that intensive flaking of cores and nodules was carried out well away from the most central zones of occupation, either close to the talus region of rock-shelters (as at La Ferrassie) or adjacent to large stone blocks around the edges of the most centrally occupied areas (as at the Grotte du Lazaret and apparently in certain levels at Combe Grenal).

4. Some further general patterns seem to emerge from the distribution of the larger and more complete fragments of animal bones on sites. In general, most kinds of faunal debris tend to be fairly widely distributed over occupation surfaces, and are usually less tightly concentrated than those of lithic flaking debitage – with the possible exception of bone splinters and heavily fragmented parts of skulls and jaws. At several sites, however, most of the larger bone fragments seem to be distributed mainly in the more marginal areas of occupation, well away from the most intensively used parts of the site. At both Les Canalettes and Grotte de l'Hyene (Arcy-sur-Cure) it is reported that most of the large bone elements were concentrated mainly towards the walls of the shelters (Meignen 1994; Farizy 1990a; de Lumley & Boone 1976b), while at Grotte Vaufrey and Grotte du Lazaret it seems that they are fairly dispersed over most of the marginal areas. Two possible explanations could be advanced to account for these patterns. One possibility is that it was these marginal areas of the occupation surfaces that were reserved for the 'primary' or heavy duty butchery of animal carcases, from which certain selected parts of the carcases were removed for more intensive processing in adjacent areas. The alternative is that these peripheral areas served simply as general refuse disposal zones, where the bulkier and cumbersome elements from the butchered animal carcases were deliberately discarded well away from the central 'domestic' or industrial areas. Whatever the explanation, there does seem to be a significant contrast between the distribution of larger bone elements as opposed to fragmented bones and teeth in many Middle Palaeolithic sites.

5. Finally, what patterns can be documented in the distribution of different categories of retouched stone tools on the occupation surfaces? Unfortunately, most of the individual occupation levels in which these distributions have been recorded in detail (as at Vaufrey and Lazaret) have yielded far too small samples of the individual tool categories to provide any secure basis for analysis. Binford (1992) has suggested that some

general patterns of this kind can be detected in various occupation levels at Combe Grenal, with a tendency for most of the simple notched and denticulated forms to be concentrated mainly in the central zones of the occupation surfaces, while various forms of racloirs, points and related types appear to be distributed more frequently in the more peripheral parts. Until the detailed distribution plots to support these generalizations have been published it will inevitably remain difficult to evaluate these patterns. As noted above, however, there may be hints of a similar pattern at the Grotte Vaufrey; according to the spatial plots provided by Rigaud and Geneste (1988: 608–9), it seems that the overall ratios of denticulates and notches to racloirs are perhaps slightly higher in the central zones of the occupation (towards the south wall of the cave) than in the marginal areas towards the north and east (Figs 9.6, 9.7). Even so it would almost certainly be premature to attach too much significance to these patterns until the critical supporting data from Combe Grenal and other sites have been fully analysed and published.

Discussion

The impression which emerges is that Middle Palaeolithic sites are by no means lacking in certain kinds of internal organization, or even clear 'structure', when viewed in spatial terms. The patterns documented at the Grotte du Lazaret, Grotte Vaufrey, Les Canalettes, Arcy-sur-Cure etc. leave no doubt that the various economic, technological and social activities carried out on Middle Palaeolithic sites were patterned and regulated in various ways by certain spatial constraints which, despite the inevitable blurring effects of occupational palimpsests and post-depositional disturbance, have left some discernible traces in the archaeological record. There seems little doubt that certain activities were carried out predominantly in relatively close

proximity to hearth areas, while other activities were located primarily in more marginal areas of the sites. Overall, the evidence points to at least some broad regularities in the ways in which certain activities were organized and separated within the available occupation space on the sites.

The existence of such patterns is of course hardly surprising. Similar patterns of use of different areas of occupation surfaces for different economic or technological activities can now be documented from some of the earliest hominid sites, such as those at Olduvai Gorge or Koobi Fora in East Africa, as well as in many of the later Acheulian sites such as Isimila and Terra Amata (Isaac 1984; Howell *et al.* 1962; de Lumley & Boone 1976a). Indeed, even in some primate groups, the existence of such behavioural patterns can be documented in the use of different activity locations – for example in recent studies of chimpanzee communities in West Africa (Sept 1992). The question of how far we can attach any deeper conceptual or cognitive significance to these patterns is of course far more debatable. To take a minimalist perspective, one could argue that effectively all of the documented spatial patterning recorded in Lower and Middle Palaeolithic sites could be seen essentially as a purely pragmatic response to the simple functional requirements and constraints involved in the organization of various economic, technological and social activities within specific, restricted areas. Thus the concentration of certain activities in close proximity to fire places need reflect nothing more than either the understandable desire for greater comfort during long periods of food processing or industrial activities (particularly under conditions of glacial climate) or alternatively the need for fires in the essential processing of the materials being worked – for example in the cooking of foods, the heat-treatment of flint to improve its flaking properties (Meignen 1982) or the fire-hardening of tips of spears or digging sticks. Similarly, the

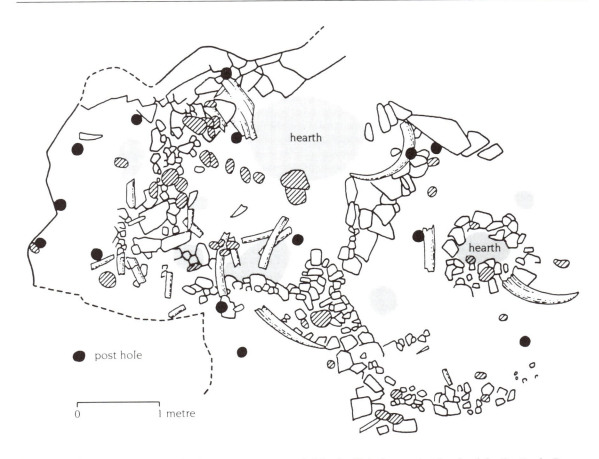

hearth

hearth

post hole

0   1 metre

Figure 9.31 *Plan of two circular hut structures recorded in the Châtelperronian levels of the Grotte du Renne Cave at Arcy-sur-Cure (central France), dated by radiocarbon to ca 33–34,000 BP. The huts are defined by settings of stones, post-holes, and centrally located hearths. After Farizy 1990b.*

tendency to process and discard some of the larger components of animal carcasses well away from the more central zones of domestic or industrial activities need reflect nothing more than the understandable inclination to keep bulky, cumbersome or smelly parts of domestic refuse away from the most heavily used areas. Granted these factors, it is difficult to see the need to invoke a deeper cognitive or conceptual structure in the organization of activities on Middle Palaeolithic sites which goes beyond such simple functional or pragmatic considerations. Even

the practice of deliberately laying down localized areas of pebble paving to provide drier or more stable surfaces need hardly go beyond basic pragmatic considerations – well within the capacity of groups who were habitually accustomed to importing large quantities of lithic raw materials for use in occupation sites.

Clearly, this is a minimal view of the nature of spatial patterning in Middle Palaeolithic sites, for which more imaginative interpretations would no doubt be possible. Nevertheless, very few if any of the spatial patterns so far documented in Middle Palaeolithic sites exhibit the same degree of complexity, organization or obvious structure which can now be documented in many Upper Palaeolithic settlements. A detailed analysis of all the relevant evidence from Upper Palaeolithic sites goes well beyond the scope of this discussion. Briefly, however, one can point to at least three major features now documented from a wide range of Upper Palaeolithic settlements which seem to go significantly beyond any of the patterns so far documented in Middle Palaeolithic sites:

1. First, there can be no doubt that many Upper Palaeolithic sites show far clearer and more sharply defined evidence for deliberate living structures than anything so far documented from Middle Palaeolithic sites. Some of the most impressive and best documented of these structures are those recorded from the so-called 'East Gravettian' open-air sites in central and eastern Europe, such as Dolní Věstonice, Pavlov, Kostienki, Mezhirich, Mezin etc. (Klein 1969, 1973; Gamble 1986; Soffer 1985a). In western Europe, however, we now have evidence for equally clear living structures extending back to the earliest stages of the Upper Palaeolithic sequence – most notably in the well defined hut foundations, marked by circular settings of stones and associated post-hole arrangements, documented in the Châtelperronian levels at Arcy-sur-Cure (Fig. 9.31) (Leroi-Gourham & Leroi-Gourhan 1964; Farizy 1990a) and, rather more speculatively, in the apparently rectangular structure documented in the roughly contemporaneous Aurignacian levels at the Cueva Morín in northern Spain (Freeman & Echegaray 1970). Later examples of similar structures are recorded from the Upper Perigordian levels at Villerest (Loire) (Fig. 9.32) and at Corbiac (Dordogne) (Combier 1982, 1984; Bordes 1968b). However much weight one may attach to the slight evidence for possible occupation structures recorded in a few Middle Palaeolithic sites (Păunescu

1989), none of this can compare with the substantial and explicit evidence for living structures documented from many Upper Palaeolithic sites.

2. Second, there would seem to be evidence for some kind of clearly structured, preconceived form in the design and construction of many Upper Palaeolithic living structures. Arguably the most significant feature in this context is the conspicuously circular plans apparent in many of these structures – as for example in the two juxtaposed hut plans recorded in the Châtelperronian levels at Arcy-sur-Cure (Fig. 9.31) or in the similar plans visible at many of the later Gravettian or Upper Perigordian sites such as Villerest, Dolní Věstonice, Mezhirich etc. (see Fig. 9.32: Leroi-Gourhan & Leroi-Gourhan 1964; Leroi-Gourhan 1976; Combier 1982, 1984; Gamble 1986). The element of clear form and structure visible in these settlement plans goes beyond anything at present demonstrable from Middle Palaeolithic sites – with the possible (though disputed) exception of the apparently circular ring of mammoth bones documented at Molodova (cf. Soffer 1989a).

3. Finally, one of the most striking features of many documented Upper Palaeolithic settlements is the way in which the principal areas of occupation can usually be seen to be centred around one major and centrally located hearth. Numerous examples could be quoted going back to the very early structures at Arcy-sur-Cure (Fig. 9.31) and continuing through all later stages of the Upper Palaeolithic sequence (cf. Gamble 1986: 251–272). Binford (1988: 559–60) has recently commented on this point in relation to the Grotte Vaufrey excavations, and points out that this tendency to concentrate activities around a central hearth seems to be characteristic not only of most Upper Palaeolithic settlements but also of most of the ethnographically documented settlements of modern hunter-gatherers. As Gamble (1986: 264) has pointed

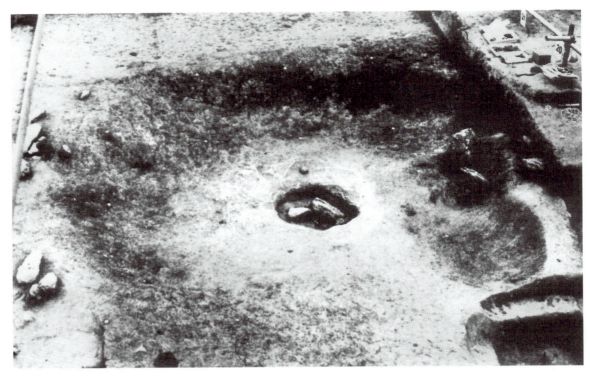

Figure 9.32 *Foundations of circular hut structure, with a centrally located hearth, recorded at the Gravettian site of Vigne-Brune (Villerest) in eastern France, dated to ca 27,000 BP. After Combier 1984.*

out, this clear centralization of economic and social activities could possibly reflect several factors, ranging from patterns of mutual cooperation and sharing in nuclear family units to more general patterns of communication between groups of individuals engaged in particular communal tasks. Whatever the explanation, it seems that there is at present much less evidence for this kind of patterning in Middle Palaeolithic sites. With a few significant exceptions, most of the hearths documented in Middle Palaeolithic sites seem to show an irregular, widely dispersed pattern which bears little resemblance to those recor-

ded in most of the documented Upper Palaeolithic sites. These points will be pursued further in later chapters, in the context of more general comparisons of the social and cognitive organization of Middle and Upper Palaeolithic groups. The general implications nevertheless seem evident. Whatever significance one may attach to some of the apparent elements of spatial organization in Middle Palaeolithic sites, the character of this patterning remains in at least certain fundamental respects clearly different from that documented in many of the settlements of the Upper Palaeolithic period.

CHAPTER 10

The Significance of Industrial Variability

We now come to the issue which has largely dominated studies of the Middle Palaeolithic for the last thirty years – namely the challenge of explaining the bewildering array of variation documented in the stone-tool assemblages from Middle Palaeolithic sites. Several aspects of this question have been touched on in earlier chapters, including the effects of different raw materials on some of the technical and morphological aspects of the tools, or the possible effects of spatial localizations of different tool types on occupation surfaces on the composition of artefact assemblages recovered during excavations. The most critical debates of the past three decades, however, have centred not so much on these small-scale variations in tool morphology or assemblage composition but on the very gross contrasts in the relative frequencies of the major retouched tool forms which collectively define the chief industrial variants of the Bordes taxonomy – i.e. the Ferrassie, Quina, Denticulate, Typical and Mousterian of Acheulian tradition (MTA) variants. Three radically contrasting explanations for these patterns have been advanced: first, the 'functional variability' models, associated mainly with the work of Lewis and Sally Binford; second, the 'tool reduction' models proposed by Nicholas Rolland, Harold Dibble and others; and third, the more traditional 'cultural' models, advocated throughout the publications of François Bordes.

These debates over the character and behavioural significance of industrial variability in the Middle Palaeolithic are of interest not merely for their own sake but because of their critical bearing on almost all aspects of our current understanding of the basic patterns of technological adaptation, economic organization, settlement strategies and indeed social and demographic structure of Middle Palaeolithic groups. The various theoretical issues raised by these debates also have implications which extend far beyond the immediate scope of the Mousterian problem and impinge on the interpretation of technological variation throughout all stages of the Palaeolithic. The complex, technical, and seemingly endless debates on Mousterian variability in so much of the literature during the last 30 years may be tedious but they are not trivial or peripheral to any attempt to understand the overall structure or organization of Middle Palaeolithic society. The following sections will take each of the major models in turn and examine their general implications and testability in terms of the evidence from the southwestern French sites.

The functional variability Models of Lewis Binford

Few debates had a greater impact on Palaeolithic archaeology during the post-war period than those surrounding the 'func-

tional' models of industrial variability proposed in several publications by Lewis
Binford (initially in collaboration with Sally
Binford) (Binford & Binford 1966, 1969; Binford 1973, 1983a; see also Freeman 1966).
These interpretations were formulated initially during the early years of the 1960s and,
as Binford has emphasized, were presented
largely as an explicit reaction against the
idea of contemporaneous, parallel cultural
phylla within the Mousterian proposed in
the interpretations of François Bordes. This
essentially reactive nature of the functional
interpretations was stressed by Binford:

> In view of the demonstrated alternation of
> industries, one must envision a perpetual
> movement of culturally distinct peoples, never
> reacting to or coping with their neighbors. Nor
> do they exhibit the typically human
> characteristics of mutual influence and
> borrowing. Such a situation is totally foreign,
> in terms of our knowledge of *sapiens* behavior.
> The purpose of this paper is to present an
> alternative set of testable hypotheses as
> possible explanations for the observed
> variation and alternation of Mousterian
> industries demonstrated by Bordes.
> (Binford & Binford 1966: 240)

Some of the more general implications of the
Binford models have been discussed in earlier chapters. Without attempting an exhaus-
tive review of all the publications by Binford
on these issues, it can be said that these
models reduce, in essence, to three basic
propositions (Binford & Binford 1966: 267–9,
1969: 70–2; Binford 1983a: 66–9):

1. That Middle Palaeolithic social groups
   were divided internally (probably according to age and sex) into separate activity or
   foraging units, which carried out largely
   discrete subsistence and other economic
   activities, either in different spatial locations or in different seasons of the year.

2. These activities were patterned largely in
   response to the variable distribution of
   different economic resources and led to
   specific combinations of particular economic or technological activities at different
   site locations within the group's foraging
   territories (Binford 1982a).

3. In the course of these activities, different
   forms of tools were employed for certain
   specific activities, related directly to the
   resources being exploited. Since most
   tools employed by Middle Palaeolithic
   groups are assumed to have been used in
   an 'expedient' way (i.e. manufactured,
   used and discarded over a short time span)
   the particular associations of tool forms
   encountered at each site are assumed to
   reflect, fairly accurately and directly, the
   varying character of the activities carried
   out in the different locations (Binford
   1973: 242–53, 1977: 266–7, 1983a: 66).

The development of these central and basic
assumptions can be traced through Binford's
publications over thirty years (Binford & Binford 1966, 1969; Binford 1972, 1973, 1977,
1979, 1982a,b, 1983a,b, 1989, 1992). Initially,
the models were applied specifically to the
various Levallois-Mousterian industries
from the Middle Eastern region (notably
those of Jabrud and Shubbabiq in Israel: Binford & Binford 1966, 1969) and were only
extended later to the classic sequence of
Mousterian industries from the southwestern
French sites (Binford & Binford 1969; Binford
1973). Their application to the French industries has, as yet, been focused almost exclusively on the long succession of Mousterian
levels (55 layers) recorded during Bordes'
excavations at Combe Grenal – involving (as
noted in Chapter 9) detailed analyses not
only of the varying composition of the lithic
assemblages, but also the associated patterns
of faunal material and the overall spatial
distribution of the lithic and faunal remains
in the different occupation levels. Detailed

discussion of these results is hampered by the fact that as yet only brief and partial results have been published (see for example Binford 1978, 1981, 1982b, 1983a, 1984, 1992). Clearly, a full analysis of Binford's interpretations of the Combe Grenal material must await his forthcoming monograph with full supporting analytical and quantitative data.

From results published so far, however, and from further information provided personally by Binford, it is clear that he visualizes effectively the whole spectrum of typological and technical variation documented in the Combe Grenal sequence – and by implication from other southwestern French sites – as primarily reflecting the various kinds of functional mechanisms outlined above. Whilst not denying the possibility of certain broadly adaptive shifts in basic economic and settlement patterns in response to climatic and ecological oscillations during the earlier part of the last glaciation (Binford 1973: 231–2; 1982a) he nevertheless rejects the notion that these adaptations can be used to explain the major and most distinctive industrial facies of the Mousterian documented in the studies of François Bordes. In essence, Binford suggests that these classic variants of the Mousterian (Ferrassie, Quina, Denticulate, Typical, MTA etc.) must be seen as reflections of activities carried out by segments of essentially the same human groups in a range of different environmental or activity locations (Binford 1973: 231–2, 1982a: 369–77, 1983a: 65–9, 1983b: 157–8). Thus, his model corresponds, in other words, to what I defined in an earlier publication (Mellars 1970: 74) as the 'inter-locational' (or 'functional facies') model of industrial variation within the Mousterian complex.

Beyond these broad statements of principle, we have as yet only a limited indication of exactly how Binford visualizes the different industrial variants in specific functional or activity-related terms. As Binford is aware, we are at present very poorly equipped with reliable data on the original functions of the major Mousterian tool types (racloirs, denticulates, backed knives, hand axes etc.) and any suggestions about how the different types may have functioned remain at best highly speculative. The closest which Binford has come to proposing specific functions for the different hypothetical tool kits is the suggestion that the major categories of sharp-edged tools (i.e. the various forms of racloirs and related point forms) were probably associated primarily with butchery and processing of animal carcases and the notched or denticulated tools are more likely to relate either to woodworking activities or certain plant processing tasks (Binford & Binford 1966: 249–59, 1969: 79). He then suggests that the industries which show a heavy predominance of racloirs (i.e. the various Ferrassie and Quina assemblages) probably reflect mainly activities relating to the primary processing of animal carcases, while the various denticulate-dominated industries may relate to either the processing of plant foods or perhaps to more general food-processing activities, possibly involving the intensive extraction of brains, tongues and bone marrow from the residues of hunted or scavenged animals. He envisages that the former would have been carried out mainly by the males and the latter mainly by the females (Binford 1992). Beyond these very general speculations, there are virtually no specific suggestions as to how the individual industrial variants of the Mousterian should be interpreted in particular functional terms, nor how the different patterns of activity were related either to the occupation of different site locations or to different seasons. Expressed in such general terms, it is inevitably difficult to subject the Binford models to any systematic testing against the available archaeological data, or indeed to formulate how they could possibly be tested in the course of future research. Hopefully, these issues will be outlined more fully in the forthcoming monograph on the results of the Combe Grenal research.

Evaluation of the Binford models

Even the harshest critics of the Binford models (e.g. Jelinek 1988a: 200) would not deny the fundamental theoretical importance of his functional approach to the study of industrial variability, nor the profound impact it has had on Palaeolithic research during the last thirty years. The inherent rationale of the functional approach is self-evident. Even a cursory acquaintance with the ethnographic literature on hunter-gatherer behaviour (e.g. Lee & DeVore 1968, Damas 1969, Bicchieri 1972, etc.) confirms that virtually all hunters and gatherers practise highly variable, mobile patterns of economic and social behaviour, which must inevitably leave an imprint on the character and composition of the tool assemblages and associated manufacturing residues abandoned at different locations. There can be no doubt that *some* component of variability of this kind must be reflected in the surviving archaeological residues of the Middle Palaeolithic – as indeed those of the Upper Palaeolithic, Mesolithic and later periods (cf. Mellars 1976). The only question is the scale on which this kind of variability can be recognized. In other words can we invoke these models to provide an almost complete explanation for the bewildering range of typological and technical variation apparent in the archaeological records of the Middle Palaeolithic, or do they account in practice for only a very small part of this variation?

In the present context there are two separate issues posed by the Binford models. On the one hand there are the various theoretical or methodological assumptions which in one form or another underlie the functional models. On the other hand there is the question of how far these models can be applied to some of the hard facts of the archaeological record, specifically those from the southwestern French sites. These two issues are addressed successively in the following sections.

Theoretical assumptions

1. A central assumption which underlies the whole of the Binford models is that the basic typological systematics employed in the analysis of Middle Palaeolithic assemblages measure, primarily if not entirely, functional contrasts in the use of different tool forms. Thus, the Binford models imply a largely direct equation between form and function for most documented Middle Palaeolithic tool types. However cautiously expressed, this assumption is ultimately central to the functional variability models, since the recognition that certain tool types were not related directly to specific functional requirements would automatically be a strong argument for some other, explicitly *non*-functional variation in the lithic assemblages.

As discussed earlier, this assumption has been questioned by many aspects of research into both the morphology and specific functional orientation of Middle Palaeolithic stone tools. One of the most striking contributions of recent use-wear studies is that in many contexts, different morphological forms of Middle Palaeolithic tools seem to have been used for broadly similar functions, which evidently overlapped between the morphologically contrasting types in complex and enigmatic ways. This has now been clearly documented for the various categories of sidescrapers, backed knives, points and apparently hand-axe forms, and may even have extended to some of the more apparently functionally specialized forms such as notches and denticulates (cf. Keeley 1980; Beyries 1987, 1988a; Plisson 1988; Anderson-Gerfaud 1981, 1990). To assume a simple, direct equation between the forms of different types and the specific economic or technological activities for which they were employed would run counter to all recent research in these fields. The recent tool-reduction models of Dibble and Rolland, discussed in the following section, are also significant in this context. Leaving aside some of the more controversial aspects of these models, Rolland and Dibble (1990; also Rolland 1977, 1981; Dibble & Rolland 1992) have argued persuasively that many common and widespread activities in Middle Palaeolithic sites are likely to have been undertaken with either unretouched or very lightly retouched flakes (cf. Beyries 1987, 1988a; Beyries & Boëda 1983; Keeley 1980; see also Binford 1979: 277 and Gould 1968: 47 for ethnographic accounts of the use of unretouched flakes in different activities). Clearly, any notion that all the recognized typological variants of Middle Palaeolithic tools were produced with differentiated functions in mind, can be questioned from several aspects of the current data.

Viewed from this perspective, it is clearly debatable how far many of the hypothetically distinct Mousterian variants in the original Bordes taxonomic scheme can be seen as significantly different from a functional viewpoint. This is most obviously true of Bordes' distinction between the Ferrassie and Quina variants, which in his terms was always defined primarily by contrasts in the basic production techniques rather than by variations in the overall tool-type composition of the assemblages (Bordes 1953a, 1961a, 1968a, 1981). The same may be true of the Typical Mousterian industries, many of which differ from the Ferrassie/Quina group only in relatively slight differences in racloir percentages (see Chapter 6). If we assume that many of the functions of hand axes and backed knives could have overlapped with those of the various racloir categories, as much of the present evidence suggests (Keeley 1980; Beyries 1987, 1988a; Plisson 1988; Anderson-Gerfaud 1990), then there could be equally little functional contrast between most of the MTA industries and those of the other groups (Ferrassie, Quina and Typical Mousterian) referred to above. Arguably the only variants of the Mousterian which could suggest any obvious contrasts in overall functional orientation are those of the Denticulate group, and perhaps some of the later

(Type B) MTA industries, which show frequencies of racloirs versus notched and denticulated forms which are (by definition) different from those of the other industrial groups. But as noted earlier, this would effectively ignore the possible role in these industries of unretouched flakes (cf. Rolland and Dibble 1990; Mellars 1970: 85) and would in any case account for only a small part of the total industrial variability within the Middle Palaeolithic.

None of this is meant to imply that some of the industrial variations discussed above do not have some significance in broadly functional terms, but until we have more reliable information on the functions of the different tool forms, any suggestion that the varying tool inventories must reflect clear functional contrasts remains purely an assumption on the part of the Binford models, without any supporting data from the archaeological record itself.

2. A second curious assumption of the Binford models is that effectively all the documented contrasts between the different industrial variants of the Mousterian lie in gross variations in the overall frequencies of the different tool forms, rather than in the presence or absence of specific, idiosyncratic tool forms (Binford 1973: 246, 1982b: 180). Binford attaches some significance to this, since he argues:

> It seems more reasonable to recognize that social and ethnic considerations do not seem to be operative at the level of tool design, *since typologically identical forms are demonstrably distributed over half the inhabited world during the Middle Palaeolithic.* To demand that they are being expressed in a form which is not cognitively apparent, e.g. proportional frequencies varying between locations *in identical tool forms,* clearly goes beyond any understanding of the functions of culture and its dependence on cognitively recognizable expediently assigned meanings to form – i.e. symbols.
>
> (Binford 1973: 246; italics added)

Exactly where this idea of the absence of distinctive morphological tool forms in certain Mousterian variants may have arisen is something of a mystery. As Bordes (1961b, 1968a, 1977, 1981, 1984), Peyrony (1920, 1930) and others emphasized on numerous occasions, the whole definition of the MTA variant in southwestern French sites has always rested on the presence in these industries of two idiosyncratic tool forms – typical cordiform hand axes and typical (i.e. extensively retouched) backed knives (Figs 4.19, 4.22) – both of which seem to be lacking in well documented occurrences of all other Mousterian variants in southwestern French sites (see below and Chapter 6). Similarly, Bordes frequently pointed out that several other distinctive forms, such as typical limaces (Fig. 6.7) and large, bifacially retouched tranchoirs (Figs 6.5, 6.6), can be regarded virtually as type-fossils of the broad Charentian (i.e. Ferrassie plus Quina) group and again seem to be unknown (at least in characteristic form) in the other industrial variants (Bordes 1961b: 805, 1968a: 101, 1981: 78, 1984: 160; Turq *et al.* 1990: 62–3). Seen in a wider geographical focus there are equally diagnostic typological signatures in many Middle Palaeolithic industries of central and eastern Europe (notably various Micoquian hand-axe forms and distinctive bifacial leaf points: Figs 6.8, 6.9) and in the well known Vasconian industries (characterized by idiosyncratic flake-cleaver forms: Fig. 4.28) from the Pyrenees and Cantabrian Spain (Bordes 1977: 39, 1984: 174–209; Bosinski 1967; Cabrera Valdés 1988 etc.). Any suggestion that the documented range of variability in Middle Palaeolithic assemblages lies entirely in varying proportional frequencies of uniform, universally distributed tool forms, is therefore hardly in keeping with the available data from either southwestern French sites or many other parts of Europe.

There are similar contradictions in Binford's comments on the apparent lack of clustering of technological patterns in the Middle Palaeolithic. Again, he seems to attach great significance to this, since he regards the evidence for chronological and geographical clustering of specific typological or technological patterns as a main criteria for the recognition of explicitly cultural groupings in the classic Upper Palaeolithic sequence (1982b: 180). Here again, the available data appear to be in direct conflict with these suggestions. In the French sequence, there is strong evidence for a tight chronological clustering of three of the major industrial variants (i.e. the Ferrassie, Quina and MTA variants – as discussed in Chapter 6 and as Binford seems to acknowledge (1973: 231)). Geographically, equally clear clustering of distinctive technological and typological patterns has been documented in the distribution of the Yabrudian and Levallois-Mousterian industries in the Middle East, in some of the Micoquian and Blattspitzen industries in central and eastern Europe and in the characteristic Aterian industries of north-west Africa (Bordes 1984; Rolland & Dibble 1990: 492; Kozlowski 1992). If clumping or clustering of technological patterns is a possible criterion for the presence of non-functional variation in the Middle Palaeolithic, then there is ample evidence for this in many of the industrial sequences from both Europe and western Asia.

3. Finally, the most puzzling and possibly self-contradictory aspect of the whole of the Binford interpretations lies in his discussion of the respective roles of 'expediency' versus 'curation' in the manufacture and use of stone tools (Binford 1973: 242–53, 1977: 264–8, 1979: 283–6). Central to his model of functional variability in the Middle Palaeolithic is the assumption that Middle Palaeolithic tools were used in what he refers to as a highly expedient way – i.e. that the tools were manufactured mainly in response to immediate economic and technological needs as these arose in day-to-day foraging activities

and, equally if not more significant, that they were normally discarded almost immediately after use, at the actual location of manufacture and use (Binford 1973: 242–4, 251, 1977, 1979). In this way Binford envisages an essentially direct relationship between the activities carried out in particular locations and the tools discarded on the sites. This is in direct opposition to the idea of deliberate curation of stone tools, where it is assumed that tools would be manufactured well in advance of when they were needed (i.e. that the technology would anticipate the needs for future economic and technological activities) and that the tools would only be discarded at the end of a relatively long and complex use-life when no further use for them could be envisaged. In this case there need be virtually no direct connection between the particular activities undertaken on specific sites and the tools discarded there (Binford 1977: 266–7).

This conflict between expedient and curated technologies has become central to the Binford model of industrial variation in the Middle Palaeolithic, and to the postulated contrasts between this patterning and that documented not only in the later stages of the Palaeolithic but also in the observed technological behaviour of recent hunting and gathering groups. As Binford discusses at some length (1973: 242–4, 249–53, 1977: 267), one of the most striking features which has emerged from recent ethnoarchaeological studies of modern hunter-gatherers is that in most cases there does not seem to be any simple or direct relationship between the activities undertaken in particular locations and the tools left behind on the sites. Binford has documented this pattern for various Nunamiut groups in Alaska and quotes similar observations from studies of other recent groups such as the north Australian aborigines and the Kalahari Bushmen (Binford 1973: 244; White & Petersen 1969). The pattern documented by Binford among the Alaskan groups has been summed up as follows:

Taking a strict archaeological perspective the contents of these sites are grossly similar, in that the most common items on all are generally similar and are those items most expendable, of least importance in the ongoing technological system. Plotting these assemblages by means of cumulative graphs shows that they exhibit, in general, less variability in content than might be currently demonstrable between many different assemblages from the Upper Palaeolithic assignable to a given 'cultural phase'. Locations which are demonstrably very different from the behavioural standpoint are only differentiated archaeologically by items which are rare and almost always broken and modified heavily through use.

(Binford 1973: 242)

He goes on to suggest that essentially the same patterns seem to be reflected in much of the archaeological records of both the European Upper Palaeolithic and the various Palaeoindian and early postglacial Archaic groups in North America (1973: 250–1). In these contexts, he argues, there seems to be remarkably little evidence for clear functional variability in the overall composition of the lithic tool assemblages from different sites, which could be held to reflect variations in the activities carried out on the sites. In terms of both the ranges of tool forms represented on the sites and the varying frequencies of the different types, the industries appear in most cases to be remarkably uniform or from site to site (see also Bordes & de Sonneville-Bordes 1970; Mellars 1970).

The whole of this pattern stands in stark contrast to the situation which Binford envisages for industrial variation in the Middle Palaeolithic. In other words, the mechanisms to which he attributes almost all documented variation in Middle Palaeolithic contexts seems to be largely lacking in the observed technological behaviour of more recent hunter-gatherer groups. His explanation for this anomaly is that this reflects a profound shift in the whole of the organizational and cognitive aspects of technology, away from the purely expedient technologies of the Middle Palaeolithic towards the curated (i.e. more anticipatory) technologies of the Upper Palaeolithic and later periods (Binford 1973: 250–1, 1979: 283, 1982a: 373). But of course this entire scenaro is pure hypothesis and speculation. Why one should envisage such a profound shift in these fundamental aspects of technological organization is entirely unclear. By proposing this pattern Binford is forced to invoke a massive shift in the associated cognitive aspects of technology, which then becomes the basis for postulating even more far-reaching changes in the conceptual and planning abilities of human groups over the period of the conventional Middle-Upper Palaeolithic transition (Binford 1982b: 178, 1989). Even if Binford may eventually prove to be right in some of these speculations (see Chapter 12), this could be seen as an extreme case of effectively assuming what one should be attempting to prove.

There are striking contrasts in other aspects of the technological records of the Middle and Upper Palaeolithic, as visualized in the current Binford models. As Binford points out (1973), it is now clear that many of the documented contrasts between the major industrial and cultural stages of the Upper Palaeolithic sequence are reflected not only in the presence or absence of particular type fossil forms (generally assumed to be distinctive stylistic features of different cultural groups) but also in the overall quantitative composition of many of the shared types in the different assemblages. Thus relative frequencies of end-scrapers, burins, perforators, and other apparently functional types in many Upper Palaeolithic industries vary widely during different chronological stages of the Upper Palaeolithic sequence. Sharply defined patterns of this kind have been documented, for example, during the different stages of the Aurignacian sequence in southwestern France and also in the Solutrian and Magdalenian industries – often in a way which allows the industries to be seriated into clear and independently verifiable chronological sequences on the basis of these quantitative features alone (de Sonneville-Bordes 1960; Collins 1965). To the best of my knowledge, neither Binford nor anyone else has ever suggested that all, or even most, of these patterns can be explained in purely functional terms – i.e. in response to basic shifts in the range and relative importance of different activities carried out during separate stages of the Upper Palaeolithic sequence. And yet it is precisely these patterns – i.e. variations in the relative frequencies of discrete, morphologically distinct tool forms – which are assumed in the case of Middle Palaeolithic industries to be explicable almost entirely in functional terms. Exactly why there should be such a radical contrast between patterns of industrial variability in the Middle and Upper Palaeolithic remains unexplained.

The role of functional variability within the Mousterian succession of southwestern France

Before attempting to evaluate how far the Binford models may or may not be applicable to some of the specific industrial variants of the Mousterian in the southwestern French sites, there are a number of specific aspects of the available evidence from these sites which are possibly highly relevant to these interpretations and which should be recognized at the outset. The question is, in essence, how far – granted the particular nature and sources of the available samples of archaeological material – it is reasonable to expect to find evidence of the kind of variability which Binford has proposed. In this context, three aspects are especially significant:

1. It is now generally recognized that almost all the available samples of lithic assemblages recovered from the various caves and rock shelters of southwestern France are likely to represent compound composite entities, resulting from many separate episodes of occupation or activity at the same locations. This was emphasized by Bordes (1961b: 806, 1968a: 144; Bordes & de Sonneville-Bordes 1970: 66–8), and has been clearly recognized by Binford himself (1973: 241, 1982a, b: 179). The clear implication is that most of the available archaeologically documented assemblages of artefacts are likely to represent palimpsests of activity, in which the effects of any relatively short-term (e.g seasonal) variations in economic activities or tool manufacture on the sites will inevitably have been largely if not entirely obscured by the artificial conflation of many different occupational horizons into a single excavated sample. Only if certain sites were reserved primarily or exclusively for some very specific activities over long periods (probably amounting to several decades if not centuries) would these different activities be clearly isolated and possibly recognizable in the surviving archaeological records from the sites (Binford 1982a; see also Dibble & Rolland 1992: 6).

2. Virtually all of the well documented Mousterian assemblages from southwestern France, and certainly those on which the Binford interpretations are based, derive from a single, relatively specialized type of settlement location – namely various cave and rock-shelter sites located in or closely adjacent to the major river valleys of the region. As emphasized in Chapter 8, it is now clear that these sites have many basic features in common. All offer considerable protection from local climatic and weather conditions; all provide easy access to fresh water resources and (in most cases) local, abundant supplies of lithic raw materials; almost all face directly to the south or southeast; and (possibly most important) most are located in classic ecotone situations, offering quick and easy access to a wide range of contrasting ecological and economic resources, extending over both the adjacent river valley environments and the sharply contrasting habitats of the neighbouring limestone plateaux (see also Geneste 1985; Turq 1988a, 1989a; Duchadeau-Kervazo 1984, 1986). The implications of this pattern for the functional variability models seem self-evident. Since almost all the available samples of artefacts derive from these relatively specialized and similar locations, the question inevitably arises as to how far one would *expect* to observe sharply contrasting patterns of activities in this range of sites. By analogy with the behaviour of modern hunter-gatherer groups, one would expect evidence for specialized economic or technological activities to be encountered in contrasting settlement locations, suited either to the exploitation of very different economic resources, or to occupation in different seasons. To find evidence for so many specialized functions within the same topographic and ecological locations would provide a curious contrast with the patterns of behaviour observed amongst all modern, ethnographically documented hunter-gatherers.

The same point is underscored by many aspects of archaeological assemblages recovered from these sites, discussed in Chapters 8 and 9. Almost all occupation levels documented in the cave and rock-shelter sites appear to reflect a relatively varied spectrum of activities – indicated by the wide range of animal species introduced into the sites; the invariable evidence for long and complex patterns of tool manufacture, use and discard; the importation of lithic raw materials from a variety of sources; and frequent evidence for importation of other raw materials, such as manganese dioxide or iron ochre. As Geneste (1985: 516), Turq (1988a: 104, 1989a: 196) and others have frequently emphasized,

most assemblages recovered from cave and
rock-shelter sites appear more suggestive of
very generalized patterns of occupation, than
of specialized, possibly very short-term episodes of activity. My own view is that most
such locations were probably selected deliberately by Middle Palaeolithic groups to
serve in some way as 'central places' from
which diverse economic and domestic activities could be pursued (see also Duchadeau-
Kervazo 1984: 48).

These observations become even more significant when one recalls that the major variants of the Mousterian complex are not in fact
found in separate settlement locations but in
most cases in precisely the same sites. Thus,
Charentian (i.e. Ferrassie and Quina) industries have been found in the same sites as
MTA industries in at least 16 sites (see Table
6.2); Denticulate assemblages in the same
sequences as Charentian industries in at least
eight sites; MTA industries with Denticulate
industries in five or six sites; and similar
superimposed occurrences of Ferrassie,
Quina and Typical Mousterian industries in
at least seven sites (Mellars 1970). This pattern reaches its most extreme expression at
Combe Grenal, where multiple, superimposed levels of all five of Bordes major industrial variants occur within the same
stratigraphic sequence (Figs 6.16, 6.22). Any
idea that the different industrial variants are
distributed in significantly different site
locations therefore would not be in keeping
with available evidence from southwestern
France.

3. Finally, it is worth recalling some of the
observations made in an earlier very
perceptive study by Campbell (1968) on
the settlement and mobility patterns of
Tuluaqmiut reindeer-hunting groups in
northern Alaska. After stressing the complex
patterns of activities and contrasting settlement types generated by these groups, he
concluded:

We must recognize that the archaeological
record is likely to be selective not only in
respect to the kinds of artefacts that survive,
but also in respect to the kinds of sites that can
be recognized. Tuluaqmiut settlements of
Types II, III, IV and VI are fundamental
components of the way of life, and information
derived from Types I and V alone gives a
biased picture. However, only the latter types
normally possess sufficient cultural debris to
permit discovery by archaeologists centuries
or millennia after their abandonment. This fact
is not always recognized by specialists in
Palaeoindian or Palaeolithic cultures; on the
contrary, known sites are likely to be taken as
representative of settlement patterns of these
early hunting groups.

(Campbell 1968: 18)

The implications are again largely self-
explanatory: even if Middle Palaeolithic
groups in the Perigord region did engage in
complex and differentiated economic and
technological activities at different times and
locations in the landscape, how many would
one expect to be clearly represented and
archaeologically identifiable in the surviving
traces of these settlements? This becomes
even more pertinent if we take into account
both the complex palimpsest nature of the
archaeological assemblages recovered from
most documented sites and the special character of the various cave and rock-shelter
sites from which they have been recovered.

The archaeological perspective

There remains the question how far the Binford models can be reconciled with some of
the hard archaeological evidence for the
character, chronology and associations of the
major industrial variants of the Mousterian
in the southwestern French sites. These
issues have been debated at length in earlier
papers (e.g. Mellars 1969, 1970; Bordes & de
Sonneville-Bordes 1970; Binford 1973 etc.),
and many of the most pertinent issues are no
less relevant now than they were 20 years

ago. My own view, as explained in earlier papers (1970, 1988) is that for the three most distinctive of the Bordes industrial variants – i.e. Quina, Ferrassie and MTA groups – the available evidence provides an almost categorical refutation of any idea that these industries can be regarded as simple, functionally differentiated facies of a single cultural or behavioural system. The evidence relating to the various occurrences of taxonomically Denticulate and Typical Mousterian assemblages remains more enigmatic and no doubt open to a wider range of interpretations. Briefly, the most critical observations in this context can be summarized as follows.

1. The most categorical argument against a purely functional interpretation of the Ferrassie, Quina and MTA industries is provided by the evidence bearing on the relative and absolute chronology of these variants discussed in Chapter 6. A direct and inescapable assumption of the functional-facies model is that the separate, functionally differentiated facies of the Mousterian were manufactured over essentially the same time spans – i.e. that they were broadly if not entirely contemporaneous (cf. Binford 1983b: 158). Stated bluntly, unless the different industrial variants are essentially synchronous, then they can hardly be products of the same human groups. As we have seen in Chapter 6, the evidence against this pattern of large-scale synchronism in the case of the Ferrassie, Quina and MTA variants in the southwestern French sites is now virtually conclusive (Mellars 1969, 1986a,b, 1988, 1989c). To recapitulate the evidence set out more fully in Chapter 6, the most critical observations in this context are as follows:

(a) The existence of at least 16 well documented sites in which levels of MTA are stratified above those of Ferrassie or Quina Mousterian and the lack of any site where this situation is demonstrably reversed (see Table 6.2).

(b) The invariable tendency for MTA industries to occur in the upper levels of stratified Mousterian successions, in at least 14 cases stratified immediately beneath levels with early Upper Palaeolithic industries (see Table 6.3).

(c) The sharp localization and complete stratigraphic separation of multiple levels of Ferrassie, Quina and MTA industries throughout the 55 levels of Mousterian occupation at Combe Grenal (Fig. 6.16) – arguing categorically against the idea that all five variants were manufactured continuously in this region throughout the whole of the 50–60,000 year period spanned by these deposits.

(d) The remarkable consistency of the technological and typological trends which can be documented in all the stratified sequences of Ferrassie and Quina assemblages at Combe Grenal, Abri Chadourne, Abri Caminade, La Ferrassie, Roc de Marsal, Petit-Puymoyen and Roc-en-Pail, strongly suggesting a gradual technological development from one variant to the other during the course of the Würm II period (see Figs 6.17, 6.18).

(e) Other geological, climatic, faunal and radiometric dating evidence, all of which appears to reinforce the evidence for the relative sequence of the Ferrassie, Quina and MTA industries outlined above. The most significant observations are the consistent associations of classic Quina Mousterian industries with evidence of rigorous climatic conditions associated with oxygen isotope stage 4 (Delpech 1990; Guadelli 1990) and the similar association of MTA industries with the milder conditions of isotope stage 3 (Figs 2.23, 6.20, 6.21). The results of absolute dating (both TL and ESR) at Le Moustier, Pech de l'Azé II, La Quina and Fonseigner (Valladas *et al.* 1986, 1987; Grün *et al.* 1991; Mellars & Grün 1991) provide further confirmation that the MTA industries date largely if not entirely from the terminal stages of the Mousterian sequence.

The implications are evident. Clearly, the evidence as a whole is categorically opposed to an extensive overlapping of the time-ranges occupied by the Ferrassie, Quina and MTA industries in the southwestern French sites and suggests that these industries most probably formed a simple chronological succession in these sites. If this conclusion is valid, then any notion of these three variants as purely functionally differentiated products of the same human groups must inevitably be rejected.

2. A further powerful argument against the functional-facies model is provided by the tightly restricted and exclusive distribution of a number of distinctive type-fossil forms in certain industrial variants. This is illustrated most strikingly by the distribution of typical cordiform hand axes in the southwestern French sites. As Bordes has emphasized (1953a: 462, 1961b: 804–5, 1968a: 98–105, 1977: 39, 1981: 77–9, 1984: 137–64; Bordes & de Sonneville-Bordes 1970: 61–3), these forms can be regarded in every sense as a classic *fossile directeur* of the MTA assemblages in the southwestern French sites, and appear to be totally lacking from well documented and well excavated occurrences of all other Mousterian variants in this region. The clearest illustration of this is provided by the sequence at Combe Grenal (Bordes 1972). The point to be emphasized here is not merely that the industries attributed by Bordes himself to the MTA variant are confined exclusively to the uppermost five levels in this sequence, but that typical (or even atypical) specimens of cordiform hand axes were found to be entirely lacking throughout all the lower 50 Mousterian levels in the sequence – i.e. throughout a sequence of levels comprising multi-layered occurrences of all four of the non-hand axe variants of the Mousterian, and with a total of over 10,000 retouched tools. The question is how this observation can be reconciled with the idea that throughout the entire time span of the

Combe Grenal deposits, the human groups who occupied the site were producing assemblages comprising large numbers of typical cordiform hand axes at other, closely adjacent sites in the same region. So why did these groups never produce, import or discard a single cordiform hand axe in this location during a period of over 50,000 years? The answer of course is that the production of hand axes was simply not a part of the technological or cultural repertoire of the successive human groups who occupied the Combe Grenal site throughout the first 50,000 years of the last-glacial succession.

Similar observations could be made for some of the other distinctive type-fossil forms in the Mousterian. As Bordes (1953a, 1961b, 1968a, 1977, 1981, 1984), Peyrony (1920, 1930), Bourgon (1957) and others have stressed, fully typical (i.e. extensively retouched) backed-knife forms (Fig. 4.19) seem to be almost as distinctive a type-fossil of the MTA variant as are cordiform hand axes, and are virtually lacking in all the other industrial variants. At Combe Grenal, for example, Bordes recorded only three possibly typical specimens of backed knives amongst over 10,000 retouched tools recovered from the entire sequence of layers 8 to 55. The same could be said for some of the more distinctive forms of thick, convex, heavily retouched Quina-type racloirs (Fig. 6.10). It is now clear that these forms are not merely scarce but in many cases entirely lacking in most other industrial variants of the Mousterian (Fig. 6.11), for example in most of the well excavated occurrences of Denticulate assemblages and in at least the majority of the classic MTA industries (Bordes 1968a: 102, 1981: 78–9, 1984: 158; Bordes & de Sonneville-Bordes 1970: 71). Again, the question arises as to how this can be reconciled convincingly with the idea that at other sites in the region, the same human groups (or segments of the same groups) were manufacturing assemblages heavily dominated by typical Quina racloir forms.

3. Arguments concerning the precise forms of stone tools extend to the techniques by which they were manufactured. As Bordes (1953a, 1961b, 1977, 1980, 1981), Turq (1988b, 1989b, 1992a), Geneste (1985) and others have documented, the detailed character of the techniques employed for primary flake production in the southwestern French industries varies dramatically between the different industrial variants, with the Ferrassie-type industries and many of the earlier MTA assemblages relying heavily on typical Levallois techniques, while the classic Quina industries were dependent on technologically simpler but equally distinctive procedures oriented towards the production of thicker, more cortex-covered, non-Levallois flakes (Turq 1988b, 1989b, 1992a). As yet, no clear explanation has been offered as to how these technological contrasts can be accommodated convincingly within the functional-facies models. These contrasts are illustrated most strikingly when attention is focused on a single morphological (and presumably functional) category of tools, such as simple, single-edged racloir forms. As I pointed out in an earlier paper (Mellars 1970: 79–80) if one were to analyse a sample of, say, 100 typical racloirs in a characteristic Ferrassie-Mousterian assemblage, a high proportion (ca 20–40 percent) would have been manufactured from typical Levallois flakes and an even higher proportion (ca 50–60 percent) would show extensively faceted striking platforms. If one were to analyse a typical Quina assemblage, by contrast, virtually none of the tools (at most ca 2–3 percent) would be on Levallois flakes and only ca 10–20 percent would show faceted striking platforms. In this case one is making comparisons between groups of tools which, in Binford's terms, are assumed to be direct functional equivalents, oriented towards essentially the same economic and technological goals. Even if some broadly functional significance could be attached to the use of Levallois versus non-Levallois flakes for

racloir production, it is difficult to see how the presence or absence of facetted striking platforms could be explained in such terms. The evidence, again, is strongly opposed to the idea that these diverse technological strategies were employed in different activity locations by precisely the same human groups – especially since in most cases these contrasting strategies can be documented in different occupation levels of the same site (for example in the various levels of Ferrassie and Quina Mousterian at Combe Grenal, Abri Chadourne, Abri Caminade etc.) (cf. Rolland 1988b). In the case of the Ferrassie/Quina dichotomy discussed above, it has already been pointed out that this contrast in flaking strategies almost certainly reflects a pattern of gradual technological evolution over time, marked by a progressive phasing-out of the use of Levallois techniques over a period of perhaps 5,000–10,000 years (Figs 6.17, 6.18).

More general observations can be made. As already discussed in Chapter 6, the precise forms of side scrapers documented in MTA industries are very different from those in either the Quina or Ferrassie-type assemblages, generally with lower frequencies of heavily convex rather than straight retouched edges, and (by comparison with Quina industries) much lower frequencies of transverse forms and racloirs with typical Quina retouch (cf. Bordes & de Sonneville-Bordes 1970: 72). The racloirs in Denticulate-type assemblages are even more distinctive and are usually characterized by very light, discontinuous retouch applied to much smaller flakes (Table 10.1, Fig. 4.5) than those used for racloir manufacture in either the Charentian or MTA industries (Bordes 1968a: 102, 1977: 38, 1981: 79; Bordes & de Sonneville-Bordes 1970: 63, 71; also Rolland 1988b: 173, Table 9.4B). These contrasts between the precise forms and techniques of manufacture of specific tool-forms within the different industrial variants are some of the most obvious objections to any interpretation of the

Table 10.1
*Mean lengths of racloirs and unretouched flakes (mm)
recorded in levels of Ferrassie/Quina Mousterian and
Denticulate Mousterian at Combe Grenal*

	Racloirs	Unretouched flakes
Ferrassie/Quina		
Layer 17	49.6	45.4
Layer 21	52.2	45.5
Layer 23	56.9	48.4
Layer 25	53.6	45.8
Layer 27	55.7	47.6
Layer 35	56.2	47.9
Denticulate		
Layer 11	52.5	47.3
Layer 13	40.7	40.8
Layer 14	39.5	40.9
Layer 16	46.4	52.6
Layer 20	48.5	43.1

According to Rolland (1988): Table 9.4B.

total spectrum of Mousterian variability in simple functional or activity-facies terms.

4. The last set of arguments relate to some of the more direct evidence for clear functional or activity specialization in the sites under review. To what extent can we identify evidence of highly specialized technological or economic activities associated with the different industrial variants of the Mousterian in the southwestern French sites, which might provide a clear rationale for the kind of functional-facies model which Binford has proposed?

The main point to recall is the remarkable scale of the industrial variation to be accounted for in terms of the functional models and the number of separate technological dimensions in which this variation is reflected in

the different Mousterian variants. Thus the frequencies of side scrapers in different assemblages can range from less than 5 percent in Denticulate assemblages and some of the later (Type B) MTA industries, to more than 80 percent in some of the Ferrassie and Quina industries (Figs 6.1, 6.2); frequencies of denticulates can vary similarly from less than 5 percent to over 40 percent; frequencies of hand axes from zero to more than 30 percent; backed knives from zero to 25 percent; and so on. The same scale of variation is apparent in the various technical parameters of the industries, with Levallois indices ranging from effectively zero to over 30 percent, facetting indices from 10 to 60 percent, blade indices from zero to 30 percent, and so on (see Bordes 1977: 37, 1984: 1327–9).

The direct and inescapable implication of the functional models is that all these variations must be accounted for in some way by commensurate variations in the nature, range or relative importance of the economic or technological activities undertaken in the different sites. The question therefore arises as to how far we can identify evidence for extreme functional specialization of this kind in the available data from the southwestern French sites?

It must be acknowledged that we are still very poorly equipped with information on many of the potentially relevant aspects of behaviour in Mousterian sites – most notably on the seasonality of occupation and activity patterns, or the relative importance of exploitation of plant versus animal resources in different sites. Nevertheless, the point has been emphasized in the preceding chapters that where we do appear to have evidence for apparently specialized economic and technological activities in different sites, these usually fail to show any direct or obvious correlations with specific industrial variants of the Mousterian (see for example Mellars 1970; Bordes & de Sonneville-Bordes 1970; Chase 1986a,b). Some of the most relevant observations in this context are as follows:

1. First, there is surprisingly little evidence for clear patterning or segregation in the specific locations occupied by the different industrial variants. As noted earlier, MTA industries have been found in the same stratigraphic sequences as Ferrassie or Quina assemblages in at least 16 sites (Table 6.2); Denticulate industries in the same sequences as Charentian industries in at least eight sites; Quina industries with Ferrassie industries in at least five sites; and so on. The classic illustration of this is provided by the sequence at Combe Grenal, where multiple, superimposed levels of all five of the Bordes Mousterian variants occur within a single stratigraphic succession (Figs 6.16, 6.22). The apparent predominance of MTA industries at open-air sites in the Perigord region may conceivably be an exception to this pattern, but as noted earlier, this is by no means invariable and as discussed in Chapter 8 has almost certainly been greatly distorted by past generations of selective surface collecting. The fact that most of the MTA industries appear to date from a period of temperate climate (coinciding with the earlier part of isotope stage 3) may provide a simple explanation for this proliferation of open-air sites in purely climatic terms.

2. The association of particular industrial variants with particular faunal species shows a similar lack of any clear or simple correlations (Mellars 1970; Chase 1986a,b). There are of course some broad associations of this kind – for example in the frequent association of Quina-type industries with high percentages of reindeer remains or the tendency for MTA industries to be associated predominantly with bovids or red deer remains (Fig. 6.21). However both patterns are subject to several well documented exceptions (for example in the high frequencies of horses and bovids associated with the Quina assemblages at Mas-Viel and Puycelsi, and the high frequencies of reindeer associated with the MTA industries at the Gare de Couze and La

Quina: Niederlender et al. 1956; Tavoso 1987a; Peyrony 1932; Jelinek, personal communication) which clearly demonstrate that there is no simple or direct connection between the specific character of the lithic assemblages and the particular species of animals exploited. This lack of correspondence between faunal assemblages and industrial patterns is even more apparent in the succession of Quina and the immediately overlying Denticulate Mousterian levels in layers 23 to 11 at Combe Grenal (see Fig. 10.2; cf. Mellars 1970; Chase 1986a, b; Bordes & Prat 1965). Any broad correlations between particular faunal species and particular industrial variants in the southwestern French sites probably have more to do with basic climatic and ecological changes during the course of the Mousterian sequence than with any functional specialization in the activities carried out in the different sites.

3. Finally, there seems to be a remarkable lack of correlation between different industrial variants of the Mousterian and the overall abundance of faunal remains – in relation to the associated lithic artefacts – recorded in different sites and occupation levels. As I pointed out in 1970, this pattern is illustrated most strikingly in the sequence at Combe Grenal, where one can observe essentially the same range of variations in the ratios of identifiable faunal remains to retouched stone tools in each of the four major variants of the Mousterian (Ferrassie, Quina, Denticulate and Typical) and where the average values of this ratio recorded for the four variants are remarkably similar (see Fig. 10.1). Clearly, if there were any significant differences in the degree to which systematic processing of animal resources rather than plant food resources were carried out in the different industrial horizons, this is not reflected in any consistent way in the overall abundance of faunal remains found in association with different industrial variants at Combe Grenal.

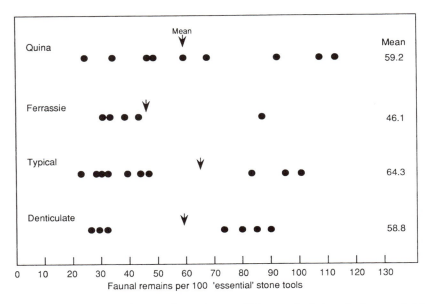

Figure 10.1 *Frequencies of identifiable animal remains per 100 retouched stone tools recorded in association with different Mousterian variants at Combe Grenal. Note the very wide range of the frequencies recorded for each variant, and the close similarity in the means. Data on the numbers of identifiable faunal remains in the different levels are from Bordes & Prat 1965 (see also Mellars 1970: 79).*

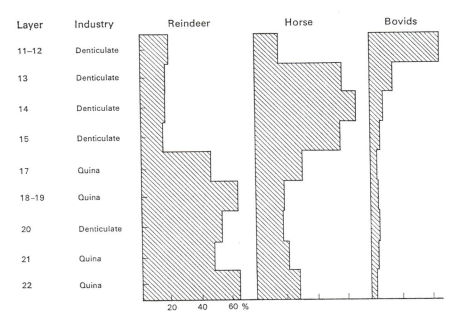

Figure 10.2 *Frequencies of main faunal species associated with different levels of Denticulate and Quina Mousterian in levels 22–11 at Combe Grenal – indicating a lack of any simple correlation between the type of industry and the associated fauna. From Mellars 1970.*

Whether we can identify any clear patterns in the particular skeletal elements of animal species associated with particular industrial variants remains to be seen. At Combe Grenal, there are some possible hints of such patterns – as for example in the apparently stronger representation of reindeer head and teeth remains in several levels of Denticulate Mousterian in the upper levels of the sequence than in those of the immediately preceding Charentian (Ferrassie and Quina) horizons (Fig. 7.5; Chase 1986a). However, these patterns are by no means simple and in fact one of the highest ratios of reindeer heads and jaws to postcranial bones was recorded not in a level of Denticulate Mousterian but in one of the richest and most typical Ferrassie Mousterian levels, in layer 27 (Chase 1986a: Fig. 8). Any suggestion of a direct, causative link between the character of faunal exploitation and the composition of the associated lithic industries seems to be contradicted by the detailed evidence from Combe Grenal.

Conclusions

The implication of the evidence discussed in the preceding section is that at present, any direct, pragmatic support for the kind of functional variability models which Binford has proposed to account for the major industrial variants of the Mousterian complex is still remarkably difficult to identify within the southwestern French sites. For the most distinctive industrial variants in this region, i.e. the Ferrassie, Quina and MTA variants, the combined evidence against a simple functional explanation is, in my view, effectively conclusive. The idea that these industries represent synchronous, functionally differentiated facies manufactured by the same human groups is flatly contradicted not only by a wealth of direct stratigraphic and chronological evidence but also by many of the specific technical and typological features of the industries themselves.

The status of the various occurrences of taxonomically Denticulate and Typical Mousterian assemblages is clearly more enigmatic. As emphasized in Chapter 6, the whole integrity of these two variants of the Bordes system as discrete, well defined entities is open to serious doubt (as Bordes himself acknowledged on several occasions: Bordes 1977: 38, 1981: 79; Bordes et al. 1954: 249), and there is now clear stratigraphic evidence that both these variants can occur in at least two separate stages of the Mousterian succession. In certain respects, therefore, these industries do exhibit some of the features one might anticipate from functional or activity variants. There remain, however, serious difficulties in attempting to see these industries as simple functional facies of the other, better defined variants. Why, for example, are the occasional racloirs encountered in Denticulate Mousterian assemblages so obviously different from those in the immediately preceding Quina Mousterian levels (Fig. 4.5: Bordes & de Sonneville-Bordes 1970: 63, 71; Bordes 1961b: 805, 1977: 38, 1984: 158; Rolland 1988b: 173)? Why are typical hand axes conspicuously lacking from these rich levels of Denticulate and Typical Mousterian at Combe Grenal and elsewhere? And why is there such a conspicuous absence of well documented cases of interstratification of Denticulate and Typical Mousterian assemblages *within* the numerous, rich and multi-layered sequences of Ferrassie, Quina and MTA industries in the Perigord sites – with the one notable exception of the isolated level of Denticulate Mousterian encountered in the uppermost Quina levels (layer 20) at Combe Grenal? Even for these enigmatic and poorly characterized variants of the Bordes taxonomic scheme, there are severe difficulties in attempting to account for the recorded assemblages in purely functional terms.

The fact remains nevertheless that some component of purely functional patterning of the kind envisaged by Binford is totally predictable in theoretical terms and must inevitably account for some of the documented variation in the Mousterian assemblages from southwestern France – as indeed in those of the Upper Palaeolithic and later periods (Mellars 1970: 84, 1976). My own suggestion, as discussed in Chapter 8, is that clear evidence for such patterning is more likely to be found in some of the less intensively occupied and potentially more functionally specialized open-air sites in the Perigord region, rather than in the much more intensively occupied and seemingly more generalized cave and rock-shelter sites. Possible examples of functionally specialized open-air sites have already been documented at several locations in southwestern France, for example at the bovid-butchery sites of La Borde, Coudoulous and Mauran some way to the south of the Perigord and perhaps also at some of the smaller open-air sites in the northern Perigord region (for example, at the isolated mammoth-butchery site at the Carrière Thomasson, and other small and apparently ephemerally occupied sites in the Euche valley: Geneste 1985: 76–105; Duchadeau-Kervazo 1986). Further research on these sites may reveal a much more complex picture of Neanderthal activities and associated patterns of industrial variation than that provided by the classic sequence of industries from the better known cave and rock-shelter sites.

The tool-reduction models of Dibble and Rolland

A central assumption of both the Binford functional variability models and the original cultural interpretations of François Bordes is that the majority of the typologically or morphologically contrasting tool forms encountered in Middle Palaeolithic industries can be seen as deliberate products that were conceived by their manufacturers as discrete and intentional forms. Both models imply, at least to some degree, the notion of discrete 'mental

templates' in the conception and production of the tools. In both cases the primary determinants of the different tool forms are assumed to have been closely related to the intended functions of the tools, although in the case of the Bordes model, the existence of additional 'cultural' variants is assumed to have been superimposed on the primarily functional forms (Bordes & de Sonneville-Bordes 1970). Both the models of Binford and Bordes, in other words, assume that the major retouched tool types in Middle Palaeolithic assemblages were intended from the start as discrete products and, crucial to these interpretations, were discarded in the archaeological record in essentially the same form in which they were originally produced.

It is this central idea of intentionality and immutability in the forms of retouched stone tools which has been challenged, vigorously and emphatically, in the tool reduction models of Nicholas Rolland and Harold Dibble (Rolland 1977, 1981, 1988a,b, 1990; Dibble 1984a,b, 1987a,b,c, 1988a,b, 1989, 1991a,b; Rolland & Dibble 1990; Dibble & Rolland 1992; see also Jelinek 1988a). The basic premise underlying their studies is that the majority of tool types encountered in Middle Palaeolithic industries were never conceived from the outset as discrete and morphologically separate forms, and only attained the form in which they are encountered in the archaeological record as a result of a long and complex process of reworking and resharpening in the course of use. The assumption, in other words, is that most retouched tool types started their effective use-lives simply as unretouched flakes, which were then progressively resharpened, reworked and in some cases substantially remodelled, as the edges of the original flakes became blunted or damaged in the course of heavy use (Rolland & Dibble 1990: 485–6; Rolland 1981: 28). As Dibble and Rolland point out, these ideas are by no means new to the archaeological literature and have been raised in earlier

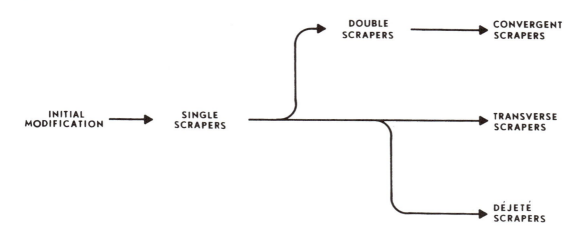

Figure 10.3 *Summary of Dibble's model for the progressive reduction of unretouched flakes into either double and convergent racloirs (upper) or transverse and* déjeté *racloirs (lower) (see also Figs 3.8, 3.12). The original shape and nature of the edges on the parent flakes are assumed to dictate which particular reduction pattern is followed. After Dibble 1987a.*

publications by Leroi-Gourhan (1956, 1966), Combier (1967: 194–6), Jelinek (1976) and indeed by Bordes himself (1953b, 1984: 166–9). Nevertheless, the recent publications of Dibble and Rolland carry these interpretations to a much more sophisticated and systematic level than any of the earlier flirtations with the same ideas.

As discussed in Chapter 4, there are two basic paths along which these processes of reduction and resharpening are assumed to have proceeded (see Fig. 10.3). One starts with the use of simple unretouched flakes as cutting tools which were then progressively retouched into simple, single-edged racloir forms as the edges of the original flakes became blunted. The second involved various subsequent stages of repeated resharpening and remodelling of these simple, single-edged racloirs into a variety of more complex racloir forms. Depending on the shape of the original flake, these might lead to typical transverse racloirs (Fig. 4.8) or (where the tools were manufactured from more elongated, thinner flakes) to various forms of either double or convergent racloirs

(Fig. 4.12: Rolland 1977, 1981, 1988a; Dibble 1984a, 1987a,b,c, 1988b; Rolland & Dibble 1990). According to this model therefore, almost all racloir forms in Middle Palaeolithic industries can be attributed to the progressive reduction and resharpening of pieces which started their initial lives as unretouched, primary flakes. Broadly similar paths have been envisaged for the production of denticulate tools, which are assumed to have started life as simple notch forms, subsequently transformed by the addition of further, adjacent notches along the same working edge (Dibble 1988a: 190; Rolland & Dibble 1990: 485–7).

If the models of Dibble and Rolland are accepted, therefore, we have a situation in which a number of radical transformations could have taken place in the character and composition of the retouched tool component in Middle Palaeolithic industries. If – as they suggest – the edges of cutting tools were likely to become more rapidly damaged in use than those of more robust notched forms, then the effects of intensive resharpening and reduction could easily lead to major changes

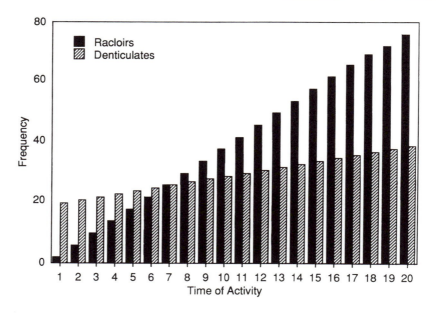

Figure 10.4 *Dibble & Rolland's hypothetical model of how progressive tool reduction and resharpening can transform an industry from one dominated by notched and denticulated forms to one heavily dominated by racloirs. The model assumes that for each unit of time four flakes are reduced into racloirs, while only one flake is transformed into a denticulate. The model clearly infers a much longer duration of occupation for a racloir-dominated industry than for one dominated by denticulates. After Dibble & Rolland 1992.*

in the relative frequencies of racloirs versus notches and denticulates in particular assemblages (Fig. 10.4), possibly creating a shift from a typical Denticulate industry into an assemblage of either Typical Mousterian or even Charentian form (Rolland & Dibble 1990: 485–6; Dibble & Rolland 1992; Rolland 1981: 28). Similarly, further reduction and resharpening of racloirs could lead to progressive increases in the frequencies of transverse or double and convergent racloirs, in relation to simpler, single-edged lateral racloirs (Dibble 1984a, 1987a etc.). The cumulative impact of such transformations could lead to dramatic shifts in the typological structure and composition in which particular lithic assemblages eventually found their way into the archaeological record.

Two basic mechanisms are envisaged by Dibble and Rolland for these transformations

(see Fig. 10.5). The first is the availability of local flint supplies within the immediate catchment area of the sites. According to them, a significant scarcity of local flint supplies would lead to deliberate economizing in the use of available resources, which would inevitably encourage maximum reuse and resharpening of lithic artefacts to extend their use-life (Rolland & Dibble 1990: 486–7; Dibble 1984b; Dibble & Holdaway 1989). The second relates to the intensity or duration of occupation on particular sites. In this case, it is assumed that relatively long and continuous episodes of occupation would encourage frequent reuse and resharpening of tools immediately available on the sites, as opposed to the manufacture of entirely new tools, which could involve additional time and work securing additional flint supplies (Rolland & Dibble 1990: 487–8; Jelinek 1988a:

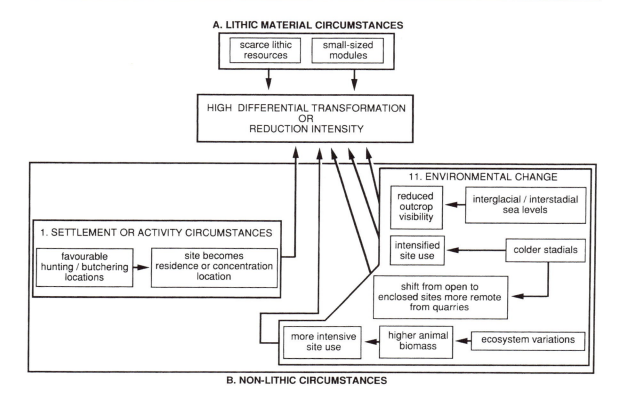

A. LITHIC MATERIAL CIRCUMSTANCES

| scarce lithic resources | small-sized modules |

HIGH DIFFERENTIAL TRANSFORMATION
OR
REDUCTION INTENSITY

11. ENVIRONMENTAL CHANGE

reduced outcrop visibility ← interglacial / interstadial sea levels

intensified site use ← colder stadials

shift from open to enclosed sites more remote from quarries

1. SETTLEMENT OR ACTIVITY CIRCUMSTANCES

favourable hunting / butchering locations → site becomes residence or concentration location

more intensive site use ← higher animal biomass ← ecosystem variations

B. NON-LITHIC CIRCUMSTANCES

Figure 10.5 *Rolland's model of how various environmental and other factors can influence the degree of tool reduction in Middle Palaeolithic sites. After Rolland 1990: Fig. 13.1.*

208). A further contributary factor would be the variable mobility of the groups, which would similarly influence the ease with which new flint supplies could be collected in the course of routine foraging activities (Rolland & Dibble 1990: 488–9; Rolland 1981: 22). In short, heavily reduced lithic industries would be likely to reflect either inherent scarcities in lithic raw materials or simply varying degrees of permanence in the occupation patterns of certain site locations.

Critique of the tool-reduction models

Outlined in these terms, the tool reduction models of Rolland and Dibble are coherent,

explicit and in certain respects highly plausible. The basic elements of the models are in good accord with expectations derived from ethnographic observations and with many features of the archaeological record itself. Ethnographically, the extensive resharpening of stone tools has been documented in many different contexts and in some cases (most notably the Australian tula adzes) led to radical changes in the form and dimensions of discarded tools (e.g. Hayden 1979). Similarly, there is now evidence from several archaeological contexts not only that many totally unretouched flakes were employed for a range of functions (Keeley 1980; Beyries 1987; 1988a: 216–8; Beyries & Boëda 1983) but also that several tool types (including typical racloir forms) were systematically resharpened at one if not several points in their total use-life (Figs 4.3, 4.4: Cornford 1986; Lenoir 1986; Meignen 1988; Schlanger 1989). These observations leave no doubt that both the systematic use of unretouched flakes and resharpening of tools in the course of use was a regular part of stone tool production apparently during all stages of the Palaeolithic sequence.

The question here, therefore, is not whether the tool reduction models of Dibble and Rolland have some significant role to play in explaining the total range of variation in Middle Palaeolithic assemblages, but simply the scale on which the models can be applied. Some of the basic problems raised by the reduction models concerning the production of different racloir types in Middle Palaeolithic industries have been discussed in Chapter 4 and need not be repeated here. The critical question in the present context is how far the Dibble and Rolland models can be invoked to account for some of the much more radical contrasts in lithic assemblages which have traditionally been used to differentiate between the major Mousterian variants – notably the large variations in racloir frequencies which distinguish principally the Denticulate, Typical and Charentian variants of Bordes' taxonomy, and the occurrence of other taxonomically diagnostic features such as hand axes or backed knives (Bordes 1953a, 1961b, 1968a, 1981, 1984). Clearly, these are fundamental issues to any assessment of industrial variation in the Middle Palaeolithic and deserve close examination.

1. The first and most important point to emphasize is that the tool reduction models have never been seen as a way of accounting for all the differences which define the main industrial variants of the Bordes taxonomy. As discussed above, these models have been focused on two specific aspects: first, the varying production of racloirs in different assemblages, as a result of systematic reshaping of unretouched flakes; and second, on some of the finer variations in the specific forms of racloirs (i.e. the varying percentages of lateral, transverse, double and convergent forms). While these variations could in principle be used to explain many of the documented contrasts between industries of the Denticulate, Typical and Charentian groups, alone, they have no bearing whatever on the basic typological and technological features which have always been used to differentiate between the three most distinctive and sharply characterized variants of the Bordes taxonomy – i.e. the Ferrassie, Quina and MTA variants. As discussed earlier, the fundamental distinction between the Ferrassie and Quina variants is based not on the overall typological composition of the tool assemblages but on the basic primary flaking techniques by which the tools were produced – i.e. on the predominance of Levallois techniques in the Ferrassie assemblages and simpler, non-Levallois techniques in the Quina industries (Bordes 1953a, 1961b, 1981, 1984). Clearly, the tool reduction models of Dibble and Rolland have no bearing whatsoever on these underlying technological contrasts between the Ferrassie and Quina industries and can in no way be used to explain this particular dichotomy within the Bordes system. The same is equally true of the specific hallmarks which have always been used to define the MTA variant – i.e. the presence of characteristic cordiform hand axes and distinctive backed knife forms (Peyrony 1920, 1930; Bourgon 1957; Bordes 1953a, 1961b, 1981, 1984). Again, these features are totally beyond the range of the current tool reduction models and stand as unique, idiosyncratic features of the MTA industries as a whole. All of these points have been recognized by Dibble and Rolland, and have never been presented as a part of their general tool reduction models. Nevertheless, it is important to recognize that whatever credence may eventually be given to the tool reduction

models, they leave some of the most basic and striking aspects of industrial variation in the French Mousterian totally unexplained.

2. The suggestion that tool reduction models could account for the varying frequencies of racloirs and notches/denticulates which distinguish the Denticulate, Typical and Charentian industries is in principle much more plausible and this no doubt stands as the main potential contribution of the Dibble and Rolland models. There is little doubt that tool reduction patterns have some significant role to play in accounting for some of the documented industrial variations within these three groups, at least when seen from a broad geographical and chronological perspective. The question once again is on what scale? In their most recent publications Dibble and Rolland have argued that even the most dramatic contrasts between racloir and notch/denticulate frequencies – i.e. those of the Ferrassie and Quina industries versus the classic Denticulate-type industries – can be explained entirely in these terms (Rolland & Dibble 1990: 485–6; Dibble & Rolland 1992; see also Rolland 1981: 28). As a specific example they quote the contrast between the later Quina-type Mousterian levels and the immediately overlying Denticulate horizons recorded in the later Würm II levels (layers 26–11) at Combe Grenal. They argue that a shift in the intensity of occupation patterns in these levels could potentially account for all the documented contrasts between the racloir versus notch/denticulate frequencies in these levels – i.e. for the shift from racloir-to-denticulate ratios of around 10 to 1 in the late Quina levels to less than 0.2 to 1 in the immediately overlying Denticulate levels (Dibble & Rolland 1992: Fig. 1.3; see also Rolland & Dibble 1990: 487–8, Table 2). As noted above, their suggestion is that during episodes of intensive, semi-permanent occupation, frequencies of various racloir forms would increase more rapidly in the assemblages as a result of progressive retouching of

unretouched flakes than those of the notch/denticulate group, eventually leading to a massive shift in the relative frequencies of these two tool forms in the tool assemblages (Fig. 10.4). Exactly why these major shifts in the intensity or duration of occupation patterns should have taken place between the Quina and Denticulate levels has not yet been explained. Nevertheless the implication that such changes could have led to the dramatic contrasts between the tool-type composition of the Denticulate and Quina assemblages in this part of the Combe Grenal sequence is clearly set out in Dibble and Rolland's paper.

However elegant and coherent these models may appear in a theoretical sense, there are a number of problems in attempting to reconcile them with some of the hard archaeological data in this particular case. Three observations are especially relevant:

(a) First, the basic mathematics underlying the Dibble and Rolland models lead to some surprising conclusions. If we assume, as they suggest, that all the observed contrasts between frequencies of racloirs versus notches and denticulates recorded in these levels can be attributed to the varying intensity of occupation patterns (i.e. the length of time during which each occupation horizon was occupied) then we must assume that the occupation intensity factor was at least 50 times higher during occupation of the various Quina Mousterian levels than during that of the adjacent Denticulate levels. This conclusion follows directly from their own mathematics (as illustrated in Table 2 of Rolland & Dibble 1990; see also Dibble & Rolland 1992: Fig. 1.3) and makes the minimal assumption that whilst the production (and therefore relative frequencies) of racloirs increased progressively with the duration of occupational episodes, the production of notched and denticulated forms showed no comparable increase over these intervals. If we assume that both racloir and denticulate production would have increased progressively – though at different rates – over these intervals, as Dibble and Rolland concede is a more likely scenario (see Fig. 10.4), then the contrasts in the occupational-intensity index needed to produce the observed dichotomy between the Denticulate and Quina assemblages becomes even more dramatic – probably in the region of 100:1. Exactly why there should be such massive contrasts between the intensity or duration of occupation patterns in these particular levels of the Combe Grenal sequence remains as yet unexplained.

(b) Presumably, if one were to envisage such massive contrasts between the relative intensity and duration of occupation episodes in the Denticulate and Quina levels, one would reasonably expect to see some reflection of this in certain other features of the archaeological material – for example in the total quantities of faunal remains introduced into the site, or in evidence for intensive use and re-use of hearths and associated hearth refuse in the different occupation levels. In fact, the available evidence from Combe Grenal yields few indications of such contrasts. As discussed in the preceding section, the ratios of identifiable faunal remains to stone tools documented in the various levels of Denticulate Mousterian at Combe Grenal cover broadly similar ranges to those recorded in the adjacent levels of Quina Mousterian and, when calculated as average values, show effectively no discernible contrasts between the Quina and Denticulate levels (Fig. 10.1). If the ratios are calculated to include the unretouched flake components of the assemblages (on the assumption that these could have functioned as simple, non-reduced tool forms) then admittedly the figures do shift in favour of rather higher overall faunal ratios in the Quina than in the Denticulate levels – potentially by a factor of up to ×5. Even this, however, is far removed from the contrast in occupation-intensity ratios implied by the basic mathematics of the Dibble and Rolland models.

Similar observations could be made for the evidence of hearths and other indications of burning in the Denticulate and Quina levels. Clearly, these features are difficult to quantify objectively, but from data provided by Bordes (1972) and from additional observations of Binford (1992, and personal communication) and Laquay (1981), it seems evident that evidence for the use and reuse of hearths is if anything even more conspicuous in the Denticulate levels at Combe Grenal than in the adjacent Quina levels. Binford (1992, and personal communication), for example, has pointed out that many Denticulate levels seem to include not only relatively large, frequently re-used hearths but also often associated evidence for large-scale 'sheet burning' affecting the occupation levels as a whole. Rolland himself (1988b: 168) has drawn attention to the evidence for extensive hearths and ash bands in the single Denticulate level at Pech de l'Azé II. Similar observations are suggested by the frequencies of burnt animal bones and teeth recorded by Binford (1984) and Laquay (1981: 420–2) in the occupation levels, which again suggest significantly higher frequencies of burning in the Denticulate than in the Quina levels. As in the case of faunal remains therefore, there is little evidence to suggest that the intensity or duration of occupation patterns was significantly greater in the Quina levels than in the Denticulate Mousterian at Combe Grenal – and certainly not on the dramatic scale which the current tool-reduction models imply.

(c) Finally, Rolland has quoted some interesting data on the dimensions of tools and unretouched flakes in the various Denticulate and Quina Mousterian levels at Combe Grenal, which appear to conflict directly with the stated implications of the tool reduction models (Rolland 1988b: 173, Table 9.4B). As shown in Table 10.1, these reveal that the average lengths of both racloirs and unretouched flakes are appreciably smaller in the

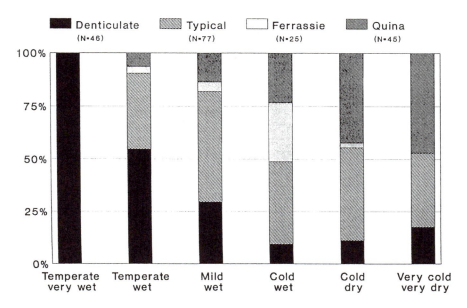

Figure 10.6 *Frequencies with which different types of Mousterian assemblage are associated with different climatic regimes in western Europe, according to Dibble & Rolland 1992 (see also Rolland 1990). Note that with the exception of the 'temperate, very wet' category characteristic of the west Mediterranean coastal zone, all types of industry are found in association with all climatic regimes.*

majority of Denticulate Mousterian levels at Combe Grenal than in those of the adjacent Quina levels (see also Fig. 4.5). Both of these observations would seem to be in direct conflict with the patterns of tool and flake reduction envisaged for the Denticulate and Quina industries. Clearly, if the largest available flakes were being deliberately selected for retouching into racloirs in the Quina assemblages (as Rolland and Dibble imply: 1990: 485; also Dibble 1988a: 193) then the remaining unretouched flakes in these assemblages should presumably be significantly smaller than in the typologically non-reduced Denticulate assemblages. By the same token it should follow that various racloir forms documented in the Quina-type assemblages should be significantly smaller than in the Denticulate industries, since these tools are (by definition, in the Dibble and Rolland

models) more heavily reduced than in the Denticulate industries. In both respects, the recorded artefact dimensions seem to reveal the reverse of the pattern that would be predicted from the tool reduction models. Seemingly the most contradictory feature is the much smaller average size of the racloirs in the Denticulate than in the Quina industries, since this would imply that intensive reduction of the tools (in the Quina assemblages) somehow made the tools larger!

Similar conflicts are apparent in some of the other data cited by Dibble and Rolland for the Denticulate and Quina levels at Combe Grenal. Two features referred to repeatedly in earlier publications as general criteria for identifying intensive reduction patterns in Middle Palaeolithic industries are sharp reductions in the sizes of discarded cores (to achieve maximum economy in the use of

available raw materials) and a progressive
increase in the frequencies of denticulates
compared with notches in tool assemblages,
reflecting repeated resharpening of single
notches into more complex, multiple-
notched forms (Dibble 1988a: 190; Rolland &
Dibble 1990: 485–7). Again, the available data
on these two features from the Denticulate
and Quina levels at Combe Grenal seem to
show the reverse of the patterns predicted by
the lithic reduction models. Rolland has
pointed out (1988b: 196) that the average
sizes of cores recovered from the Quina levels
at Combe Grenal are substantially larger than
those from the adjacent Denticulate levels
(hardly an index of heavy reduction or economizing on raw materials in the Quina levels),
while according to data provided by Dibble
(1988b; Fig. 10.6) ratios of denticulates to
notches are in general somewhat higher in
the hypothetically unreduced Denticulate
levels than in those of the hypothetically
heavily reduced Quina horizons (Fig. 4.11).
Again, these apparently contradictory observations are passed without comment or
explanation in the recent tool-reduction
explanations for the Quina/Denticulate
Mousterian dichotomy at Combe Grenal.

3. Potentially the most interesting correlations proposed by Dibble and Rolland are
those which relate the varying intensity of
reduction of particular industries to certain
features of the environmental or ecological
contexts of the sites. They suggest that heavily reduced industries tend to be encountered
primarily either in areas where lithic raw
material supplies are in short supply, or
where climatic or other environmental factors would have reduced the mobility of the
human groups and therefore the frequency of
access to available raw material sources
(Rolland & Dibble 1990: 484–90; Dibble &
Rolland 1992). In both cases there is assumed
to be a close correlation between the degree
of reduction and the ecological or environmental context of the sites.

Some of the correlations proposed by Dibble and Rolland – such as the tendency for
more lightly reduced industries to occur predominantly during mild climatic episodes
(Fig. 10.6), or the occurrence of heavily
reduced industries in areas far removed from
raw material supplies, such as the Zagros
region of southwest Asia – are unquestionably interesting and worth pursuing further
(Rolland 1981: 29–34; Dibble 1984b, 1991a,b;
Dibble & Holdaway 1989; Rolland & Dibble
1990: 486–90). However there are again problems in reconciling these patterns with the
totality of the available data from the French
sites. As Rolland has pointed out (1988b)
there is no very simple correlation between
the intensity of tool reduction and the availability of raw materials at specific site locations. At many sites in southwestern France
(most notably at Combe Grenal) the complete
spectrum of variation from heavily reduced
(i.e. Ferrassie and Quina) to almost totally
unreduced (i.e. Denticulate) assemblages can
be seen to occur in a single location, where
the availability of local raw materials must
presumably be seen as an effective constant
throughout the different occupation episodes
(see Bordes 1984: 169). Equally significant is
the occurrence of a number of heavily
reduced industries in contexts where abundant and high quality flint supplies were
apparently readily available on, or closely
adjacent to, the occupation sites. Striking
illustrations of this are provided, for example, at the site of Combe-Capelle Bas in the
Couze valley (Peyrony 1943; Bourgon 1957),
or at sites such as Biache-Saint-Vaast,
Champvoisy and Riencourt-les-Bapaume in
the flint-rich regions of northern France (Tuffreau & Sommé 1988; Tuffreau 1988b, 1993).
Why it should have been necessary to employ
heavily economizing tool-resharpening strategies in these and similar contexts remains
unclear. Conversely, other *non*-reduced
(Denticulate) industries have been recorded
from contexts (such as Arcy-sur-Cure in the
Yonne and Roc-en-Pail in Maine-et-Loire)

where local flint supplies were apparently scarce (Girard 1978; Gruet 1976).

There are similar problems in attempting to pin-point specific associations between the degree of reduction and particular combinations of climatic and ecological conditions (Fig. 10.6). As Dibble and Rolland suggest, it may well be significant that many hypothetically reduced industries from the southwestern French sites seem to correlate primarily with the extremely cold climatic conditions of the Würm II (i.e. isotope stage 4) phase (as reflected by the various occurrences of Quina and Ferrassie industries), while the bulk of the industries from the milder, Würm I phase (= isotope stages 5a–d) belong to either the Typical or Denticulate variants (Rolland & Dibble 1990: 488–90; Rolland 1981: 29–34, 1988a: 177–8; Jelinek 1988a: 20–8). However, these correlations are by no means exact. Thus several of the Denticulate Mousterian horizons in the later Würm II levels at Combe Grenal demonstrably date from a period of very cold climate (Bordes *et al.* 1966; Laville *et al.* 1980), while in other contexts heavily reduced, Charentian-like industries have been recorded from periods of relatively mild climate – for example the Ferrassie-type assemblages from Rescoundudou (Aveyron) and Biache-Saint-Vaast (Jaubert 1983, 1989; Tuffreau & Sommé 1988). Some of the proposed correlations between the relatively unreduced industries and the occupation of open-air sites as opposed to cave and rock-shelter sites are equally ambiguous. The discovery of a number of characteristic Charentian industries at a range of open-air sites in southern France (Chinchon, Puycelsi, Rescoundudou, Plateau Baillard, Champlost etc.) is particularly significant in this regard and shows clearly that the production of heavily reduced industries was by no means confined either to enclosed cave/rock-shelter sites, or locations where local flint sources were in short supply (Sireix & Bordes 1972; Tavoso 1987a; Jaubert 1983, 1989; Le Tensorer 1973; Farizy 1985). The pattern of correlation between lithic reduction intensity and the associated environmental context of the assemblages is therefore at best rather vague and hardly shows the kind of close, direct correlations which the tool reduction models would seem to require.

Summary

In conclusion, none of these observations are intended to deny the inherent theoretical importance of the Dibble and Rolland models of industrial variability. The systematic resharpening, reworking and progressive reduction of stone tools clearly did take place in many contexts, and this must inevitably have contributed to some degree to the spectrum of variability within the Middle Palaeolithic as a whole. The central question is simply the scale on which these models can be applied. My own view is that they may perhaps help explain some of the more limited variations in tool percentages, such as those reflected in the dichotomy between the Typical and Denticulate assemblages in the Würm I levels at Combe Grenal, and perhaps also in the later Würm II levels there (Bordes 1972). Similar factors could possibly have led on occasions to fluctuations in racloir percentages which transgressed the critical threshold set by Bordes between industries of the Ferrassie group and those of the Typical Mousterian group – as reflected for instance in layers 28–30 at Combe Grenal, or in layers G-I2 at Pech de l'Azé IV (Bordes 1975a). Finally, there are some potentially important implications in these models for more long-term adaptive patterns of development in Mousterian technology, to be discussed in the following section. The only objection to the Dibble and Rolland models lies in their claim that 'these features alone account for many, if not most, aspects of assemblage composition variability seen in the Middle Palaeolithic of Western Eurasia' (Rolland & Dibble 1990: 490). I would suggest that this is a seriously exaggerated claim, for all the specific reasons outlined above. The tool reduction models clearly have some role to play in explaining the total industrial variability in the Middle Palaeolithic, but on a much more limited scale than the recent publications would imply (see also Close 1990; Kuhn 1992).

Overview: the nature of industrial variation in the Middle Palaeolithic

As emphasized frequently in the preceding sections, the most striking feature of Middle Palaeolithic technology – especially when viewed over long time spans and wide geographical areas – is its great variability. This is seen in many parameters: in the primary flaking techniques; in the dramatic variations in the relative frequencies of different tool forms; in some conspicuous variations in the forms of retouched tools (e.g. side scrapers, hand axes and other bifacial forms) and in the highly localized distribution of certain, idiosyncratic type-fossil forms (cordiform hand axes, backed knives, limaces, bifacial leaf-points etc.) (Bordes 1961a, b, 1968a, 1981, 1984; Bordes & de Sonneville-Bordes 1970). Whatever models or interpretations we eventually invoke to account for this bewildering range of technical and typological variation, the massive scale of this variation must be kept in mind.

From the results of research during the last twenty years there can be no doubt that at least some component of this variation is explicable in relatively simple functional terms. Quite clearly, the nature and availability of local raw materials must have had a significant impact on certain aspects of the industries – such as the size of tools, the relative finesse achieved in retouching the tools, or the choice of different core-reduction strategies for use on materials with sharply differing flaking properties (such

as coarse-grained quartz or quartzite as opposed to fine-grained flint) (see Chapter 4). Similarly, there can be no doubt that certain variations in the activities undertaken in different locations must have influenced the relative frequencies of particular tool types abandoned on the sites (along the lines suggested by Binford) and that certain other variations are attributable to the amount of reduction and resharpening of tools in different occupational or environmental contexts (as suggested by Rolland and Dibble). These points can be readily conceded and must be given due weight in any attempt to explain the total range of industrial variation within the Mousterian complex (Mellars 1992a).

The central question remains that of the scale on which these interpretations can be invoked. In the preceding sections it has been accepted that functional differences may well account for some of the more small-scale variations in the industries from the southwestern French sites – such as those which distinguish between some of the assemblages of the Typical and Denticulate Mousterian groups or between certain Ferrassie-type industries and those of the so-called 'racloir-enriched' Typical Mousterian group. The possibility that these models can be invoked to account for the more dramatic variations between the assemblages – particularly those which characterize the Ferrassie, Quina and MTA assemblages – must be rejected firmly on all the grounds outlined in the preceding sections.

If these points are conceded, then we are forced to look for some other, predominantly *non*-functional mechanisms by which the variety of technological patterns documented in the Middle Palaeolithic could have arisen. The answer, I would suggest, lies in a combination of four separate but inevitably interacting factors: the first is the role of social distance or social isolation in fostering separate patterns of technological development at different times and places within the Middle Palaeolithic time range; the second is the role of various demographic factors, leading to alternating periods of population decline or expansion, in some cases accompanied by major shifts in the geographical ranges occupied by the different groups; third, the role of more long-term technological adaptation under the influence of changing environmental or ecological conditions; and fourth, the possible role of convergence in patterns of technological development, constrained by the limited range of technological options available in the Middle Palaeolithic repertoire.

Social distance

The idea of social distance or social isolation is inherent in the time and space dimensions of the Middle Palaeolithic. Even defined in its most narrow sense (i.e. as coinciding essentially with the earlier part of the last glaciation) the Middle Palaeolithic covers an immense period of time, amounting to at least 80,000 years and extending to two or three times this length if the definition is allowed to embrace all the more complex forms of Levallois and related technologies which extend back at least to the period of isotope stage 7 (Rolland 1988b, 1990). In a spatial sense the Middle Palaeolithic covers the whole of Europe and western Asia and (in the form of the 'Middle Stone Age') the whole of Africa (Klein 1989a; Clark 1992). Thus the scope for social distance and social isolation arising from chronological or geographical factors alone is immense. To these space and time dimensions, one must add the effects of major geographical barriers to communication created by rivers, deserts or mountain ranges and the effects of possibly intermittent, discontinuous patterns of occupation in certain regions under the impact of sharply oscillating climatic and environmental conditions (Gamble 1984, 1986; Whallon 1989).

The effects of social distance in fostering separate and diverging patterns of techno-logical development in the Palaeolithic have been discussed frequently in the earlier literature (e.g. Isaac 1969, 1972a,b; Binford 1973: 245–7; Sackett 1982 etc.). All authors seem agreed that under conditions of social separation, certain divergences in technological patterns are not only plausible, but in most contexts almost inevitable. Isaac (1969, 1972a), for example, has stressed the potential impact of purely stochastic shifts in patterns of tool production, resulting from nothing more than inaccurate replication of tool-production patterns, accumulated and transmitted over many successive generations of stone workers – a process which might perhaps be referred to as 'technological drift'. He illustrates how such variations could well lead in some cases to apparently directional changes in patterns of technological development. Thus, small-scale divergences in tool or flake production could easily accumulate over several millennia as quite dramatic shifts in the overall patterns of tool technology, without necessarily involving any underlying technological or environmental constraints to promote technological change along specific adaptive or functionally-oriented lines. The same point seems to have been accepted in principle in several of Binford's publications (e.g. 1973, 1982b).

If these points are accepted, then we have an immediate possible explanation for many of the technical and typological variations documented throughout the enormous space and time range of the Middle Palaeolithic. Any interpretation in these terms would of course need to assume that many of these variations were largely neutral in an adaptive sense, in the sense of representing equally viable, essentially alternative ways of achieving the same technological and economic goals. But from all the available information on both the specific and general functions of Middle Palaeolithic tool and flake forms, there is no reason to doubt that this was largely true. As discussed earlier, it is possible that many forms such as racloirs, hand axes, backed knives, bifacial leaf points as well as simple unretouched flake forms, could have been utilized for very similar purposes (see Keeley 1980; Beyries 1987, 1988a; Anderson-Gerfaud 1990; Plisson 1988). If so, then the effects of social distance and social isolation factors alone might be sufficient to account for much of the documented range of technological and typological variation within the Middle Palaeolithic.

Clearly, the whole of this discussion leads us back to the time-honoured notion of 'technological traditions' in tool manufacture – i.e. that there were indeed certain basic patterns of tool production that can be understood only as a result of long-term, cumulative patterns of technological divergence which persisted in some cases over long periods of time (see Isaac 1969, 1972a,b; see also Boyd & Richerson 1986 and Gibson 1990 for theoretical discussions of the concepts of 'behavioural traditions' and related 'social learning processes'). My own view is that this represents the only viable explanation at present for many of the observed variations in Middle Palaeolithic technology for which strongly patterned distributions can now be seen in space or in time. In the case of the French industries this seems the only plausible explanation for the striking technological contrasts between the long succession of Charentian (i.e. Ferrassie and Quina) industries and those of the succeeding MTA group. On a broader geographical scale the same pattern may be reflected in the restricted spatial distributions of industries such as the Vasconian of the Franco-Cantabrian region, the various Micoquian and related leaf-point industries of central and eastern Europe and no doubt also in the clear chronological separation of the Yabrudian and Levallois-Mousterian industries of the Middle East. As discussed earlier, some of the strong contrasts in techniques of primary flake production (such as those apparent in some of the Levallois-point-dominated industries of the Near East or some specialized Levallois and

blade technologies of the north European plain: Meignen & Bar-Yosef 1988; Boëda 1988a, b; Boëda *et al.* 1990) must almost certainly be explained in similar terms. If a name is required for these patterns, I would suggest that the term 'techno-traditions' would serve as both a convenient label and as reflecting the essential nature of the industrial patterning in both a technological and social or demographic sense.

How far such variations can be described under the rubric of 'stylistic' patterning has been deliberately avoided in this discussion, to obviate some of the endless debates over the dialectics and semantics of this term which have continued for the last twenty years (see for example Sackett 1973, 1982, 1986a,b, 1988; Weissner 1983, 1984; Close 1980, 1989; Chase & Dibble 1987; Chase 1992). If the notion of 'style' is understood to apply to almost any variation in material technology which cannot be explained in purely functional or adaptive terms (roughly equivalent with Sackett's notion of 'isochrestic style'; see also Close 1980), then no doubt most of the variations discussed above could be described in these terms. If on the other hand 'style' is taken to imply some kind of almost conscious ethnic symbolism in the patterns of tool manufacture (what Chase and Dibble (1987) have described as 'iconological style'), then the whole notion of style in Lower and Middle Palaeolithic contexts becomes so elusive and controversial as to rank more as a red herring than as a positive aid to discussion. Even if some of the documented variations in Middle Palaeolithic tool forms can be argued to correspond with major demographic or spatial divisions in Middle Palaeolithic society, it is difficult to see how one could ever demonstrate that these were socially perceived patterns, rather than a result of purely long-term, stochastic divergences in technology along the lines discussed above (Binford 1973: 235–7). To introduce the notion of 'style' into discussions of Middle Palaeolithic technology would almost certainly do more to confuse than to clarify discussions of the nature and meaning of industrial variability, in the present state of research.

Demographic fluctuations

Arguably the least understood and most neglected aspect of Middle Palaeolithic adaptation is that of possible fluctuations in the demography and spatial distribution of human populations. Most of the earlier discussions of economic and social adaptation in the Middle Palaeolithic and the patterns of technological change have tended to assume essentially static, largely stable, populations, which are thought to have persisted in certain regions over immense spans of time (Bordes 1953a: 465–6, 1959: 101,1961b: 806–7; Binford & Binford 1966, 1969; Rolland 1988b; Rolland & Dibble 1990; Jelinek 1988a). I would suggest that these models ignore some of the most central and critical dimensions of potential variability in Middle Palaeolithic populations. The main points to emphasize are as follows:

1. All of the current evidence suggests that in most areas of Eurasia, Middle Palaeolithic groups were living in relatively low population densities, fairly thinly distributed over the available habitats (Bordes 1968a; Mellars 1973; Klein 1973; Foley 1989; Whallon 1989; Soffer 1989; Binford 1982b). There can be little doubt that in many regions, including the Perigord, most Middle Palaeolithic groups were living in much lower populations than those during the ensuing Upper Palaeolithic phase (Bordes 1968a; Mellars 1973, 1982). How far these populations may or may not have been isolated demographically, and exactly what forms of breeding networks were maintained between the populations in different regions (cf. Wobst 1974, 1976) remains speculative. However, it seems unlikely that the individual demographic units in most areas were very large – especially under the conditions of last-glacial

Europe, where the combination of harsh climatic conditions and natural geographical barriers may well have set clear limits on the extent of inter-communication possible between communities across wide geographical areas (Whallon 1989; Gamble 1982, 1984, 1986).

The demographic implications of small population units of this kind are well known. Under these conditions, sharp fluctuations in population numbers can easily arise for several reasons, including purely stochastic fluctuations in the birth or death rates of the population the effects of disease, or periods of economic or social stress resulting for example from a succession of exceptionally severe winters, periods of unusually heavy and prolonged snowfall, or occasional unpredictable fluctuations in the numbers or local migration patterns of animal populations (cf. McArthur & Wilson 1967; McArthur et al. 1976; Birdsell 1968; David 1973; Dunbar 1987; Zubrow 1989). Whether such fluctuations would ever be sufficient to totally eliminate local populations in certain regions is more debatable. Nevertheless, ethnographic and ethnohistorical data leave no doubt that under certain combinations of economic, environmental or social stress, relatively small, isolated communities of hunter-gatherers can be affected by fluctuations in natural population numbers to the point where the viability of the population – in both economic and social terms – is effectively undermined (e.g. Lee & Devore 1968; Damas 1969; Dumont 1977; David 1973: 296–7). The possibility that such fluctuations could have occurred at many different times and places during the 80,000 years or more of the Middle Palaeolithic sequence should be kept in mind.

2. A second, equally if not more significant point to emphasize is the impressive scale and frequency of the climatic and ecological oscillations which are now known to have occurred at many different times during the

Middle Palaeolithic period. Even if we confine attention purely to the earlier stages of the last glaciation (i.e. ignoring the dramatic climatic fluctuations which marked the period of the last glacial/interglacial cycle, between isotope stages 6 and 5) it is now evident that ecological and vegetational changes in many parts of western and central Europe shifted at different periods from almost treeless periglacial tundra or steppe (with year-round temperatures at least 10°C lower than those of the present day) to episodes of dense deciduous forest, not very different from those of fully interglacial conditions (notably during isotope stages 5a and 5c and conceivably at certain points during stage 3) (see Chapter 2, Figs 2.6, 2.7, 2.21 etc.). These climatic and ecological fluctuations were on a massive scale, sharply oscillatory in character and demonstrably associated with major changes in the composition of the associated animal populations. For groups who were heavily dependent on the exploitation of animal resources, it may be difficult to over-emphasize the impact of the changes which occurred for example at the transition from isotope stage 5a to 4 (marked in the Perigord region by a rapid shift from predominantly red deer to reindeer populations) or the equally abrupt changes which occurred at the transition from stage 4 to stage 3 (marked by a rapid decline in reindeer, in favour of bovids, horse and red deer) (see Fig. 2.22; Bordes & Prat 1965; Guadelli 1987; Laville & Guadelli 1990; Raynal & Guadelli 1990). As discussed further below, it is likely that some of these changes were dealt with by the human groups by rapid adaptations in local subsistence strategies, probably accompanied by corresponding adaptations in the patterns of lithic technology. But it is equally likely that these ecological shifts would have led in certain contexts to major shifts or displacements in the territories exploited by the human groups, leading to major changes in the geographical ranges occupied by particular populations. More simply, one could

say that if Middle Palaeolithic populations are assumed to have been specifically adapted in some way (economically, socially or technologically) to specific environmental and ecological conditions, then any major displacements in the boundaries of these different zones would almost inevitably be accompanied by corresponding shifts in the ranges occupied by the groups. With small-scale, highly mobile populations, demographic shifts of this kind are not only plausible but probably inevitable in response to the major climatic fluctuations of the last-glacial period.

In this context it is important to remember that virtually all documented hunter-gatherers tend to be highly mobile in their seasonal and annual foraging patterns (cf. Lee & DeVore 1968; Binford 1982a,1983c) and also that in the case of Middle Palaeolithic communities we would seem to have an explicit indication of this mobility in the remarkably long distances travelled by several raw materials (see Chapter 5). As Binford has emphasized (1982a) such mobility invariably ensures that human groups are familiar with the economic and environmental resources of extensive areas, extending well beyond the immediate foraging territories of individual groups and often extending into adjacent territories of neighbouring groups. It is equally significant that very few hunter-gatherers are known to be strictly territorial in their behaviour (in the sense of attempting to defend exclusive rights to certain geographical territories or ecological resources) and indeed are usually incapable of maintaining an effective defence of territorial boundaries under conditions of low overall population densities (Lee & DeVore 1968; Woodburn 1982; Layton 1986; Dyson-Hudson & Smith 1978). Presumably, the combination of these factors would have made a process of gradual, intermittent shifts in the specific geographical ranges occupied and exploited by different Middle Palaeolithic groups not only a relatively easy adaptation, but almost inevitable in the context of the rapidly oscillating climatic and ecological conditions of the late Pleistocene.

3. The final point to be kept in mind is the geographical diversity and complexity of most regions of Europe – characterized by a complex juxtaposition of many ecologically and topographically contrasting regions, each with its own particular range of ecological and economic resources. The Perigord region provides a classic illustration of this. To the north of the region one moves fairly rapidly (within a space of 300–400 km) into the sharply contrasting environments of the north French Plain, characterized by topographic, climatic, vegetational and even raw material patterns unlike those in the Perigord area. To the south one moves equally rapidly into the very different environment of the Languedoc region, and eventually into the even more contrasting habitats of the Pyrenees and northern Spain. To the east and southeast lie the Massif Central, the Rhône valley and the Mediterranean coastal zone. These strongly contrasting regions all lie within at most 300–400 km of the Perigord area. Presumably these areas are likely to have supported at least partially separate human populations, with economic, social and possibly technological adaptations which could have differed significantly from those in the Perigord region (see de Lumley 1976; Bordes 1984). Potentially, all of these are regions from which, during periods of demographic flux or rapid climatic change, new human populations could have moved into the southwestern French zone. In addition, it should be kept in mind that the Perigord region was evidently a very favourable habitat for hunter-gatherer groups throughout the last glacial sequence, both in terms of subsistence resources and the availability of naturally sheltered living sites and abundant raw materials – as the exceptional density and concentration of Upper Palaeolithic sites in the region clearly indicates. The possibility of di-

rect social or economic competition for the use of this region throughout the various stages of the Middle and Upper Palaeolithic sequence should therefore not be too lightly dismissed.

My own view is that this scenario of population shifts, displacements and eventually replacements should now be seen as one of the central elements in any attempt to account for overall patterns of human occupation in the different regions of Europe throughout the long time span of the Middle Palaeolithic succession. As I have argued elsewhere (Mellars 1969: 148–50, 1992a) this almost certainly accounts for the abrupt appearance of the classic MTA industries in southwestern France, following the long period occupied by the preceding sequence of Charentian and Denticulate Mousterian industries. Significantly, this can be seen to coincide with the period of rapidly improving climatic conditions around the transition from isotope stage 4 to stage 3 (Mellars 1986a, 1988). There is at least a possibility that some of the other industrial changes documented around the same point in the climatic succession could reflect similar population displacements – for example the curious intrusion of the isolated level of Denticulate Mousterian (layer 20) within the uppermost Quina Mousterian levels at Combe Grenal and perhaps some of the subsequent shifts between Denticulate and Typical levels in the immediately overlying levels (Bordes 1972). In other regions, such as northern France, there can be no doubt whatever that major population movements occurred in direct response to local climatic and ecological changes at several points during the Middle Palaeolithic succession (Bordes 1954a; Farizy & Tuffreau 1986). Demographic oscillations of this kind are not merely plausible or acceptable in a general theoretical sense but to a large extent inevitable and predictable, given our current understanding of human demography and the pattern of climatic and ecological fluctuations throughout the course of the Middle Palaeolithic period.

As a final point it should be recalled that demographic displacements and replacements of this kind have now been accepted as an inescapable component of the classic succession of Upper Palaeolithic occupations in both southwestern France and several other regions of Europe. In the Perigord, for example, it is now generally accepted that the alternation of Aurignacian and Châtelperronian industries during the initial stages of the Upper Palaeolithic sequence (reflected stratigraphically at Roc de Combe and Le Piage) can only be plausibly explained in these terms (Hublin & Demars 1989; Demars 1990; Harrold 1989 etc.) The same is almost certainly true of the sudden replacement of the Aurignacian by the earliest Upper Perigordian (Gravettian) industries, and perhaps by the equally abrupt appearance of the Noaillian industries towards the end of the Upper Perigordian sequence (de Sonneville-Bordes 1960, 1973; David 1973, 1985; Bricker & David 1984). Similar population replacements have been documented from both archaeological and ethnohistorical evidence among many recent hunter-gatherer populations – for example between various Inuit and Indian groups in the Canadian arctic. Population fluctuations, displacements and replacements must now be seen as a largely inevitable and predictable part of human demographic and adaptative patterns throughout the whole of the Palaeolithic sequence.

Technological adaptations

The discussion in the preceding sections has focused on some of the possible mechanisms by which significant changes in lithic technology could have taken place. Throughout, the emphasis has been on variables which could generate change without necessarily involving any associated changes in the specific economic, technological or 'functional' contexts in which the tool assemblages were employed.

None of this discussion is intended to dismiss the idea that certain aspects of lithic technology were related closely to functional or adaptive mechanisms of some kind, nor that these features could, over extended periods of time, have constrained and influenced the patterns of technological development in significant ways. My own view is that these attempts to find functional or adaptive explanations for the documented variations in technology are at present seriously handicapped by the scarcity, if not total absence, of data bearing on some of the most crucial behavioural parameters of the assemblages, such as the specific functions of different tool types, variations in patterns of seasonal occupation in different sites, or indeed the size and character of the human groups who occupied the sites. Nevertheless these practical limitations of the current data base should not prevent some of the basic questions of functional or adaptive interpretations from being raised.

In the southwestern French sites, an obvious challenge is presented by the clear patterns of technological development apparent in the stratified sequences of Charentian (i.e. Quina and Ferrassie) assemblages discussed in Chapter 6. From the sequences recorded at Combe Grenal, Abri Chadourne, Abri Caminade, La Ferrassie, Roc de Marsal, Petit-Puymoyen and several other sites, it would now seem that throughout the entire time span of these Charentian sequences, there were certain gradual, progressive changes in basic techniques of flake production, leading initially from the use of relatively sophisticated Levallois techniques, towards much simpler techniques, evidently oriented towards the production of thicker, heavier flakes (see Figs 6.17, 6.18). Interestingly, this sequence does not seem to have been accompanied by major changes in the overall composition of the tool assemblages (as reflected in the relative frequencies of racloirs, notches, denticulates etc.), and almost all of the documented changes in the 'typological' character of the industries (such as the progressive increase in the frequencies of transverse as opposed to lateral racloirs, increased frequencies of tools shaped by typical Quina retouch, and the associated decrease in frequencies of more complex forms of double and convergent racloirs) seem to be attributable largely if not entirely to the underlying shifts in the character of basic flaking strategies by which the flake blanks for tool manufacture were produced (Fig. 4.7: Bordes 1961b: 806, 1968a: 101–2, 1977: 38, 1981: 78–9, 1984; Bordes & de Sonneville-Bordes 1970: 61, 71). Chronologically, this entire sequence of development seems to have taken place over about 10,000 years, coinciding with the period of rapid climatic deterioration over the transition from isotope stage 5 to stage 4 (Mellars 1986a, 1988). The obvious question therefore arises as to how far this technological development can be seen as a direct adaptation to changing environmental conditions throughout this particular segment of the last glacial sequence. What follows is an attempt to explore some of the alternative adaptive scenarios that could potentially be advanced to account for this pattern of technological change (Mellars 1992a).

1. One of the most obvious and seductive explanations would be to account for long-term patterns of technological change primarily in terms of variable constraints on the character or accessibility of local raw material supplies. In the case of Levallois technology, for example, it has often been suggested that for effective, large-scale application of these techniques it is necessary to have access not only to relatively fine-grained materials, in which the flaking strategies can be accurately controlled, but also large, fairly regular nodules (Fish 1981: 377–9; Dibble 1985: 392). Both these generalizations can be questioned from specific archaeological contexts (e.g. Turq 1989b; Tavoso 1984), but in general terms may well be valid (cf. Fig. 3.26). If they are valid, then it would be reasonable to look

for explanations of long-term reductions in Levallois techniques in terms of continuing constraints on the quality or local availability of raw material supplies. Two main scenarios could be visualized in this context. One would be to suggest that the initial, high-Levallois phases of Charentian technology were developed in some contexts where raw materials were relatively abundant and of high quality (such as northern France) from which the groups subsequently shifted to areas of generally poorer quality and/or less abundant materials (for example on most of the cretaceous and jurassic outcrops of the southwestern French region: Geneste 1985, Demars 1982). Under these conditions it could be argued that the groups would adapt to the generally poorer raw materials by gradually phasing out the reliance on Levallois techniques in favour of technologically simpler or more economical non-Levallois techniques (Turq 1988b, 1992a). The alternative possibility suggested by Rolland and Dibble (1990) is that a decline in Levallois techniques could be related more simply to a progressive over-exploitation of the available raw material sources in the Perigord region. Whilst the latter possibility may be difficult to visualize on a broad, regional scale, it could conceivably have been a significant factor in some more intensively occupied locations such as Combe Grenal, which are known to have been occupied and presumably continuously exploited for raw materials over the greater part of the Mousterian period. What this model cannot explain is why some of the demonstrably later Mousterian industries in the same locations should have been able to revert (apparently rapidly) to the use of much more extensive Levallois techniques (see for example Rolland 1988b: Fig. 9.2).

2. A second alternative discussed by Rolland and Dibble (1990) is that a progressive decline in the use of Levallois techniques could have been related to increasingly in-

tensive or sedentary occupation patterns in particular sites, leading to increased pressures for the most economical use of available raw materials. In the case of the Ferrassie/ Quina industries, for example, they argue that the sharp deterioration in climatic conditions towards the peak of the Würm II (i.e. isotope stage 4) climatic episode could have discouraged large-scale mobility and tended to concentrate foraging activities within a short distance of the major occupied sites – perhaps related in part to the exploitation of more aggregated resources such as migrating reindeer herds (see also Jelinek 1988a). This in turn would place further constraints on access to raw materials and would have provided a further incentive towards the use of the most economical, arguably non-Levallois, flaking strategies.

3. A totally different approach would be to emphasize not so much the character of the primary flaking techniques but rather the specific forms and functions of different retouched tools, produced by different techniques (e.g. Turq 1989b). For example, it could be argued that typically Quina side scrapers (i.e. relatively thick, convex forms, shaped by heavy, invasive retouch) might well be more appropriate for heavy-duty hide or skin-scraping than were the thinner, more lightly retouched forms encountered in Ferrassie-type industries, and could reflect simply a shift towards the increased use of animal skins (for clothing, rugs, shelters etc.) during the increasingly cold conditions of isotope stage 4. As discussed earlier, the available micro-wear evidence can hardly be used to provide any conclusive test for this particular hypothesis. If this pattern were to be substantiated, however, it would provide an entirely separate line of interpretation for the documented shifts from predominantly Levallois to non-Levallois flaking techniques over the period of the Ferrassie-Quina transition. As Turq has pointed out (1988b, 1989b, 1992a) the characteristically thick, triangular-

sectioned flakes employed for the production of Quina racloirs (Fig. 3.14) could be seen as deliberate, preconceived products, designed specifically for the production of these tools. The thrust of this hypothesis would therefore reverse the generally assumed relationship between technology and typology and imply that in at least some contexts it was the changing morphology of stone tools that dictated shifts in the associated primary flaking strategies, and not vice versa. More explicit data on the actual functions of Quina versus non-Quina racloirs would be needed to evaluate this hypothesis further.

4. Finally, the recent tool-reduction models of Dibble and Rolland could be invoked, in a slightly different guise, to account for the documented shifts between Levallois and non-Levallois flaking techniques. If Dibble (1984a, 1987a) is correct in suggesting that typical Quina-type racloirs are in general more heavily reduced and resharpened than most of the other side-scraper forms, one could ask how far the relatively thick, steep-edged flakes used for the production of Quina scrapers may have been deliberately selected with this specific procedure in mind. In other words, if the need for repeated resharpening of side scrapers became more pronounced during the course of the Ferrassie-Quina development (possibly as a result of increased sedentism on the sites: Jelinek 1988a: 208) this could have provided an incentive for the production of thick, heavy flakes, particularly well adapted for such repeated resharpening. Here again, the thrust of the argument would reverse the normally assumed normal relationships between technology and typology and suggest that changes in tool morphology, and/or function, might serve in some cases as the primary determinants of change in associated flaking strategies.

The above example illustrates some of the inherent problems encountered in attempting to explain long-term patterns of change in lithic technology in convincing functional or adaptive terms. The central problem is the difficulty of testing adequately the various alternative models for change against the available archaeological data. Various possible scenarios for long-term technological adaptation can be suggested, and some apparent implications of these different scenarios can be spelled out. The task of choosing between the different alternatives remains, as ever, extremely difficult.

There are similar problems in attempting to explain certain other, apparently well documented patterns of change in other segments of the Middle Palaeolithic sequence in southwestern France. As discussed earlier, the long-term development of the MTA industries in the Perigord seems to reveal a chronological pattern no less sharply defined and well documented than that of the Charentian industries. In this case the development seems to have been marked again partly by a general reduction in the Levallois component of the industries, accompanied by a parallel decrease in frequencies of typical hand axes and various racloir forms (Bordes 1954–55, 1975a: 301–3, 1981: 77–8, 1984: 137–49; Pelegrin 1990; see Chapter 6).

To account for the change in Levallois techniques one could invoke similar mechanisms to those discussed above for the development of the Charentian industries – i.e. a gradual shift in core-reduction methods in response to the declining availability of local raw material supplies. However, in this case the specific patterns of technological change followed a very different path from that documented in the Charentian industries, and seem to have been marked by increasing use of smaller, simpler forms of primary flakes rather than by developing specialized techniques to produce large, thick, cortical-covered flakes of the kind encountered in the later Charentian (i.e. Quina-type) industries (Pelegrin 1990). If raw material constraints were involved in this sequence, then the specific patterns of adaptation were very different from those during the development of the Charentian industries. How one should account for the major shifts in the frequencies of different tool types during the course of the MTA development remains equally enigmatic. Both Rolland (1981: 26–31, 1990: 371–3; Dibble & Rolland 1992: 17) and Binford & Binford (1966: 256, 259, 1969: 79) have suggested that a move away from the production of large, sharp-edged tools towards various notched and denticulated forms might reflect either a shift towards a greater use of plant-food resources or alternatively an increasing emphasis on wood-working activities. Since the chronological development of the MTA industries appears to have been associated with a period of mainly temperate climatic conditions, when both wood and plant food resources would presumably have become more abundant, one might well visualize some of these changes within a broadly adaptive framework (Mellars 1992a).

What these models cannot explain is the initial, very abrupt appearance of the MTA industries in the Perigord region, nor some of the distinctive typological and technological features (such as the sudden appearance of typical hand axes and backed knives and the exceptionally high ratios of Levallois techniques) which differentiate the earliest stages of the MTA sequence from the final stages of the preceding Ferrassie-Quina succession (Mellars 1965; Rolland 1988b). Even if some component of long-term ecological adaptation can be identified in these sequences, therefore, this cannot provide a satisfactory explanation for the detailed patterns of technological development reflected in the industrial succession as a whole. There seems little doubt that the only convincing explanation for the sudden appearance of the MTA industries in the Perigord sequence lies in a replacement of population. Even if this was precipitated in some way by the contemporaneous climatic and ecological changes (Mellars 1969: 148–50), it seems evident that such changes alone can hardly account for the detailed character of changes in lithic tool kits or for their occurrence at this particular time in the Mousterian succession.

The role of technological convergence

The final factor which must be kept in mind in any general study of Middle Palaeolithic variability is the possible role of technological convergence in promoting certain combinations of technical and typological features at different times and places in the Middle Palaeolithic universe. The central idea of convergence in technological development is by no means new. Indeed, this was inherent in Bordes' original notion of the *évolution buissonante* of Lower and Middle Palaeolithic industries, published over 40 years ago (Bordes 1950). The same idea is equally implicit, in different ways, in the recent functional models of Binford and in the tool-reduction models of Dibble and Rolland (Binford 1973, 1982a; Rolland 1981: 23; Rolland & Dibble 1990). Whether phrased in primarily cultural terms (as visualized by Bordes) or in functional or adaptive terms (as visualized by Binford and by Dibble and Rolland) this notion remains central to an understanding of industrial variation documented throughout the extensive time and space range of the Middle Palaeolithic.

However this notion of technological convergence is expressed, the idea reduces in essence to what might be termed the 'principle of limited possibilities' in technological expression (cf. Rolland 1981: 19). The entire structure of basic taxonomic divisions within Middle Palaeolithic industries set out in the publications of François Bordes rested primarily on a relatively small number of quantitatively defined typological and technical features (Bordes 1953a, 1961b, 1981, 1984; Bordes & Bourgon 1951). The most critical parameters were those based on overall percentages of racloirs in different industries (which collectively defined and differenti-

ated the Charentian, Typical and Denticulate variants of his taxonomy) and the gross distinction between industries defined as predominantly 'Levallois' or 'non-Levallois' in a technological sense (i.e. with Levallois indices respectively above or below ca 10 percent). The various possible permutations of these basic typological and technical parameters provided the basis for almost all his taxonomic divisions within the French industries – with the single notable exception of the MTA variant, which was always defined primarily on the basis of two distinctive type-fossil forms – typical cordiform hand axes and typical backed knives (Bordes 1981, 1984).

The implications of this approach to taxonomy are largely self evident. With such a small and restricted range of possible typological and technological variables at our disposal, the probability that certain specific combinations of these variables will arise at several different times and places becomes not merely possible but almost inevitable. This is true whether one attributes the determinants of these different patterns to specific cultural mechanisms, as proposed by Bordes, or to broadly functional mechanisms, as proposed by Binford and in the tool-reduction models of Dibble and Rolland. The 'law of limited possibilities' when applied to the main quantitative aspects of the industries, dictates that certain recurrent associations of these basic features must inevitably recur at several different times and places, regardless of how these variations are interpreted in specific behavioural terms.

The archaeological records of the Middle Palaeolithic would seem to provide several obvious examples of such convergence. The classic illustration perhaps is provided by the spatial and chronological distribution of industries which have generally been described as having a broadly 'Quina Mousterian' aspect. Industries included under this broad rubric have been identified from geographical contexts as widely separated as western

France and the Yabrudian industries of Israel and Syria, and from sites extending chronologically from the earlier stages of the penultimate glacial (as apparently at Les Tares and La Micoque) to the classic sequence of Quina-type industries dating from isotope stage 4 (ca 60–70,000 BP) of the last glaciation (see Bordes 1968a: 98–120, 126–30, 1981; 1984: 57–64; Rigaud & Texier 1981; de Lumley 1971: 351–65). As soon as one turns to the specific features of these industries, however, it becomes clear that the only significant feature which they share is a relatively high overall frequency of various racloir forms (generally greater than ca 50–55 percent of the total retouched tool component) manufactured from relatively thick, unprepared non-Levallois flakes. All other common features among the different industries, such as the occasional production of thick, heavily retouched Quina-type racloirs, or increased ratios of transverse to lateral racloir forms, can be seen as an effectively automatic result of the combination of these two basic typological and technological features (see Chapter 6 and Bordes 1961b: 806, 1968a: 101–2, 1977: 38, 1981: 78–9). To regard the industries defined in these terms as having any necessary connection in either cultural or any other behavioural terms would make little sense from the evidence at present available (see also Rolland 1981: 23).

The reality of this dilemma in the case of these taxonomically Quina-type industries has now been recognized by the majority of French workers. In discussing the industry from the Rissian site of Les Tares, for example, Rigaud and Texier (1981) have emphasized that the assemblage contrasts in several obvious respects with most of the industries of the classic Quina-Mousterian group – for example in the much higher frequencies of notched and denticulated tools, in the exceptionally low percentage of transverse as opposed to lateral racloirs, in the total absence of any trace of either Levallois or disc-core flaking techniques, in the curiously denticulated character of the retouched edges on many of the racloirs, in the complete absence of classic limaces or large, bifacially worked racloir forms, and several other features (see also Geneste 1990). In addition to these strictly typological and technological contrasts, it is now clear that these so-called 'Proto-Charentian' or 'Proto-Quina' industries as represented at Les Tares, La Micoque and elsewhere are separated from the sequence of classic (i.e. last-glacial) Quina-Mousterian industries by an enormous span of time, possibly amounting to 100,000 years or more, without any clear evidence for the continuity of similar typological or technological patterns throughout the intervening period. The total absence of characteristic Quina-type assemblages from all of the long, stratified sequences of 'Würm I' Mousterian levels at Combe Grenal, Le Regourdou, La Chaise, Pech de l'Azé etc. is especially significant in this regard. Finally and most significantly, we now have what seems to be almost conclusive evidence for the gradual re-emergence of classic Quina technology during the course of the last glacial sequence, by a process of gradual technological development from industries of the Ferrassie type – as reflected in the long stratified sequences recorded at Combe Grenal, Abri Chadourne, Abri Caminade, La Ferrassie, Petit-Puymoyen, Roc de Marsal, Roc-en-Pail and elsewhere (Fig. 6.17). Confronted by this battery of observations, the case for clear convergence in the patterns of development of certain 'Quina-type' industries during at least two entirely separate and independent points in the Middle Palaeolithic succession of western France would seem to be beyond question. The same can be said for the supposedly Quina-like 'Yabrudian' and related industries of the Middle East, which are again linked with those of the classic Quina group purely by the combination of high overall frequencies of racloirs manufactured on thick, non-Levallois flakes.

Similar observations can be made for some

of the recent discoveries of 'Ferrassie-type' assemblages recovered from penultimate-glacial age sites such as Biache-Saint-Vaast in northern France and the Cotte de Saint-Brelade in Jersey (Tuffreau & Sommé 1988; Callow & Cornford 1986). Again, the only features to link these industries with those of the classic Ferrassie group seem to be the combination of high racloir frequencies with relatively high frequencies of various Levallois flaking techniques. Any other features shared in common between the assemblages, such as moderately high frequencies of double and convergent racloir forms and the low frequencies of transverse racloirs, would seem to be explicable largely in terms of the common use of thin, elongated forms of flake blanks for tool manufacture in the different industries (see Dibble 1987a,b). The same arguments apply with even greater force to the various occurrences of taxonomically Denticulate and Typical assemblages. As noted in Chapter 6, Bordes acknowledged clearly in his later publications that these two variants carried little conviction either taxonomically or culturally and almost certainly represented the products of several, independent lines of convergent technological development at different times and places in the Mousterian succession (e.g. Bordes 1977: 38, 1981: 79; Bordes *et al.* 1954: 249).

Summary

As discussed above, this kind of technological convergence within certain very broadly defined taxonomic groupings of Middle Palaeolithic industries must be accepted as a predictable feature of Middle Palaeolithic technology, regardless of how this variability is interpreted in specific cultural or behavioural terms. This pattern is expectable and inevitable purely in terms of the principle of limited technological alternatives discussed above (Rolland 1981: 19). But to leap from this conclusion to the suggestion that there is no significant patterning recognizable within

the overall time and space framework of the Mousterian complex would be not merely irrational but demonstrably in conflict with many explicit features of the archaeological evidence. Three aspects should be re-emphasized in this context:

(a) The existence of clearly defined spatial patterns in the distribution of certain distinctive industrial variants – for example the MTA industries of western France, the Vasconian industries of the Vasco-Cantabrian region, the various Micoquian and leaf-point industries of central and eastern Europe or the equally distinctive Yabrudian and Levallois-Mousterian technologies of the Middle East.

(b) The existence of clear and demonstrable chronological patterns in the assemblages from particular regions – for example in the chronological separation of the Ferrassie, Quina and MTA industries within the last-glacial sequence of western France and the equally well defined chronological and technological trends demonstrable *within* the various stratified sequences of Ferrassie and Quina industries in the same region. Beyond the Perigord, equally clear chronological patterns have been documented in the relative positions of the Yabrudian and Levallois-Mousterian industries of the Middle East, in several well defined stages in the development of the Micoquian and leaf-point assemblages in central Europe and apparently in similar patterns in certain Mousterian industries of eastern and southeastern Europe (Jelinek 1982; Bar-Yosef 1989; Bosinski 1967; Kozlowski 1992).

(c) Finally, the existence of sharply defined patterns in some of the more specific typological and technological features of the industries, which go far beyond the broad quantitative features employed by Bordes in his generalized taxonomy of the industries. Several examples of these idiosyncratic features of particular industries have been given in preceding sections: the restricted distribution of typical cordiform hand axes and typical backed knives within the MTA industries of western France; some distinctive type-fossils of the classic Ferrassie and Quina-type industries (typical limaces, large, bifacially worked tranchoirs, and other forms of bifacially worked points); the various 'prodniks', Micoquian type hand axes and bifacial leaf points of the central European industries; the flake cleavers of the Vasco-Cantabrian Vasconian industries; the distinctive tanged and stemmed points of the North African Aterian industries and so on (Bordes 1984; Gamble 1986; Klein 1989a). It is these distinctive, essentially qualitative rather than quantitative features, which characterize some of the specific industrial variants of the Middle Palaeolithic, and which exhibit the clearest chronological and geographical patterns. My own view is that these features are in no way a product of broad convergence mechanisms and cannot at present be explained entirely in any simple or convincing functional or raw material related terms. The only plausible explanation for these patterns lies in the notion of separate technological traditions – i.e. separate patterns of technological development, fostered by the variable degrees of social distance maintained between the human populations involved. In this sense, the existence of a real element of 'cultural' patterning in the character of technological variation within the Middle Palaeolithic seems difficult, if not impossible, to deny.

CHAPTER 11

Neanderthal Society

Social organization is central to the adaptive strategies of all animal communities. The size and structure of local groups, the particular roles and relationships of individuals within these groups, the existence of any systematic alliances between groups and their integration into larger, regional populations are all in one way or another critical to the survival prospects of communities under varying ecological and demographic conditions. All of these factors impinge on the ability of groups to secure a reliable and predictable food supply, their capacity to protect themselves from attacks by predators or members of their own species, and their ability to form viable, stable breeding populations in the short and long term. It is now generally recognized that understanding the nature and structure of social organization is fundamental not only to studies of human and primate communities, but to those of all animal populations (e.g. Hinde 1983; Smuts *et al.* 1987; Standen & Foley 1989).

The central issue here is how the social structure and organization of Neanderthal communities may have differed from that of behaviourally and biologically modern populations. Was it essentially the same as in modern communities or was there, as White and others have argued, a fundamental 'restructuring in social relationships' coinciding with the transition from the Middle to the Upper Palaeolithic (White 1982; see also Binford 1992; Soffer 1994)?

What we mean by 'modern' patterns of social organization has been discussed fully in a recent article by Lars Rodseth and others in *Current Anthropology* ('The human community as a primate society': 1992). Rodseth *et al.* provide a systematic comparison of various aspects of social organization across a broad range of modern (principally hunter-gatherer) human communities, set against those of other primate species. What emerges is that while the groups of non-human primates display a great variety of social structures, from the solitary, monogamous families of Asian gibbons to the male-dominated harem groups of mountain gorillas (Smuts *et al.* 1987), there are a number of specific patterns which are characteristic of the great majority of present-day human communities but which occur much less commonly, if at all, in groups of non-human primates. These include the strong tendency to form permanent relationships between particular pairs of males and females throughout life; the tendency for several of these closely bonded family units to form relatively stable interacting and residential groupings over substantial periods of time; the almost universal tendency in these contexts for food to be shared extensively between members of the two sexes, together with the dependent children; the equally universal tendency for these multi-family groups to practise strict rules of exogamy, most commonly involving the movement of females away from parental groups; and the establishment of at least temporary 'home bases', which usually provide the focus for many different activities (sleeping, eating, socializing, food preparation, craftwork etc.) over periods of at least several days, if not much longer. A further and particularly diagnostic feature of modern human communities is their tendency to maintain long-term social relationships between individuals who are related by birth, but who habitually reside in separate social groups, sometimes at great distances from their original birthplace (Rodseth *et al.* 1992: 237–40). It is this feature, described by Foley & Lee (1989) as 'combined kinship and lineage', which provides one of the most striking contrasts between modern human communities and all the recorded groups of non-human primates. Other factors, such as the importance and rigidity of various social or economic roles ascribed to individual members, the division of economic or other activities between the sexes, and the degree of territoriality maintained between adjacent groups, all vary fairly widely in both human and primate communities and therefore cannot be used to provide any firm criteria for the definition of characteristically human as opposed to primate behaviour. Most of these features are familiar components of the classic 'home-base' model of distinctively human behaviour which has been discussed in the anthropological and archaeological literature since the 1960s (e.g. Isaac 1969, 1978, 1984; Binford 1984, 1985). What is new about Rodseth *et al.*'s analysis is that it puts these specific features of characteristically 'human' or 'modern' behaviour firmly within the framework of broader comparative studies of human and other primate societies.

The value of Rodseth *et al.*'s study therefore is that it provides a systematic yardstick for identifying what we mean by modern social organization. There seems to be widespread agreement among archaeologists and anthropologists that the main features identi-

fied in Rodseth's study would be equally valid for most culturally Upper Palaeolithic communities throughout Eurasia, extending back at least to the earlier stages of the Aurignacian (White 1982). The critical question is how the behaviour and organization of the preceding Middle Palaeolithic and Neanderthal communities matches this model. Since there are currently very differing viewpoints on this issue (compare for example White 1982 and associated comments; Binford 1984, 1985, 1992; Soffer 1994; Clark & Lindly 1989), the various lines of evidence which have been advanced require careful consideration.

The Binford model

Recently, a stark alternative to the idea of essentially modern patterns of social organization among Neanderthal and earlier populations has been argued forcefully in the publications of Lewis Binford, based largely on his studies of the detailed spatial organization of the various occupation levels at Combe Grenal (Binford 1992, and personal communication). In essence, Binford believes that it is possible to identify several clear and repeated patterns in the spatial distribution of hearths, animal remains, stone tools, flaking debris etc. throughout most of the separate occupation horizons at Combe Grenal, which to him carry evident sociological implications. As outlined in Chapter 9, he believes he can identify the following recurrent patterns:

1. A well defined area generally located towards the centre of the rock-shelter deposits, marked by an irregular scatter of ashes and burned bones, and usually lying on soft, fine-grained, sediments. The most frequent items associated with these central, ashy zones tend to be predominantly notched and denticulated tools, with concentrations of flaking debris resulting from their manufacture and many heavily fragmented splinters of long-bones, evidently reflecting inten-

sive smashing of bones for extraction of marrow. Also found in these central zones are concentrations of teeth or skull fragments of large animals (especially bovids and horse) often with traces of burning. Binford therefore believes that this represents an area of intensive processing of certain selected parts of animal carcasses, mainly to get at the rich supplies of bone marrow or brains and tongues from the long bones or jaws and skulls of the animals respectively. Recently, Binford has referred to these areas as representing the 'nest' of the site (1992: 48).

2. Binford identifies a second zone of more marginal activities, generally located towards the front of the rock-shelter, and often associated with scatters of fallen limestone blocks. In these peripheral parts of the occupation area, Binford claims to detect localized concentrations of what he interprets as primary butchery tools (principally various forms of side-scrapers and points) associated predominantly with the detached articular ends of long bones, and in some cases with fragmented mandibles. In certain levels he claims that these are associated with small, concentrated hearths, which show evidence for more intensive burning than the diffuse scatters of ash and burned bone in the centre of the site. He sees these therefore as more general, primary butchery or carcase-processing areas, essentially preliminary to the intensive processing of selected carcase parts in the central parts of the site.

3. Finally, Binford claims to have detected some clear contrasts between the kinds of lithic raw materials associated with these two main activity zones. The central areas, he suggests, seems to be associated predominantly with very local and generally rather poor quality raw materials, usually obtained from a maximum of 2–3 km from the site. The peripheral areas, by contrast, tend to produce greater quantities of more varied and fine-grained materials, apparently from better

quality flint outcrops at distances of several kilometres from the site.

The most critical aspect of Binford's analysis of these patterns lies in his interpretation of the spatial distributions in social terms. As noted in Chapter 9, he sees the patterns primarily as reflecting a sexual division of activities in the site. Thus, he sees the central 'nest' areas as specifically female activity areas, characterized by the intensive extraction of bone marrow and brains or tongues and probably also by the processing of certain plant food resources, such as cat-tail or other aquatic plants collected from waterside habitats close to the Combe Grenal site. He interprets the peripheral areas, by contrast, as specifically male zones, associated with the heavy-duty butchery of complete animal carcases or at least substantial limb segments from the scavenging of carcases. He sees the contrasts in raw materials associated with these two locations (mostly local flint in the central area, versus more distant materials in the marginal zones) as a further, direct reflection of this, with the males exploiting more distant and generally better quality flint sources in the course of wider-ranging foraging activities than those engaged in by the females (Binford 1992: 49).

None of this patterning is particularly surprising and could no doubt be explained by a number of different economic or social models. However, Binford has chosen to interpret this as indicating a radically different pattern of social organization among the Mousterian occupants of Combe Grenal from anything which he regards as characteristic of fully modern human communities, and one which he believes reflects a fundamental separation between males and females of Neanderthal groups. Thus, he suggests that Combe Grenal reflects mainly the activities of closely integrated female groups – probably representing the activities of 5–10 adult females with dependent children – who were only occasionally visited by the males, specifically during brief periods of mating. In his recent

Discover interview (intriguingly entitled 'Hard Evidence') he suggests that these 'visiting firemen' may only have been present on the site for a few days per month – the rest of the time presumably forming separate male foraging units, effectively isolated from the female groups (Binford 1992: 50). The scenario visualized by Binford is thus radically different from the classic home-base model, generally assumed to be typical not only of all present-day hunter-gatherers, but also of most Upper Palaeolithic groups. In these respects Binford's model would make the social organization of Neanderthal communities much closer to that of most present-day ape communities than to that of behaviourally and biologically modern populations (Smuts *et al.* 1987; Wrangham 1987b; Foley & Lee 1989; Rodseth *et al.* 1992).

Critique of the Binford model

Binford's model for the Combe Grenal occupations – and by implication for the social organization of Neanderthal groups in general – has all the classic Binfordian hallmarks of originality, ingenuity and radical creativity, and in the final analysis of course he could be right. The difficulties of identifying the sexual composition of social groups from the character and organization of occupation residues are self-evident – and indeed from this perspective we might well be pessimistic of ever presenting a secure reconstruction of changing patterns of social organization during the successive stages of human evolution. However, it is possible to advance here a number of specific objections to the central components of the Binford model. Two are concerned with Binford's analyses of the archaeological data from Combe Grenal and the interpretations he draws from them. The other objections arise from more general theoretical approaches to the analysis and reconstruction of social patterns in earlier human populations, derived from some of the recent work in the field of so-called comparative

socioecology. Briefly, the relevant considerations are as follows.

1. All the specific patterns which Binford claims to have identified within the Combe Grenal deposits rest at present on unpublished data. Until Binford has published the analytical data on which these spatial patterns are based it will be impossible to assess exactly how strong – or weak – the patterns may be in quantitative terms, and whether any are repeated consistently in the different occupation levels. There are already reasons to think that many of the patterns are more diffuse, more variable and more questionable than his preliminary reports would imply. The main point to emphasize, however, is that even if the patterns do have some validity in broad statistical terms, it is possible to suggest a number of much simpler and more economical explanations for these distributions than the heavily sociological interpretations which he has advanced. For example, any slight disparity between the overall spatial distribution of notches and denticulates, versus side scrapers and points, could be related not to a sexual division of activities but to simpler functional factors. If side scrapers were employed mainly for butchery of large segments of animal carcases, as Binford suggests, then it would hardly be surprising if such activities were carried out mainly in the peripheral areas, if only to avoid the accumulation of bulky and cumbersome segments of long bones, ribs, vertebrae etc. – together with other more unpleasant residues of butchering activities – in the central areas of the site. Conversely, it would hardly be surprising if the intensive processing of the less bulky parts of carcases, such as the extraction of marrow from fragmented long bones, was carried out preferentially in the more central areas, perhaps immediately adjacent to the main eating and socializing areas. It is presumably there that one would also expect most of the craftwork (such as wood working) to be carried out – which

could perhaps explain why notches and denticulates appear to be slightly more concentrated in these areas than in the peripheral zones. Similar explanations could be offered for any disparities between the raw materials used for different tool forms. As discussed in Chapter 4, any such patterns seem to be on a relatively minor scale, and can be attributed more economically to the varying demands placed on different raw materials for the manufacture of morphologically contrasting tool forms (e.g. for the manufacture of relatively large, complex tools, versus smaller, simpler tools) than to a sexual division between the individuals who manufactured and used them. Until Binford's analyses of the Combe Grenal data have been published in detail, it will be difficult to comment further on any of these reported patterns. Meanwhile I see no reason to interpret any of these claimed patterns in explicitly sexual terms rather than as an indication of more basic functional activity patterns on the site.

2. How far one can cite any direct archaeological evidence against Binford's interpretations is perhaps more debatable. However, there are several features of the archaeological data which are at best difficult to harmonize with his model. For example, if there was indeed a sharp, consistent separation between the economic activities of male and female foraging groups in the Middle Palaeolithic, then where is the archaeological evidence for the *synchronous* presence of these two contrasting patterns of occupation – and associated stone tool assemblages – within the documented Mousterian sequences in southwestern France (see Chapter 6)? Similarly, if we assume that assemblages dominated by notched and denticulated tools were strictly female tool kits (devoted principally to the intensive processing of either plant foods, or marrow and brains from the remains of scavenged animal carcasses) then why do we find industries consisting pre-

dominantly of notches and denticulates in certain sites which quite obviously reflect large-scale, primary butchery of large animal carcasses – for example at the open-air sites of Mauran, La Borde and Le Roc (Farizy *et al.* 1994; Jaubert *et al.* 1990; Geneste 1985: see Chapter 7). Again, if we are to interpret assemblages dominated by various forms of side scrapers and points etc. as strictly male tool kits, then why do we find the remains of very young children frequently associated with industries of this kind, as at Rescoundudou, Monsempron, Combe Grenal, La Ferrassie and La Quina (Vandermeersch 1965; Jaubert 1989). Finally, it should be kept in mind that all the spatial patterns which Binford claims to have identified at Combe Grenal show at best simply statistical tendencies in the distribution of these tool forms, with a massive degree of spatial overlap in the actual distribution of different forms. If we were to assume (as Binford seems to imply) that these two tool forms were associated specifically with one sex rather than the other, then we would have to accept that the areas utilized by both sexes on the individual occupation floors at Combe Grenal showed a much greater degree of overlap and spatial integration than his model would suggest. Indeed, it is interesting to see that in level VIII at the Grotte Vaufrey, which Binford himself has interpreted as an all male processing site, there appears to be a broadly similar pattern in the spatial distribution of notches and denticulates versus side scrapers and points to that which he has reported – and claimed as evidence of sexually segregated activities – at Combe Grenal (see Chapter 9, and Binford 1988; Rigaud & Geneste 1988). How far any of these observations could be said to refute the Binford model is open to debate. It is evident, however, that none of these observations accords very easily with his basic sociological models for the spatial patterns at Combe Grenal, or indeed Middle Palaeolithic social organization in general.

Perspectives from comparative socioecology

How far we can make reliable inferences about patterns of social and demographic organization of earlier human populations from a comparison of patterns recorded in other primate species is currently one of the most controversial issues in palaeoanthropology. This field of so-called comparative socioecology has generated a massive literature over the past decade, stimulated largely by a spate of field and laboratory studies of different primate species (e.g. Smuts *et al.* 1987; Kinzey 1987; Standen & Foley 1989). There are unfortunately a number of obvious problems and pitfalls in these studies, which have been emphasized by Wrangham (1987a,b) and others – including for example the limited amount of field research on some of the most potentially relevant species, the difficulties of demonstrating that particular aspects of social organization are indeed related to specific ecological variables, and the extent of variation in behaviour and social patterns recorded between separate populations of the same species in different ecological and demographic contexts. The problems inherent in these approaches have been spelled out emphatically by Dunbar:

> In some respects we are now forced to argue that each species, perhaps each population, is a historically unique entity, and that its behaviour can only be understood in terms of the particular constellation of factors that came together in that particular time and place. This is not to suggest that there are no universal rules, but rather that the level at which these rules exist in biology is one step further removed than we had previously thought ... The same theoretical principle will predict quite contradictory optimal strategies in different environments.
>
> (Dunbar 1991: 215)

Despite these caveats, there seems to be increasing agreement that at least some of the patterns and correlations observed in these recent socioecological studies are sufficiently well documented and robust to provide some reasonable working guidelines for the patterns of social development during the later stages of human evolution. Two factors seem especially relevant to an analysis of the social organization of European Neanderthal populations: first, the demands imposed by an increasing reliance on mainly animal food resources in the diet, particularly where this involved a major component of hunting; second, the demands imposed by the general delay in rates of growth and maturation of young children which characterized the later stages of human evolution, coupled with a rapid increase in the size (and therefore the nutritional demands of early growth) of the human brain.

In this context it is arguable that the second factor is more important than the first. The rapid increase in human brain size which characterized the past million years or so (see Fig. 12.1) would have had the effect of significantly retarding the overall rate of growth and maturation of young children – and thereby increasing their period of maternal dependence – and also of making heavy demands on nutrition levels of females during pregnancy, in order to sustain the very heavy metabolic costs involved in rapid brain growth (Foley 1989; Foley & Lee 1989, 1991; Parker 1990; Aiello & Wheeler 1994). The combined effect of the two factors, it is argued, would have made human females less mobile to engage in frequent and far-reaching searches for food and also dependent on an especially abundant, nutritious, and reliable day-to-day food supply to support the heavy energy cost of child rearing. The obvious solution, both socially and ecologically, would be for a substantial part of the food requirements of females in local groups to be provided directly by the males. Thus, the situation would virtually demand that males and females in these local groups should be involved in integrated and mutually cooperative food sharing activities, if only to ensure the well being – and hence the evolutionary survival prospects – of the dependent children (see for example Lovejoy 1981; Tooby & DeVore 1987; Foley & Lee 1989; Aiello 1995). Further arguments sometimes advanced are that increased male involvement in these child-rearing groups would be needed to protect the women and children from attacks either by predators or by members of other groups, and that this increased parental investment on the part of the males might lead to increasingly strong and permanent bonds between specific pairs of males and females within local groups (Lovejoy 1981; Foley & Lee 1989; Aiello 1995; see also Binford 1985: 312–14). In other words, all these ecological, evolutionary and social pressures would converge towards encouraging the formation of essentially nuclear family groups, in which extensive food sharing between males, females and children and the occupation of 'home-base' locations – where integrated food-sharing and child-rearing activities could take place – would become essential to ensure the long-term survival and evolutionary continuity of the social groups.

The arguments concerning the increased reliance on animal food resources during the later stages of human evolution relate more specifically to the patterns of integration and cooperation between males in local groups. The argument here is that survival within the extreme periglacial and highly seasonal environments of northern Eurasia would have demanded substantial and deliberate hunting of large animal species (for reasons discussed in Chapter 7) and that this in turn would dictate the formation of mutually cooperating and interacting groups of males not only to effectively locate, pursue and kill the animals, but above all to ensure a reliable and predictable distribution of food resources between different group members day-to-day (Washburn & Lancaster 1968; Tooby & DeVore 1987). As many authors

have pointed out, it would make no sense whatever, in either social or ecological terms, for hunters of large animals to exploit resources on a purely individual basis, and cooperation in both killing and sharing of animal resources would seem an inevitable response (e.g. Isaac 1984; Binford 1985; Tooby & DeVore 1987). Whether this communal sharing of food resources would have extended to females and children in local groups is more debatable, but all the reasons discussed in the preceding paragraph suggest that this would almost certainly have been the case.

Again, all the ecological arguments point strongly to a pattern of mutual cooperation and food sharing within Neanderthal communities – at least within the more extreme, periglacial environments of northern Europe – which would almost certainly have led to a very close integration of males and females in local groups. Clearly, none of this reasoning is in harmony with Binford's (1992: 48–50) suggestion that male and female communities at Combe Grenal and elsewhere led 'strangely separate' lives, with only perhaps a few days each month when the two sexes came together for strictly reproductive purposes.

Archaeological perspectives

Finally, what observations can be made about more general patterns in the archaeological evidence? How far does the archaeological evidence support the suggestion of a broadly similar pattern of social and demographic organization over the period of the Middle-to-Upper Palaeolithic transition, as opposed to the alternative hypothesis of a dramatic restructuring of social and residential patterns over this interval?

Arguably the most striking feature of the archaeological evidence is the remarkable similarity of general residential and occupation patterns in the Middle and Upper Palaeolithic periods – at least in the core region of southwestern France. This is observable in several different aspects of the data:

1. In the first place, both the overall spatial distribution and specific locations of Middle and Upper Palaeolithic sites are in most respects broadly similar. As discussed in Chapter 8, Middle and Upper Palaeolithic cave and rock-shelter sites in southwestern France tend to concentrate not only in the same general areas of major river valleys (especially those of the Dordogne, Vézère, Couze, Isle and Dronne) but very often in precisely the same site locations. Of the total of 50 or so documented Mousterian cave and rock-shelter sites in the Perigord, well over half are overlain directly by substantial Upper Palaeolithic occupation levels. Of the remainder, almost all are sites where the sedimentary sequence of deposits had effectively filled up the occupation space in shelters by the end of the Mousterian sequence – for example at Combe Grenal, Abri Chadourne, Pech de l'Azé sites I and II, La Quina etc. If there are any significant contrasts between the spatial distributions of Middle and Upper Palaeolithic sites, this is reflected mainly in the apparently greater frequency of Mousterian open-air locations, for which the probable reasons (principally climatic) have been discussed in Chapter 8. The apparently greater concentration of Upper Palaeolithic sites in certain restricted stretches of the Vézère valley (see Fig. 8.2) seems to be related to a shift towards a more sharply focused and anticipatory exploitation of migrating reindeer populations which emerged during the earliest stages of the Upper Palaeolithic sequence (Mellars 1989a: 358). In general there seems no reason to think that any very different criteria were involvedintheselectionofmostMiddlePalaeolithic sites, than those which controlled the selection and use of Upper Palaeolithic sites.

2. A further feature of most Mousterian cave and rock-shelter sites in southwestern France

is the diversity of the economic and technological activities which seem to have taken place in the sites (Mellars 1989a: 360). Most tool inventories, for example, include a wide range of different tool forms, and also clear evidence that most of the tools were produced directly on the sites – usually through long and complex sequences extending from the importation of original unworked nodules, through production of various forms of primary flakes, to the systematic retouching, use and resharpening of finished tools (Geneste 1985, 1988). The raw materials employed for tool production were usually derived from many sources, extending in different directions from the site locations (Geneste 1988; Turq 1989a). The associated faunal assemblages indicate a similar diversity: most include remains of many different species, which were apparently processed through the different stages of skinning, butchering, marrow and brain extraction etc. within the different sites (Chase 1986a: see Chapter 7). Similarly, many sites produced substantial quantities of one or two colouring pigments (manganese dioxide or iron ochre), presumably reflecting other social or technological activities (Bordes 1952, 1972; Demars 1992). Nor should we forget the existence of deliberate and in some cases multiple human burials on several Mousterian sites (Vandermeersch 1976). Overall, there seems no reason to suspect that the range and diversity of economic, technological and social activities undertaken at most Mousterian sites in the Perigord area was any less than in most Upper Palaeolithic sites. Indeed, it has been argued that the specific locations of most Mousterian cave and rock-shelter sites were deliberately chosen to allow exploitation of the widest possible range of economic resources, from a variety of contrasting habitats and ecological zones within the overall catchment areas.

3. Evidence for the duration and intensity of occupation episodes in archaeological sites is

notoriously difficult to interpret. For most Mousterian cave and rock-shelter sites in the Perigord region, however, there seems no reason to suspect that the intensity of occupation was significantly less than in the majority of Upper Palaeolithic sites. In the lower shelter at Le Moustier, for example, it is reported that over 90 percent of the total sediments in certain levels was composed of either faunal fragments or flint artefacts, suggesting a general density of occupation residues hardly less than that documented in some of the classic and most intensively occupied Upper Palaeolithic sites such as Laugerie Haute, Abri Pataud or La Madeleine (Laville *et al.* 1980: 177). Equally dense concentrations of occupation debris seem indicated in many other sites, such as the MTA levels at Tourtoirac and Pech de l'Azé I, the Denticulate Mousterian levels at Saint-Césaire and La Quina, or the Ferrassie and Quina Mousterian levels at La Ferrassie, La Chapelle-aux-Saints and Combe Grenal.

Equally significant is the evidence for deliberate improvement of occupation surfaces in certain Mousterian sites. At La Quina, for example, there can be little doubt that the surface of one of the uppermost occupation levels was deliberately levelled off by the Mousterian occupants, almost certainly to produce a more regular and roomy living area in the diminishing space available below the rock overhang (Jelinek *et al.* 1988). From several sites in southeastern France (Baume-Bonne, Aldène) there seems to be evidence for the deliberate paving of occupation surfaces with large numbers of cobbles carried laboriously into the sites from adjacent river deposits (Lumley & Boone 1976a, b). Evidence for apparently constructed hearths has now been recorded from many Middle Palaeolithic sites, and there is also evidence of substantial and deliberately excavated pit features in at least three sites (Combe Grenal, Le Moustier and La Quina), together with one clearly defined post-hole in the later Mousterian levels at Combe Grenal

(Bordes 1972; Jelinek *et al.* 1988). Finally, as Binford argued for the Combe Grenal occupations, it seems that the use of available activity space in many Mousterian sites was organized and structured in some way, even if on a less elaborate scale than in many Upper Palaeolithic sites. In some sites there also seems to be evidence that the disposal of certain kinds of refuse, such as large animal bones or piles of lithic flaking debris, was controlled in a way which may imply the anticipation of extended episodes of occupation on the sites (Leroi-Gourham 1976; Meignen 1993).

The point of these observations is simply to demonstrate that many Middle Palaeolithic cave and rock-shelter sites in the Perigord region do seem to reflect episodes of relatively intensive occupation, in which many different economic, technological and probably social activities were carried out. From the evidence cited here, there seems no reason to reject the notion that many of these sites represent typical 'home base' occupations, with all the social and economic connotations which this term implies. If we are looking for evidence that Middle Palaeolithic sites reflect radically different patterns of social organization and group structure from those documented in the archaeological records of the Upper Palaeolithic societies, or indeed the archaeological signatures of recent hunter-gatherer groups (Gamble & Boismier 1991), it will not be easy to identify this with any confidence in the documented Mousterian sites in southwestern France.

This is not to suggest that the social organization of Middle Palaeolithic groups was in all respects analogous to that of Upper Palaeolithic communities. I have argued on several occasions that the sudden appearance and proliferation of personal ornaments in early Upper Palaeolithic sites is likely to reflect either an increase of emphasis on individual personal roles and social relationships in Upper Palaeolithic societies, or a much clearer definition and symbolic expression of these roles, based on more complex and highly structured language patterns (Mellars 1973, 1989a; see also White 1989, 1993; Gellner 1989; Donald 1991). I have also pointed out that most of the documented Mousterian sites in the Perigord region are relatively small and never appear to attain the dimensions recorded for the largest Upper Palaeolithic sites, such as Laugerie Haute, Abri Pataud, Laussel, La Madeleine, etc. (Mellars 1973, 1982, 1989a). This would argue for relatively small social units which rarely came together as large, multi-family social aggregations (cf. David 1973; Mellars 1985). It is also possible that the generally simpler patterns of spatial organization and associated structural features recorded in most Mousterian sites may reflect more short-term and transitory episodes of occupation than those documented in the larger and more structured Upper Palaeolithic sites (Mellars 1973: 266–7). The question is to what degree the simpler patterns of Middle Palaeolithic sites were related not to the character of social and economic organization on the sites but rather to fundamental contrasts in the ways in which use of living space was conceptualized by Middle and Upper Palaeolithic groups (Mellars 1989b, 1991; Binford 1989). Until these points are clarified, I believe it would be premature to accept some of the more dramatic contrasts between the social structure of Middle and Upper Palaeolithic groups which have been visualized in the recent publications of Binford (1992), Soffer (1994) and others.

CHAPTER 12

The Neanderthal Mind _____

The preceding chapters have provided a review of the evidence for different aspects of behaviour of Middle Palaeolithic populations in western Europe – their technology, subsistence strategies, patterns of mobility and spatial, social and demographic organization. What these have in common is that they are all the product of purposeful and intelligent behaviour on the part of Neanderthal groups, reflecting the mental and cognitive capacities of the individuals concerned. At this point, therefore, the question arises as to the essential character of Neanderthal mentality, and in particular to the issue of how it may have differed from that of biologically and behaviourally modern populations.

The difficulties of attempting to draw firm conclusions about the character of cognition or intelligence from surviving archaeological records are self-evident. My own approach to this question hinges on two basic assumptions. The first is that as an initial working premise we should assume that there were no significant contrasts between the mental capacities of Neanderthal and modern human populations, and that such contrasts should be inferred only after a rigorous and essentially sceptical analysis of the available archaeological evidence makes this conclusion at least highly plausible if not inescapable. There is an obvious danger of making simplistic equations between 'simplicity of behaviour' and 'simplicity of mind' which

somehow short-circuit scientific analysis, and effectively assume what one should be attempting to find out. Obviously, the less complex patterns of technological or economic behaviour of modern communities such as Australian Aborigines, Bushmen or sixteenth century Europeans in no sense imply that that they had an inherently inferior intelligence or simpler cognitive capacities than those of modern industrial societies, and we should beware of applying such simplistic thinking in our analysis of the behaviour and mentality of biologically pre-modern populations (Mellars 1989a: 377). In my view the onus of proof rests firmly on those who argue that the cognitive or intellectual abilities of Neanderthal groups were significantly inferior to those of modern populations.

My second assumption, however, is that we cannot rule out the possibility that there were significant differences in the mental or cognitive capacities of Neanderthal populations. There are several reasons why this must be recognized at the start of this discussion. First, it is inherent in evolutionary theory that the brain has been no less subject to adaptive processes over the time-span of human evolution than have other physical and biological features of early populations. There is no doubt that modern human brains are significantly more complex, highly structured, efficient and intelligent than those of even the most intelligent great ape (Passingham 1975, 1989; Gibson 1990) and this

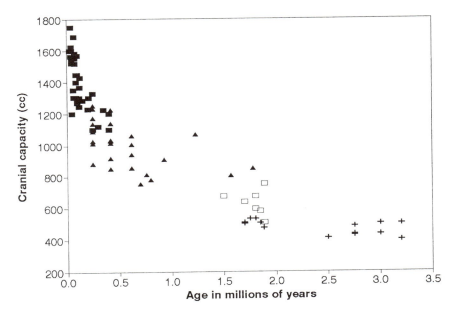

Figure 12.1 *Increase in hominid cranial capacities over the past three million years, according to Aiello 1995. (See also Aiello & Dunbar 1993.)*

increase in intelligence has clearly taken place at various stages during the last five million years or so (see Fig. 12.1). Once this is accepted it would be irrational to assume that there could have been no significant changes in the structure, complexity or intelligence of the brain over the 200–300,000 years since the emergence of the earliest taxonomically and anatomically Neanderthal populations (Gibson 1985, 1988; Parker & Milbraith 1993).

This conclusion would not only be unwarranted in biological and evolutionary terms but would be equally difficult to reconcile with most recent thinking about the mechanisms for the development of increased intelligence during the course of human evolution. Following the work of Humphrey (1976) and others, it is now accepted that major stimulus to the development of increased intelligence lay not only in the challenge for pure biological survival in the face of competing species and changing environmental conditions, but also the challenge of competing in social and personal terms with members of the same species (see for example Alexander 1974, 1989; Byrne & Whiten 1988, 1992). The obvious corollary is that the tempo of mental development is likely to have increased sharply with the increasing complexity of human behaviour during the course of the Pleistocene. It would be remarkable if mental evolution had come to an end at precisely the point when we know that all forms of human technological, economic and social development – and therefore intraspecies social competition – were becoming most intense. If we accept the implications of recent genetic studies that the evolutionary trajectories of the Neanderthal and modern populations are likely to have been separated over a period of at least 300,000 if not almost a million years (Cann *et al*. 1987; Stoneking & Cann 1989; Wilson & Cann 1992; Stringer & Gamble 1993) then the basis for such an assumption would be even more absurd.

How far we can make reliable inferences

about intelligence or other aspects of mentality from the size and morphology of endocranial casts of human brains has been discussed frequently in the literature during the last century. Initially, suggestions were made for a significantly simpler brain structure among Neanderthals, based on features such as the supposedly reduced size of the frontal lobe, the apparently simpler patterns of the sylvian and parieto-occipital fissures and the more forward position of the lunate sulcus when compared with those of modern populations (e.g. Boule & Anthony 1911; Connolly 1950). More recently, all these features have been subjected to critical review, particularly in the publications of Ralph Holloway (1969, 1976, 1983, 1985). Holloway's most recent conclusion is:

> I have no confidence in unambiguously identifying any convolutional patterns that are suggestive of a 'primitive condition' ... I believe Neanderthal brains were fully *Homo*, with no significant differences in their organization compared to our own.
> (Holloway 1985: 320–1)

He points to further ambiguities in attempts to use estimates of hominid brain sizes for comparing intelligence. While, as he points out, Neanderthal brains appear to be slightly larger than those of most modern populations when measured in terms of gross volume, much of this variability in size is clearly related to the body size of the different populations, and probably to other environmental factors such as the metabolic costs of maintaining brain functions under varying climatic regimes. In any case, brain size alone is a notoriously poor indicator of overall intelligence levels in different species and we are of course almost totally ignorant of the detailed internal structure and organization of the brain tissue itself (Whitcombe 1995). Dunbar and others have recently argued that a better reflection of varying intelligence in different primate species may be provided by the so-called 'neocortical

ratio' – i.e. the ratio of the outer neocortical region of the brain to the total brain volume (Dunbar 1992; Aiello & Dunbar 1993). Measured in these terms Neanderthal brains may perhaps be slightly less complex than modern brains (Foley 1995) (see Fig. 12.1, Table 12.1). But these calculations rest at present on such limited evidence for Neanderthal populations that it would probably be premature to press the comparisons further (see Byrne 1995).

Table 12.1

'Neocortex ratios' (= ratio of neocortex region of brain to total brain volume) calculated for different hominid taxa over the past two million years

Taxon	Neocortex ratio
Homo habilis	3.44
Homo erectus (Africa)	3.66
Homo erectus (Europe)	3.79
Pre-*Homo sapiens*	3.95
Homo neanderthalensis	4.06
Homo sapiens	4.12

After Foley (1995); see also Aiello & Dunbar (1993).

This brings us back to the surviving archaeological evidence as the only reliable basis for making any direct comparisons between mental and cognitive abilities of Neanderthal as opposed to anatomically and behaviourally modern populations. The following sections will focus on three specific issues which have generated most debate in the literature during the last decade:

1. Evidence for various forms of symbolic expression or behaviour in the Middle Palaeolithic.
2. Evidence for the more general intelligence or cognitive capacities of Neanderthals.
3. Evidence for the presence or absence of language prior to the emergence of anatomically modern populations.

Symbolism

The topic of Middle Palaeolithic symbolism has generated a large and lively literature over the past few years (Chase & Dibble 1987, 1992; Chase 1991; Dibble & Chase 1990; Dibble 1989; Lindly & Clark 1990; Marshack 1972, 1981, 1989, 1990; Byers 1994). The first and most basic question is one of terminology: what exactly do we mean by symbolism? The most widely accepted definition of a symbol is anything, be it object, sign, gesture or vocal expression which in some way refers to or represents something beyond itself (Chase 1991; Hodder 1982). This leads to the fundamental distinction between the symbol itself (the 'signifier') and the object or idea signified (the 'referent' or 'signified'). Most workers go on to make distinctions between different forms of symbols, at increasing levels of abstraction between the symbol and the referent. Thus Chase (1991: 195) differentiates between what he refers to as 'iconic' symbols, which relate in a very direct and obvious way between the symbol and the referent (as a road sign of a jumping deer might refer to the presence of wild game in the vicinity), 'index' symbols (which might use, say, the symbol of smoke to indicate the notion of fire) and totally abstract or arbitrary symbols which have no obvious relationship to the idea being symbolized beyond that attached arbitrarily by the individuals sharing this particular pattern of symbolic expression. The classic illustrations of the latter are most words, which can only be related to the objects or ideas being symbolized in essentially abstract, arbitrary ways.

Most discussions of symbolism go on to make direct connections between the existence of explicit symbolic expression and at least rudimentary forms of language (e.g. Holloway 1969, 1983; Falk 1987; Dibble 1989; Davidson & Noble 1989, 1993). Whether these correlations are as direct and obvious as many authors seem to assume (in the sense of necessarily implying the existence of language) is clearly debatable. Even so, no one would question that the whole of present-day culture and behaviour is highly symbolic in character, and that the heavily symbol-laden nature of culture is ultimately fundamental not only to modern language patterns but to effectively all other aspects of society, technology, communication etc. (Gellner 1989; Alexander 1989; Binford 1987, 1989). There is also widespread agreement that the same patterns of complex symbolic, and probably linguistic, expression can be traced back to the beginning of the Upper Palaeolithic sequence and arguably provide the most dramatic contrast with behaviour and organization of preceding Middle Palaeolithic communities (Pfeiffer 1982; Chase & Dibble 1987, 1992; White 1989; Binford 1987, 1989; Mellars 1989a, b, 1991; Davidson & Noble 1989).

When viewed in these terms the existence and character of symbolic behaviour among Middle Palaeolithic communities must be seen as central to any understanding of both the mentality and cognition of these groups and to the nature and significance of the documented behavioural contrasts between the Middle and Upper Palaeolithic. In the following sections I will review the principal areas where the existence of symbolic expression in Middle Palaeolithic/Neanderthal contexts has often been claimed and then assess what significance can be attached to this evidence.

Use of pigments

The occurrence of what are almost certainly colouring materials in Middle Palaeolithic contexts is now beyond dispute (Bordes 1952, 1972; Wreschner 1980; Marshack 1982; Demars 1993). The evidence comes in two main forms: first fragments of iron oxide or red ochre which, depending on the source, can provide a range of colours from yellow to deep maroon or red-brown; and black manganese dioxide. Fragments of ochre have now been recorded from at least a dozen different

Figure 12.2 *Fragments of manganese dioxide apparently used as pigments, from the MTA levels of Pech de l'Azé I. The pieces show traces of either facets or smoothing (suggesting their applicaton to a smooth surface) or scraping of the surface to produce powder. After Bordes 1972.*

Middle Palaeolithic sites in southwestern France, while the occurence of manganese dioxide is even more frequent (Bordes 1952; Demars 1992). Evidence that the materials were used as pigments seems difficult to dispute. Many individual fragments show either clear signs of scraping (presumably to yield a powder) or well developed facets on one or more surfaces which suggest that they were applied directly to a hard or soft surface (see Fig. 12.2: Bordes 1952, 1972). Significantly, both minerals are known to have been used extensively as pigments for the production of cave art throughout the European Upper Palaeolithic sequence. Whilst the possibility of other uses has occasionally been suggested, such as the use of ochre in tanning hides, the status of both ochre and manganese dioxide as colouring pigments in many Middle Palaeolithic contexts seems undeniable.

The critical issue here is whether the simple use of colouring materials can be regarded as an explicitly symbolic act. What is conspicuously lacking from the Middle Pal-

aeolithic is any indication of exactly how these colours were employed. There are a number of reports of fragments of stone apparently stained with ochre, and traces of ochre smeared on either stone tools or occasional fragments of bone or bone artefacts (e.g. Marshack 1988, 1990). But in none of these cases is there any evidence that the pigments were used to produce designs or other identifiable markings. Thus, the possibility remains that colour was employed in Middle Palaeolithic contexts purely to change the surface appearance of things, much as one might paint a piece of furniture or dye clothing. Captive chimpanzees have been taught to use paint and in many cases obviously enjoy – or at least are intrigued by – the visual effects of the paint on the canvas or other objects provided. It therefore remains highly debatable whether the mere existence and use of pigments in Neanderthal contexts need reflect anything more than natural curiosity about transforming the appearance of objects, or at best perhaps a rudimentary form of aesthetic appreciation

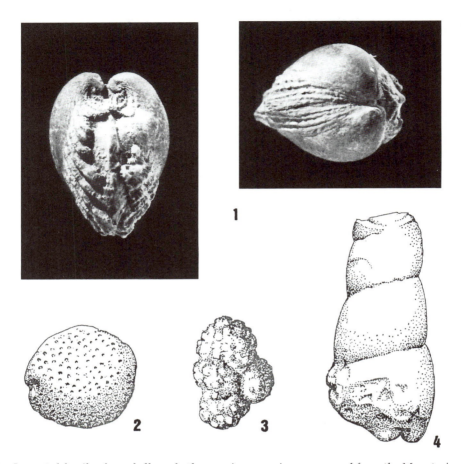

Figure 12.3 *Imported fossils of sea shells and other marine organisms recovered from the Mousterian levels of Chez-Pourrez (Corrèze: no. 1) and the Grotte de l'Hyène (Arcy-sur-Cure, Yonne: nos. 2–4). After Lhomme & Freneix 1993; Poplin 1988.*

(Chase & Dibble 1987). The same kind of curiosity may perhaps be reflected in the occasional specimens of fossil shells recovered from Mousterian sites (e.g. Fig. 12.3). Whether the pigments were applied to human skin, or other items such as skin clothing, wooden artefacts etc. remains equally enigmatic from the archaeological evidence.

Symbolic artefacts

Most arguments about explicit symbolism in Middle Palaeolithic contexts have centred on a small number of bone artefacts or bone fragments which are held to show either deliberate patterns of incisions on their surfaces or occasionally deliberately perforated holes (Chase & Dibble 1987, 1992; Marshack

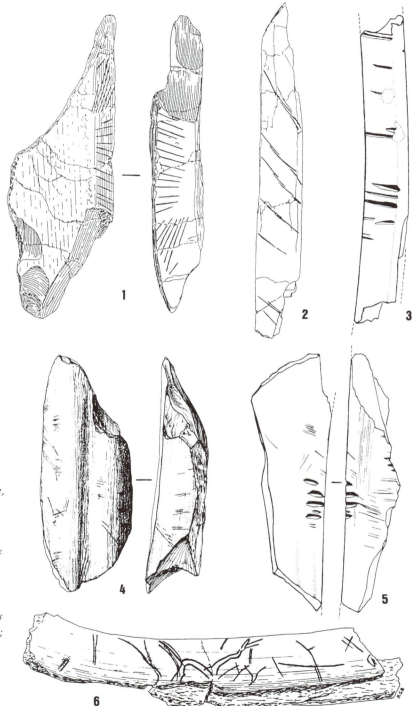

Figure 12.4 *Various incised bones from Lower and Middle Palaeolithic contexts in Europe, often claimed as evidence for 'decorative' or 'symbolic' artefacts in pre-Upper Palaeolithic contexts. 1, 2 from the Middle Pleistocene levels of Bilzingsleben (eastern Germany); 3, 5 from the Mousterian levels of the Abri Suard (La Chaise, Charente); 4 from the Mousterian level 17 of Cueva Morín (northern Spain); 6 from the Acheulian levels of Pech de l' Azé II (Dordogne). After Bednarik 1992 (nos 1, 2, 6); Debénath & Duport 1971 (nos 3, 5); Gonzalez Echegaray 1988 (no. 4). (Various scales)*

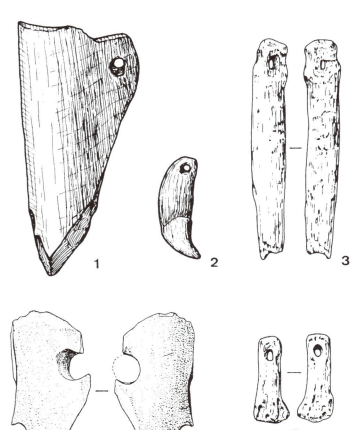

Figure 12.5 *Perforated bones and teeth from the Middle Palaeolithic levels of Repolosthöhle, Austria (nos 1, 2); the Micoquian levels of Bocksteinschmiede, Germany (nos 3, 5 – a wolf metapodium and swan's vertebra respectively); and the Acheulian levels of Pech de l'Azé II, France (no. 4). After Bednarik 1992 (nos 1, 2); Marshack 1990 (nos 3, 5); Bordes 1969 (no. 4).*

1972, 1988, 1990; Lindly & Clark 1990; Bednarik 1992). The most controversial pieces are some of the supposedly decorated bones, such as those recovered from one of the Rissian levels at Pech de l'Azé II (Bordes 1969, 1972), the later Mousterian levels at Cueva Morín in northern Spain (Gonzalez Echegaray 1988), and the much earlier, Middle Pleistocene levels at Bilzingsleben in east Germany (Mania & Mania 1988) (see Fig. 12.4). Among the smaller group of perforated objects are the swan's vertebra and wolf metapodial from the Micoquian levels at Bocksteinschmiede in west Germany, the perforated reindeer phalanges from the Quina-Mousterian levels at La Quina,

the flattened bone fragment with a small perforation from Repolusthöhle, and the two typical but remarkably isolated specimens of perforated animal teeth recovered from supposedly Mousterian levels at La Quina and Repolusthöhle respectively (Fig. 12.5). Finally, and most significantly, there is the intriguing nummulite fossil from Tata in Hungary, on which a natural crack running through the centre of the fossil has been supplemented by a further, finely incised line running almost exactly at right angles to form a regular, symmetrical cross (Fig. 12.6: see Marshack 1988, 1990 and Bednarik 1992 for details of these pieces and full references to the relevant literature).

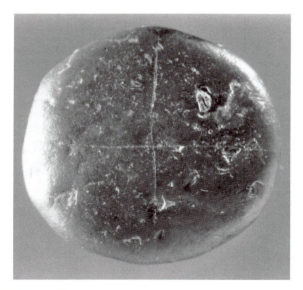

Figure 12.6 *Fossil nummulite from the Mousterian levels of Tata (Hungary), on which a cross has been formed by incising a single line at right angles to a natural crack running through the fossil. After Marshack 1990.*

The problems of accepting these items as convincing examples of symbolic or even decorative objects have been spelled out fully by Chase, Dibble and others (Chase & Dibble 1987, 1992; Dibble & Chase 1990; Davidson & Noble 1989, 1993). The most obvious ambiguities concern the grouping of supposedly engraved bones. As Chase and Dibble (1992) point out, deep and in some cases quite regularly spaced incisions on bone fragments can be produced easily when skinning and butchering animal carcases and are frequent occurrences on most Palaeolithic and Mesolithic sites with well preserved faunal material. In my own opinion – and that of Chase and Dibble – none of the markings so far described from Lower and Middle Palaeolithic sites in Europe shows anything approaching the degree of clarity, regularity, organization, or obvious intentionality that one would expect from deliberately and consciously symbolic engravings. In this respect

they contrast strikingly with the great majority of engravings recovered from Upper Palaeolithic sites, where the obvious patterning and clear intentionality of the engravings cannot be questioned (e.g. Fig. 13.2) (Marshack 1972). The status of these and similar pieces has been summed up aptly by Chase as follows:

> (Most) archaeological bone assemblages contain very large numbers of specimens with tool marks, many of them with many such marks. This means a large number of patterns of marks. It is all too easy to pick out a few specimens from the eye-catching end of the natural variation among these patterns and to attribute them to intentional symbol making rather than to the laws of probability.
> (Chase 1991: 210)

The occasional specimens of perforated bones and teeth reported from Middle and Lower Palaeolithic contexts pose similar problems. It is now generally accepted that the claimed perforation on the bone fragment from Pech de l'Azé II (Fig. 12.5), and the perforated reindeer phalanges from La Quina are almost certainly the results of natural damage, caused by chemical erosion of the bones or by carnivore gnawing (Chase 1990). The two very characteristic but extremely isolated specimens of perforated animal teeth described from the Mousterian levels at La Quina and Repolusthöhle (a fox canine and a wolf incisor respectively: Fig. 12.5) on the other hand clearly cannot be explained in such terms. As White (1989: 386) has pointed out, the critical question is whether these pieces were truly associated with the Mousterian levels on the sites or, as seems more likely, they were misplaced during the original excavations from overlying Upper Palaeolithic levels on the sites. We are left with the intriguing perforated swan's vertebra and wolf metapodial from the Bocksteinschmiede cave (Fig. 12.5) which, if genuinely associated with the Micoquian levels, do seem to represent deliberately perforated objects (Marshack 1990). Whether these two, isolated specimens can be taken to represent explicitly symbolic or even decorative items is still an open question. Chase and Dibble (1992) have recently suggested the possibility of a purely utilitarian function, for example as parts of toggles, thongs or other items of clothing.

The main problem in accepting these objects as convincing evidence for symbolic expression in the Middle Palaeolithic lies in their extreme rarity – especially when seen in the context of the long time-span of the Middle Palaeolithic (over 150,000 years) and the large numbers of sites and occupation levels of this period which have been investigated throughout Europe. As Chase & Dibble (1987, 1992) and others have emphasized, for any pattern of behaviour to be regarded as clearly symbolic there must presumably be evidence that the behaviour was shared amongst members of the societies in question and served as a medium for communication between group members. How such significance can be attached to objects such as the two perforated bones from Bocksteinschmiede, or the two highly questionable perforated teeth from La Quina and Repolusthöhle, which are effectively unique in the European Middle Palaeolithic, is by no means clear. Again, the contrast with the Upper Palaeolithic could hardly be more dramatic. From the earlier stages of the Aurignacian, for example, we not only have large numbers of perforated animal teeth recovered from many individual occupation levels (such as the 53 teeth recovered from the Aurignacian I levels at La Souquette, or the 29 specimens from the similar levels at the Grotte de Spy: White 1989; Otte 1979) but also a *repetition* of this pattern from many early Aurignacian sites distributed across the length and breadth of Europe, from the Balkans to Belgium, southwestern France, Italy and northern Spain (Han 1972, 1977; Otte 1979; Kozlowski & Klima 1982; White 1989, 1993). Regardless of the significance attached to any individual objects, such as the enigmatic

engraved cross (Fig. 12.6) on the Tata pebble, the scarcity and isolation of these objects within the Middle Palaeolithic universe as a whole makes it difficult to see symbolic expression as a significant component of Neanderthal behaviour.

Burial practices

The long-debated topic of Middle Palaeolithic burials and associated burial rituals raise issues similar to those discussed above. Is the archaeological evidence reliable? If it is, what can we legitimately infer from it about the symbolic or other nature of the behaviour represented?

The issue of Middle Palaeolithic burial practices has been discussed at length in a *Current Anthropology* review article by Robert Gargett (1989). Gargett sets out the case systematically against the acceptance of most reported incidences of deliberate burials in Neanderthal contexts, followed by a variety of positive and negative responses from other commentators. What emerged is that while the case for the existence of intentional burial practices in Neanderthal contexts remains strong, evidence for possible rituals or symbolic offerings associated with these burials (at least in European contexts) is very weak.

Arguments for accepting the majority of claimed burials were set out persuasively in the comments by Bricker, Trinkaus, Frayer, Montet-White and others on Gargett's article. As they point out, it seems impossible to dismiss the evidence for deliberately excavated graves in at least three of the southwestern French sites – at La Chapelle-aux-Saints, La Ferrassie and Le Moustier (Figs 12.7–12.9). At La Chapelle-aux-Saints there are accounts of a rectangular burial pit at the base of the Mousterian deposits (Fig. 12.7) which is difficult if not impossible to attribute of any natural geological processes on the site (Bouyssonie *et al.* 1908, 1913). The same is suggested by the original accounts of

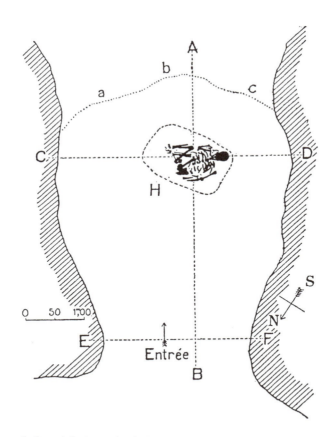

Figure 12.7   *Section and plan of the Neanderthal grave recorded by Boule at La Chapelle-aux-Saints (Corrèze). The basal level containing the burial produced a rich and typical Quina Mousterian assemblage. After Boule 1909.*

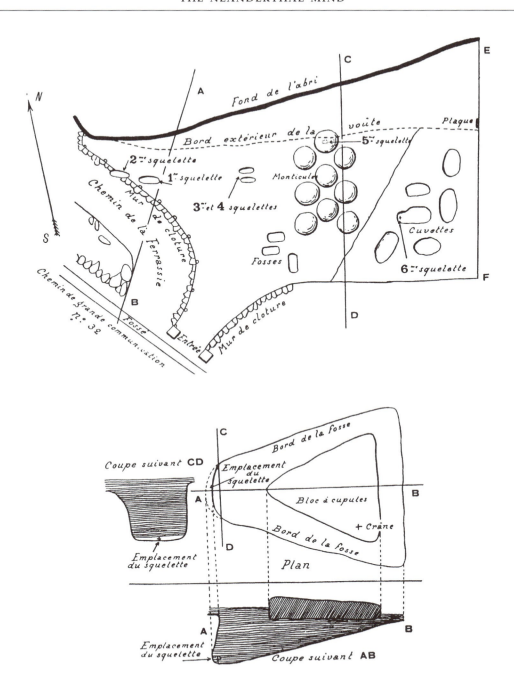

Figure 12.8 *Upper: Plan of the six Neanderthal burials recorded by Peyrony in the Mousterian levels at La Ferrassie. Lower: Plan and section of burial no. 6 from La Ferrassie, showing the position of the triangular stone – possibly with artificial 'cup' depressions – overlying the grave. After Peyrony 1934.*

Figure 12.9 *Photograph of the skeleton of Neanderthal burial no. 2 at La Ferrassie, taken at the time of the discovery. Note the flexed position of the skeleton. After Peyrony 1934.*

at least one of the Neanderthal burials at La Ferrassie (Fig. 12.8: Peyrony 1921, 1934, 1939; see especially the comments by Bricker on these burials in Gargett 1989) and by similar accounts of an apparently deep grave pit containing the remains of a young child recorded in the later Mousterian levels at Le Moustier (Fig. 9.29: Peyrony 1930; Bordes 1959b).

Perhaps the most persuasive arguments in favour of the intentionality of these Neanderthal burials, however, is provided by purely taphonomic considerations. As Trinkaus and others pointed out, it seems inconceivable that so many individual Neanderthal remains could have survived as largely intact, articulated skeletons unless the bodies had been deliberately protected from both predators and natural decay processes shortly after death. This is especially true in view of the relatively large numbers of individual corpses recorded from certain sites (such as the remains of at least seven individuals at La Ferrassie and nine at Shanidar) and

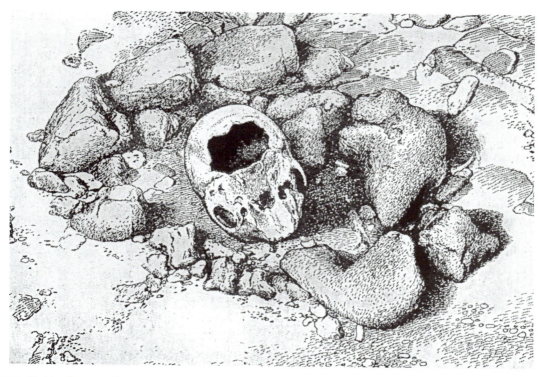

Figure 12.10 *Supposedly 'ritual' placement of the Neanderthal skull from the Grotta Guattari (Monte Circeo, western Italy), according to Blanc 1958. Recent research suggests that both the position and damaged condition of the skull are the result of carnivore activity in the site (see Stiner 1991c).*

the fact that many are remains of very young children whose bones are especially delicate. Combined with the stratigraphic and other observations discussed above, the case for deliberate interment of most of these skeletons appears virtually beyond dispute.

The arguments for deliberate grave offerings or other forms of ritual associated with these burials are very different. As Chase & Dibble (1987) and others have pointed out, most of the supposedly deliberate grave-goods which have been reported in association with Middle Palaeolithic burials in Europe can be interpreted more convincingly as objects which were incorporated accidentally into grave infillings at the time of interment. This applies most obviously to stone tools and animal bones, which are of course

some of the commonest objects encountered in Mousterian occupation levels and would almost inevitably become incorporated into the infill of grave pits from material which was simply lying on the surface at the time of burial. Indeed, since most of the burial pits were excavated through Mousterian occupation deposits, for example at La Ferrassie and Le Moustier, the occurrence of stone tools and faunal remains in the grave fillings is almost inevitable. At best, all the reported occurrences of these grave offerings from European sites must be regarded as unproven. Whether the supposed ring of Ibex horns recorded in association with the Neanderthal burial at Teshik Tash in Uzbekistan (Movius 1953) or the curious concentration of flower pollen recorded in one of the Neanderthal

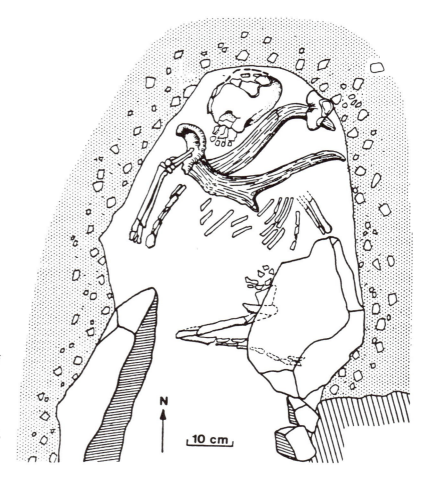

Figure 12.11 *Burial of one of the anatomically modern hominids from the Djebel Qafzeh (Israel) associated with a large pair of fallow deer antlers. After Vandermeersch 1970. The burial is dated by TL and ESR dating to around 90–100,000 BP.*

graves at Shanidar in Iraq (Leroi-Gourhan 1975) can be interpreted in similar terms is perhaps more controversial (see for example the comments by Arlette Leroi-Gourhan on Gargett's paper). Equally controversial is the supposedly ritualistic setting of the isolated Neanderthal skull from the Grotta Guattari Cave in central Italy (Fig. 12.10), which has been reinterpreted recently as almost certainly a product of hyena activity in the site (Stiner 1991c). If these conclusions are accepted, we are left with only two convincing examples of intentional and potentially symbolic grave offerings recorded from well documented Middle Palaeolithic contexts –

the large fallow-deer antlers (Fig. 12.11), and the complete boar's jaw, associated respectively with the Middle Palaeolithic burials in the Qafzeh and Skhul caves in Israel (Vandermeersch 1970). The critical interest of these finds is that they were both associated with skeletons claimed to be of essentially anatomically modern, as opposed to Neanderthal, physical type (Vandermeersch 1981, 1989).

The final and no doubt most controversial question is how far the existence of intential burials can be taken as a reliable reflection of symbolic behaviour among Neanderthal groups. Does the mere act of human burial need to be seen as explicitly symbolic in nature, or could it be seen as a reflection of much simpler and more basic human motivation? The argument that Neanderthal burials were a purely functional response to the need to dispose of dead bodies seems rather slim – especially in view of the relatively large numbers of such burials recorded at La Ferrassie and Shanidar and the fact that there are of course much simpler and less labour-intensive ways of disposing of corpses, as the ethnographic record demonstrates. At the very least we must assume that the act of deliberate burial implies the existence of some kind of strong social or emotional bonds within Neanderthal societies, which dictated that the remains of relatives or other close kin should be carefully protected and perhaps preserved in some way after death. But to suggest that the burial act must be seen as necessarily symbolic seems difficult to sustain. In the absence of either clear ritual or unambiguous grave offerings associated with the documented range of Neanderthal burials in Europe, it must be concluded that the case for a symbolic component in burial practices remains at best unproven.

Symbolism and style in tool manufacture

The possibility that evidence for symbolism may be reflected in the morphology and patterns of manufacture of stone tools has also been much discussed. Holloway (1969, 1983) argued that the mere existence of clearly differentiated forms of stone tools at Olduvai Gorge and elsewhere could be taken as reasonable evidence for the existence of essentially linguistic mental concepts two million years ago. The same idea has been repeated in many other publications (Isaac 1969; Parker & Gibson 1979; Gibson 1988; Foster 1990; see Hewes 1993 for a review).

Recently, these rather simplistic notions of a symbolic element in early Palaeolithic stone tools have come under attack from different directions. Dibble (1987, 1989) has argued that almost all documented variation in Lower and Middle Palaeolithic stone tools can be attributed either to the different tasks for which the tools were intended, or to successive modifications in the shape and apparent design of the tools as a result of repeated episodes of resharpening and reworking in the course of use (see Chapter 4, Figs 4.8, 4.12). He believes that there were few, if any, basic concepts for the different tool forms beyond a simple mental categorization of tools in terms of the particular functions for which they were intended (see also Chase & Dibble 1987; Chase 1991).

Broadly similar arguments have been expressed in my own publications. In discussing the range of side-scraper forms from the Greek Mousterian site of Kokkinopilos, for example, I commented that 'the edge alone is the important part of the tool; the overall shape is of little importance' (Mellars 1964: 231). I went on to add: 'despite this variety of form, it is difficult to identify any clearly intentional sub-types amongst the racloirs, since so much of the secondary working applied to a piece depends on the original form of the flake chosen and the natural edges on it'. In a later paper I argued that the forms of most Mousterian tools, in contrast to those of the Upper Palaeolithic, showed a high degree of fluidity and lack of morphological standardization, with the result that Middle Palaeolithic tool assemblages were far more difficult to classify into discrete, clearly separate forms than were those of the Upper Palaeolithic, and that any attempt to categorize Middle Palaeolithic industries would require a much more limited range of discrete forms than would taxonomies of Upper Palaeolithic industries (see Chapter 4 and Mellars 1989b, 1991)

As discussed in Chapter 4, the essence of the contrasts between the morphology of Middle versus Upper Palaeolithic tools lies in the notion of deliberately 'imposed form' (Mellars 1989b, 1991; see also Chase 1991).

The central proposition is that in many if not the majority of Upper Palaeolithic tools the artisan seems to have invested much effort to control and modify the shapes of the original flake blanks, to achieve a distinctive and in most cases relatively standardized appearance in the form of the finished tools. In the majority of Lower and Middle Palaeolithic tools, by contrast, this element of imposed form seems to be lacking; most attention during tool manufacture seems to have been paid to the strictly functional properties of the working edges, with little attention paid to the overall shape or appearance of the resulting tool (Fig. 4.30). I have summed up this contrast by suggesting that Upper Palaeolithic artisans adhered to the maxim 'this is an endscraper: I use it as an endscraper, I call it an endscraper, and it must therefore *look* like an endscraper', whereas Middle Palaeolithic groups seem to have been content with the notion 'who cares what this tool looks like, so long as it performs adequately for the job in hand' (Mellars 1991: 66).

This same element of tightly imposed form and standardization in visual morphology is apparent in many other aspects of Upper Palaeolithic technology. As discussed elsewhere (Mellars 1989b, 1991) it is arguably even more explicit in the form and design of various Upper Palaeolithic bone and antler tools (which again contrast dramatically with the much simpler forms of Middle Palaeolithic bone tools) and may also be discernible in such features as the highly 'structured' organization of many Upper Palaeolithic living sites (e.g. Figs 9.31, 9.32: Combier 1988; Soffer 1985a), as well as in the similar regularity and standardization of Upper Palaeolithic art and decorative motifs (Figs 13.2, 13.4, 13.5: Mellars 1989b, 1991).

The central assumption is that this kind of morphological standardization and imposed form in tool manufacture does indeed have some clear symbolic significance (see for example discussions in Chase 1991 and Byers 1994). The argument is that this element of standardization and imposed form goes beyond any purely functional or utilitarian requirements of the tools, and must necessarily imply some symbolic concept of individual tool forms, which in turn was reflected either consciously or unconsciously in their specific shapes and visual appearance (see also Wynn 1991: 203–4). Exactly what symbolic significance or associations this patterning might have inevitably remains more speculative. Two obvious possibilities are that it could be related to the social and ethnic identity of the groups producing the tools – reflecting what Wiessner (1983, 1984, 1990) and others have referred to as 'emblemic style' in tool manufacture – or alternatively that it could be related to the specific economic, environmental and social contexts in which the different tool forms were employed. In an earlier paper I commented:

> Eskimo groups, for example, make sharp conceptual distinctions between tools used for processing different forms of plant versus animal products, between tools used on different social occasions or at different seasons of the year, or between tools used by males and females (e.g. Birket-Smith 1945; McGhee 1977). Potentially, therefore, formalized, perceptually defined differences in the forms of stone and bone artefacts could have been tied into a much wider framework of symbolism and symbolically defined behaviour embracing many different aspects of the social and economic organization of Upper Palaeolithic groups.
>
> (Mellars 1989b: 359)

Whether this element of imposed form in tool manufacture is entirely lacking from Middle Palaeolithic artefacts is more debatable. As discussed in Chapter 4, a fairly strong case for the existence of clear 'mental templates' could be argued for some of the distinctive forms of bifacial tools recorded in certain Middle Palaeolithic industries, such as the sharply triangular or 'bout coupé' forms of certain hand axes, or the symmetrical, bifacial leaf-point forms (see Figs 4.22,

4.25, 4.26, 4.27; Bordes 1961a, 1984). Despite some recent protests (e.g. Dibble 1989), it seems difficult to escape the conclusion that these particular forms were produced with a clear template in the minds of the artisans, which dictated that the tools should not only perform adequately for some particular tasks, but also conform to specific, predetermined and presumably preconceived forms (see also Gowlett 1984, 1995).

In this context there would seem to be two possibilities: either that the occurrence of these pieces does reflect an incipient measure of symbolic patterning in Middle Palaeolithic stone-tools, directly heralding that which emerged on a greatly increased scale during the ensuing Upper Palaeolithic. Or alternatively that the regularity, symmetry and standardization apparent in these particular tool forms can be attributed to a more basic and innate motivation, reflecting either simple aesthetic appreciation (i.e. curiosity or satisfaction in the production of regular and symmetrical tool shapes), or perhaps even (as Binford has suggested: 1989: 28) to a more long-term adaptation to the production of these symmetrical, bifacial tool forms extending, in the case of the Acheulian, for over a million years (see also Wynn 1979, 1985; Gowlett 1984, 1995; Schick & Toth 1993). At present I can see no way of choosing between these alternatives. What must be accepted is that whatever degree of standardization, symmetry, imposed form etc. we can recognize in the morphology of Middle Palaeolithic tools emerges on a much greater, more complex and more rapidly changing scale in the ensuing tool inventories of Upper Palaeolithic groups. It is this dramatic and well defined shift in tool production patterns that suggests that there was indeed a major change in the symbolic and cognitive properties of tool manufacture between the Middle and Upper Palaeolithic periods, which may have equally significant implications for the general mental and cognitive dimensions of the populations involved.

Intelligence

Attempts to evaluate the intelligence of early humans is one of the most challenging tasks to confront students of human evolution. At least some of the problems in this context stem from the difficulties of defining exactly what 'intelligence' implies in a psychological or cognitive sense, and the complexities of the current arguments among cognitive psychologists about how to measure and evaluate varying degrees of intelligence, in different individuals and animal species (Parker & Gibson 1979; Gibson 1990; Wynn 1991). There are equal problems in attempting to relate specific aspects of intelligence (visual skills, communication, memory etc.) to specific areas or neurological structures in the brain (Passingham 1989; Gibson 1988, 1990; Whitcombe 1995).

Leaving aside these theoretical issues, some of the possible criteria by which one might evaluate and compare relative degrees of intelligence amongst present-day primate species – and also during the course of human evolution – have been discussed in a series of useful papers by Kathleen Gibson (e.g. 1988, 1990, 1993). Basically, Gibson argues that increasing intelligence in the course of primate and human evolution was linked to increasing levels of complexity in the organization of the neocortical region of the brain, which led to capacities for increasingly complex linkages between each processing area. This in turn led to an increasing capacity for processing several types of information (visual, auditory, tactile etc.) along parallel and simultaneous lines. Briefly, she argues that the effects of such increasingly complex neurological linkages are likely to be reflected in various aspects of behaviour of different species, such as the ability to engage in increasingly long and complex sequences of thought involving multiple, sequential cause and effect relationships; the ability to integrate several kinds of behaviour (e.g. stacking, hammering, throwing etc.)

into single sequences of action; the ability to form complex mental maps and images of, for example, the spatial and temporal distribution of different economic resources within the environment; the ability to coordinate the behaviour of individuals with that of increasingly larger numbers of individuals; and, above all, the capacity to visualize and construct simultaneous, alternative mental scenarios for the behaviour and movements of both other animal species in the environment and members of one's own species. While the precise relationships of these patterns of behaviour to specific neurological systems are still unclear (Passingham 1989; Gibson 1990, 1993; Whitcombe 1995), the features discussed by Gibson provide some useful working criteria for comparing the relative complexity of intelligent behavioural patterns among different primate species and potentially for those which emerged during the various stages of human evolution (see also Wynn 1991; Tooby & Cosmides 1992).

The central issue is how we can use specific aspects of the archaeological records of human behaviour to evaluate possible shifts in intelligence during the course of human evolution – and specifically during the transition from archaic (i.e. Neanderthal) to fully modern humans? In other words how far can we use the archaeological evidence for different aspects of human behaviour as a structured test to evaluate changing levels of intelligence or mental evolution (Gibson 1985; Parker 1985; Gibson & Ingold 1993; Gowlett 1984, 1985)? Surprisingly few attempts have been made to examine the archaeological record in these terms. The two most relevant studies are Thomas Wynn's (1979, 1985, 1989, 1991) attempts to use a Piagetian perspective in analysing the development of early tool-making abilities and some of the more recent speculations of Lewis Binford (1987, 1989, 1992) regarding evidence for increasing 'time depth' in the planning and structuring of activities over the period of the Middle-to-Upper Palaeolithic transition.

Wynn's studies are based on the classic investigations of Jean Piaget into patterns of cognitive development in human children – using this as a possible analogue for the development of similar mental abilities during the course of human evolution (Piaget 1960, 1970). Piaget identified three major stages of cognitive development, which he referred to respectively as the 'sensorimotor' stage, extending from birth to about 18–24 months; the 'pre-operational' stage, from 18–24 months to around 6–7 years; and the fully 'operational' stage, from 6–7 years onwards. Although each stage was defined by a number of different cognitive aspects (levels of communication skills, emotional reactions, social relationships etc.), Piaget argued that each stage was reflected equally in the ways in which children visualize and manipulate physical objects. Piaget was convinced that in recognizing these sequential stages in childhood development he had also identified the major stages in the mental evolution of early human populations – i.e. that ontogeny could be assumed broadly to mirror (or 'recapitulate') phylogeny – an issue which has generated endless controversy in the subsequent literature (see for example Longuet-Higgins *et al.* 1973; Parker & Gibson 1979, with following discussion; Lock 1993).

It is this particular viewpoint which has been taken up and developed in Wynn's studies of the development of early stone tool technology (Wynn 1979, 1985, 1989, 1991, 1993). To summarize Wynn's results, he believes that when stone tools are evaluated in terms of the level of complexity and conceptual vision implied by their sequence of manufacture, the whole spectrum of stone-tool production documented during the last two million years can be compared closely with the two final stages of Piaget's developmental scheme – i.e. with the stages of pre-operational intelligence and full operational intelligence. The critical conclusion from an archaeological perspective is that the major threshold between pre-operational and operational stages seems to be reflected in the patterns of tool manufacture during the middle or later stages of the Acheulian – i.e. by around 300,000 years BP. By this stage, Wynn argues, there is evidence that the manufacture of regular, carefully retouched hand-axe forms involved a perception of both bilateral symmetry (apparent in the plan of the tools and in the less readily-perceived cross sections), together with closely related ideas of 'reversibility' in the patterns of flaking applied alternately to the upper and lower faces of the tools (Wynn 1979, 1985, 1991). Seen from a Piagetian perspective, this symmetrical, carefully controlled tool manufacture could not have been achieved by any modern individual who had not attained the fully operational stage of spatial and visual competence – normally achieved in modern children at an age of around 6–7 years. Wynn goes on to argue (significantly, in terms of the present discussion) that he can see nothing in the subsequent patterns of tool manufacture during either the Middle or Upper Palaeolithic to suggest that there were any further significant advances in basic cognitive abilities beyond those reflected in the production of hand axes (Wynn 1985: 41).

How far Wynn's conclusions in this regard are justified could be debated from several different perspectives. Intelligence is by definition a complex, multi-facetted phenomenon, and it is possible to argue that there could be several other dimensions in which basic cognitive abilities could have developed (for example in the spheres of memory, communication, social awareness etc.) which need hardly be reflected directly in the manufacture of stone tools (Gibson 1985, 1988, 1990, 1993). More significantly perhaps, Piaget recognized two major substages within his final stage of fully operational intelligence – those of 'concrete operations' and 'propositional operations' respectively – which are not discussed directly in Wynn's analysis, and which could be highly signifi-

cant for some of the later cognitive developments over the period of the Middle-Upper Palaeolithic transition (see Parker & Gibson 1979; Gibson 1985; Parker & Milbraith 1993). Finally, as argued earlier, one could contest strongly on theoretical grounds the notion that all forms of human cognitive evolution are likely to have come abruptly to a halt as long ago as 300,000 years – i.e. precisely at the time when most of the social, technological and perhaps ecological stimuli to human development are likely to have become most intense (Alexander 1989; Gibson 1985, 1988; Parker & Milbraith 1993; Aiello & Dunbar 1993). Nevertheless, Wynn's application of Piaget's basic perspective is refreshingly articulate and stands as a seminal contribution to the study of human mental and cognitive development.

Binford's speculations on cognitive development during the Lower, Middle and Upper Palaeolithic periods are more intuitively based than those of Wynn, and relate mainly to his perception of the varying degrees of 'time depth' and long-range planning implied by various activities reflected in the archaeological records of the Palaeolithic succession (Binford 1985, 1987, 1989, 1992). Binford argues that in the behavioural records of Upper Palaeolithic groups – as well as in those of all recent hunter-gatherers – we can document an impressive capacity to plan and structure activities over a relatively long time-span and with a clear perception of ultimate goals. Thus, tools and equipment tend to be produced long before they are actually needed for use, and with an anticipation of extended periods of use. Occupation sites are similarly planned and structured with a view to relatively long periods of occupation and with an anticipation of the needs that will arise during these periods of extended use. Systematic movements of camp sites are made with a view to intercepting or harvesting food resources at long distances from the earlier camps, and so on. These capacities for long-term planning, forethought, anticipation etc. are seen by Binford as central to the behavioural adaptations of all present-day hunter-gatherers, and he believes they are equally explicit in the behavioural records of most Upper Palaeolithic groups (Binford 1989; see also Gibson 1985, 1988; Parker & Milbraith 1993).

The main thrust of Binford's argument is that this dimension of long-term planning, forethought, prediction etc., appears to be largely lacking from the archaeological records of the Lower and Middle Palaeolithic periods (Binford 1987, 1989, 1992). He argues, for example, that Lower and Middle Palaeolithic tools were normally manufactured, used and discarded within a short span of time and, even if occasionally transported across the landscape, rarely remained within the behavioural system for more than a few days at most (Binford 1973). He similarly argues that most food resources exploited by Middle Palaeolithic groups were processed and consumed close to, if not actually on, the point of procurement, with little attempt to anticipate regular shifts in distribution or location of resources at different times of the year. In the same vein he argues that almost all claimed occupation sites of the Lower and Middle Palaeolithic appear to reflect very short-term activities, mostly by small-scale human groups, with little evidence for the organization or structuring of occupation surfaces with a view to long periods of occupation.

Many of Binford's conclusions could no doubt be contested from several aspects of the archaeological data. As discussed in Chapter 5, there is ample evidence that certain varieties of high-quality raw materials were systematically transported across the landscape by Middle Palaeolithic groups (see Figs 5.5, 5.8, 5.18), in some cases apparently from specialized quarry or procurement sites (Geneste 1985, 1989a; Turq 1989a). There is also strong evidence that some specialized quarry or workshop sites were devoted to the production of one or more specific artefact

forms (such as Levallois flakes or cordiform hand axes) which were subsequently carried away from the sites for further use or shaping elsewhere (Geneste 1985). And as Tavoso (1984) has persuasively argued, there can be no doubt that fully shaped tools were frequently carried for long distances into areas where lithic raw material supplies were entirely lacking, with an anticipation of their use far from the original point of production. All this evidence could therefore be held up as proof that Neanderthal groups did have a clear perception of time-structured technologies and of potential future needs, in at least certain aspects of their activities.

Nevertheless the central point of Binford's argument is that whatever degree of forethought, prediction, long-term planning etc. can be identified in the behaviour of Middle Palaeolithic groups, would seem to be greatly exceeded by the patterns that one can document in the ensuing Upper Palaeolithic. In this respect, my own perception of the archaeological evidence coincides closely with that of Binford. In the Upper Palaeolithic, for example, we have evidence not only that high-quality lithic raw materials were transported over the landscape in much more substantial quantities than in the Middle Palaeolithic (Fig. 5.20: Geneste 1989a, 1990) but also that certain specific commodities (most notably marine shells) were in some cases transported over much greater distances (Fig. 13.6), in some cases 500 km or more (Roebroeks et al. 1988; Gamble 1986; Bahn 1982; Geneste 1992; Taborin 1993). In the same vein there is no doubt that certain Upper Palaeolithic occupation sites (particularly those of the Upper Perigordian and later periods) do show much clearer evidence for highly structured internal organization than any of the sites so far documented from the Lower and Middle Palaeolithic (Figs 9.31. 9.32), in a way which almost certainly reflects the anticipation of much longer periods of occupation (Mellars 1973; Soffer 1985a, b; Gamble 1986; Combier 1988). Most significant of all perhaps is the apparent evidence for deliberate storage of food resources on certain Upper Palaeolithic sites – if indeed some of the large pit-like features recorded at open-air sites in central and eastern Europe can be reliably interpreted in these terms (Soffer 1985a, b). Finally, I have argued elsewhere that some of the strategies of hunting documented in the Upper Palaeolithic sites of southwestern France appear to reflect a more sharply focused exploitation of reindeer resources than anything so far documented in the preceding Mousterian sites of the same region (see Chapter 7, and Mellars 1989a: 357–8). This seems to be indicated not only by the extremely specialized composition of the faunal assemblages recovered from many sites (often with up to 99 percent of reindeer remains – even in sites dating from the earlier stages of the Aurignacian: cf. Fig. 7.4) but also by the conspicuous clustering of these specialized reindeer-hunting sites along certain restricted stretches of the Vézère and Dordogne valleys (Fig. 8.2), which are likely to have formed the major and most predictable migration routes for reindeer herds. I would argue that it is this aspect of the evidence which demonstrates most directly the capacity of Upper Palaeolithic groups to plan and organize their subsistence activities in a more efficient and strategic way than the preceding Middle Palaeolithic groups in the same region.

The final question is what does this apparently clear evidence for increased time perspectives, long-range planning etc. imply for the relative intelligence of Upper and Middle Palaeolithic populations? The answer could be argued in two different ways. Following the criteria outlined by Gibson (1990), one could certainly argue that the appearance of more complex technological operations and long-term planning – as well as the integration within these activities of increased numbers of cooperating individuals – is fully consistent with a significant increase in human cognitive abilities over the Middle-

Upper Palaeolithic transition (cf. Gibson 1985, 1988; Parker & Milbraith 1993). But as Binford has emphasized (1987, 1989), there are potentially other, less dramatic explanations for the same patterns of change. Binford's suggestion is that this major shift could indicate the emergence not of increased intelligence but simply of new cognitive structures, marked specifically by the appearance of more highly structured forms of language (Binford 1987: 692). It is this issue which raises some of the most intriguing questions in current studies of the character and significance of the Middle-Upper Palaeolithic behavioural transition.

Language

Discussions on the origins and development of language have filled several books over the past decade (e.g. Chomsky 1986; Landsberg 1988; Lieberman 1990; Bickerton 1990; Parker & Gibson 1990; Gibson & Ingold 1993) and also spawned a society devoted exclusively to the study of language origins (Wind et al. 1989). It is tempting to suggest that there are almost as many views of the possible origins of language as there are linguists, psychologists and palaeoanthropologists who have written on the issue. No one doubts that even the earliest hominids must have possessed a rudimentary form of language – since several species of great apes and monkeys have been shown to employ relatively complex forms of vocal communication between individuals, for a number of different functions (e.g. Smuts et al. 1987; Cheney & Seyfarth 1990; Parker & Gibson 1990). The critical question is how these primitive forms of communication were transformed into fully developed language as observed in all present-day human populations.

The essence of most of these debates lies in the dichotomy between the 'evolutionary' views of language origins (which envisage a fairly gradual, step-by-step increase in the complexity and structure of language over the course of human evolution) and the alternative of a relatively abrupt or 'catastrophic' transition at some point in language development – reflecting either a major genetic shift in the neurological organization of the human brain, or at least a radical cognitive 'invention' which would have revolutionized conceptual thought and associated communication patterns (e.g. Ragir 1985; Bichakjian 1989; Foster 1990; Lieberman 1990; Bickerton 1981, 1990; Pinker, 1994). The most explicit account of the 'catastrophist' hypothesis has been set out in the publications of Derek Bickerton (1981, 1990). Bickerton argues for a fundamental distinction between what he refers to as 'proto-language' and fully developed, 'true' language. The theoretical basis for this distinction derives partly from studies of language development in young children and partly from similar studies of documented transitions between 'pidgin' and more complex 'creole' languages observed in many contact situations between European and non-European populations – all of which appear to reflect a relatively abrupt transition from simple, rudimentary forms of language to more complex and highly structured language within a remarkably short time. To Bickerton, these studies suggest that there are probably no intermediate patterns between these two forms (Bickerton 1990: 140, 164–74, 190). The crucial distinction between the two language patterns lies in the structure of complex syntax or grammar which is the diagnostic hallmark of all present-day language patterns. Proto-languages by contrast (for example the language of very young children and various forms of pidgin) are characterized by a virtual lack of grammatical structure or syntax and consist essentially of short sequences of words, usually referring to immediate here-and-now situations and relationships or at best to events in the immediate past or future. While the structure of proto-languages does not necessarily rule out the possibility of occasional reference to objects or events at some distance in time or space (i.e. some notion of 'displacement'), Bickerton argues that the lack of clear syntactical structure would dramatically reduce the potential complexity of any such discussions – as well as introducing potential ambiguities – and would therefore have a drastically limiting effect on the efficiency of the language. He argues, more controversially, that this could influence not only the actions of the human groups, but also their capacity for structured coherent thought:

> Thinking of the kind that humans do is at best extremely difficult in the absence syntax, since it depends crucially on the existence of structures like: x happened because y happened; whenever x happens, y happens; unless x happens, y will not happen … Complex sentences of any kind are impossible in protolanguage because there is no way in which one clause can be inserted into another … Without any reliable way of determining 'who did what to whom', ambiguities will quickly accumulate until they are too numerous to process. This would pose almost as much of a problem for coherent thought as it would for speech.
>
> (Bickerton 1990: 162–3)

Whether the emergence of language during the course of human evolution did take this dramatic punctuated form remains the major issue in language origins (e.g. Ragir 1985; Chomsky 1986; Bichakjian 1989; Foster 1990; Bickerton 1981, 1990; Pinker 1994). The argument which has been advanced by Binford (1987, 1989), Whallon (1989), Clark (1981, 1992), myself (1989b, 1991) and several others is that if there were a relatively abrupt shift in the overall complexity and structure of language patterns associated with the transition from archaic to anatomically modern populations (even if on a less dramatic scale than that envisaged by Bickerton) this might help explain some of the radical transformations in human behavioural patterns over this period, documented in the archaeological records. The arguments are as follows:

1. First, Lieberman, Laitman and others have presented a series of purely biological arguments for questioning the linguistic capacities of Neanderthals (e.g. Lieberman 1989, 1990; Lieberman & Crelin 1971; Laitman *et al.* 1979). Their argument, in essence, is that detailed studies of the form of the mandible and basi-cranial region in several Neanderthal skulls (most notably that from La Chapelle-aux-Saints) seem to rule out the possibility that these hominids possessed a fully developed vocal tract and the associated ability to form a full range of clear vowel sounds. These conclusions have been contested (for example by Arensburg [1989] based on his studies of the recently discovered Neanderthal hyoid bone from Kebara in Israel; see also Falk 1975, Dubrul 1977), and Lieberman (1989) admits that more highly developed capacities for speech seem to be indicated in some earlier, pre-Neanderthal populations (e.g. Petralona and Steinheim). Even if the anatomical arguments are accepted, it could be argued that the mere inability to form a full range of vowel sounds need not automatically rule out the possibility of relatively advanced language among the Neanderthals (Pinker 1994: 354) – although as Leiberman points out it would inevitably restrict the complexity of vocal communications and might introduce serious ambiguities into language patterns.

To take a slightly different perspective, it would be possible to reverse the basic line of Lieberman's argument and see the lack of full anatomical ability for highly developed speech as an indication of the limited *need* for this capacity among Neanderthals – i.e. as a reflection of the much simpler patterns of vocal communication in these groups. Either way, the evidence cited by Lieberman and Laitman could provide an initial argument against the presence of fully developed language among the European Neanderthals.

2. In terms of the archaeological evidence, the most significant observation is the virtual lack of convincing evidence for symbolic behaviour or expression in Neanderthal contexts. As discussed earlier in this chapter there is a lack of well documented decorative or artistic items in Mousterian contexts; a lack of any obvious symbolic component in most Middle Palaeolithic tools, and a lack of convincing evidence for ceremonial burials. The possible links between language and various forms of visual or other symbolism can be argued in different ways (e.g. Chase & Dibble 1987; Davidson & Noble 1989; Donald 1991; Gibson & Ingold 1993). No one would question, however, that elaborate symbolic thought and expression is one of the defining hallmarks of all fully developed languages. The virtual lack of convincing evidence for symbolism in Mousterian contexts is at least consistent with the lack of highly developed language in Neanderthal communities, even if it cannot be taken as concrete proof of this.

3. Potentially the most significant aspect of 'symbolic' expression was discussed earlier in relation to the detailed morphology and taxonomy of Middle and Upper Palaeolithic tool forms. The critical point is that Middle Palaeolithic tools not only exhibit far less diversity in shape than those in Upper Palaeolithic contexts, but also a much weaker degree of morphological standardization and deliberately imposed form (Fig. 4.30). As I have argued in more detail elsewhere (1989b, 1991), this would suggest that the inferred 'mental templates' which lay behind the production of most Middle Palaeolithic tool forms were far less sharply defined than for those of the Upper Palaeolithic (see also Dibble 1989). This would fall naturally into place if the mental and associated *linguistic* categorization of different tool forms in the Middle Palaeolithic was much less tightly structured than in the Upper Palaeolithic, and could suggest that even if Neanderthal groups did have discrete mental and linguistic categories for certain groups of tools, the

total number of these categories was very much less than in the Upper Palaeolithic – arguably reflecting a general increase in the overall scale and complexity of vocabularies over this transition.

4. Finally, we can return to the arguments advanced by Binford and others concerning the degree of long-range planning and associated strategic organization of economic and social activities in the Middle Palaeolithic. As discussed above, a significant shift in the complexity of these planning strategies over the Middle-Upper Palaeolithic transition can be argued from several aspects of the archaeological record – the patterns of procurement and distribution of raw materials, evidence for increased storage of food resources, more highly structured patterns of seasonal mobility and settlement strategies, and a more specialized and sharply focused pattern of animal exploitation, apparently based, in the Upper Palaeolithic, on an increased ability to predict and anticipate the movements of animal herds. As Binford points out (1987, 1989) such complex, logistical planning and organization would virtually *demand* the existence of relatively advanced and structured language, in which the ability to discuss events far removed in space and time would be indispensable to the effective organization and coordination of economic and social strategies. Essentially the same point has been made by Whallon (1989), Soffer (1994) and others, who argue that the increased ability to share information about the distribution of essential economic resources and to coordinate and integrate the social and economic activities of both individuals and widely dispersed local groups – which only developed language could provide – could well have been the critical factor which allowed the colonization of some of the more extreme and unpredictable periglacial environments in central and eastern Europe which seem to have been occupied for the first time during the Upper Palaeolithic:

The capacity to plan ahead certainly exists on an individual basis even among primates. The capacity to do this on a group basis, however, requires an ability to communicate among the members of the group, and in so doing refer to and discuss anticipated events in the future ... Groups lacking such language capabilities would have been incapable of regularly and reliably organizing themselves for logistically exploiting their environments, and therefore would have operated with less and less efficiency in environments of increasing differential resource distribution and decreasing resource variety.

(Whallon 1989: 442–3)

As Whallon (1989: 443) and Bickerton (1990: 158–60) point out, complex language would be equally indispensable to the long-term storage of information about past events and experiences, both economic or social, which could be equally critical to the survival prospects of human groups in many environments (see also Donald 1991; Pinker 1994: 368).

On present evidence the arguments about the character of Neanderthal language can hardly be taken further. No one would seriously question that Neanderthals, as well as much earlier hominids, must have possessed a reasonably effective form of vocal communication and, as Bickerton has argued at some length, this must have been far more complex and structured than that of any of the living primates. The question is simply whether Neanderthal language was of essentially modern form, or, as Bickerton and many others have argued, fundamentally simpler in its basic grammatical and syntactical structure. In common with many other prehistorians I hold the view that a radical restructuring of language patterns would not only be consistent with the available archaeological records of behavioural changes over the period of the Middle-Upper Palaeolithic transition but might provide the most economical single explanation for these dramatic changes. Whether such changes were associated with equally fundamental changes in the neurological organization of the human brain is of course an entirely separate question.

As a final point we should perhaps return to the fundamental proposition advanced by Chomsky (1986), Bickerton (1981, 1990) and others that the emergence of essentially modern language must by its nature have been a relatively sudden, 'catastrophic' event, rather than a gradual process of mental and linguistic evolution. If this conclusion is valid then we must presumably expect to see a fairly dramatic reflection of this transition in the available behavioural records of human development, and arguably in the whole spectrum of behaviour ranging from technology, through subsistence and social patterns, to the more overtly symbolic domains of the human groups (Pinker 1994: 367–9). The question is where, in the available archaeological records of Europe, might we identify such a watershed, if not over the period of the Middle-to-Upper Palaeolithic transition?

Note: *An earlier version of this chapter was presented to the Conference: 'Modelling the Early Human Mind' (Cambridge, 1993) and is reproduced with the permission of the McDonald Institute for Archaeological Research.*

CHAPTER 13

The Big Transition

Perhaps the most intriguing and enigmatic aspects of the Middle Palaeolithic period is how and why it came to an end, after a period of around 200,000 years of remarkable stability. From the preceding chapters it has emerged that while there were significant shifts in the precise morphology and technology of stone tool production, subsistence patterns, site distributions etc. at different stages of the Middle Palaeolithic, very few if any of these seem to reflect any radical reorganization or restructuring of technological, economic or social patterns. Most of the documented changes appear to be more cyclical than directional in character – e.g. oscillations between the use or non-use of blade technology, the appearance and disappearance of bifaces, shifts between specialized and more general animal exploitation, and so on. To what extent such shifts can be seen as simple adaptations to the rapidly oscillating climatic and ecological conditions during the course of the Upper Pleistocene period remains a major question for future research. The point to be emphasized is that none of these changes at present suggests more than a reshuffling of basic cultural and behavioural patterns which, in one form or another, can be traced back into the time range of the penultimate glaciation (Mellars 1973: 256–7, 261; Rolland 1990; Rigaud 1993).

The dramatic break in this pattern of behavioural stability occurs at the time of the classic Middle-to-Upper Palaeolithic transi-

tion, dated in most regions of Europe at around 35–40,000 BP. So much has been written on the character and significance of this transition that it is difficult in a single chapter to do more than highlight the central elements in these debates (see for example papers in Hoffecker & Wolf 1988; Trinkaus 1989; Kozlowski 1989; Mellars & Stringer 1989; Mellars 1990; Farizy 1990; Aitken *et al.* 1992; Cabrera Valdés 1993; Knecht *et al.* 1993; Nitecki & Nitecki 1994). Most recent discussions on this topic have focused on four major issues:

1. What is the precise character of the behavioural changes which define the conventional Middle-Upper Palaeolithic transition, and how reliably can they be documented from the archaeological record?

2. To what extent were these changes associated with a major dispersal of new human populations across Europe, i.e. populations of anatomically modern, as opposed to the preceding populations of 'archaic' or Neanderthal, form?

3. If there is persuasive evidence for a dispersal of new populations over Europe at this point in the archaeological sequence, how far can we identify evidence for any contact or interaction between the expanding populations and the local Neanderthal/Middle Palaeolithic populations in the different regions?

4. Finally, why should we encounter this particular pattern of combined biological and behavioural change at this specific point in the archaeological sequence, approximately midway during the last glaciation and at a time when large parts of Europe were still in the grip of a severe periglacial climate?

Behavioural change

A criticism sometimes made against earlier discussions of the character of the Middle-Upper Palaeolithic transition is that these have relied on very broad characterizations of archaeological evidence from the two periods, in a way which tends to exaggerate the true character of behavioural changes over the period of the hypothetical transition itself (e.g. Simek & Price 1990: 243; Clark & Lindly 1989: 634). Thus, the charge is levelled that proponents of a behavioural 'revolution' at this point in the archaeological sequence are contrasting average patterns of behaviour in the two periods, or at worst, in the words of Lawrence Straus, comparing 'the Mousterian with the Magdalenian' (Straus 1983).

Whether there is more than a grain of truth in these allegations could be debated. Nevertheless it is evident that any attempt to document a radical shift or 'revolution' in behavioural patterns must focus initially on a specific and preferably relatively brief time span. The aim of this section is to adopt precisely this focus by concentrating on the time range of the earlier Aurignacian industries in Europe, i.e., on industries which can now be dated securely in radiocarbon terms to well before 30,000 BP and in many cases to before 35,000 BP (Fig. 13.13: Mellars 1992b; Bánesz & Kozlowski 1993). The critical issue is how far we can identify significant innovations in the archaeological records of human behaviour over this period, which are demonstrably lacking from the archaeological records of the preceding Middle Palaeolithic in the same regions. To summarize results of

earlier syntheses by Kozlowski (1990), White (1989), myself (1973, 1989a,b, 1991) and others, the answer lies in a combination of the following features:

1. *Improved blade technology.* The existence of blade technology in a basic morphological sense is by no means confined to the Upper Palaeolithic – as discussed fully in Chapter 3. However, there are indications that the scale of blade technology increased sharply not only in quantitative terms in the earlier Aurignacian industries of Europe, but also in the specific techniques of blade production – particularly those involving the use of soft as opposed to hard hammer techniques and probably also indirect 'punch' techniques (Boëda 1989, 1990; Kozlowski 1990; Pelegrin 1990; Rigaud 1993). In most areas of Europe there is a striking contrast between the relative poverty of blade technology in the final stages of the Mousterian and the rapid proliferation of blade production in the earliest stages of the Aurignacian (Kozlowski 1990).

2. *New forms of stone tools.* In an earlier paper I commented 'the generation of qualitatively new artefact forms seems to be specifically associated with the earliest phases of the Upper Palaeolithic succession throughout all regions of Western Eurasia' (Mellars 1989b: 341; see also Mellars 1973: 257). In the case of the earlier Aurignacian industries, this is reflected in the appearance of forms such as elegantly fluted carinated and nosed scrapers, new and more complex types of burins (including multifaceted and eventually classic *busqué* forms), extensively edge-retouched blades (including distinctively strangulated forms) and – above all – a proliferation of small, carefully retouched bladelets of both 'Font Yves' smaller 'Dufour' forms (Fig. 13.1: de Sonneville-Bordes 1960; Bosinski 1990; Bánesz & Kozlowski 1993). While it could be argued that some of these features (especially the proliferation of bladelets) might be attributed to the emergence of new forms of blade technology, several of

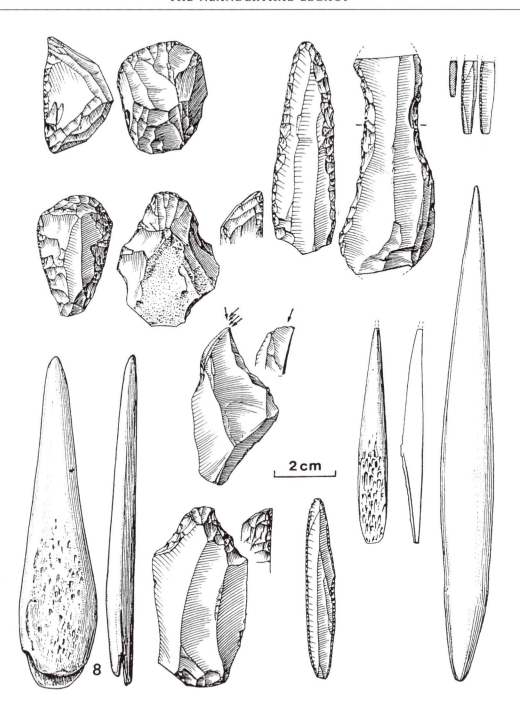

2 cm

Figure 13.1 *Aurignacian stone and bone artefacts from sites in western France. After Bordes 1968a.*

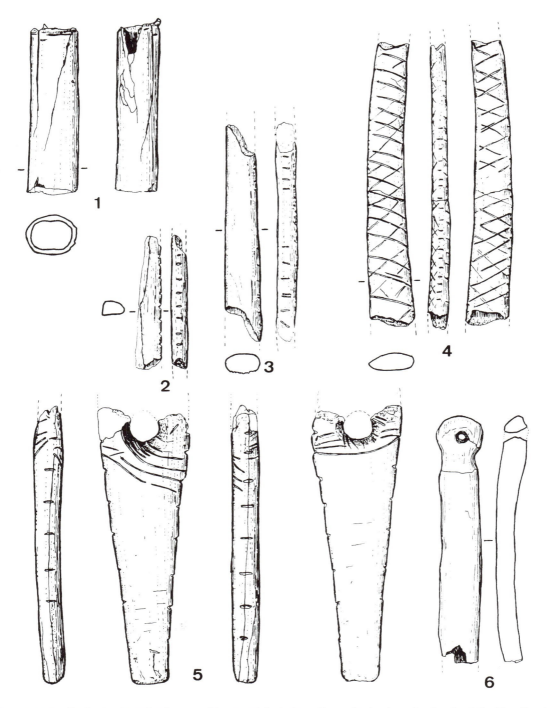

Figure 13.2 *Early Aurignacian bone and ivory artefacts from the early Aurignacian levels of the Vogelherd cave, southern Germany (nos 1–5) and Mladeč, Czechoslovakia (no. 6). After Hahn 1977.*

the other distinctively Aurignacian forms (classic nosed and carinated scrapers, *busqué* burins etc.) are rarely, if ever, manufactured from blade as opposed to flake blanks (Mellars 1989b). How far the appearance of typical end-scrapers and burins can be regarded as an explicitly Upper Palaeolithic feature is more debatable (see Chapter 4). Both of these types, however, appear in much larger numbers and usually in much more classic forms in the early Aurignacian industries of Europe than in any of the preceding Middle Palaeolithic industries in the same regions (Mellars 1989b; Anderson-Gerfaud 1990; Rigaud 1993).

3. *Bone, antler and ivory technology*. The appearance of complex, standardized and extensively shaped bone, antler and ivory artefacts is a striking feature of early Aurignacian sites throughout Europe (Hahn 1977; Bánesz & Kozlowski 1993). The tools appear in a remarkable diversity of forms, ranging from the classic split-base and biconical or lozangic bone/antler/ivory points, to more enigmatic types such as perforated antler batons, delicate, perforated bow-like forms, bone tubes, etc. (Figs 13.1, 13.2). The technology involved in the production of these tools was equally complex, ranging from initial cutting and grooving of the original blanks (in some cases demonstrably involving groove and splinter techniques), through to finer sawing, grinding and polishing of the finished tools (Mellars 1973: 258–9; 1989b; Knecht 1993). The element of imposed form is arguably even more explicit in these bone and antler tools than in the shapes of the associated stone tools. No one would seriously question that the proliferation of this complex, tightly controlled bone, antler and ivory technology is a radical innovation in the technological records of Europe, without parallel in the preceding Middle Palaeolithic industries of the region (Camps-Fabrer 1976; Vincent 1988; Rigaud 1993).

Figure 13.3 *Perforated deer tooth fom the early Aurignacian levels of La Souquette, southwestern France. After White 1989.*

4. *Personal ornaments*. The appearance of various types of personal ornaments is a further hallmark of early Aurignacian sites across Europe (Hahn 1972, 1977; White 1989, 1993). As discussed in the preceding chapter, these vary from simple perforated animal teeth (Fig. 13.3) and marine shells, to more elaborate beads laboriously manufactured from either ivory (as at Spy and Vogelherd) or hard stones such as steatite and serpentine (as in several of the French sites). As White (1989, 1993) and others have emphasized, these personal ornaments seem to be an invariable component of early Aurignacian sites across Europe and frequently occur in remarkable quantities – for example the 800 or so beads of stone and ivory recovered from the three adjacent sites of Abri Blanchard, Castanet and La Souquette in the Castelmerle valley of southwestern France. With the exception of the few dubious and extremely

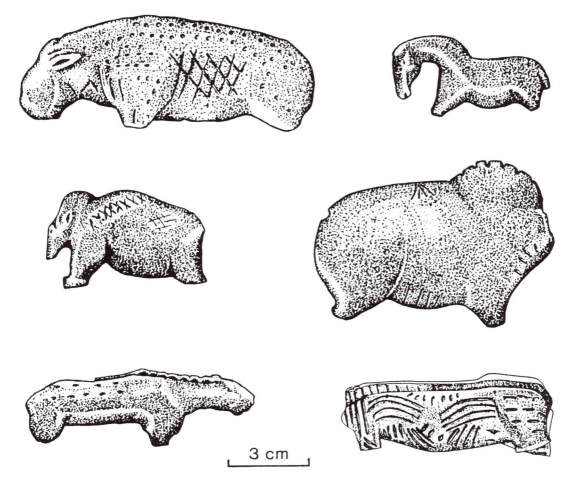

Figure 13.4 *Animal figurines carved from mammoth ivory from the early Aurignacian levels in the Vogelherd cave, southern Germany.*

isolated occurrences discussed in Chapter 12, such decorative items appear to be lacking from well documented Middle Palaeolithic contexts throughout Europe.

5. *Art and decoration.* The appearance of complex and sophisticated representational art provides the most dramatic reflection of the 'explosion' of symbolic expression associated with the earliest stages of the Upper Palaeolithic in Europe. The art is now documented from a range of Aurignacian sites in both western and central Europe, and varies from simple engraved outlines of either animals or female vulvar symbols, as at La Ferrassie, Abri Blanchard, Belcayre, Abri Cellier etc. in western France, to the more elaborate and beautiful statuettes of animals meticulously carved out of mammoth ivory from the sites of Vogelherd (Fig. 13.4), das Geissenklösterle and Höhlenstein-stadel in southern Germany (Delluc & Delluc 1978; Hahn 1972, 1977, 1984,

Figure 13.5 *Lion-headed human figure of mammoth ivory from the early Aurignacian levels in the Höhlenstein-Stadel cave, southern Germany.*

1993a; Bosinski 1990). Equally striking are the phallic carvings recorded from the Abri Blanchard and Vogelherd and the recently discovered ivory statuette of a male human figure equipped with a lion's head from the earlier Aurignacian levels of the Hohlenstein-Stadel cave (Fig. 13.5). Viewed as a whole, this art could hardly be described as crude, uniform, or proto-typical. Most of these discoveries were associated securely with the earlier Aurignacian levels in the different sites and can hardly be younger than ca 32–34,000 BP.

A proliferation of more abstract decorative motifs is a further widespread feature of early Aurignacian sites across Europe (Hahn 1972, 1977, 1993a). Again, the forms vary from relatively simple, regularly-spaced notches along the edges of bone or ivory fragments (e.g. Fig. 13.2), to more complex arrangements of dots, lines, crosses or other motifs carefully incised on larger fragments of bone or stone. Marshack (1972, 1985 etc.) has seen these engravings as an indication of either sophisticated numerical notation systems or even simple lunar calendars. Regardless of interpretations, these artefacts are not only abundant in Aurignacian contexts but still without convincing parallels from the preceding Middle Palaeolithic sites of Europe (cf. Fig. 12.4; Chase & Dibble 1987; Bednarik 1992, with comments).

6. *Expanded distribution and trading networks.* The most impressive indication of wide distribution networks for scarce or valued commodities in early Upper Palaeolithic contexts is provided by the occurrence of far-travelled marine shells in early Aurignacian sites throughout Europe – for example the frequent occurrence of shells from the Atlantic and Mediterranean coasts in early Aurignacian levels in the Perigord region (Fig. 13.6: Mellars 1973: 267; Taborin 1990, 1993), or the shells from the Black Sea coast at sites in the Kostenki region of south Russia (Hahn 1977: map 5), in some cases indicating their trans-

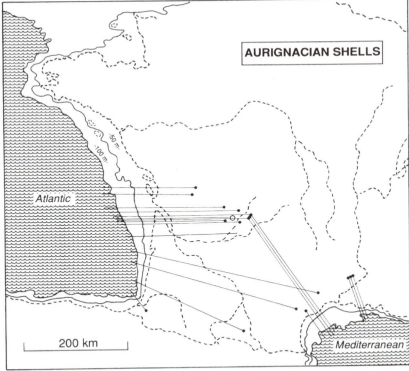

Figure 13.6 *Upper: Perforated sea shells, from the early Aurignacian levels of La Souquette, southwestern France. After White 1989. Lower: sources of sea shells found in Aurignacian levels in southern France, deriving from both the Atlantic and Mediterranean coasts. After Taborin 1993.*

portation of over distances of 500 km or more. Once again, movement of either shells or any other materials on this scale is without parallel from Mousterian sites in Europe.

Whether there was a similar increase in the range or efficiency of distribution networks for flint supplies in the early Upper Palaeolithic is perhaps more debatable. As discussed in Chapter 5, there is certainly evidence that especially high-quality raw materials were transported across the landscape in substantially larger quantities, and apparently on a more organized scale, in the Aurignacian sites of southwestern France than in any of the Mousterian sites of the same region (Fig. 5.20) – especially apparent in the distribution of high-quality Bergerac flint to sites in the Brive region, or to sites in the southern Perigord region between the Lot and the Dordogne (Demars 1982, 1990; Turq 1993). A similar sharp increase in frequencies of imported, high-quality flints has been recorded in early Aurignacian levels in Bulgaria, southern Germany and northern Spain (Kozlowski 1982; Hahn 1977; Hahn & Owen 1985; Straus & Heller 1988: 115–18). There can be little doubt that the overall scale on which lithic raw materials were transported, if not the maximum distances, increased significantly during the earlier stages of the Upper Palaeolithic compared with earlier Mousterian contexts.

The preceding paragraphs have focused on areas where major changes in human behavioural patterns can be documented most easily and directly from the 'hard' archaeological evidence. As I have discussed elsewhere (1973, 1982, 1989a) one could argue for equally significant changes in several other aspects of human organization, such as the densities of populations in some of the more ecologically favourable areas such as southwestern France, central Europe and northern Spain (marked by a sharp increase in numbers of occupied cave and rock-shelter sites), an increase in the maximum sizes of human residential groups (indicated by an increase

in the spatial extent and intensity of occupation residues in some of the larger sites) and almost certainly a shift towards a more sharply focused pattern of animal exploitation – reflected in both the highly specialized composition of the faunal assemblages recovered from several early Aurignacian sites (Abri Pataud, La Gravette, Roc de Combe, Le Piage etc.: see Fig. 7.4) and the apparent tendency for sites to be more densely clustered along some of the major migration routes of reindeer and other species such as the valleys of the Vézère and the Dordogne (see Chapter 7). Since the interpretation of these social and economic aspects of the evidence has generated more controversy than the direct artefactual evidence discussed above (e.g. Chase 1987; Clark & Lindly 1989), it is perhaps best to leave these issues for the present discussion.

Several aspects of the evidence outlined above should be emphasized. First is the wide range of the different types of behaviour involved. The earlier stages of the Aurignacian in Europe provide explicit evidence for radical changes in at least five or six separate aspects of behaviour: the technology and typology of stone tool production; the production of bone, antler and ivory artefacts; the emergence of varied and complex personal ornaments; the 'explosion' of representational art and associated abstract ornamentation; and the emergence of greatly expanded networks for the distribution of marine shells and in some cases high-quality raw materials. All these innovations are in place and well documented in the archaeological records of some of the earliest Aurignacian sites in Europe, certainly well before 30,000 BP, and probably before 35,000 BP. In short, we have evidence for a remarkable package of cultural and behavioural innovations which appear relatively suddenly within the archaeological records of Europe over a time span of at most 5–10,000 years and – within individual regions – probably over a much narrower time range of ca

1–2000 years. Regardless of how these developments are interpreted in broader social or cognitive terms, it is this package of innovations which justifies speaking of a major revolution in human behaviour coinciding closely with the conventional transition from the Middle to the Upper Palaeolithic stages in Europe.

Population dispersal

The issue of population dispersal and ultimately population replacement is central to all of the current debates over modern human origins in Europe. To demonstrate that there was a revolution in cultural and behavioural patterns associated with the appearance of anatomically modern anatomy is one thing. To argue that this reflects a dispersal of entirely new populations over the different regions of Europe is an entirely separate and far more contentious issue.

Throughout this century opinions have tended to polarize between those who saw the appearance of anatomically modern populations (i.e. *Homo sapiens sapiens*) in terms of colonization deriving initially from one specific geographical centre and subsequently dispersing throughout all areas of the world – the so-called 'Garden of Eden' hypothesis; and those who saw this event as a much more gradual and localized process of more long-term evolutionary development within each region of the world, without any significant dispersal or replacement of populations – the 'regional continuity' or 'multiregional evolution' hypothesis (Spencer 1984; Smith 1991; Stringer & Gamble 1993). A spate of recent publications on this issue shows that opinions remain no less divided now than 20 or even 50 years ago (e.g. Stringer & Andrews 1988; Stringer 1990, 1994; Wolpoff 1989; Thorne & Wolpoff 1992; Wolpoff *et al*. 1994; Wilson *et al*. 1992).

There is little doubt that a number of discoveries over the past 15 years have tended to shift the balance of evidence in favour of the population dispersal hypothesis. Two developments have been especially critical. On the one hand has been the work on genetic finger-printing of present-day human populations in different regions of the world, based on patterns of variation of nuclear and mitochondrial DNA. In particular, research by Allan Wilson, Rebecca Cann, Mark Stoneking and others on the patterns of mitochondrial DNA (known to be inherited exclusively through the female line of descent, and subject to an unusually rapid rate of genetic mutation) points to a surprisingly recent point of origin for present-day world populations, probably reaching back no more than ca 200–300,000 years (Cann *et al*. 1987, 1994; Stoneking & Cann 1989; Wilson & Cann 1992; Stoneking *et al*.1992). Combined with results of similar studies of nuclear DNA (inherited through both the male and female lineages: Lucotte 1989; Wainscoat *et al*. 1989; Mountain *et al*. 1992) the genetic evidence as a whole points strongly to Africa as the most likely point of origin of these genetically modern populations – although other potential homelands further to the north and east in Asia have occasionally been debated in the literature (cf. Maddison 1991; Templeton 1992). The second crucial development has come from recent advances in absolute dating techniques, which now make it possible to attribute secure relative and absolute ages to several human skeletal remains whose ages had previously been extremely controversial (Aitken *et al*. 1992). Perhaps the most important development has been the recent dating of large samples of anatomically modern skeletal remains from the sites of Mugharet-es-Skhul and Djebel Qafzeh in Israel to around 90–110,000 BP (Bar-Yosef 1992, 1994). The clear implication is that human populations that were essentially modern in at least most anatomical respects (Fig. 13.7) had become established in the Middle Eastern region at least 50–60,000 years before their appearance in the more northern regions of Europe and Asia, and must therefore have

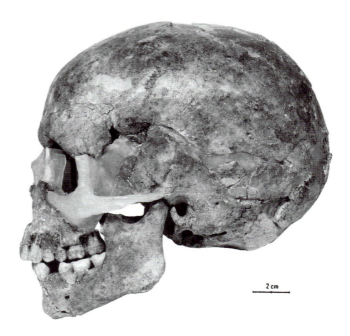

Figure 13.7 *Skull of anatomically modern form from Djebel Qafzeh, Israel, dated to ca 90–100,000 BP. After Tiller 1989.*

coexisted with various Neanderthal and other archaic populations of these regions throughout this period (Vandermeersch 1989, 1993a). Combined with the dating to a similar period of a range of essentially anatomically modern hominids at several sites in southern Africa (e.g. Omo, Border cave, Klasie's River Mouth) this gives strong support to the 'Out of Africa' model and is much less consistent with the implications of the multiregional evolution hypothesis (Stringer & Andrews 1988; Stringer 1990, 1994; Bräuer 1989; Howell 1994).

One final development which has a particularly critical bearing on the interpretation of evidence from western Europe is the discovery of human skeletal remains at the site of Saint-Césaire in the Charente-Maritime Department of western France (Fig. 13.8). Although fragmentary, reconstruction of these remains produced a skull now accepted as being of essentially classic Neanderthal type (Lévêque & Vandermeersch 1980; Stringer *et al.* 1984; Vandermeersch 1993b). The importance of this discovery lies in its dating. From a variety of evidence (archaeological associations, stratigraphy, pollen analysis, as well as direct thermoluminescence measurements on associated burnt flint samples) it is clear that this skull must date from no earlier than 35–38,000 BP (Mercier *et al.* 1991). As such, the remains can hardly be more than 3000–5000 years older than the earliest well documented specimens of fully 'Cro-Magnon' forms in western Europe, represented for example at Vogelherd in Germany, Les Rois in western France and Cro-Magnon itself (Fig. 13.9) (Gambier 1989, 1993). The argument, quite simply, is that the human population represented by the Saint-Césaire remains could not have evolved into populations of fully modern skeletal form within the time span available without a massive component of external gene-flow,

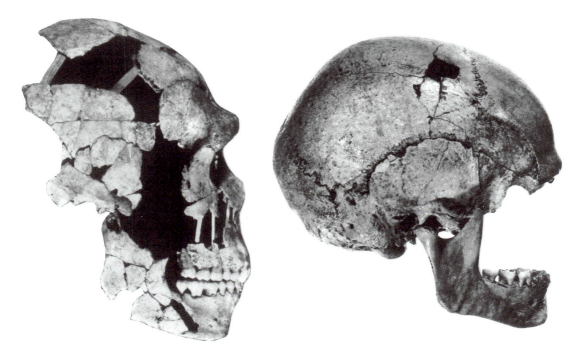

Figure 13.8   *Left: Late Neanderthal skull from the Châtelperronian levels (ca 35,000 BP) at Saint-Césaire (Charente-Maritime, southwestern France) Right: Anatomically modern skull from the early Aurignacian levels at Vogelherd (Stetten, southern Germany). After Bräuer 1989.*

which would have effectively swamped the genetic and anatomical features of the local Neanderthal populations. Even the most ardent proponents of the population continuity hypothesis seem to accept that the characteristics of the Saint-Césaire skeleton point strongly to a major population influx into these extreme western fringes of Europe (e.g. Smith 1991).

Most of these discoveries have been challenged in various ways by proponents of the multiregional evolution school, at least regarding the specific interpretations drawn from the genetic and skeletal evidence. Wolpoff, Thorne and others, for example, have challenged the chronological interpretations of mitochondrial DNA evidence and argued that by adopting a rather different rate of genetic divergence in DNA patterns one could redate the inferred dispersal of anatomically modern populations from the presumed African homeland to around 900,000 BP, that is close to the generally accepted date for the initial dispersal of *Homo erectus* populations into northern latitudes in the early Pleistocene (Wolpoff 1989; Thorne & Wolpoff 1992; Wolpoff *et al.* 1994). Wolpoff, Thorne, Smith and others have also contested the interpretation of the skeletal evidence, arguing that many of the supposed dichotomies between anatomically archaic and anatomically modern populations (for example in Africa, the Middle East and indeed parts of Europe) fail to make due allowance for the probable scale of individual anatomical variation within the local populations (e.g. Smith 1991, 1994). They also argue that in certain other regions, such as southeast Asia, Aus-

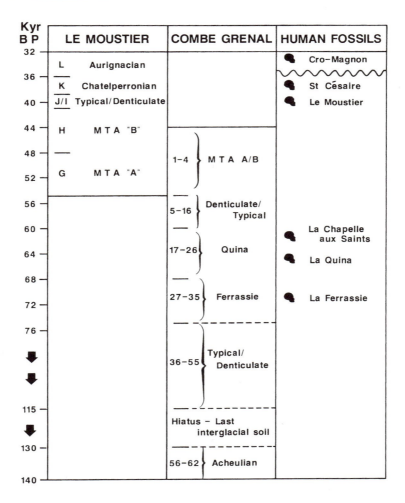

Figure 13.9 *Suggested chronology of the principal Neanderthal remains from southwestern France, compared with the archaeologically stratigraphies at Le Moustier and Combe Grenal. After Mellars 1986a.*

tralasia and parts of central and eastern Europe, there are strong indications in the skeletal evidence for some direct genetic continuity between the latest archaic and earliest anatomically modern population (Thorne & Wolpoff 1992). Possibly more seriously, there have been a number of criticisms of the statistical calculations which underlay some of the original DNA studies, and which are claimed to cast doubt on a specifically African origin for modern humans (Maddison 1992; Templeton 1993). Most of the recent studies of both the mitochondrial and nuclear DNA evidence, however, continue to favour some version of the population dispersal model, with Africa still remaining the most likely point of origin for the earliest biologically modern populations (Harpending *et al.* 1993; Ruvolo *et al.* 1993; Hasegawa *et al.* 1993; Cann *et al.* 1994; Stringer & Gamble 1994).

Population dispersal in Europe: the archaeological perspective

There is little doubt that many of the current controversies in the interpretation of available genetic and anatomical evidence stem from the attempt to adopt a single, unified view for the emergence of anatomically modern populations which is applicable to all areas of the world, regardless of the character of local geographical and environmental circumstances, or the particular patterns of demographic and evolutionary development within each region. Fortunately, the issues here are simpler and concern specifically the European evidence, and particularly that from the western zone of Europe. Having already looked briefly at the basic biological and skeletal arguments, I will now focus on the bearing of the available archaeological evidence on the character of this transition. The question is how far the archaeological evidence can be used to argue forcibly either for or against a rapid dispersal of entirely new human populations across different regions of Europe, associated with the earliest appearance of anatomically modern morphology in these areas?

As I have discussed in more detail elsewhere, all the current arguments in this context hinge on one critical correlation, the assumption that the earliest and most securely documented specimens of fully modern anatomy in Europe are associated with one specific archaeological entity, the grouping of so-called Aurignacian industries (Mellars 1992b). Leaving aside some of the more controversial specimens, well documented associations of this kind have now been recorded from at least four or five separate localities in Europe – notably from Vogelherd (i.e. Stetten) in Germany, Mladeč in Czechoslovakia, Velika Pečina in Yugoslavia and Les Rois and (perhaps less certainly) Cro-Magnon in western France (Fig. 13.8: Smith 1984; Stringer *et al*. 1984; Gambier 1989, 1993; Hublin 1990). Certainly, no serious claim has been made for an association between typically Aurignacian assemblages and anatomically Neanderthal remains in Europe. If this critical correlation is accepted, then the whole of the archaeological aspect of this particular debate hinges on the specific origins and mutual interrelationships of these Aurignacian industries within the different regions of Europe. Specifically, do these industries reflect the dispersal of entirely new human populations over the different parts of the continent? Or do they reflect simply a diversity of essentially local patterns of technological and demographic development, stemming directly from the immediately preceding Middle Palaeolithic/ Neanderthal populations within each region? As a generalization it is probably fair to say that most European archaeologists who have recently expressed a view on this issue have opted strongly for the population-dispersal hypothesis, based on the following range of observations (Allsworth-Jones 1986, 1990; Kozlowski 1988, 1990, 1993; Demars & Hublin 1989; Demars 1990; Hublin 1990; Harrold 1989; Mellars 1989a, 1992; Mussi 1990; Goia 1990; Farizy 1990; Bischoff *et al.* 1989; Broglio 1993; Djindjian 1993):

1. Archaeologically, the most striking feature is the remarkable uniformity of Aurignacian technology, extending across almost all eastern, central and western Europe and into the northern parts of the Middle East – a span of over 4000 kilometres (Fig. 13.10). As François Bordes (e.g. 1968: 200) and others have emphasized, industries recovered from sites such as Ksar Akil in Lebanon and Hayonim and Kebara in Israel are virtually indistinguishable from those in many of the classic Aurignacian sites in western Europe, reflected not only in the detailed typology of the stone tools (Fig. 13.11) but also in several idiosyncratic forms of bone and antler tools, such as typical split-base and biconical bone points (Fig. 13.12: Bar-Yosef & Belfer-Cohen 1988). At no other point in the Upper Palaeo-

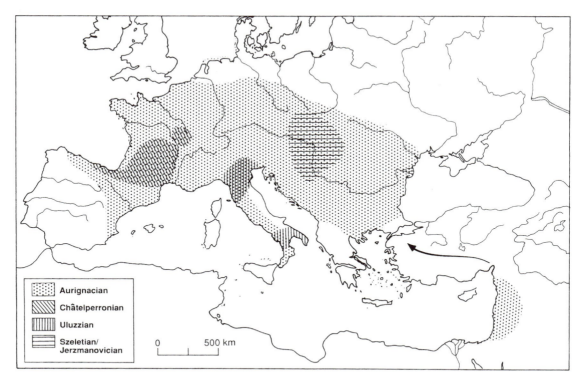

Figure 13.10 *Geographical distribution of Aurignacian industries in Europe and the Middle East, compared with the distribution of Châtelperronian, Szeletian and Uluzzian industries. In addition to the distribution shown, further occurrences of Aurignacian technology have been reported from Portugal, Britain, Sicily, southern Russia and Afghanistan. From Mellars 1992b.*

lithic sequence can one demonstrate such a uniform technology extending over such a wide diversity of contrasting environmental and ecological zones. Whether this would have been possible without a similar uniformity in language patterns across this region remains an interesting point for speculation (cf. Cavalli-Sforza 1991; Renfrew 1987).

2. This uniformity in technology of the earlier Aurignacian industries contrasts sharply with the diversity of the immediately preceding Middle Palaeolithic technologies in different regions of Europe. As Kozlowski (1992) and others have emphasized, the final stages of the Middle Palaeolithic seem to have been characterized by a variety of technological patterns: typical Mousterian of Acheulian tradition (MTA) industries on the extreme western fringes of the continent; various forms of either leaf-point or 'eastern Charentian' industries in central and eastern Europe; Denticulate industries in parts of Italy and northern Spain; and a variety of either Levallois or Levallois-point dominated technologies in the Balkans and south-eastern Europe. It is difficult to see how a technology as uniform and widespread as the Aurignacian could have sprung, rapidly and essentially independently, from such a diversity of technological roots.

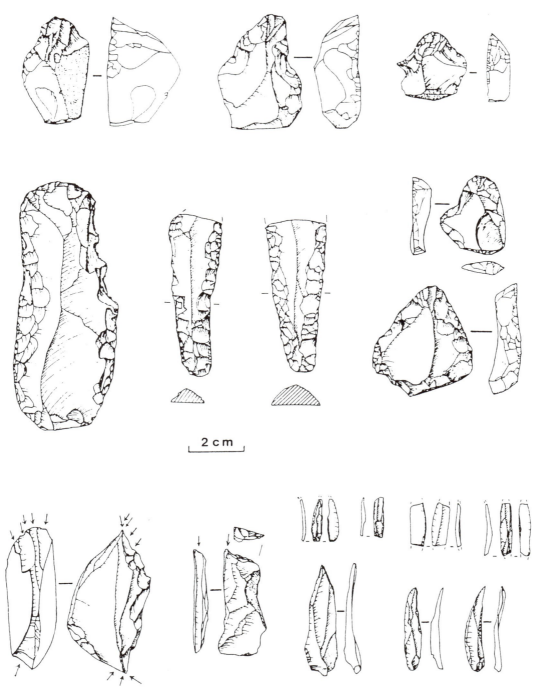

2 cm

Figure 13.11 *Aurignacian stone tools from the Hayonim cave (Israel) showing a range of forms closely similar to those from Aurignacian sites in western and central Europe (see Fig. 13.1). After Belfer-Cohen & Bar-Yosef 1981.*

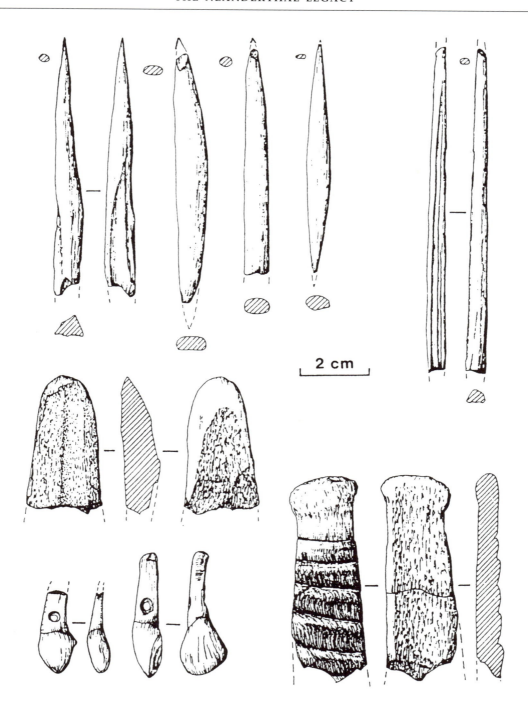

Figure 13.12   *Bone artefacts and animal-tooth pendants from the Aurignacian levels of the Hayonim cave, Israel. After Belfer-Cohen & Bar-Yosef 1981.*

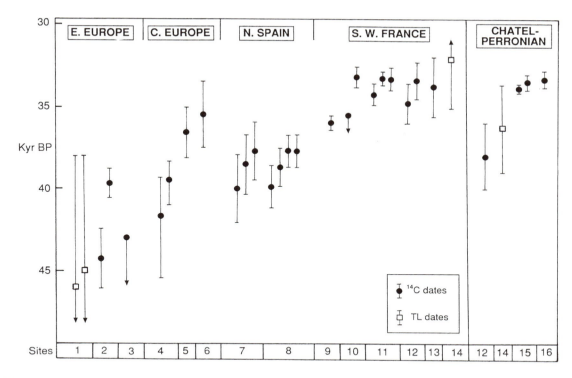

Figure 13.13 *Absolute age measurements for early Aurignacian industries in eastern, central and western Europe, and for Châtelperronian industries in France. For the radiocarbon dates (indicated by circles) the graph includes only the oldest dates available from each region, on the assumption that these are likely to show patterns least affected by problems of contamination with more recent, intrusive carbon. Vertical bars indicate one standard deviation; vertical arrows indicate 'greater than' ages. The sites shown are: 1. Temnata (Bulgaria); 2. Istállóskö (Hungary); 3. Bacho Kiro (Bulgaria); 4. Willendorf (Austria); 5. Geissenklosterle (Germany); 6. Krems (Austria); 7. Castillo (northwestern Spain); 8. L'Arbreda (northeastern Spain); 9. La Rochette (France); 10. La Ferrassie (France); 11. Abri Pataud (France); 12. Roc de Combe (France); 13. Le Flageolet (France); 14. Saint-Césaire (France); 15. Arcy-sur-Cure (France); 16. Les Cottés (France). The dates are taken from the following sources: Allsworth-Jones 1986; Bischoff et al. 1989; Cabrera-Valdés & Bischoff 1989; Delibrias & Fontugne 1990; Haesaerts 1990; Kozlowski 1982, 1992; Leroi-Gourhan & Leroi-Gourhan 1964; Mellars 1990a; Mellars et al. 1987; Mercier et al. 1991; Movius 1975. Note that radiocarbon dates in this age range are likely to be systematically younger than those produced by other dating techniques, perhaps by ca 3000 years (Bard et al. 1990a). The precise taxonomy of the industries from Willendorf (site 4) remains to be clarified. From Mellars 1992b.*

3. It now seems to be generally accepted that it is extremely difficult to find convincing origins for the distinctive patterns of Aurignacian technology within most regions of Europe. This has been emphasized by Kozlowski (1982, 1992) for the Balkans and southeast Europe; by Allsworth-Jones (1986, 1990), Otte (1990) and others for central Europe; by Mussi (1990), Goia (1990) and Broglio (1993) for Italy; by Bordes (1968), de Sonneville-Bordes (1960), Demars (1990), Rigaud (1993) and others for western France; and by Bischoff *et al.* (1989) for northern Spain. In all these areas the earliest Aur-

ignacian industries appear as an abrupt break in the local patterns of technological development, with no apparent links with the immediately preceding Mousterian industries in the same regions. Only rarely has the possibility of purely local origins been suggested for Aurignacian technology in Europe, for example by Cabrera Valdés and Bernaldo de Quirós (1990) for the succession at El Castillo in northern Spain and by Valoch (1983) for some of the Czechoslovakian industries. Both suggestions, however, have been contested by other workers (e.g. Kozlowski 1993; Djindjian 1993) and have since been withdrawn by Valoch himself (1990) for the Czechoslovakian industries.

At present the most plausible origins for Aurignacian technology seem to lie in some of the Middle Eastern industries, most notably in the long sequence of Aurignacian, proto-Aurignacian and so-called transitional industries recorded at Ksar Akil in the Lebanon (Copeland 1976; Marks & Ferring 1988; Ohnuma & Bergman 1990). Significantly, it is there, in contrast to the various regions of Europe, that the earliest Aurignacian industries were preceded by a long succession of demonstrably earlier Upper Palaeolithic technologies, apparently extending back to 45–50,000 BP (Marks & Ferring 1988; Mellars & Tixier 1989; Bar-Yosef 1994).

4. The relative and absolute chronology of the earliest stages of the Aurignacian within the different regions of Europe still remains to be documented in secure terms – largely owing to the inherent limitations of radiocarbon dating within this age range. The overall pattern of available dates nevertheless appears to suggest a pattern of successively younger dates extending from east to west across the continent, from around 43–45,000 BP in eastern and southeastern Europe, through to ca 40,0000 BP in central Europe, northern Spain and the Mediterranean coast, to around 35,000 BP in the classic region of southwestern France (Fig. 13.13: Kozlowski 1992, 1993; Mellars 1992b; Hahn 1993b). There is an urgent need for more dates to confirm this pattern, preferably with methods other than radiocarbon. As the evidence stands at present, however, it could be seen as at least consistent with the hypothesis of a progressive spread of Aurignacian technology from east to west across the continent.

5. Finally, the character and scale of the various technological and other behavioural innovations associated specifically with the earlier stages of the Aurignacian in different regions of Europe should be re-emphasized – ranging from innovations in the technology and typology of stone tool production, through extensively shaped bone, antler and ivory artefacts, to the effective explosion of symbolic artefacts in the form of notched and incised bonework, various forms of personal ornaments and remarkably varied and sophisticated representational art. Even if we set aside evidence for apparent shifts in the densities of human population, patterns of animal exploitation and the sizes of local social and residential groups, this is an impressive range of behavioural innovations which, as argued earlier in Chapters 11 and 12, almost certainly reflects equally radical changes in the social, cognitive and most probably linguistic patterns of the associated populations. Of course, radical and wide-ranging innovations of this kind cannot be taken as an automatic indication of episodes of population dispersal or replacement in the archaeological record, since it is clear that under certain conditions episodes of rapid behavioural change can occur through processes of either cultural diffusion or purely internal cultural change. Nevertheless the close association of all these behavioural innovations with the first appearance of Aurignacian technology – and apparently fully modern skeletal anatomy – in the different regions of Europe, is at least consistent with the hypothesis of a population dispersal at

this point in the archaeological sequence, even if archaeological evidence alone cannot be held up as conclusive proof.

Viewed as a whole, therefore, the archaeological evidence for the Aurignacian in the different regions of Europe coincides closely with the patterns that one would predict from the implications of current population-dispersal scenarios of modern human origins. Whether the same data could be held to be equally consistent with the population-continuity or multiregional evolution hypothesis is much more doubtful. How in this case would one account for the striking uniformity of Aurignacian technology over such a vast area of Europe and the Middle East, superimposed on so much diversity in the technology of the immediately preceding Middle Palaeolithic populations in the same regions? How would one explain the sudden and abrupt appearance of this technology in so many different regions, without any convincing origins or antecedents in preceding technologies in the same areas – or the sheer range, diversity and magnitude of the various cultural and behavioural innovations involved? In the classic region of western France there can be no serious doubt that the appearance of the Aurignacian reflects the intrusion of an essentially new human population, not only in the sudden and abrupt appearance of this technology (clearly *later* than in the neighbouring areas of northern Spain and the Mediterranean coast: Fig. 13.13), but also in the clear evidence that the earliest Aurignacian communities in this area persisted and apparently coexisted for some time alongside the latest Neanderthal populations in the same region (see below: cf Mellars 1989a; Demars & Hublin 1989; Rigaud 1993). If we accept such population intrusion in southwestern France, we should be prepared to give the same hypothesis equal consideration in the other regions of Europe where the overall spectrum and character of the archaeological evidence appears to show a similar pattern.

Population interaction

One of the most intriguing questions posed by current studies of the origins and dispersal of biologically modern human populations is how far we can identify evidence for any contact or interaction between the final Neanderthal populations and the earliest, hypothetically intrusive populations of anatomically modern humans in the different regions of Europe? If the population-dispersal scenario is valid, then this particular issue cannot be avoided. The inescapable implication of this model is that some kind of contact and interaction between the intrusive, expanding populations of anatomically modern hominids and the local, indigenous populations of archaic Neanderthals must have occurred repeatedly, and over the whole of the geographical range occupied by the expanding modern populations. This scenario has provided the inspiration for several popular novels, such as William Golding's *The Inheritors* and Jean Auel's *The Clan of the Cave Bear*, but remains surprisingly poorly studied from the perspective of the archaeological evidence.

Over the last decade, evidence for such chronological overlap, contact and apparent interaction between the final archaic and earliest anatomically modern populations has been claimed from several different regions of Europe (e.g. Allsworth-Jones 1986, 1990; Kozlowski 1988, 1990, 1993; Harrold 1989; Otte 1990; Mussi 1990; Goia 1990; Valoch 1990; Demars 1990; Demars & Hublin 1989; Hublin 1990; Mellars 1989a, 1991; Rigaud 1993; Djindjian 1993). The clearest evidence once again comes from the extreme western fringes of Europe, centred on the Perigord and adjacent provinces of southwestern France and resides in the demonstrable contemporaneity of two quite distinct and sharply contrasting archaeological assemblages, represented by the classic Aurignacian industries on the one hand and those of the Châtelperronian or Lower Perigordian group on the other. The juxtaposition of these two industries raises a number of intriguing issues which are worth examining closely.

1. On the basis of simple technological and geographical criteria, there can be no doubt that the Aurignacian and Châtelperronian industries were products of separate human populations in the southwestern French sites. The distinctive type-fossils which define the two industries (Châtelperron points for the Châtelperronian and various forms of nosed and carinate scrapers, Aurignacian blades, Dufour and Font Yves bladelets, split-base bone points etc. for the Aurignacian: cf. Figs 13.1, 13.14) show mutually exclusive distributions in material from the most recently excavated sites, and there is evidence that both the basic techniques of flake and blade production and the specific sources exploited for lithic raw materials in the two variants were different (de Sonneville-Bordes 1960; Harrold 1989; Demars 1990; Demars & Hublin 1989; Pelegrin 1990; Lévêque et al. 1993; Rigaud 1993). Most significantly, the geographical distributions of the two industries are radically different: whereas the Aurignacian has a distribution extending over virtually the whole of western, central and eastern Europe, the Châtelperronian is restricted to a small zone confined entirely to the western and central parts of France (to the west of the Rhône valley) and penetrating for a short distance into the adjacent parts of the Pyrenees and northern Spain (Fig. 13.10).

2. The existence of a substantial period of overlap between the Aurignacian and Châtelperronian populations can be demonstrated from several different aspects of the chronological data. In addition to correlations based on detailed climatic and vegetational sequences recorded at different sites (Leroyer & Leroi-Gourhan 1983; Leroyer 1988) we now have evidence from three sites in southern France and northern Spain where discrete

levels of Châtelperronian and Aurignacian industries occur directly interstratified within the same stratigraphic sequences – notably at the Roc de Combe and Le Piage in southwestern France and at El Pendo in Cantabria (Bordes & Labrot 1967; Champagne & Espitalié 1981; Harrold 1989; Demars 1990). The available radiocarbon evidence admittedly remains rather sparse and potentially ambiguous for the southwestern French sites. From the immediately adjacent areas of the Mediterranean coast and Cantabria, however, there is now clear radiocarbon evidence that typically Aurignacian industries were being manufactured by at least 38–40,000 BP, preceding by at least 4000–5000 years dates for typical Châtelperronian industries at sites such as Les Cottés and Arcy-sur-Cure in western and central France (Fig. 13.13: Bischoff et al. 1989; Cabrera Valdes & Bischoff 1989; Harrold 1989; Farizy 1990). From the combined palaeoclimatic, stratigraphic and radiocarbon evidence, there can be no doubt that the time ranges of the Aurignacian and Châtelperronian industries must have overlapped within these extreme western parts of Europe for several thousand years.

3. The critical importance of this chronological overlap lies in the fact that there is now almost conclusive evidence that these two technologies were the product of contrasting biological populations. All the available skeletal evidence from France and other regions of Europe suggests that the Aurignacian industries were the product of fully anatomically modern populations (Howell 1984, 1994; Stringer et al. 1984; Smith 1984; Gambier 1989, 1993; Demars & Hublin 1989; Hublin 1990). By contrast, there is explicit evidence from the hominid remains recovered from Saint-Césaire (Fig. 13.8), as well as from the series of human teeth recovered from earlier excavations at Arcy-sur-Cure, that the populations responsible for the Châtelperronian industries were of distinctively archaic, essentially classic Neanderthal type (Lévê-

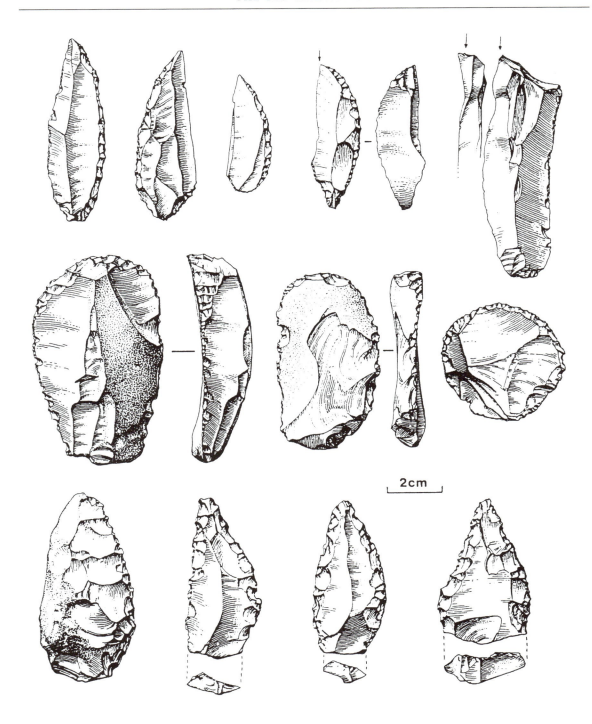

Figure 13.14 *Châtelperronian stone tools from the Grotte du Renne, Arcy-sur-Cure, central France, showing a combination of Upper Palaeolithic and Mousterian forms. After Leroi-Gourhan & Leroi-Gourhan 1964.*

que & Vandermeersch 1980; Vandermeersch 1993b; Stringer *et al.* 1984; Leroi-Gourhan 1958). If this evidence is accepted, then it confirms the coexistence of these two biologically contrasting populations within the western fringes of Europe over a very substantial span of time.

What has not always been so clearly recognized in the earlier literature is that these archaic associations of the Châtelperronian industries had already been predicted, several decades before the discovery of the Saint-Césaire skeleton, purely on the basis of the technology of the industries. As long ago as 1954 François Bordes argued that many of the distinctive technological features of the Châtelperronian industries, such as the character of the steeply backed Châtelperron points, as well as the occurrence of typical side scrapers, denticulates, and even small, bifacial hand-axe forms (Fig. 13.14), showed obvious links with the preceding Mousterian industries of the same region, especially with those of the MTA group (Bordes 1954–55, 1958, 1968, 1972). Later, I added further strands to these arguments, by pointing to the closely similar geographical distributions of the Châtelperronian and MTA industries (both confined to areas to the west of the Rhône valley in France and both extending into the adjacent areas of northern Spain) and arguing that the MTA industries appeared from several lines of evidence to represent the final stages of the local Mousterian sequence in southwestern France, immediately preceding the emergence of the Châtelperronian industries (Mellars 1969, 1973). In short, the arguments for believing that the Châtelperronian industries are the product of entirely indigenous, Neanderthal, populations in western Europe can be supported strongly on the basis of both the direct skeletal associations of the industries (at Saint-Césaire and Arcy-sur-Cure) and the basic technology, chronology and spatial distribution of the industries themselves.

4. The final and in many ways most interesting point is that this period of overlap between the Aurignacian and Châtelperronian populations seems to be reflected in the various forms of interaction or acculturation between the two populations. As discussed in detail elsewhere (e.g. Harrold 1989; Mellars 1989a, 1991; Farizy 1990, 1994; Lévêque 1993) it is now clear that while the basic technological roots of the Châtelperronian industries lie in the immediately preceding Mousterian industries, many of their specific features are of distinctively Upper Palaeolithic type. This applies not only to the strong component of typical blade technology apparent in most Châtelperronian assemblages but also to the presence of highly typical and abundant forms of end scrapers and burins and – in at least some sites – simple but extensively shaped bone and antler tools and even personal ornaments, in the form of carefully grooved and perforated animal teeth (Fig. 13.15: Harrold 1989; Farizy 1990; Lévêque 1993; Leroi-Gourhan & Leroi-Gourhan 1964). The crucial point to recognize is that all these specifically Upper Palaeolithic elements in the Châtelperronian appear to have developed at a relatively late stage, certainly long after the initial appearance of fully Aurignacian industries in northern Spain (Fig. 13.13) and probably while Aurignacian populations were already present in the southeastern parts of France (Leroyer & Leroi-Gourhan 1983; Leroyer 1988; Cabrera Valdés & Bischoff 1989; Mellars 1992b). Exactly how such interaction and apparent acculturation between the final Neanderthal and earliest anatomically modern populations should be visualized remains more controversial (see Graves 1991 and associated comments for further discussion). There seems little doubt, however, that the emergence of typically Upper Palaeolithic technological features amongst the final Neanderthal populations of western Europe can be explained more economically by vari-

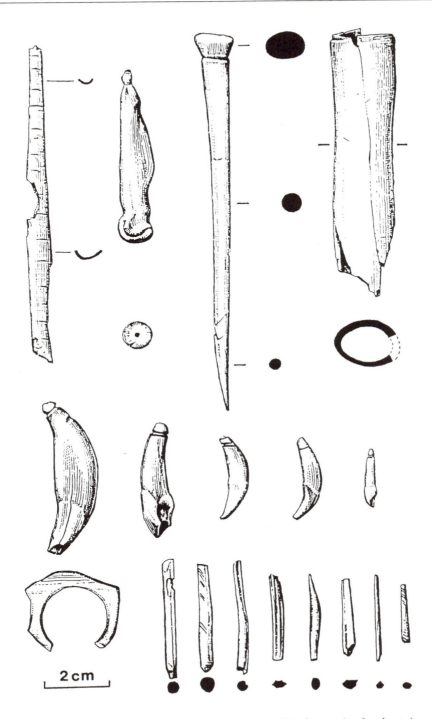

Figure 13.15 *Bone artefacts and animal-tooth pendants from the Châtelperronian levels at Arcy-sur-Cure, central France (ca 33–34,000 BP). After Leroi-Gourhan & Leroi-Gourhan 1964.*

ous contact and acculturation processes than by a purely spontaneous invention of Upper Palaeolithic technology by the final Neanderthal communities themselves.

Exactly how this kind of coexistence between the two populations could be maintained is more difficult to answer from the available archaeological evidence. We are still extremely ignorant about many of the most basic adaptive and organizational features of the Châtelperronian populations, mainly due to the poverty of faunal material recovered from most of the sites and the lack of detailed studies of the available economic data. One possibility is that the Châtelperronian and early Aurignacian groups were adapted to significantly different foraging and subsistence strategies – with the Aurignacian perhaps focusing mainly on specialized hunting of reindeer herds along major migration trails (such as the valleys of the Dordogne and Vézère) while Châtelperronian groups were adapted to more generalized animal exploitation, perhaps still dependent partially on scavenging rather than on deliberate and strategic hunting of game. The very generalized faunal assemblages recovered from Châtelperronian levels at Saint-Césaire, Roc de Combe, Châtelperron, Trou de La Chèvre etc. (in each case showing fairly balanced frequencies of horse, red deer, bovids, reindeer etc.) could be taken to support this suggestion (Delpech 1983, 1993; Patou-Mathis 1993). Another possibility is that the overall levels of population density and highly mobile patterns of seasonal and annual foraging strategies practised by the two groups were such that there was rarely any direct competition between them for particular economic resources or for the simultaneous occupation of the same territories. Evidence for close interstratification of Aurignacian and Châtelperronian levels documented at the Roc de Combe and Le Piage (Bordes & Labrot 1967; Champagne & Espitalié 1981; Demars 1990) might perhaps be seen as a direct reflection of these highly

mobile foraging patterns. Most probably it was only when the population density of the Aurignacian groups built up to relatively high levels during the middle and later stages of the Aurignacian that any strong economic and social competition for the use of particular resources or social territories would have emerged in some of the more ecologically favoured areas such as the Dordogne and Vézère valleys (de Sonneville-Bordes 1960; Mellars 1989a; Demars 1990, Rigaud 1993). It is at this time, significantly, that evidence for Châtelperronian occupation seems to be restricted mainly to the peripheral zones of western and central France, such as the Arcy-sur-Cure caves and some of the areas to the north and south of the Perigord region, in the sites of Les Cottés, Fténioux and Quinçay in the Vienne, or at Roc de Combe and Le Piage in the Lot (Leroyer & Leroi-Gourhan 1983; Leroyer 1988; Lévêque 1993). Seen in these terms it is reasonable to suggest that the process of eventual population replacement of the Châtelperronian by the Aurignacian groups was a relatively gradual and progressive phenomenon, probably reflecting more of a gradual shift in population numbers and the occupation of specific territories rather than any outright confrontation between the two groups (Zubrow 1989).

How far similar interaction and acculturation patterns between final Neanderthal and earliest anatomically modern populations can be recognized in other regions of Europe remains a topic of continuing debate. Allsworth-Jones (1986, 1990), Kozlowski (1988, 1990, 1993), Valoch (1990) and several others have put forward this argument for the emergence of the Szeletian and related leaf-point industries of central and eastern Europe (Figs 13.10, 13.16), pointing out that the time-range occupied by these industries almost certainly overlaps with that of the, apparently intrusive, Aurignacian industries in the same regions, and that strictly local roots for these distinctive industries can be seen in both the

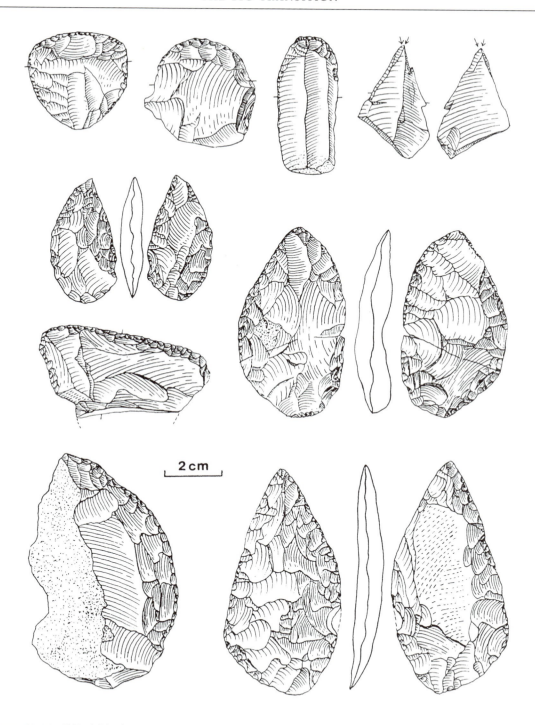

Figure 13.16 *Bifacial leaf points and associated tools from the Szeletian site of Vedrovice V, Czechoslovakia. After Valoch 1990.*

technology and spatial distribution (Fig. 13.10) of the archaeological assemblages. Mussi (1990), Goia (1990) and others have presented similar arguments for the emergence of the Uluzzian industries of the Italian peninsula – again almost certainly contemporaneous with the presence of typical Aurignacian industries in the adjacent areas of the Mediterranean coast, and again showing a restricted geographical distribution in the Italian sites (Fig. 13.10). Further to the east, similar patterns may be reflected in the dichotomy between the Streletskaya and Spitsinskaya industries of the south Russian Plain (Soffer 1985; Hoffecker 1988).

To summarize, recent research into the earliest stages of the Upper Palaeolithic now seems to reveal a broadly similar pattern in different regions of Europe. In each area there is evidence for the presence of typical and seemingly intrusive Aurignacian industries, apparently associated with fully anatomically modern hominids, and appearing in most regions between ca 40,000 and 35,000 BP. Alongside these industries, and at a broadly similar date, there is evidence for the emergence of a range of sharply contrasting forms of early Upper Palaeolithic technology, each restricted to a limited geographical area and each showing strong and obvious links with the latest Middle Palaeolithic technologies in the same regions. It is only in western Europe that these local technologies have been found in association with well documented human skeletal remains but in this particular case (i.e. the Châtelperronian) the skeletal remains are of distinctively archaic, Neanderthal form (Fig. 13.8). Proponents of the population dispersal hypothesis would argue that this pattern coincides closely, if not exactly, with the pattern one would predict from the scenario of a rapid dispersal of new human populations over different regions of Europe, combined with varying degrees of chronological overlap, contact and eventually acculturation with the local, indigenous Neanderthal populations.

Colonization scenarios

How and why a major episode of population dispersal occurred at this particular point in the Upper Pleistocene has been much discussed (e.g. Zubrow 1989; Mellars 1989a, 1992b; Kozlowski 1993; Bar-Yosef 1994; Gamble 1993). It is now clear from recent dating of the large samples of skeletal remains from the sites of Skhul and Qafzeh in Israel that human populations that were essentially modern in most anatomical respects (Fig. 13.7), had become established in the Middle East by around 100,000 BP and must therefore have coexisted with Neanderthal populations in the immediately adjacent areas of Europe over a period of at least 50–60,000 years (Bar-Yosef 1992, 1994). Possible reasons for such prolonged coexistence may not be too difficult to discern. If, as most scenarios still suggest, the anatomically modern populations had evolved initially in the tropical and subtropical environments of southern Africa, then they could hardly be expected to possess the appropriate range of either biological or cultural adaptations to allow rapid colonization of the severe periglacial environments which made up the greater part of Europe during Upper Pleistocene times. By contrast, the Neanderthal populations had evolved and evidently flourished in these particular environments for at least 100,000 if not 200,000 years (Stringer et al. 1984; Stringer & Gamble 1993; Hublin 1990). As I have discussed elsewhere (Mellars 1989a, 1992b), it was almost certainly the complex technological and cultural changes inherent in the 'Upper Palaeolithic Revolution', whatever its ultimate causes, which eventually gave a strong adaptive advantage to the anatomically modern populations in the Middle Eastern area, and equipped them both to colonize a range of entirely new glacial environments and to compete effectively with local Neanderthal populations in these regions. Significantly, the initial stages of this technological and cultural revolution seem to have occurred several thousand years earlier in the Middle East (at sites such as Boker Tachtit and Ksar Akil, both dated to ca 45–50,000 BP) than in the adjacent zones of Europe (Mellars 1989a; Mellars & Tixier 1989; Marks 1993; Bar-Yosef 1994). The precise mechanisms and initial stimulus for this cultural transformation remain enigmatic. As discussed earlier, however, we should give serious consideration to the possibility that it was the emergence of more complex, essentially modern forms of language – with its attendant consequences for almost all aspects of human behaviour and organization – which played a major if not primary role in this transformation (Mellars 1989a, 1991; Clark 1992).

Regardless of the initial stimulus, the process of population expansion may have been greatly facilitated by climatic and ecological events around the middle of the last glaciation. It is now evident that the period between 50,000 and 30,000 BP (i.e. the later part of stage 3 of the oxygen-isotope sequence) was marked by a series of major climatic fluctuations, during which average temperatures in many regions rose by at least 5–6°C and which allowed the expansion of temperate woodland into many areas of Europe previously dominated by periglacial tundra or steppe (see Figs. 2.8, 2.14, 2.16: Guiot *et al.* 1989, 1993; Dansgaard *et al.* 1993). To groups who were ecologically adapted, biologically and culturally, to the temperate environments of the east Mediterranean zone, these ecological changes would inevita-

bly have made the process of population expansion into areas lying to the north and west easier to achieve. It has been argued recently that this population dispersal could have occurred along two major routes: one along the Mediterranean coast and one through the Danube valley, perhaps in a roughly analogous way to the much later dispersal of the earliest farming communities across Europe (Kozlowski 1993; Djindjian 1993). It may well be that the same ecological changes served to destabilize some of the ecological and cultural adaptations of local Neanderthal populations in these regions, leading either to significant shifts in the geographical ranges occupied by individual groups or perhaps even to major episodes of population decline. Zubrow (1989) has argued that it would require little more than a minor shift in birth/death rates between the two populations to lead to an effective replacement of one population by the other in specific regions of Europe within a span of at most 1,000 years.

Whether or not such a process of total demographic and biological replacement did occur in Europe, or any other part of the world, remains controversial. It is now clear, however, that such replacement is by no means inconceivable in either cultural or demographic terms and could have been achieved without dramatic confrontation, let alone mass genocide, which have been envisaged in some of the more fanciful discussions of the origins and dispersal of modern humans.

References _____

Adam, K. 1951. Der waldelefant von Lehringen, ein jagdbeute des diluvialen menschen. *Quärtar* **5**: 79–92.

Aharon, P. 1983. 140,000-yr isotope climatic record from raised coral reefs in New Guinea. *Nature* **304**: 720–723.

Aharon, P. & Chappell, J. 1986. Oxygen isotopes, sea level changes and the temperature history of a coral reef environment in New Guinea over the last 10^5 years. *Palaeogeography, Palaeoclimatology, Palaeoecology* **56**: 337–379.

Aiello, L.C. 1995. Hominine preadaptations for language and cognition. In: *Modelling the Early Human Mind* (eds P. Mellars & K. Gibson). Cambridge: McDonald Institute for Archaeological Research (in press).

Aiello, L.C. & Dunbar, R.I.M. 1993. Neocortex size, group size and the evolution of language. *Current Anthropology* **34**: 184–193.

Airvaux, J. & Chollet, A. 1975. Le site moustérien de la Fontaine à Scorbé-Clairvaux (Vienne). *Bulletin de la Société Préhistorique Française* **72**: 209–217.

Aitken, M.J. 1985. *Thermoluminescence Dating*. London: Academic Press.

Aitken, M.J. 1990. *Science-based Dating in Archaeology*. London & New York: Longman.

Aitken, M.J., Huxtable, J. & Debenham, N.C. 1986. Thermoluminescence dating in the Palaeolithic: burned flints, stalagmitic calcite and sediment. In: *Chronostratigraphie et Faciès Culturels du Paléolithique Inférieur et Moyen dans l'Europe du Nord-Ouest* (eds A. Tuffreau & J. Sommé). Paris: Association Française pour l'Etude du Quaternaire, pp. 7–14.

Aitken, M.J., Stringer, C.B. & Mellars, P.A. (eds) 1992. *The Origin of Modern Humans and the Impact of Chronometric Dating*. London: Royal Society (Philosophical Transactions of the Royal Society, series B, **337**, no. 1280).

Alexander, R.D. 1974. The evolution of social behavior. *Annual Review of Ecology and Systematics* **5**: 325–383.

Alexander, R.D. 1989. The evolution of the human psyche. In: *The Human Revolution: behavioural and biological perspectives on the origins of modern humans* (eds P. Mellars & C. Stringer). Princeton: Princeton University Press, pp. 455–513.

Allsworth-Jones, P. 1986. *The Szeletian and the Transition from Middle to Upper Palaeolithic in Central Europe*. Oxford: Oxford University Press.

Allsworth-Jones, P. 1990. The Szeletian and the stratigraphic succession in Central Europe and adjacent areas: main trends, recent results, and problems for solution. In: *The Emergence of Modern Humans: an archaeological perspective* (ed. P. Mellars). Edinburgh: Edinburgh University Press, pp. 160–243.

Ameloot-van der Heijden, N. 1993a. L'industrie laminaire du niveau CA. In: *Riencourt-les-Bapaume (Pas-de-Calais): un Gisement du Paléolithique Moyen* (ed. A. Tuffreau). Paris: Maison des Sciences de l'Homme, pp. 26–52.

Ameloot-van der Heijden, N. 1993b. L'industrie laminaire du niveau CA du gisement Paléolithique moyen de Riencourt-les-Bapaume

(Pas de Calais). *Bulletin de la Société Préhistorique Française* **90**: 324–327.

Ameloot-Van der Heijden, A. & Tuffreau, A. 1993. Les industries lithiques de Riencourt-les-Bapaume dans le contexte de l'Europe du Nord-Ouest. In: *Riencourt-les-Bapaume (Pas-de-Calais): un Gisement du Paléolithique Moyen* (ed. A. Tuffreau). Paris: Maison des Sciences de l'Homme, pp. 107–111.

Anderson-Gerfaud, P. 1980. A testimony of prehistoric tasks: diagnostic residues on stone tool working edges. *World Archaeology* **12**: 181–194.

Anderson-Gerfaud, P. 1981. *Contribution Methodologique à l'Analyse des Microtraces d'Utilisation sur les Outils Préhistoriques*. Doctoral dissertation, University of Bordeaux I.

Anderson-Gerfaud, P. 1990. Aspects of behaviour in the Middle Palaeolithic: functional analysis of stone tools from southwest France. In: *The Emergence of Modern Humans: an archaeological perspective* (ed. P. Mellars). Edinburgh: Edinburgh University Press, pp. 389–418.

Antoine, P. 1993. L'environnement des occupations humaines au Paléolithique moyen récent dans la France septentrionale. *Bulletin de la Société Préhistorique Française* **90**: 320–323.

Arensburg, B. 1989. New skeletal evidence concerning the anatomy of Middle Palaeolithic populations in the Middle East: the Kebara skeleton. In: *The Human Revolution: behavioural and biological perspectives on the origins of modern humans* (eds P. Mellars & C. Stringer). Princeton: Princeton University Press, pp. 165–171.

Ashton, N.M. 1983. Spatial patterning in the Middle-Upper Palaeolithic transition. *World Archaeology* **15**: 224–235.

Ashton, N.M. 1985. Style et fonction dans le moustérien français. *Bulletin de la Société Préhistorique Française* **82**: 112–115.

Ashton, N. & Cook, J. 1986. Dating and correlating the French Mousterian. *Nature* **324**: 113.

Assassi, F. 1986. *Recherches Sedimentologiques sur la Climatologie du Würm Ancien et de l'Interstade Würmien en Périgord*. Doctoral dissertation, University of Bordeaux I.

Audouze, F. 1987. The Paris Basin in Magdalenian times. In: *The Pleistocene of the Old World: regional perspectives* (ed. O. Soffer). New York: Plenum Press, pp. 183–200.

Audouze, F. 1988. Des modèles et des faits: les modèles de A. Leroi-Gourhan et de L. Binford confrontés aux résultats récents. *Bulletin de la Société Préhistorique Française* **84**: 343–352.

Bánesz, L. & Kozlowski, J.K. (eds) 1993. *Aurignacien en Europe et au Proche Orient*. Bratislava: Acts of 12th International Congress of Prehistoric and Protohistoric Sciences.

Bar-Yosef, O. 1992. The role of western Asia in modern human origins. In: *The Origin of Modern Humans and the Impact of Chronometric Dating* (eds M.J. Aitken, C.B. Stringer & P.A. Mellars). London: Royal Society (Philosophical Transactions of the Royal Society, series B, **337**, no. 1280), pp. 193–200.

Bar-Yosef, O. 1994. The contributions of southwest Asia to the study of the origin of modern humans. In: *Origins of Anatomically Modern Humans* (eds M.H. Nitecki & D.V. Nitecki). New York: Plenum Press, pp. 24–66.

Bar-Yosef, O. & Belfer Cohen, A. 1988. The early Upper Palaeolithic in the Levantine Caves. In: *The Early Upper Palaeolithic: Evidence from Europe and the Near East* (eds J.F. Hoffecker & C.A. Wolf). Oxford: British Archaeological Reports International Series 437, pp. 23–42.

Bar-Yosef, O. & Meignen, L. 1992. Insights into Levantine Middle Paleolithic cultural variability. In: *The Middle Paleolithic: adaptation, behavior, and variability* (eds H.L. Dibble & P.A. Mellars). Philadelphia: University of Pennsylvania University Museum Monographs No. 72, pp. 163–182.

Bard, E., Hamelin, B., Fairbanks, R.G. & Zindler, A. 1990a. Calibration of the ^{14}C timescale over the past 30,000 years using mass spectrometric U-Th ages from Barbados corals. *Nature* **354**: 405–410.

Bard, E., Hamelin, B., & Fairbanks, R.G. 1990b. U-Th ages obtained by mass spectrometry in corals from Barbados: sea level during the past 130,000 years. *Nature* **346**: 456–458.

Beaulieu, J.-L. de & Reille, M. 1984. A long Upper Pleistocene pollen record from Les Echets, near Lyon, France. *Boreas* **13**: 111–131.

Bednarik, R.G. 1992. Palaeoart and archaeological myths. *Cambridge Archaeological Journal* **2**: 27–57.

Behre, K.-E. 1990. Biostratigraphy of the last glacial period in Europe. *Quaternary Science
Reviews* **8**: 25–44.

Behre, K.-E. & van der Plicht, J. 1992. Towards an
absolute chronology for the last glacial period
in Europe: radiocarbon dates from Oerel,
northern Germany. *Vegetation History and
Archaeobotany* **1**: 111–117.

Belfer-Cohen, A. & Bar-Yosef, O, 1981. The Aurignacian at Hayonim Cave. *Paléorient* **7**: 19–42.

Bernaldo de Quirós, F. & Cabrera Valdés, V. 1993.
Early Upper Palaeolithic industries of Cantabrian Spain. In: *Before Lascaux: the complex record of
the early Upper Paleolithic* (eds H. Knecht, A.
Pike-Tay & R. White). Boca Raton: CRC Press,
pp. 7–70.

Beyries, S. 1983. Etude technologique et traces
d'utilisation des 'éclats débordants' de Corbehem (Pas-de-Calais). *Bulletin de la Société Préhistorique Française* **80**: 275–279.

Beyries, S. 1986. Approche fonctionnelle de l'ou-
tillage provenant d'un site paléolithique moyen
du Nord de la France: Corbehem. In: *Chronostratigraphie et Faciès Culturels du Paléolithique
Inférieur et Moyen dans l'Europe du Nord-Ouest*
(eds A. Tuffreau & J. Sommé).Paris: Association
Française pour l'Etude du Quaternaire, pp.
219–224.

Beyries, S. 1987. *Variabilité de l'Industrie Lithique au
Moustérien: approche fonctionelle sur quelques
gisements Français.* British Archaeological
Reports International Series S328.

Beyries, S. 1988a. Functional variability of lithic
sets in the Middle Paleolithic. In: *Upper Pleistocene Prehistory of Western Eurasia* (eds H. Dibble
& A. Montet-White). Philadelphia: University
of Pennsylvania, University Museum Monographs No. 4, pp. 213–224.

Beyries, S. 1988b. Etude tracéologique des racloirs
du niveau IIA. In: *Le Gisement Paléolithique
Moyen de Biache-Saint-Vaast (Pas-de-Calais).* Vol.
I: *Stratigraphie, environnement, études archéologiques* (eds A. Tuffreau & J. Sommé). Paris:
Mémoires de la Société Préhistorique Française
21, pp. 215–230.

Beyries, S. 1990. Problems of interpreting the functional results for ancient periods. In: *The Interpretative Possibilities of Microwear Studies* (ed. K.

Knutson). Aun 14. Uppsala: Societas Archae-
ologica Upsaliensis, pp. 71–76.

Beyries, S. 1993. Analyse fonctionelle de l'industrie lithique du niveau CA: rapport préliminaire et directions de recherche. In: *Riencourt-
les-Bapaume (Pas-de-Calais): un Gisement du
Paléolithique Moyen* (ed. A. Tuffreau). Paris:
Maison des Sciences de l'Homme, pp. 53–61.

Beyries, S. & Boëda, E. 1983. Etude technologique
et traces d'utilisation des 'éclats débordants' de
Corbehem (Pas-de-Calais). *Bulletin de la Société
Préhistorique Française* **80**: 275–279.

Bicchieri, M.G. (ed.) 1972. *Hunters and Gatherers
Today.* New York: Holt, Rinehart & Winston.

Bichakjian, B. 1989. Origine du langage et évolution linguistique. *Société d'Etudes et de Recherches Préhistoriques des Eyzies* **39**: 93–111.

Bickerton, D. 1981. *Roots of Language.* Ann Arbor:
Karoma.

Bickerton, D. 1990. *Language & Species.* Chicago:
University of Chicago Press.

Binford, L.R. 1972. Contemporary model building:
paradigms and the current state of Palaeolithic
research. In: *Models in Archaeology* (ed. D.L.
Clarke). London: Methuen, pp. 109–166.

Binford, L.R. 1973. Interassemblage variability:
the Mousterian and the 'functional' argument.
In: *The Explanation of Culture Change* (ed. C.
Renfrew). London: Duckworth, pp. 227–254.

Binford, L.R. 1977. Forty-seven trips: a case study
in the character of archaeological formation
processes. In: *Stone Tools as Cultural Markers*
(ed. R.V.S. Wright). Canberra: Australian Institute of Aboriginal Studies, pp. 24–36 (reprinted
in Binford 1983c, pp. 243–268).

Binford, L.R. 1978. *Nunamiut Ethnoarchaeology.*
New York: Academic Press.

Binford, L.R. 1979. Organization and formation
processes: looking at curatedtechnologies. *Journal of Anthropological Research* **35**: 255–273
(reprinted in Binford 1983c, pp. 269–286).

Binford, L.R. 1981. *Bones: ancient men and modern
myths.* New York: Academic Press.

Binford, L.R. 1982a. The archaeology of place.
Journal of Anthropological Archaeology **1**: –31
(reprinted in Binford 1983c, pp. 357–378).

Binford, L.R. 1982b. Comment on R. White 'Rethinking the Middle/Upper Palaeolithic transition'. *Current Anthropology* **23**: 177–181.

Binford, L.R. 1983a. Working at archaeology: the Mousterian problem – learning how to learn. In: *Working at Archaeology* (ed. L.R. Binford). New York: Academic Press, pp. 65–69.

Binford, L.R. 1983b. Working at archaeology: the Mousterian debate, arguments of relevance, and the generation gap. In: *Working at Archaeology* (ed. L.R. Binford). New York: Academic Press, pp. 157–167.

Binford, L.R. 1983c. (ed.) *Working at Archaeology.* New York: Academic Press.

Binford, L.R. 1984. *Faunal Remains from Klasies River Mouth.* New York: Academic Press.

Binford, L.R. 1985. Human ancestors: changing views of their behavior. *Journal of Anthropological Archaeology* **4**: 292–327.

Binford, L.R. 1987. An interview with Lewis Binford (edited by A.C. Renfrew). *Current Anthropology* **28**: 683–694.

Binford, L.R. 1988. Etude taphonomique des restes fauniques de la Grotte Vaufrey, couche VIII. In: *La Grotte Vaufrey: paléoenvironnement, chronologie, activités humaines* (ed. J.-P. Rigaud). Mémoires de la Société Préhistorique Française 19, pp. 35–564.

Binford, L.R. 1989. Isolating the transition to cultural adaptations: an organizational approach. In: *The Emergence of Modern Humans: biocultural adaptations in the later Pleistocene* (ed. E. Trinkaus). Cambridge: Cambridge University Press, pp. 18–41.

Binford, L.R. 1991. Review of P. Mellars & C. Stringer (eds.) The Human Revolution. *Journal of Field Archaeology* **18**: 111–115.

Binford, L.R. 1992. Hard evidence. *Discover*, February 1992: 44–51.

Binford, L.R. & Bertram, J. 1977. Bone frequencies and attritional processes. In: *For Theory Building in Archaeology* (ed. L.R. Binford). New York: Academic Press, pp. 77–153.

Binford, L.R. & Binford, S.R. 1966. A preliminary analysis of functional variability in the Mousterian of Levallois facies. *American Anthropologist* **68** (no. 2, part 2): 238–295.

Binford, S.R. & Binford, L.R. 1969. Stone tools and human behavior. *Scientific American* **220**(4): 70–84.

Birket-Smith, K. 1945. *Ethnographical Collections from the Northwest Passage.* Copenhagen: Report of the Fifth Thule Expedition 1921–1924 (Vol. 6, no. 2).

Bischoff, J.L., Soler, N., Maroto, J. & Juliá, R. 1989. Abrupt Mousterian/Aurignacian boundary at c. 40 ka bp: accelerator ^{14}C dates from l'Arbreda Cave (Catalunya, Spain). *Journal of Archaeological Science* **16**, 63–576.

Blanc, A.C. 1958. Torre in Pietra, Saccopastore, Monte Circeo. On the position of the Mousterian culture in the Pleistocene sequence of the Rome area. In: *Hundert Jahre Neanderthaler (Neanderthal Centenary) 1856–1956* (ed. G.H.R. Von Koenigswald). Utrecht: Kemink en Zoon, pp. 167–174.

Blumenschine, R. 1986. *Early Hominid Scavenging Opportunities: implications of carcass availability in the Serengeti and Ngorongoro ecosystems.* British Archaeological Reports International Series 283.

Blumenschine, R. 1987. Characteristics of an early hominid scavenging niche. *Current Anthropology* **28**: 383–407.

Boëda, E. 1982. Approche technologique de la variabilité de la méthode Levallois: industries de Bagarre et de Corbehem (Pas-de-Calais). *Bulletin de l'Association Française pour l'Etude du Quaternaire* 2–3: 63–66.

Boëda, E. 1984. Méthode d'étude d'un nucléus Levallois à éclat préférentiel. *Cahiers de Géographie Physique* : 95–133.

Boëda, E. 1986. *Approche Technologique du Concept Levallois et Evaluation de son Champs d'Application: étude de trois gisements saaliens et weichseliens de la France septentrionale.* Doctoral dissertation, University of Paris X.

Boëda, E. 1988a. Le concept Levallois et évaluation de son champ d'application. In: *L'Homme de Néandertal*, Vol. 4: *La Technique* (ed. M. Otte). Liège: Etudes et Recherches Archéologiques de l'Université de Liège, pp. 13–26.

Boëda, E. 1988b. Le concept laminaire: rupture et filiation avec le concept Levallois. In: *L'Homme de Néandertal*, Vol. 8: *La Mutation* (ed. M. Otte).

Liège: Etudes et Recherches Archéologiques de l'Université de Liège, pp. 41–59.

Boëda, E. 1988c. Analyse technologique du débitage du niveau IIa. In: *Le Gisement Paléolithique Moyen de Biache-Saint-Vaast (Pas-de-Calais)*. Vol. I: *Stratigraphie, environnement, études archéologiques* (eds A. Tuffreau & J. Sommé). Paris: Mémoires de la Société Préhistorique Française 21, pp. 185–214.

Boëda, E. 1990. De la surface au volume: analyse des conceptions des débitages Levallois et laminaire. In: *Paléolithique Moyen Récent et Paléolithique Supérieur Ancien en Europe* (ed. C. Farizy). Nemours: Mémoires du Musée de Préhistoire d'Ile de France No. 3, pp. 63–68.

Boëda, E. 1991. Approche de la variabilité des systèmes de production lithique des industries du Paléolithique inférieur et moyen: chronique d'une variabilité attendue. *Techniques et Culture* 17–18: 37–79.

Boëda, E. 1993a. Le débitage discoïde et le débitage Levallois recurrente centripète. *Bulletin de la Société Préhistorique Française* 90: 392–404.

Boëda, E. 1993b. *Le Concept Levallois: variabilité des méthodes*. Paris: Centre National de la Recherche Scientifique (Centre de Recherches Archéologiques, Monograph 9).

Boëda, E. & Pelegrin, J. 1983. Approche technologique du nucléus Levallois à éclat. *Etudes Préhistoriques* 15: 41–48.

Boëda, E., Geneste, J.-M., & Meignen, L. 1990. Identification de chaînes opératoires lithiques du paléolithique ancien et moyen. *Paleo* 2: 43–80.

Boesch, C. 1993. Aspects of transmission of tool use in wild chimpanzees. In: *Tools, Language and Cognition in Human Evolution* (eds K. Gibson & T. Ingold). Cambridge: Cambridge University Press, pp. 171–183.

Bohmers, A. 1951. Die Höhlen von Mauern. *Palaeohistoria* 1: 1–107.

Bond, G., Broeker, W., Johnsen, S., McManus, J., Labeyrie, L., Jouzel, J. & Bonani, G. 1993. Correlations between climate records from North Atlantic sediments and Greenland ice. *Nature* 365: 143–147.

Bonifay, E. 1964. La grotte du Regourdou (Montignac, Dordogne): stratigraphie et industrie lithique moustérienne. *L'Anthropologie* 68: 49–64.

Bonifay, E. 1988. Fréquence et signification des sépultures néandertaliennes. In: *L'Homme de Néandertal*, Vol. 5: *La Pensée* (ed. M. Otte). Liège: Etudes et Recherches Archéologiques de l'Université de Liège, pp. 31–36.

Bordes, F. 1947. Etude comparative des différentes techniques de taille du silex et des roches dures. *L'Anthropologie* 51: 1–29.

Bordes, F. 1948. Les couches moustériennes du gisement du Moustier (Dordogne); typologie et technique de taille. *Bulletin de la Société Préhistorique Française* 45: 113–125.

Bordes, F. 1950a. Principes d'une méthode d'étude des techniques de débitage et de la typologie du Paléolithique ancien et moyen. *L'Anthropologie* 54: 19–34.

Bordes, F. 1950b. L'évolution buissonnante des industries en Europe occidentale: considérations théoriques sur le Paléolithique ancien et moyen. *L'Anthropologie* 54: 393–420.

Bordes, F. 1952. Sur l'usage probable des peintures corporelles dans certaines tribus moustériennes. *Bulletin de la Société Préhistorique Française* 49: 169.

Bordes, F. 1953a. Essai de classification des industries moustériennes. *Bulletin de la Société Préhistorique Française* 50: 457–466.

Bordes, F. 1953b. Levalloisien et Moustérien. *Bulletin de la Société Préhistorique Française* 50: 226–235.

Bordes, F. 1954a. *Les Limons Quaternaires du Bassin de la Seine*. Paris: Archives de l'Institut de Paléontologie Humaine, Mémoire 26.

Bordes, F. 1954b. Le Moustérien de l'Ermitage (fouilles L. Pradel): comparaisons statistiques. *L'Anthropologie* 58: 444–449.

Bordes, F. 1954–55. Les gisements du Pech de l'Azé (Dordogne). I: Le Moustérien de tradition acheuléenne. *L'Anthropologie* 58: 401–432; 59: 1–38.

Bordes, F. 1955a. La stratigraphie de la grotte de Combe-Grenal, commune de Domme (Dordogne): note préliminaire. *Bulletin de la Société Préhistorique Française* 52: 426–429.

Bordes, F. 1955b. Le Paléolithique ancien et moyen

de Jabrud (Syrie) et la question du Préaurignacien. *L'Anthropologie* **59**: 486–507.

Bordes, F. 1957. Le Moustérien de Hauteroche: comparaisons statistiques. *L'Anthropologie.* **61**: 436–441.

Bordes, F. 1958. Le passage du Paléolithique moyen au Paléolithique supérieur. In: *Hundert Yahre Neanderthaler* (ed. G.H.R. von Koenigswald), pp. 175–181. Utrecht: Kemink en Zoon.

Bordes, F. 1959a. Evolution in the Paleolithic cultures. In: *Evolution after Darwin* (ed. S. Tax). Chicago: Chicago University Press, pp. 99–110.

Bordes, F. 1959b. Le contexte archéologique des Hommes du Moustier et de Spy. *L'Anthropologie* **63**: 154–157.

Bordes, F. 1961a. *Typologie du Paléolithique Ancien et Moyen.* Bordeaux: Delmas. Publications de l'Institut de Préhistoire de l'Université de Bordeaux, Mémoire 1.

Bordes, F. 1961b. Mousterian cultures in France. *Science* **134**: 803–810.

Bordes, F. 1963. Le Moustérien à denticulés. *Archeološki Vestnik* **13–14**: 43–49.

Bordes, F. 1968a. *The Old Stone Age.* London: Weidenfeld & Nicolson.

Bordes, F. 1968b. Emplacement de tentes du Périgordien supérieur évolué à Corbiac, pres Bergerac (Dordogne). *Qartar* **19**: 251–262.

Bordes, F. 1969. Os percé moustérien et os gravé acheuléen du Pech de l'Azé II. *Quaternaria* **XI**: 1–6.

Bordes, F. 1971a. Observations sur l'Acheuléen des grottes de Dordogne. *Munibe* **23**: –23.

Bordes, F. 1971b. Physical evolution and technological evolution in man: a parallelism. *World Archaeology* **3**: 1–5.

Bordes, F. 1972. *A Tale of Two Caves.* New York: Harper & Row.

Bordes, F. 1975a. Le gisement du Pech de l'Azé IV: note préliminaire. *Bulletin de la Société Préhistorique Française* **72**: 293–308.

Bordes, F. 1975b. Sur la notion de sol d'habitat en préhistoire paléolithique. *Bulletin de la Société Préhistorique Française* **72**: 139–143.

Bordes, F. 1977. Time and space limits in the Mousterian. In: *Stone Tools as Cultural Markers* (ed. R.V.S. Wright). Canberra: Australian Institute of Aboriginal Studies, pp. 37–39.

Bordes, F. 1978. Typological variability in the Mousterian layers at Pech de l'Azé I, II, and IV. *Journal of Anthropological Research* **34**: 181–193.

Bordes, F. 1980. Le débitage Levallois et ses variantes. *Bulletin de la Société Préhistorique Française* **77**: 45–49.

Bordes, F. 1981. Vingt-cinq ans après: le complexe moustérien revisité. *Bulletin de la Société Préhistorique Française* **78**: 77–87.

Bordes, F. 1984. *Leçons sur le Paléolithique:* Vol. 2: *Le Paléolithique en Europe.* Paris: Centre National de la Recherche Scientifique (Institut du Quaternaire, Université de Bordeaux I, Cahiers du Quaternaire no. 7).

Bordes, F. & Bourgon, M. 1951. Le complexe moustérien: Moustérien, Levalloisien et Tayacien. *L'Anthropologie* **55**: 1–23.

Bordes, F. & Fitte, P. 1953. L'Atelier Commont (album de 188 dessins de Victor Commont, avec une étude de l'atelier). *L'Anthropologie* **57**: 1–44.

Bordes, F. & Labrot, J. 1967. La stratigraphie du gisement de Roc-de-Combe (Lot) et ses implications. *Bulletin de la Société Préhistorique Française* **64**: 15–28.

Bordes, F. & Lafille, J. 1962. Découverte d'un squelette d'enfant moustérien dans le gisement du Roc de Marsal, commune de Campagne-du-Bugue (Dordogne). *Comptes Rendus de l'Académie des Sciences de Paris* **254**: 714–715.

Bordes, F. & Prat, F. 1965. Observations sur les faunes du Riss et du Würm I en Dordogne. *L'Anthropologie* **69**: 31–45.

Bordes, F. & de Sonneville-Bordes, D. 1954. Présence probable de jaspe de Fontmaure dans l'Aurignacien V de Laugerie-Haute. *Bulletin de la Société Préhistorique Française* **51**: 67–68.

Bordes, F. & de Sonneville-Bordes, D. 1970. The significance of variability in Palaeolithic assemblages. *World Archaeology* **2**: 61–73.

Bordes, F., Fitte, P. & Blanc, S. 1954. L'Abri Armand Chadourne. *Bulletin de la Société Préhistorique Française* **51**: 229–254.

Bordes, F., Laville, H. & Paquereau, M.M. 1966. Observations sur le Pléistocène supérieur du gisement de Combe-Grenal (Dordogne). *Actes de la Société Linnéenne de Bordeaux* **103**: 3–19.

Bosinski, G. 1967. *Die Mittelpaläolithischen Funde im Westlichen Mitteleuropa*. Fundamenta A4. Cologne: Herman Böhlau.

Bosinski, G. 1973. Der Paläolithische fundplatz Rheindalen. In: *Neue Palaolithische und Mesolithische Ausgrabungen in der Bundesrepublik Deutchsland* (ed. H.J. Müller-Beck). Tübingen: Proceedings of the IX INQUA Congress: 11–14.

Bosinski, G. 1990. *Homo Sapiens*. Paris: Editions France.

Bouchud, J. 1966. *Essai sur le Renne et la Climatologie du Paléolithique Moyen et Supérieur*. Périgueux: Magne.

Boule, M. 1909. L'homme fossile de la Chapelle-aux-Saints (Corrèze). *L'Anthropologie* 20: 257–271.

Boule, M. & Anthony, R. 1911. L'encéphale de l'homme fossile de la Chapelle-aux-Saints. *L'Anthropologie* 22: 129–196.

Boulton, G.S. 1993. Two cores are better than one. *Nature* 366: 507–508.

Bourgon, M. 1957. *Les Industries Moustériennes et Pré-moustériennes du Périgord*. Paris: Archives de l'Institut de Paléontologie Humaine, Mémoire 27.

Bourlon, M. 1905. Une fouille au Moustier. *L'Homme Préhistorique* 7: 293–304.

Bourlon, M. 1906. L'industrie moustérienne au Moustier. *Congrès International d'Anthropologie et d'Archéologie Préhistorique* 13 (Monaco), pp. 287–322.

Bourlon, M. 1910. L'industrie des foyers supérieurs au Moustier. *Revue Préhistorique* 5: 157–167.

Bourlon, M. 1911. Industries des niveaux moyens et inférieurs de la terrasse du grand abri du Moustier. *Revue Préhistorique* 6: 283–300.

Bouyssonie, J. 1954. Les sépultures moustériennes. *Quaternaria* 1: 107–115.

Bouyssonie, A., Bouyssonie, J. & Bardon, L. 1908. Découverte d'un squelette humain moustérien à la Bouffia de la Chapelle-aux-Saints (Corrèze). *L'Anthropologie* 19: 513–518.

Bouyssonie, A., Bouyssonie, J. & Bardon, L. 1913. La station moustérienne de la 'Bouffia' Bonneval, à la Chapelle-aux-Saints. *L'Anthropologie* 24: 609–634.

Bowen, D.Q. 1978. *Quaternary Geology: a stratigraphic framework for multidisciplinary work*. Oxford: Pergamon Press.

Bowen, D.Q. 1990. The last interglacial-glacial cycle in the British Isles. *Quaternary International* 3/4: 41–47.

Bowman, S.G.E., Loosemore, R.P.W., Sieveking, G. de G. & Bordes, F. 1982. Preliminary dates for Pech de l'Azé IV. *Pact* 6: 362–369.

Bowman, S.G.E. & Sieveking, G. de G. 1983. Thermoluminescence dating of burnt flint from Combe Grenal. *Pact* 9: 253–268.

Boyle, K.V. 1990. *Upper Palaeolithic Faunas from South-west France: a zoogeographic perspective*. Oxford: British Archaeological Reports International Series 557.

Boyle, K.V. 1993. Upper Palaeolithic procurement and processing strategies in southwest France. In: *Hunting and Animal Exploitation in the Later Palaeolithic and Mesolithic of Eurasia* (eds G.L. Peterkin, H.M. Bricker & P. Mellars). Archeological Papers of the American Anthropological Association No. 4, pp. 151–162.

Bradley, B. 1977. *Experimental Lithic Technology with Special Reference to the Middle Palaeolithic*. PhD dissertation, University of Cambridge.

Bradley, B. & Sampson, C.G. 1986. Analysis by replication of two Acheulian artefact assemblages from Caddington, England. In: *Stone Age Prehistory: studies in memory of Charles McBurney* (eds G.N. Bailey and P. Callow): Cambridge: Cambridge University Press, pp. 29–45.

Bradley, R.S. 1985. *Quaternary Paleoclimatology: methods of paleoclimatic reconstruction*. Boston: Unwin Hyman.

Bräuer, G. 1989. The evolution of modern humans: a comparison of the African and non-African evidence. In: *The Human Revolution: behavioural and biological perspectives on the origins of modern humans* (eds P. Mellars & C. Stringer). Princeton: Princeton University Press, pp. 123–154.

Breuil, H. & Kozlowski, L. 1932. Etudes de stratigraphie Paléolithiques dans le nord de la France, la Belgique, et l'Angleterre. *L'Anthropologie* 41: 449–488; 42: 27–47.

Bricker, H.M. 1975. Provenience of flint used for the manufacture of tools at the Abri Pataud, Les

Eyzies (Dordogne). In: *Excavation of the Abri Pataud, Les Eyzies (Dordogne)* (ed. H.L. Movius). Cambridge (Mass.): Peabody Museum, Harvard University, pp. 194–197.

Broglio, A. 1993. L'Aurignacien au sud des Alpes. In: *Aurignacien en Europe et au Proche Orient* (eds L. Bánesz & J.K. Kozlowski). Bratislava: Acts of 12th International Congress of Prehistoric and Protohistoric Sciences, pp. 193–203.

Brugal, J.-P. & Jaubert, J. 1989. Stratégie d'exploitation et mode de vie des populations du Paléolithique moyen: exemples des sites du Sud de la France. *Congrès Préhistorique de France XIII (La Vie au Temps Préhistoriques)*, pp. 1–7.

Burke, A. 1993. Applied skeletochronology: the horse as a human prey during the pleniglacial in southwestern France. In: *Hunting and Animal Exploitation in the Later Palaeolithic and Mesolithic of Eurasia* (eds G.L. Peterkin, H.M. Bricker & P. Mellars). Archeological Papers of the American Anthropological Association No. 4, pp. 145–150.

Butzer, K. W. 1972. *Environment and Archeology*. Chicago: Aldine Atherton.

Byers, A.M. 1994. Symboling and the Middle-Upper Palaeolithic transition: a theoretical and methodological critique. *Current Anthropology* **35**: 369–399.

Byrne, R.W. 1995. Relating brain size to intelligence in primates. In: *Modelling the Early Human Mind* (eds P. Mellars & K. Gibson). Cambridge: McDonald Institute for Archaeological Research (in press).

Byrne, R.W. & Whiten, A. 1988. *Machiavellian Intelligence: social expertise and the evolution of intellect in monkeys, apes and humans*. Oxford: Oxford University Press.

Cabrera Valdés, V. 1988. Aspects of the Middle Palaeolithic in Cantabrian Spain. In: *L'Homme de Néandertal*, Vol. 4: *La Technique* (ed. M. Otte). Liège: Etudes et Recherches Archéologiques de l'Université de Liège, pp. 27–37.

Cabrera Valdés, V. (ed.) 1993. *El Origen del Hombre Moderno en el Suroeste de Europa*. Madrid: Universidad Nacional de Educacion a Distancia.

Cabrera Valdés, V. & Bernaldo de Quiros, F. 1990. Données sur la transition entre le Paléolithique moyen et Paléolithique supérieur dans la région cantabrique: révision critique. In: *Palé-*olithique Moyen Récent et Paléolithique Supérieur Ancien en Europe* (ed. C. Farizy). Nemours: Mémoires du Musée de Préhistoire d'Ile de France No. 3, pp. 185–188.

Cabrera Valdés, V. & Bernaldo de Quiros, F. 1992. Approaches to the Middle Paleolithic in northern Spain. In: *The Middle Paleolithic: adaptation, behavior, and variability* (eds H.L. Dibble & P.A. Mellars). Philadelphia: University of Pennsylvania University Museum Monographs No. 72, pp. 97–112.

Cabrera Valdés, V. & Bischoff, J.L. 1989. Accelerator ^{14}C dates for early Upper Palaeolithic (basal Aurignacian) at El Castillo Cave (Spain). *Journal of Archaeological Science* **16**, 577–584.

Cahen, D. 1984. Paléolithique inférieur et moyen en Belgique. In: *Peuples Chasseurs de la Belgique dans leur cadre Naturel* (eds D. Cahen & P. Haesaerts). Brussels: Institut Royal des Sciences Naturelles de Belgique, pp. 133–155.

Callow, P. 1986a. The flint tools. In: *La Cotte de St. Brelade 1961–1978: excavations by C.B.M. McBurney* (eds P. Callow & J.M. Cornford). Norwich: Geo Books, pp. 251–314.

Callow, P. 1986b. Raw materials and sources. In: *La Cotte de St. Brelade 1961–1978: excavations by C.B.M. McBurney* (eds P. Callow & J.M. Cornford). Norwich: Geo Books, pp. 203–211.

Callow, P. & Cornford, J.M. (eds) 1986. *La Cotte de St. Brelade 1961–1978: excavations by C.B.M. McBurney*. Norwich: Geo Books.

Callow, P. & Webb, R.E. 1977. Structure in the S.W. French Mousterian. *Computer Applications in Archaeology* **4**: 69–76.

Callow, P. & Webb, R.E. 1981. The application of multivariate statistical techniques to Middle Palaeolithic assemblages from southwestern France. *Revue d'Archéometrie* **5**: 129–138.

Cann, R.L., Stoneking, M.& Wilson, A.C. 1987. Mitochondrial DNA and human evolution. *Nature* **325**, 31–36.

Cann, R.L., Rickards, O. & Lum, J.K. 1994. Mitochondrial DNA and human evolution: our one lucky mother. In: *Origins of Anatomically Modern Humans* (eds M.H. Nitecki & D.V. Nitecki). New York: Plenum Press, pp. 135–148.

Cauvin, M.-C. 1971. L'industrie moustérienne du niveau supérieur de Hauteroche. *Mémoires de la*

Société Archéologique et Historique de la Charente, 1971: 179–188.

Cavalli-Sforza, L.L. 1991. Genes, peoples and languages. *Scientific American* **265**(11): 72–78.

Chadelle, J.-P. 1983. *Technologie et utilisation du silex au Périgordien supérieur: l'exemple de la couche VII du Flageolet I*. Dissertation, Ecole des Hautes Etudes en Sciences Sociales, Toulouse.

Chadelle, J.-P. 1989. Les gisements paléolithiques de Champ-parel à Bergerac, Dordogne, France: rapport préliminaire des opérations de sauvetage, 1985–1989. *Paléo* **1**: 125–133.

Champagne, F. & Espitalié, R. 1981. *Le Piage, site Préhistorique du Lot*. Paris: Mémoires de la Société Préhistorique Française, no. 15.

Chappell, J. & Shackleton, N.J. 1986. Oxygen isotopes and sea level. *Nature* **324**: 137–140.

Chase, P.G. 1986a. *The Hunters of Combe Grenal: approaches to Middle Paleolithic subsistence in Europe*. Oxford: British Archaeological Reports International Series 286.

Chase, P.G. 1986b. Relationships between Mousterian lithic and faunal assemblages at Combe Grenal. *Current Anthropology* **27**: 69–71.

Chase, P.G. 1987a. Spécialisation de la chasse et transition vers le Paléolithique supérieur. *L'Anthropologie* **91**: 175–188.

Chase, P.G. 1987b. The cult of the cave bear: prehistoric rite or scientific myth? *Expedition* **29**(2): 4–9.

Chase, P.G. 1988. Scavenging and hunting in the Middle Palaeolithic: the evidence from Europe. In: *Upper Pleistocene Prehistory of Western Eurasia* (eds H. Dibble & A. Montet-White). Philadelphia: University of Pennsylvania, University Museum Monographs No. 54, pp. 225–232.

Chase, P.G. 1989. How different was Middle Palaeolithic subsistence? A zooarchaeological perspective on the Middle to Upper Palaeolithic transition. In: *The Human Revolution: behavioural and biological perspectives on the origins of modern humans* (eds P. Mellars & C. Stringer). Princeton: Princeton University Press, pp. 321–337.

Chase, P.G. 1990. Sifflets du Paléolithique moyen(?). Les implications d'un coprolite de coyote actuel. *Bulletin de la Société Préhistorique Française* **87**: 165–167.

Chase, P.G. 1991. Symbols and Paleolithic artifacts: style, standardization, and the imposition of arbitrary form. *Journal of Anthropological Archeology* **10**: 193–214.

Chase, P.G. & Dibble, H.L. 1987. Middle Paleolithic symbolism: a review of current evidence and interpretations. *Journal of Anthropological Archeology* **6**: 263–296.

Chase, P.G. & Dibble, H.L. 1992. Scientific archaeology and the origins of symbolism: a reply to Bednarik. *Cambridge Archaeological Journal* **2**: 43–51.

Cheney, D.L. & Seyfarth, R.M. 1990. *How Monkeys see the World: inside the mind of another species*. Chicago: University of Chicago Press.

Cheynier, A. 1953. Stratigraphie de l'abri Lachaud et les cultures des bords abattus. *Archivo de Prehistoria Levantina* **4**: 25–55.

Chomsky, N. 1986. *Knowledge of Language: its nature, origin, and use*. New York: Praeger.

Churchill, S.E. 1993. Weapon technology, prey size selection, and hunting methods in modern hunter-gatherers: implications for hunting in the Palaeolithic and Mesolithic. In: *Hunting and Animal Exploitation in the Later Palaeolithic and Mesolithic of Eurasia* (eds G.L. Peterkin, H.M. Bricker & P. Mellars). Archeological Papers of the American Anthropological Association No. 4, pp. 11–24.

Clark, G.A. 1992. Continuity or replacement? Putting modern human origins in an evolutionary context. In: *The Middle Palaeolithic: Adaptation, Behavior and Variability* (eds H.L. Dibble & P. Mellars), pp. 183–205. Philadelphia: University of Pennsylvania, University Museum Monographs No. 72.

Clark, G.A. & Lindly, J.M. 1989. The case for continuity: observations on the biocultural transition in Europe and Western Asia. In: *The Human Revolution: behavioural and biological perspectives on the origins of modern humans* (eds P. Mellars & C. Stringer). Princeton: Princeton University Press, pp. 626–676.

Clark, J.D. 1981. 'New men, strange faces, other minds'. An archaeologist's perspective on recent discoveries relating to the origins and spread of modern man. *Proceedings of the British Academy* **67**: 163–192.

Clark, J.D. 1992. African and Asian perspectives on the origins of modern humans. In: *The Origin of Modern Humans and the Impact of Chronometric Dating* (eds M.J. Aitken, C.B. Stringer & P.A. Mellars). London: Royal Society (Philosophical Transactions of the Royal Society, series B, **337**, no. 1280), pp. 201–216.

Clarke, D.L. 1968. *Analytical Archaeology*. London: Methuen.

CLIMAP 1981. *Seasonal Reconstructions of the Earth's Surface at the Last Glacial Maximum*. New York: the Geological Society of America, Map and Chart series MC-36.

Close, A.E. 1977. *The Identification of Style in Lithic Artefacts from North East Africa*. Cairo: Mémoires de l'Institut d'Egypte.

Close, A.E. 1978. The identification of style in lithic artefacts. *World Archaeology* **10**: 223–237.

Collins, D.M. 1965. Seriation of quantitative features in late Pleistocene stone technology. *Nature* **205**: 931–932.

Combier, J. 1967. *Le Paléolithique de l'Ardèche dans son Cadre Paléoclimatique*. Bordeaux: Delmas (Publications of Institut de Préhistoire, University of Bordeaux No. 4).

Combier, J. 1984. Les habitats de plein air. *Les Dossiers Histoire et Archéologie* 87: 34–40.

Combier, J. 1988. L'organisation de l'espace habité des hommes du Paléolithique Supérieur en France. *Espacio, Tiempo y Forma, Series 1, Prehistoria* **1**: 111–124.

Commont, V. 1909. L'industrie moustérienne dans la région du nord de la France. *Congrès Préhistorique de France,* 5th session (Beauvais), pp. 115–157.

Commont, V. 1913. *Les Hommes contemporains du renne dans la vallée de la Somme*. Mémoires de la Société des Antiquaires de Picardie, 4th series, no. 7, pp. 207–646.

Conard, N.J. 1990. Laminar lithic assemblages from the last interglacial complex in northwestern Europe. *Journal of Anthropological Research* **46**: 243–262.

Conard, N.J. 1992. *Tönchesberg and its Position in the Paleolithic Prehistory of Northern Europe*. Bonn: Rudolf Habelt.

Connolly, C.J. 1950. *External Morphology of the Primate Brain*. Springfield (Illinois): C.C. Thomas.

Cook, J. 1986. A blade industry from Stoneham's Pit, Crayford. In: *The Palaeolithic of Britain and its Nearest Neighbours: recent trends*. (ed. S.N. Collcutt). Sheffield: Department of Prehistory & Archaeology, University of Sheffield, pp. 16–19.

Coon, C.S. 1962. *The Origin of Races*. New York: Knopf.

Copeland, L. 1976. Terminological correlations in the early Upper Palaeolithic of Lebanon and Palestine. In: *Deuxième Colloque sur la Terminologie de la Préhistoire du Proche Orient* (ed. F. Wendorf), pp. 35–48. Nice: International Union of Prehistoric and Proto-historic Sciences.

Copeland, L. 1983. Levallois/non-Levallois determinations in the Early Levant Mousterian: problems and questions for 1983. *Paléorient* **14**: 95–105.

Cornford, J.M. 1986. Specialized resharpening techniques and evidence of handedness. In: *La Cotte de St. Brelade 1961–1978: excavations by C.B.M. McBurney* (eds P. Callow & J. M. Cornford). Norwich: Geo Books, pp. 337–351.

Coulson, S.D. 1986. The Bout Coupé handaxe as a typological mistake. In: *The Palaeolithic of Britain and its Nearest Neighbours: recent trends* (ed. S.N. Collcutt). Sheffield: Department of Archaeology and Prehistory, University of Sheffield, pp. 53–54.

Crew, H. 1976. The Mousterian site of Rosh ein Mor. In: *Prehistory and Paleoenvironments in the Central Negev*, Vol. I (ed. A.E. Marks): Dallas: Southern Methodist University Press, pp. 75–112.

Damas, D. 1969. *Contributions to Anthropology: ecological essays*. Ottawa: National Museums of Canada, Bulletin 230.

Dansgaard, W., Johnsen, S.J., Clausen, H.B., Dahl-Jensen, D., Gundestrup, N.S., Hammer, C.U., Hvidberg, C.S., Steffensen, J.P., Sveinbjörnsdottir, A.E., Jouzel, J. & Bond, G. 1993. Evidence for general instability of past climate from a 250-kyr ice-core record. *Nature* **364**: 218–220.

Darpeix, A. 1936. Nouvelles fouilles à Tabaterie (Dordogne): gisement Sandougne (Sendonnie). *Bulletin de la Société Préhistorique Française* **33**: 417–441.

Daugas, J.-P. & Raynal, J.-P. 1989. Quelques étapes du peuplement du Massif Central français dans leur contexte paléoclimatique et paléogéographique. In: *Variations de Paléomilieux et Peuplement Préhistorique* (ed. H. Laville). Paris: Centre National de la Recherche Scientifique (University of Bordeaux, Cahiers du Qaternaire 13), pp. 67–95.

David, F. & Farizy, C. 1994. Les vestiges osseux: étude archéozoologique. In: *Hommes et Bisons du Paléolithique Moyen à Mauran (Haute-Garonne)* (eds C. Farizy, F. David & J. Jaubert). Paris: Centre National de la Recherche Scientifique (Gallia-Préhistoire Supplement 30), pp. 177–234.

David, N.C. 1973. On Upper Palaeolithic society, ecology and technological change: the Noaillian case. In: *The Explanation of Culture Change* (ed. A.C. Renfrew). London: Duckworth, pp. 277–303.

David, N.C. 1985. *Excavation of the Abri Pataud, Les Eyzies (Dordogne): the Noaillian (level 4) assemblages and the Noaillian culture in Western Europe.* Cambridge (Mass.): Peabody Museum, Harvard University. (American School of Prehistoric Research Bulletin 37.)

Davidson, I. & Noble, W. 1989. The archaeology of perception: traces of depiction and language. *Current Anthropology* **30**: 125–155.

Davidson, I. & Noble, W. 1993. Tools and language in human evolution. In: *Tools, Language and Cognition in Human Evolution* (eds K.R. Gibson & T. Ingold). Cambridge: Cambridge University Press, pp. 363–387.

Debénath, A. 1968. Le Moustérien type 'Quina' de la Vauzelle (Charente-Maritime). *Bulletin de la Société Préhistorique Française* **65**: 259–268.

Debénath, A. 1971. Note préliminaire sur la stratigraphie de la grotte Marcel Clouet à Cognac (Charente). *Bulletin de la Société Préhistorique Française* **68**: 133–135.

Debénath, A. 1973. Un foyer aménagé dans le Moustérien de Hauteroche à Chateauneuf-sur-Charente (Charente). *L'Anthropologie* **77**: 329–338.

Debénath, A. 1974. *Recherches sur les Terrains Quaternaires Charentais et les Industries qui leur sont Associées.* Doctoral dissertation, University of Bordeaux I.

Debénath, A. 1988. Recent thoughts on the Riss and early Würm lithic assemblages of La Chaise de Vouthon (Charente, France). In: *Upper Pleistocene Prehistory of Western Eurasia* (eds H. Dibble & A. Montet-White). Philadelphia: University of Pennsylvania, University Museum Monographs No. 54, pp. 85–94.

Debénath, A. & Duport, L. 1971. Os travaillés et os utilisés de quelques gisements préhistoriques charentais (Paléolithique ancien et moyen). *Mémoires de la Société Archéologique et Historique de la Charente* 1971: 189–202.

Debénath, A. & Duport, L. 1987. La Grotte de Montgaudier, commune de Montbron (Charente): Moustérien de Montgaudier (Charente). *Bulletins et Mémoires de la Société Archéologique et Historique de la Charente*, 1987: 93–104.

Delagnes, A. 1990. Analyse technologique de la méthode de débitage del'abri Suard (La-Chaise-de-Vouthon, Charente). *Paléo* **2**: 81–88.

Delagnes, A. 1992. *L'Organisation de la Production Lithique au Paléolithique Moyen: approche technologique à partir de l'étude des industries de la Chaise-de-Vouthon (Charente).* Doctoral dissertation: University of Paris X.

Delibrias, G. & Fontugne, M. 1990. Datations des gisements de l'Aurignacien et du Moustérien en France. In: *Paléolithique Moyen Récent et Paléolithique Supérieur Ancien en Europe* (ed. C. Farizy). Nemours: Mémoires du Musée de Préhistoire d'Ile de France No. 3, pp. 39–42.

Delluc, B., & Delluc, C, 1978. Les manifestations graphiques aurignaciennes sur support rocheux des environs des Eyzies (Dordogne). *Gallia Préhistoire* **21**, 213–438.

Delpech, F. 1983. *Les Faunes du Paléolithique Supérieur dans le Sud-ouest de la France.* Paris: Centre National de la Recherche Scientifique.

Delpech, F. 1988. Les grands mammifères de la grotte Vaufrey à l'exception des ursidés. In: *La Grotte Vaufrey: paléoenvironnement, chronologie, activités humaines* (ed. J.-P. Rigaud). Mémoires de la Société Préhistorique Française 19, pp. 213–290.

Delpech, F. 1990. Le milieu animal des Moustériens Quina du Périgord. In: *Les Moustériens Charentiens* (Proceedings of International Colloquium, Brive, August 1990), pp. 27–30.

Delpech, F. 1993. The fauna of the early Upper Paleolithic: biostratigraphy of large mammals and current problems of chronology. In: *Before Lacaux: the complex record of the early Upper Paleolithic* (eds H. Knecht, A. Pike-Tay & R. White). Boca Raton: CRC Press, pp. 71–84.

Delpech, F. (In press) La grotte du Regourdou (Montignac, Dordogne): les artiodactyles.

Delpech, F. & Prat, F. 1980. Les grands mammifères pléistocènes du sud-ouest de la France. In: *Problèmes de Stratigraphie Quaternaire en France et dans les Pays Limitrophes*. Supplement to *Bulletin de l'Association Française pour l'Etude du Quaternaire* No. 1, pp. 268–297.

Delpech, F., Donard, E., Gilbert, A., Guadelli, J.-L., Le Gall, O., Martini-Jacquin, A., Paquereau, M.-M., Prat, F. & Tournepiche, J.-F. 1983. Contribution à la lecture des paléoclimats Quaternaires d'après les données de la paléontologie en milieu continental. *Bulletin de l'Institut Géologique du Bassin d'Aquitaine*, Bordeaux **34**: 165–177.

Delporte, H. 1962. Le gisement Paléolithique de la Rochette. *Gallia-Préhistoire* **4**: 1–22.

Delporte, H. 1976. Les civilisations du Paléolithique Moyen en Auvergne. In: *La Préhistoire Française* (ed. H. de Lumley). Paris: Centre National de la Recherche Scientifique, pp. 1085–1088.

Delporte, H. 1984. *Le Grand Abri de la Ferrassie: fouilles 1968–1973*. Paris: Institut de Paléontologie Humaine (Université de Provence, Etudes Quaternaires, Mémoire no. 7).

Delporte, H. & David, R. 1966. L'évolution des industries moustériennes à la Rochette, commune de Saint-Léon-sur-Vézere (Dordogne). *Bulletin de la Société Préhistorique Française* **63**: 48–62.

Demars, P.-Y. 1982. *L'Utilisation du Silex au Paléolithique Supérieur: choix, approvisionnement, circulation: l'exemple du Bassin de Brive*. Paris: Centre National de la Recherche Scientifique. (Institut du Quaternaire, University of Bordeaux I, Cahiers du Quaternaire No. 5).

Demars, P.-Y. 1990a. Les interstratifications entre Aurignacien et Châtelperronien à Roc de Combe et au Piage (Lot): approvisionnement en matières premières et position chronologique.

In: *Paléolithique Moyen Récent et Paléolithique Supérieur Ancien en Europe* (ed. C. Farizy). Nemours: Mémoires du Musée de Préhistoire d'Ile de France No. 3, pp. 235–240.

Demars, P.-Y. 1990b. Les matières premières. In: *La Chapelle-aux-Saints et La Préhistoire en Corrèze* (eds J.-P. Raynal & Y. Pautrat). Bordeaux: La Nef (L'Association pour la Recherche Archéologique en Limousin) pp. 23–28.

Demars, P.-Y. 1992. Les colorants dans le Moustérien du Périgord: l'apport des fouilles de F. Bordes. *Préhistoire Ariégeoise* **47**: 185–194.

Demars, P.-Y. & Hublin, J.-J. 1989. La transition Néandertaliens/Hommes de type moderne en Europe occidentale: aspects paléontologiques etculturels. In: *L'Homme de Néandertal*, Vol. 7: *L'extinction* (ed. M. Otte). Liège: Etudes et Recherches Archéologiques de l'Université de Liège, pp. 23–27.

Dibble, H.L. 1983. Variability and change in the Middle Paleolithic of western Europe and the Near East. In: *The Mousterian Legacy* (ed. E. Trinkaus). Oxford: British Archaeological Reports International Series S164, pp. 53–71.

Dibble, H.L. 1984a. Interpreting typological variation of Middle Paleolithic scrapers: function, style, or sequence of reduction? *Journal of Field Archeology* **11**: 431–436.

Dibble, H.L. 1984b. The Mousterian industry from Bisitun Cave (Iran). *Paléorient* **10**: 323–334.

Dibble, H.L. 1985. Raw material variation in Levallois flake manufacture. *Current Anthropology* **26**: 391–393.

Dibble, H.L. 1987a. The interpretation of Middle Paleolithic scraper morphology. *American Antiquity* **52**: 109–117.

Dibble, H.L. 1987b. Reduction sequences in the manufacture of Mousterian implements of France. In: *The Pleistocene of the Old World: regional perspectives* (ed. O. Soffer). New York: Plenum, pp. 33–45.

Dibble, H.L. 1987c. Comparaisons des séquences de réduction des outils moustériens de la France et du Proche-Orient. *L'Anthropologie* **19**: 189–196.

Dibble, H.L. 1988a. Typological aspects of reduction and intensity of utilization of lithic resources in the French Mousterian. In: *Upper*

Pleistocene Prehistory of Western Eurasia (eds H. Dibble & A. Montet-White). Philadelphia: University of Pennsylvania, University Museum Monographs No. 54, pp. 181–198.

Dibble, H.L. 1988b. The interpretation of Middle Paleolithic scraper reduction patterns. In: *L'Homme de Néandertal*, Vol. 4: *La Technique* (ed. M. Otte). Liège: Etudes et Recherches Archéologiques de l'Université de Liège, pp. 49–58.

Dibble, H.L. 1989. The implications of stone tool types for the presence of language during the Lower and Middle Palaeolithic. In: *The Human Revolution: behavioural and biological perspectives on the origins of modern humans* (eds P. Mellars & C. Stringer). Princeton: Princeton University Press, pp. 415–432.

Dibble, H.L. 1991a. Local raw material exploitation and its effects on Lower and Middle Paleolithic assemblage variability. In: *Raw Material Economies Among Prehistoric Hunter-Gatherers* (eds A. Montet-White & S. Holen). Lawrence: University of Kansas Publications in Anthropology No. 19, pp. 33–46.

Dibble, H.L. 1991b. Mousterian assemblage variability on an interregional scale. *Journal of Anthropological Research* 47: 239–257.

Dibble, H.L. 1993. Le Paléolithique moyen récent du Zagros. *Bulletin de la Société Préhistorique Française* 90: 307–312.

Dibble, H.L. & Chase, P.G. 1990. Comment on J.M. Lindly and G.A. Clark 'Symbolism and modern human origins.' *Current Anthropology* 31: 241–243.

Dibble, H.L. & Holdaway, S.J. 1993. The Middle Paleolithic industries of Warwasi. In: *The Paleolithic Prehistory of the Zagros-Taurus* (eds D.I. Olszewski & H.L. Dibble): Philadelphia: University of Pennsylvania, University Museum Monographs No. 83, pp. 75–99.

Dibble, H.L. & Mellars, P.A. (eds.) 1992. *The Middle Paleolithic: Adaptation, Behavior and Variability*. Philadelphia: University of Pennsylvania, University Museum Monographs No. 72.

Dibble, H. & Montet-White, A. (eds) 1988. *Upper Pleistocene Prehistory of Western Eurasia*. Philadelphia: University of Pennsylvania, University Museum Monographs No. 54.

Dibble, H.L. & Rolland, N. 1992. On assemblage variability in the Middle Palaeolithic of Western Europe: history, perspectives, and a new synthesis. In: *The Middle Paleolithic: adaptation, behavior, and variability* (eds H.L. Dibble & P.A. Mellars). Philadelphia: University of Pennsylvania University Museum Monographs No. 72, pp. 1–28.

Djindjian, F. 1993. Les origines du peuplement Aurignacien en Europe. In: *Aurignacien en Europe et au Proche Orient* (eds L. Bánesz & J.K. Kozlowski). Bratislava: Acts of 12th International Congress of Prehistoric and Protohistoric Sciences, pp. 136–154.

Donald, M. 1991. *Origins of the Modern Mind: three stages in the evolution of culture and cognition*. Cambridge (Mass.): Harvard University Press.

Doran, J.E. & Hodson, F.R. 1966. A digital computer analysis of Palaeolithic flint assemblages. *Nature* 210: 688–689.

Doran, J.E. & Hodson, F.R. 1975. *Mathematics and Computers in Archaeology*. Edinburgh: Edinburgh University Press.

Drury, W.H. 1975. The ecology of the human occupation at the Abri Pataud. In: *Excavation of the Abri Pataud, Les Eyzies (Dordogne)* (ed. H.L. Movius). Cambridge (Mass.): Peabody Museum, Harvard University, pp. 187–196.

DuBrul, E.L. 1977. Origins of the speech apparatus and its reconstruction in fossils. *Brain and Language* 4: 365–381.

Duchadeau-Kervazo, C. 1982. *Recherches sur l'Occupation Paléolithique du Bassin de la Dronne*. Doctoral dissertation, University of Bordeaux I.

Duchadeau-Kervazo, C. 1984. Influence du substratum sur l'occupation paléolithique du bassin de la Dronne. *Bulletin de la Société Linnéenne de Bordeaux* 12: 35–50.

Duchadeau-Kervazo, C. 1986. Les sites paléolithiques du bassin de la Dronne (nord de l'Aquitaine): observations sur les modes et emplacements. *Bulletin de la Société Préhistorique Française* 83: 56–64.

Duchadeau-Kervazo, C. 1989. Role de l'architecture géologique du bassin de la Dronne (Nord Aquitaine) dans l'implantation humaine paléolithique. In: *Variations des Paléomilieux et Peuplement Préhistorique* (ed. H. Laville). Paris: Centre National de la Recherche Scientifique, pp. 161–168.

Dumond, D.E. 1977. *The Eskimos and Aleuts*. London: Thames & Hudson.

Dunbar, R.I.M. 1987. Demography and reproduction. In: *Primate Societies* (eds B.B. Smuts, D.L. Cheney, R.M. Seyfarth, R.W., Wrangham & T.T. Struhsaker). Chicago: University of Chicago Press, pp. 240–249.

Dunbar, R.I.M. 1988. *Primate Social Systems*. London: Croom Helm.

Dunbar, R.I.M. 1993 Co-evolution of neocortex size, group size, and language in humans. *Behavior and Brain Sciences* **16**: 681–735.

Falk, D. 1975. Comparative analysis of the larynx in man and the chimpanzee: implications for language in Neanderthals. *American Journal of Physical Anthropology* **43**: 123–132.

Farizy, C. 1985. Un habitat du paléolithique moyen à Champlost (Yonne, Nord-Bourgogne, France) *Archäologisches Korrespondenzblatt* **15**: 405–410.

Farizy, C. (ed.) 1990. *Paléolithique Moyen Récent et Paléolithique Supérieur Ancien en Europe*. Nemours: Mémoires du Musée de Préhistoire d'Ile de France No. 3.

Farizy, C. 1990a. The transition from Middle to Upper Palaeolithic at Arcy-sur-Cure (Yonne, France): technological, economic and social aspects. In: *The Emergence of Modern Humans: an archaeological perspective* (ed. P. Mellars). Edinburgh: Edinburgh University Press, pp. 303–326.

Farizy, C. 1990b. Du Moustérien au Châtelperronien à Arcy-sur-Cure: un état de la question. In: *Paléolithique Moyen Récent et Paléolithique Supérieur Ancien en Europe* (ed. C. Farizy). Nemours: Mémoires du Musée de Préhistoire d'Ile de France No. 3, pp. 281–290.

Farizy, C. 1994. Behavioural and cultural changes at the Middle to Upper Paleolithic transition in western Europe. In: *Origins of Anatomically Modern Humans* (eds M.H. Nitecki & D.V. Nitecki). New York: Plenum Press, pp. 93–100.

Farizy, C. & David, F. 1992 Subsistence and behavioral patterns of some Middle Palaeolithic local groups. In: *The Middle Paleolithic: adaptation, behavior, and variability* (eds H.L. Dibble & P.A. Mellars). Philadelphia: University of Pennsylvania University Museum Monographs No. 72, pp. 87–96.

Farizy, C. & Leclerc, J. 1981. Les grands chasses de Mauran. *La Recherche* **12**: 1294–1295.

Farizy, C. & Tuffreau, A. 1986. Industries et cultures du Paléolithique moyen récent dans la moitié Nord de la France. In: *Chronostratigraphie et Faciès Culturels du Paléolithique Inférieur et Moyen dans l'Europe du Nord-Ouest* (eds A. Tuffreau & J. Sommé). Paris: Association Française pour l'Etude du Quaternaire, pp. 225–234.

Farizy, C., David, J. & Jaubert, J. 1994. *Hommes et Bisons du Paléolithique Moyen à Mauran (Haute-Garonne)*. Paris: Centre National de la Recherche Scientifique (Gallia-Préhistoire Supplement 30).

Farrand, W.R. 1975. Analysis of the Abri Pataud sediments. In: *Excavation of the Abri Pataud, Les Eyzies (Dordogne)* (ed. H.L. Movius). Cambridge (Mass.): Peabody Museum of Archaeology and Ethnology, Harvard University, pp. 27–68.

Farrand, W.R. 1988. Integration of late Quaternary climatic records from France and Greece: cave sediments, pollen, and marine events. In: *Upper Pleistocene Prehistory of Western Eurasia* (eds H. Dibble & A. Montet-White). Philadelphia: University of Pennsylvania, University Monographs No. 54, pp. 305–320.

Féblot-Augustins, J.1993. Mobility strategies in the late Middle Palaeolithic of Central Europe and Western Europe: elements of stability and variability. *Journal of Anthropological Archaeology* **12**: 211–265.

Fish, P.R. 1981. Beyond tools: Middle Paleolithic debitage analysis and cultural inference. *Journal of Anthropological Research* **37**: 374–386.

Flint, R.F. 1957. *Glacial and Pleistocene Geology*. New York: Wiley.

Foley, R.A. 1989. The evolution of hominid social behaviour. In: *Comparative Socioecology* (eds R.V. Standen & R. Foley). Oxford: Blackwell, pp. 473–494.

Foley, R. 1995. Measuring cognition in extinct hominids. In: *Modelling the Early Human Mind* (eds P. Mellars & K. Gibson). Cambridge: McDonald Institute for Archaeological Research (in press).

Foley, R.A. & Lee, P.C. 1989. Finite social space, evolutionary pathways, and reconstructing hominid behavior. *Science* **243**: 901–906.

Follieri, M., Magri, D. & Sadori, L. 1989. Pollen stratigraphical synthesis from Valle di Castiglione (Roma). *Quaternary International* **3/4**: 81–84.

Fosse, G. 1989. Quelques réflexions sur l'apport du gisement moustérien de Saint-Vaast-la-Hougue (France) à l'ethnologie du Paléolithique moyen. In: *L'Homme de Néandertal*, Vol. 6: *La Subsistence* (ed. M. Otte). Liège: Etudes et Recherches Archéologiques de l'Université de Liège, pp. 63–68.

Fosse, G., Cliquet, D. & Vilgrain, G. 1986. Le Moustérien du Nord-Cotentin (département de la Manche): premiers résultats de trois fouilles en cours. In: *Chronostratigraphie et Faciès Culturels du Paléolithique Inférieur et Moyen dans l'Europe du Nord-Ouest* (eds A. Tuffreau & J. Sommé). Paris: Association Française pour l'Etude du Quaternaire, pp. 141–156.

Foster, M. L. 1990. Symbolic origins and transitions in the Palaeolithic. In: *The Emergence of Modern Humans: an archaeological perspective* (ed. P. Mellars). Edinburgh: Edinburgh University Press, pp. 517–539.

Freeman, L.G. 1986. Middle Palaeolithic dwelling remnants from Spain. In: *Les Structures d'Habitat au Paléolithique Moyen* (ed. L.G. Freeman). Nice: International Union of Prehistoric and Protohistoric Sciences (Colloquium XI), pp. 35–48.

Freeman, L.G. 1988. A Mousterian structural remnant from Cueva Morín (Cantabria, Spain). In: *L'Homme de Néandertal*, Vol. 6: *La Subsistence* (ed. M. Otte). Liège: Etudes et Recherches Archéologiques de l'Université de Liège, pp. 19–30.

Freeman, L.G. 1992. Mousterian facies in Space: New data from Morín level 16. In: *The Middle Paleolithic: adaptation, behavior, variability* (eds H. Dibble & P. Mellars). Philadelphia: University of Pennsylvania University Museum Monographs No. 72, pp. 113–126.

Freeman, L.G. & Echegaray, J.G. 1970. Aurignacian strutural features and burials at Cueva Morín (Santander, Spain). *Nature* **226**: 722–726.

Gambier, D. 1989. Fossil hominids from the early Upper Palaeolithic (Aurignacian) of France. In: *The Human Revolution: behavioural and biological perspectives on the origins of modern humans* (eds P. Mellars & C. Stringer). Princeton: Princeton University Press, pp. 194–211.

Gambier, D. 1993. Les hommes modernes du début du Paléolithique supérieur en France: bilan des données anthropologiques et perspectives. In: *El Origen del Hombre Moderno en el Suroeste de Europa* (ed. V. Cabrera Valdés). Madrid: Universidad Nacional de Educacion a Distancia, pp. 409–430.

Gamble, C. 1982. Interaction and alliance in palaeolithic society. *Man* **17**: 92–107.

Gamble, C. 1983. Culture and society in the Upper Palaeolithic of Europe. In: *Hunter-Gatherer Economy in Prehistory: a European Perspective* (ed. G.N. Bailey). Cambridge: Cambridge University Press, pp. 201–211.

Gamble, C. 1984. Regional variation in hunter-gatherer strategy in the upper Pleistocene of Europe. In: *Hominid Evolution and Community Ecology* (ed. R. Foley). London: Academic Press, pp. 237–260.

Gamble, C. 1986. *The Palaeolithic Settlement of Europe*. Cambridge: Cambridge University Press.

Gamble, C. 1987. Man the shoveler: alternative models for middle Pleistocene colonization and occupation in northern latitudes. In: *The Pleistocene of the Old World; regional perspectives* (ed. O. Soffer). New York: Plenum, pp. 81–98.

Gamble, C. 1993. *Timewalkers: the prehistory of global colonization*. Stroud: Alan Sutton.

Gargett, R.H. 1989. Grave shortcomings: the evidence for neandertal burial. *Current Anthropology* **30**: 157–190.

Garrod, D.A.E. 1956. Acheuléo-Jabrudien et 'Pré-Aurignacien' de la Grotte de Taboun (Mont Carmel): étude stratigraphique et chronologique. *Quaternaria* **3**: 39–59.

Gaussen, J. 1980. *Le Paléolithique Supérieur de Plein Air en Périgord (industries et structures d'habitat): secteur Mussidan-Saint-Astier-moyenne vallée de l'Isle*. Paris: Centre National de la Recherche Scientifique. (Gallia Préhistoire, Supplement XIV).

Gellner, E. 1989. Culture, constraint and community: semantic and coercive compensations for the genetic under-determination of *Homo sapiens sapiens*. In: *The Human Revolution: behav-*

ioural and biological perspectives on the origins of modern humans (eds P. Mellars & C. Stringer). Princeton: Princeton University Press, pp. 514–525.

Geneste, J.-M. 1985. *Analyse Lithique d'Industries Moustériennes du Périgord: une approche technologique du comportement des groupes humains au Paléolithique moyen.* Doctoral dissertation, University of Bordeaux I.

Geneste, J.-M. 1988. Les industries de la Grotte Vaufrey: technologie du débitage, économie et circulation de la matière première lithique. In: *La Grotte Vaufrey: paléoenvironnement, chronologie, activités humaines* (ed. J.-P. Rigaud). Paris: Mémoires de la Société Préhistorique Française 19, pp. 441–517.

Geneste, J.-M. 1989a. Economie des ressources lithiques dans le Moustérien du sud-ouest de la France. In: *L'Homme de Néandertal*, Vol. 6: *La Subsistence* (ed. M. Otte). Liège: Etudes et Recherches Archéologiques de l'Université de Liège, pp. 75–97.

Geneste, J.-M. 1989b. Systèmes d'approvisionnement en matières premières au paléolithique moyen et au paléolithique supérieur en Aquitaine. In: *L'Homme de Néandertal*, Vol. 8: *La Mutation* (ed. M. Otte). Liège: Etudes et Recherches Archéologiques de l'Université de Liège, pp. 61–70

Geneste, J.-M. 1990. Développement des systèmes de production lithique au cours du paléolithique moyen en Aquitaine septentrionale In: *Paléolithique Moyen Récent et Paléolithique Supérieur Ancien en Europe* (ed. C. Farizy). Nemours: Mémoires du Musée de Préhistoire d'Ile de France No. 3, pp. 205–207.

Geneste, J.-M. 1991. L'approvisionnement en matières premières dans les systèmes de production lithique: la dimension spatiale de la technologie. In: *Tecnología y Cadenas Operativas Líticas* (eds R. Mora, X. Terradas, A. Parpal & C. Plana). Barcelona: Universitat Autonoma de Bacelona, pp. 1–36.

Geneste, J.-M. & Rigaud, J.-P. 1989. Matières premières lithiques et occupation de l'espace. In: *Variations des Paléomilieux et Peuplement Préhistorique* (ed. H. Laville). Paris: Centre National de la Recherche Scientifique, pp. 205–218.

Gibson, K.R. 1985. Has the evolution of intelligence stagnated since Neanderthal Man? (Commentary on Parker). In: *Evolution and Developmental Psychology* (eds G. Butterworth, J. Rutkowska & M. Scaife). Harvester Press, pp. 102–114.

Gibson, K.R. 1988. Brain size and the evolution of language. In: *The Genesis of Language: a different judgement* (ed. M. E. Landsberg). Berlin: Mouton de Gruyter, pp. 149–172.

Gibson, K.R. 1990. New perspectives on instincts and intelligence: brain size and the emergence of heirarchical mental constructional skills. In: *'Language' and Intelligence in Monkeys and Apes: comparative developmental perspectives* (eds S.T. Parker & K.R. Gibson). Cambridge: Cambridge University Press, pp. 97–128.

Gibson, K.R. 1991. Myelination and behavioral development: a comparative perspective on questions of neoteny, altriciality and intelligence. In: *Brain Maturation and Cognitive Development: comparative and cross-cultural perspectives* (eds K.R. Gibson & A.C. Peterson). New York: Aldine de Gruyter, pp. 29–63.

Gibson, K.R. 1993. Tool use, language and social behavior in relation to information processing capacities. In: *Tools, Language and Cognition in Human Evolution* (eds K.R. Gibson & T. Ingold). Cambridge: Cambridge University Press, pp. 251–269.

Gibson, K.R. & Ingold, T. 1993. *Tools, Language and Cognition in Human Evolution.* Cambridge: Cambridge University Press.

Girard, C. 1976. L'habitat et le mode de vie au paléolithique moyen à Arcy-sur-Cure (Yonne). In: *Les Structures d'Habitat au Paléolithique Moyen* (ed. L. Freeman). Nice: Ninth Congress of International Union of Prehistoric and Protohistoric Sciences (Colloquium XI), pp. 49–63.

Girard, C. 1978. *Les Industries Moustériennes de la grotte de l'Hyene à Arcy-sur-Cure (Yonne).* Paris: Centre National de la Recherche Scientifique (Gallia-Préhistoire, Supplement 11).

Girard, C. & David, F. 1982. A propos de la chasse spécialisée au Paléolithique moyen: l'exemple de Mauran (Haute-Garonne). *Bulletin de la Société Préhistorique Française* **79**: 10–12.

Girard, C., Hoffert, M. & Miskowsky, J.-C. 1975. Contribution à la connaissance du Paléolithique moyen en Haute-Garonne: le gisement de Mauran. *Bulletin de l'Association Française pour l'Etude du Quaternaire* **1975** (3–4): 171–187.

Goia, P. 1990. La transition Paléolithique Moyen/ Paléolithique supérieur en Italie et la question de l'Uluzzien. In: *Paléolithique Moyen Récent et Paléolithique Supérieur Ancien en Europe* (ed. C. Farizy). Nemours: Mémoires du Musée de Préhistoire d'Ile de France No. 3, pp. 241–250.

Gonzalez Echegaray, J. 1988. Decorative patterns in the Mousterian of Cueva Morín. In: *L'Homme de Néandertal*, Vol. 5: *La Pensée* (ed. M. Otte). Liège: Etudes et Recherches Archéologiques de l'Université de Liège, pp. 37–42.

Gordon, B. 1988. *Of Men and Reindeer Herds in French Magdalenian Prehistory*. Oxford: British Archaeological Reports International Series S 390.

Gowlett, J.A.J. 1984. Mental abilities of early man: a look at some hard evidence. In: *Hominid Evolution and Community Ecology* (ed. R.A. Foley). London: Academic Press, pp. 167–192.

Gowlett, J.A.J. 1995. Mental abilities of early *Homo*: elements of constraint and choice in rule systems. In: *Modelling the Early Human Mind* (eds P. Mellars & K. Gibson). Cambridge: McDonald Institute for Archaeological research (in press).

Graves, P. 1991. New models and metaphors for the Neanderthal debate. *Current Anthropology* **32**, 513–541.

Grayson, D.K. 1978. On the quantification of vertebrate archaeofaunas. In: *Advances in Archaeological Method and Theory* (ed. M.B. Schiffer) Vol. 2, pp. 199–237.

Green, H.S. 1984. *Pontnewydd Cave: a Lower Palaeolithic Hominid Site in Wales: the first report*. Cardiff: National Museum of Wales.

Green, H.S. & Walker, E. 1991. *Ice Age Hunters: Neanderthals and early modern hunters in Wales*. Cardiff: National Museum of Wales.

GRIP members 1993. Climate instability during the last interglacial period recorded in the GRIP ice core. *Nature* **364**: 203–207.

Grootes, P.M., Stuiver, M., White, J.W.C., Johnsen, S. & Jouzel, J. 1993. Comparison of oxygen isotope records from the GISP2 and GRIP Greenland ice cores. *Nature* **366**: 552–554.

Gruet, M. 1976. Les civilisations du Paléolithique

Moyen dans les Pays de la Loire. In: *La Préhistoire Française* (ed. H. de Lumley). Paris: Centre National de la Recherche Scientifique, pp. 1089–1093.

Grün, R., Mellars, P. & Laville, H. 1991. ESR chronology of a 100,000-year archaeological sequence at Pech de l'Azé II, France. *Antiquity* **65**: 544–551.

Guadelli, J.-L. 1987. *Contribution à l'Etude des Zoocénoses Préhistoriques en Aquitaine (Würm ancien et Interstade würmien)*. Doctoral dissertation, University of Bordeux I.

Guadelli, J.-L. 1990. Le milieu animal et les Moustériens 'Charentiens' dans le quart sud-ouest de l'Europe. In: *Les Moustériens Charentiens* (Proceedings of International Colloquium, Brive, August 1990), pp. 31–32.

Guadelli, J.-L. & Laville, H. 1990. L'environnement climatique de la fin du Moustérien à Combe-Grenal et à Camiac: confrontation des données naturalistes et implications. In: *Paléolithique Moyen Récent et Paléolithique Supérieur Ancien en Europe* (ed. C. Farizy). Nemours: Mémoires du Musée de Préhistoire d'Ile de France No. 3, pp. 43–48.

Guichard, J. 1976. Les civilisations du Paléolithique Moyen en Périgord. In: *La Préhistoire Française* (ed. H. de Lumley). Paris: Centre National de la Recherche Scientifique, pp. 1053–1069.

Guillien, Y. & Henri-Martin, G. 1974. Croissance du renne et saison de chasse: le Moustérien à denticulés et le Moustérien de tradition acheuléenne à La Quina. *Inter-Nord* **13/14**: 119–127.

Guiot, J., Pons, A., de Beaulieu, J.-L. & Reille, M. 1989. A 140,000-yr continental climate reconstruction from two European pollen records. *Nature* **338**: 309–313.

Guiot, J., de Beaulieu, J.L., Cheddadi, R., David, F., Ponel, P. & Reille, M. 1993. The climate in Western Europe during the last Glacial/Interglacial cycle derived from pollen and insect remains. *Palaeogeography, Palaeoclimatology, Palaeoecology* **103**: 73–93.

Haddenham, E. 1980. *Secrets of the Ice Age: the world of the cave artists*. London: Heinemann.

Haesaerts, P. 1990. Nouvelles recherches au gisement de Willendorf (Basse Autriche). *Bulletin de l'Institut Royal des Sciences Naturelles de Belgique* (Sciences de la Terre) **60**: 203–218.

Hahn, J. 1972. Aurignacian signs, pendants, and art objects in Central and Eastern Europe. *World Archaeology* **3**: 252–266.

Hahn, J. 1977. *Aurignacien: Das Ältere Jungpaläolithikum in Mittel- und Osteuropa.* Köln: Fundamenta Series A9.

Hahn, J. 1993a. Aurignacian art in Central Europe. In: *Before Lacaux: the complex record of the early Upper Paleolithic* (eds H. Knecht, A. Pike-Tay & R. White). Boca Raton: CRC Press, pp. 229–242.

Hahn, J. 1993b. L'origine du Paléolithique supérieur en Europe Centrale: les datations ^{14}C. In: *El Origen del Hombre Moderno en el Suroeste de Europa* (ed. V. Cabrera Valdés). Madrid: Universidad Nacional de Educacion a Distancia, pp. 61–80.

Harpending, H., Sherry, S., Rogers, A. & Stoneking, M. 1993. The genetic structure of ancient human populations. *Current Anthropology* **34**: 483–496.

Harrold, F.B. 1989. Mousterian, Châtelperronian, and Early Aurignacian in Western Europe: continuity or discontinuity? In: *The Human Revolution: behavioural and biological perspectives on the origins of modern humans* (eds P. Mellars & C. Stringer). Princeton: Princeton University Press, pp. 677–713.

Hasegawa, M., Di Rienzo, A., Kocher, T. & Wilson, A. 1993. Towards a more accurate time scale for the human mitochondrial DNA tree. *Journal of Molecular Evolution* **37**: 347–354.

Hayden, B. 1993. The cultural capacities of Neandertals: a review and re-evaluation. *Journal of Human Evolution* **24**: 113–146.

Heinzelin, J. de & Haesaerts, P. 1983. Un cas de débitage laminaire au Paléolithique ancien: Croix-l'Abbé à Saint-Valéry-sur-Somme. *Gallia Préhistoire* **26**: 189–201.

Henri-Martin, 1907. *Recherches sur l'Evolution du Moustérien dans le Gisement de la Quina (Charente)* Vol. 1. Paris: Schleicher.

Henri-Martin, 1909. *Recherches sur l'Evolution du Moustérien dans le Gisement de la Quina (Charente)* Vol. 2. Paris: Schleicher.

Hewes, G.W. 1993. A history of speculation on the relation between tools and language. In: *Tools,*

Language and Cognition in Human Evolution (eds K.R. Gibson & T. Ingold). Cambridge: Cambridge University Press, pp. 20–31.

Hinde, R.A. (ed.) 1983. *Primate Social Relationships: an integrated approach.* Oxford: Blackwell.

Hodder, I. 1982. *Symbols in Action: ethnoarchaeological studies of material culture.* Cambridge: Cambridge University Press.

Hoffecker, J.F. 1988. Early Upper Paleolithic sites of the European USSR. In: *The Early Upper Paleolithic: Evidence from Europe and the Near East* (ed. J.F. Hoffecker & C.A. Wolf). Oxford: British Archaeological Reports International Series 437, pp. 237–272.

Holdaway, S. 1989. Were there hafted projectile points in the Mousterian? *Journal of Field Archaeology* **16**: 79–85.

Holloway, R. L. 1969. Culture: a human domain. *Current Anthropology* **10**: 395–412.

Holloway, R. L. 1976. Paleoneurological evidence for language origins. *Annals of the New York Academy of Sciences* **280**: 330–348.

Holloway, R. L. 1983. Human brain evolution: a search for units, models and synthesis. *Canadian Journal of Anthropology* **3**: 215–230.

Holloway, R. L. 1985. The poor brain of *Homo sapiens neanderthalensis*: see what you please. In: *Ancestors: the Hard Evidence* (ed. E. Delson). New York: Alan R. Liss.

Howell, F.C. 1952. Pleistocene glacial ecology and the evolution of 'classic Neandertal' man. *Southwest Journal of Anthropology* **8**: 377–410.

Howell, F.C. 1962. Isimila: a Paleolithic site in Africa. *Scientific American* **205**(4): 119–129.

Howell, F.C. 1984. Introduction. In: *The Origins of Modern Humans: a World Survey of the Fossil Evidence* (eds F.H. Smith & F. Spencer). New York: Alan R. Liss, pp. xiii–xxii.

Howell, F.C. 1994. A chronostratigraphic and taxonomic framework for the origins of modern humans. In: *Origins of Anatomically Modern Humans* (eds M.H. Nitecki & D.V. Nitecki). New York: Plenum Press, pp. 253–319.

Hublin, J.-J., 1990. Les peuplements paléolithiques de l'Europe: un point de vue géographique. In: *Paléolithique Moyen Récent et Paléolithique Supérieur Ancien en Europe* (ed. C. Farizy). Nemours:

Mémoires du Musée de Préhistoire d'Ile de France No. 3, pp. 29–37.

Humphrey, N. 1976. The social functions of intellect. In: *Growing Points in Ethology* (eds P.P.G. Bateson & R. A. Hinde). Cambridge: Cambridge University Press, pp. 303–317.

Imbrie, J., McIntyre, A. & Mix, A. 1989. Oceanic response to orbital forcing in the late Quaternary: observational and experimental strategies. In: *Climate and Geo-Sciences: A challenge for science and society in the 21st century* (eds A. Berger, S. Schneider & J.-C. Duplessy). Dordrecht: Kluwer, pp. 121–164.

Imbrie, J., Boyle, E.A., Clemens, S.C., Duffy, A., Howard, W.R., Kukla, G., Kutzbach, J., Martinson, D.G., McIntyre, A., Mix, A.C., Molfino, B., Morley, J.J., Peterson, L.C., Pisias, N.G., Prell, W.L., Raymo, M.E., Shackleton, N.J. & Toggweiler, J.R. 1992. On the structure and origin of major glaciation cycles: 1: linear responses to Milankovitch forcing. *Paleoceanography* 7: 701–738.

Isaac, G.Ll. 1969. Studies of early culture in East Africa. *World Archaeology* 1: 1–28.

Isaac, G.Ll. 1972. Early phases of human behaviour: models in Lower Palaeolithic archaeology. In: *Models in Archaeology* (ed. D.L. Clarke). London: Methuen, pp. 167–199.

Isaac, G.Ll. 1978. The food sharing behavior of proto-human hominids. *Scientific American* 238(4): 90–108.

Isaac, G.Ll. 1984. The archaeology of human origins: studies of the Lower Pleistocene in East Africa. In: *Advances in World Archaeology* Vol. 3 (eds. F. Wendorf & A. Close). New York: Academic Press, pp. 1–87.

Iversen, J. 1958. The bearing of glacial and interglacial epochs on the formation and extinction of plant taxa. *Uppsala Universitets Arsskrift* 6: 210–215.

Iversen, J. 1973. *The Development of Denmark's Nature since the Last Interglacial.* Copenhagen: Reitzels Forlag (Danmarks Geologiske Undersøgelse, V series, no. 7C).

Jacob-Friesen, K.H. 1959. *Einführung in Niedersachsens Urgeschichte I: Steinzeit.* Hildeshein: August Lax.

Jaubert, J. 1983. Le site moustérien du Rescoundudou (Sébazac-Concourés, Aveyron), présentation et problématique. *Bulletin de la Société Préhistorique Française* 80: 80–87.

Jaubert, J. 1984. *Contribution à la Connaissance du Paléolithique ancien et moyen des Causses*: Doctoral dissertation, University of Paris I.

Jaubert, J. 1987. Sébazac-Concourés: Le Rescoundudou. *Gallia-Préhistoire* 30: 91–2

Jaubert, J. 1989. Il y a 100 millenaires l'Homme de Néandertal chassait le cheval et le daim au Rescoundudou, Sébazac-Concourés. *Cahiers d'Archeologie Aveyronnaise* 2: 6–16.

Jaubert, J. 1990. Les industries lithiques: étude conventionnelle. In *Les Chasseurs d'Aurochs de La Borde: un site du Paléolithique moyen (Livernon, Lot)* (eds J. Jaubert, M. Lorblanchet, H. Laville, R. Slott-Moller, A. Turq & J.-P. Brugal). Paris: Maison des Sciences de l'Homme (Documents d'Archéologie Française 27), pp. 69–102.

Jaubert, J. 1993 Le gisement Paléolithique moyen du Mauran (Haut-Garonne): techno-économie des industries lithiques. *Bulletin de la Société Préhistorique Française* 90: 328–335.

Jaubert, J. & Brugal, J.-P. 1990. Contribution à l'étude du mode de vie au Paléolithique moyen: les chasseurs d'aurochs de la Borde. In: *Les Chasseurs d'Aurochs de La Borde: un site du Paléolithique moyen (Livernon, Lot)* (eds J. Jaubert, M. Lorblanchet, H. Laville, R. Slott-Moller, A. Turq & J.-P. Brugal). Paris: Maison des Sciences de l'Homme (Documents d'Archéologie Française 27), pp. 128–145.

Jaubert, J. & Rouzaud, F. 1985. Causses, karsts et Moustérien de tradition Acheuléene. *Bulletin de la Société Méridionale de Spéléologie et de Préhistoire* 25: 15–21.

Jaubert, J. & Turq, A. 1990. Les industries lithiques: interprétation chronoculturelle. In: *Les Chasseurs d'Aurochs de La Borde: un site du Paléolithique moyen (Livernon, Lot)* (eds J. Jaubert, M. Lorblanchet, H. Laville, R. Slott-Moller, A. Turq & J.-P. Brugal). Paris: Maison des Sciences de l'Homme (Documents d'Archéologie Française 27), pp. 117–125.

Jaubert, J., Lorblanchet, M., Laville, H., Slott-Moller, R., Turq, A. & Brugal, J.-P. 1990. *Les Chasseurs d'Aurochs de La Borde: un site du Palé-*

olithique moyen (Livernon, Lot). Paris: Maison des Sciences de l'Homme (Documents d'Archéologie Française 27).

Jaubert, J., Kervazo, B., Quinif, Y., Brugal, J-P. & Willford, O'Yl. 1992. Le site Paléolithique moyen du Rescoundudou (Aveyron, France): datations U/Th et interprétation chronostratigraphique. *L'Anthropologie* **96**: 103–112.

Jelinek, A. 1976. Form, function and style in lithic analysis. In: *Cultural Change and Continuity: essays in honor of James Bennett Griffin* (ed. C.E. Cleland). New York: Academic Press, pp. 19–76.

Jelinek, A. 1988a. Technology, typology and culture in the Middle Paleolithic. In: *Upper Pleistocene Prehistory of Western Eurasia* (eds H. Dibble & A. Montet-White). Philadelphia: University of Pennsylvania, University Museum Monographs No. 54, pp. 199–212.

Jelinek, A. 1988b. Comment on S. Beyries: Functional variability of lithic sets in the Middle Paleolithic. In: *Upper Pleistocene Prehistory of Western Eurasia* (eds H. Dibble & A. Montet-White). Philadelphia: University of Pennsylvania, University Museum Monographs No. 54, pp. 221–223.

Jelinek, A. 1990. The Amudian in the context of the Mugharan tradition at the Tabun Cave (Mount Carmel), Israel. In: *The Emergence of Modern Humans: an archaeological perspective* (ed. P. Mellars). Edinburgh: Edinburgh University Press, pp. 81–90.

Jelinek, A. 1994. Hominids, energy, environment, and behavior in the late Pleistocene. In: *Origins of Anatomically Modern Humans* (eds M.H. Nitecki & D.V. Nitecki). New York: Plenum Press, pp. 67–92.

Jelinek, A.J., Debénath, A. & Dibble, H.L. 1988. A preliminary report on evidence related to the interpretation of economic and social activities of Neandertals at the site of La Quina (Charente), France. In: *L'Homme de Néandertal*, Vol. 6: *La Subsistence* (ed. M. Otte). Liège: Etudes et Recherches Archéologiques de l'Université de Liège, pp. 99–106.

Jochim, M. 1983. Palaeolithic cave art in ecological perspective. In: *Hunter-gatherer Economy in Prehistory: a European Perspective* (ed. G.N. Bailey).

Cambridge: Cambridge University Press, pp. 212–219.

Jones, P.R. 1981. Experimental implement manufacture and use: a case study from Olduvai Gorge. *Philosophical Transactions of the Royal Society of London*, series B, **292**: 189–195.

Jones, R.L. & Keen, D.H. 1993. *Pleistocene Environments in the British Isles*. London: Chapman & Hall.

Jouzel, J., Lorius, C., Petit, J.R., Genthon, C., Barkov, N.I., Kotlyakov, V.M. & Petrov, V.M. 1987. Vostok ice core: a continuous isotope temperature record over the last climatic cycle (160,000 years). *Nature* **329**: 403–408.

Keeley, L.H. 1980. *Experimental Determination of Stone Tool Uses: a microwear analysis*. Chicago: University of Chicago Press.

Kerr, R.A. 1993. The whole world had a case of the Ice Age shivers. *Science* **262**: 1972–1973.

Kervazo, B., Turq, A. & Diot, M.-F. 1989. Le site moustérien de plein air de La Plane, commune de Mazeyrolles, Dordogne: note préliminaire. *Bulletin de la Société Préhistorique Française* **86**: 268–274.

Kinzey, W.G. (ed.) 1987. *The Evolution of Human Behavior: primate models*. Albany: State University of New York Press.

Klein, R.G. 1969. *Man and Culture in the Late Pleistocene: a case study*. San Francisco: Chandler.

Klein, R.G. 1973. *Ice-Age Hunters of the Ukraine*. Chicago: University of Chicago Press.

Klein, R.G. 1979. Stone age exploitation of animals in southern Africa. *American Scientist* **67**: 151–160.

Klein, R.G. 1982. Age (mortality) profiles as a means of distinguishing hunted species from scavenged bones in stone age archaeological sites. *Paleobiology* **8**: 151–158.

Klein, R.G. 1986. Review of L.R. Binford: Faunal Remains from Klasies River Mouth. *American Anthropologist* **88**: 494–495.

Klein, R.G. 1989a. *The Human Career: Human biological and cultural origins*. Chicago: University of Chicago Press.

Klein, R.G. 1989b. Biological and behavioural perspectives on modern human origins in southern

Africa. In: *The Human Revolution: behavioural and biological perspectives on the origins of modern humans* (eds P. Mellars & C. Stringer). Princeton: Princeton University Press, pp. 529–546.

Klein, R.G. 1989c. Why does skeletal part representation differ between smaller and larger bovids at Klasies River Mouth and other archeological sites. *Journal of Archaeological Science* 6: 363–381.

Klein, R.G. 1994. The problem of modern human origins. In: *Origins of Anatomically Modern Humans* (eds M.H. Nitecki & D.V. Nitecki). New York: Plenum Press, pp. 3–17.

Klein, R.G. & Cruz-Uribe, K. 1984. *The Analysis of Animal Bones from Archaeological Sites*. Chicago: University of Chicago Press.

Knecht, H. 1993. Splits and wedges: the techniques and technology of early Aurignacian antler working. In: *Before Lascaux: the complex record of the early Upper Paleolithic* (eds H. Knecht, A. Pike-Tay & R. White). Boca Raton: CRC Press, pp. 137–162.

Knecht, H., Pike-Tay, A. & White, R. 1993. *Before Lascaux: the complex record of the early Upper Paleolithic*. Boca Raton: CRC Press.

Knight, C.D. 1991. *Blood Relations: menstruation and the origins of culture*. New Haven: Yale University Press.

Kozlowski, J.K. 1982. *Excavation in the Bacho Kiro Cave (Bulgaria): Final Report*. Warsaw: Panstwowe Wydawnictwo Naukowe.

Kozlowski, J.K. 1988. Transition from the Middle to the early Upper Paleolithic in Central Europe and the Balkans. In: *The Early Upper Paleolithic: Evidence from Europe and the Near East* (eds J.F. Hoffecker & C.A. Wolf). Oxford: British Archaeological Reports International Series 437, pp. 193–236.

Kozlowski, J.K. 1990. A multi-aspectual approach to the origins of the Upper Palaeolithic in Europe. In: *The Emergence of Modern Humans: an archaeological perspective* (ed. P. Mellars). Edinburgh: Edinburgh University Press, 419–437.

Kozlowski, J.K. 1992. The Balkans in the Middle and Upper Palaeolithic: the gateway to Europe or a cul-de-sac? *Proceedings of the Prehistoric Society* 58: 1–20.

Kozlowski, J.K. 1993. L'Aurignacien en Europe et au Proche Orient. In: *Aurignacien en Europe et au Proche Orient* (eds L. Bánesz & J.K. Kozlowski). Bratislava: Acts of 12th International Congress of Prehistoric and Protohistoric Sciences, pp. 283–291.

Kozlowski, J.K. & Klíma, B. (eds) 1982. *Aurignacien et Gravettien en Europe*, Vol. 2. Liège: Etudes et Recherches Archéologiques de l'Université de Liège 13(2).

Kozlowski, J.K. & Otte, M. 1984. L'Aurignacien en Europe centrale, orientale et balkanique (travaux recents 1976–1981). In: *Aurignacien et Gravettien en Europe*, Vol. 3 (eds J.K. Kozlowski & R. Désbrosses), pp. 61–72. Liège: Etudes et Recherches Archéologiques de l'Université de Liège 13(3).

Kuhn, S.L. 1990. A geometric index of reduction for unifacial stone tools. *Journal of Archaeological Science* 17: 583–593.

Kuhn, S.L. 1991. 'Unpacking' reduction: lithic raw material economy in the Mousterian of west-central Italy. *Journal of Anthropological Archaeology* 10: 76–106.

Kuhn, S.L. 1992a. Blank form and reduction as determinants of Mousterian scraper morphology. *American Antiquity* 57: 115–128.

Kuhn, S.L. 1992b. On planning and curated technologies in the Middle Paleolithic. *Journal of Anthropological Research* 48: 185–214.

Kuhn, S.L. 1993. Mousterian technology as adaptive response. In: *Hunting and Animal Exploitation in the Later Palaeolithic and Mesolithic of Eurasia* (eds G.L. Peterkin, H.M. Bricker & P. Mellars). Archeological Papers of the American Anthropological Association No. 4, pp. 25–32.

Labeyrie, J. 1984. Le cadre paléoclimatique depuis 140,000 ans. *L'Anthropologie* 88: 19–48.

Laitman, J.T., Heimbuch, R.C. & Crelin, E.S. 1979. The basicranium of fossil hominids as an indicator of their upper respiratory systems. *American Journal of Physical Anthropology* 51: 15–34.

Lalanne, J.G. & Bouyssonie, J. 1946. Le gisement paléolithique de Laussel. Fouilles du Docteur Lalanne. *L'Anthropologie* 50: 1–163.

Lamb, H.H. 1977. The late Quaternary history of the climate of the British Isles. In: *British Quaternary Studies: recent advances* (ed. F.W. Shotton). Oxford: Clarendon Press, pp. 283–298.

Landsberg, M.E. (ed.) 1988. *The Genesis of Language: a different judgement of evidence*. Berlin: Mouton de Gruyter.

Laquay, G. 1981. *Recherches sur les Faunes du Würm I en Périgord*. Doctoral dissertation, University of Bordeaux I.

Larick, R.R. 1983. *The Circulation of Solutrian Foliate Point Cherts: residential mobility in the Périgord*. Doctoral dissertation, Department of Anthropology, State University of New York, Binghamton.

Larick, R.R. 1986. Périgord cherts: an analytical frame for investigating the movement of Paleolithic hunter-gatherers and their resources. In: *The Scientific Study of Flint and Chert* (eds G. de G. Sieveking & M.B. Hart). Cambridge: Cambridge University Press, pp. 111–120.

Larick, R.R. 1987. Circulation of Solutrean foliate points within the Perigord, SW France. In: *The Human Uses of Flint and Chert* (eds. G. de G. Sieveking & M.H. Newcomer). Cambridge: Cambridge University Press, pp. 217–230.

Lartet, L. & Christy, H. 1864. Cavernes du Périgord. Objets gravés et sculptés des temps Préhistoriques dans l'Europe occidentale. *Revue Archéologique* (1864): 1–37.

Lassarade, L., Rouvreau, M. & Texier, A. 1969. Le gisement paléolithique 'du lyceé', à Pons (Charente-Martitime). *Bulletin de la Société Préhistorique Française* 66: 341–354.

Laughlin, W.S. 1968. Hunting: an integrating biobehavior system and its evolutionary importance. In: *Man the Hunter* (ed. R.B. Lee and I. DeVore). Chicago: Aldine, pp. 304–320.

Laurent, P.F. 1965. *Heureuse Préhistoire*. Perigueux: Fanlac.

Laville, H. 1973. The relative position of Mousterian industries in the climatic chronology of the early Würm in the Perigord. *World Archaeology* 4: 321–329.

Laville, H. 1975. *Climatologie et Chronologie du Paléolithique en Périgord: étude sédimentologique de dépots en grottes et sous abris*. Marseilles: Université de Provence, Etudes Quaternaires, Mémoire No. 4.

Laville, H. 1988. Recent developments on the chronostratigraphy of the Paleolithic in the Périgord. In: *Upper Pleistocene Prehistory of Western Eurasia* (eds H. Dibble & A. Montet-White). Philadelphia: University of Pennsylvania, University Museum Monographs No. 54, pp. 147–160.

Laville, H. 1990. Le remplissage de l'aven: charactéristiques et signification. In: *Les Chasseurs d'Aurochs de La Borde: un site du Paléolithique moyen (Livernon, Lot)* (eds J. Jaubert, M. Lorblanchet, H. Laville, R. Slott-Moller, A. Turq & J.-P. Brugal). Paris: Maison des Sciences de l'Homme (Documents d'Archéologie Française 27), pp. 23–32.

Laville, H., Rigaud,J.-P. & Sackett, J. 1980. *Rock Shelters of the Périgord:geological stratigraphy and archaeological succession*. New York: Academic Press.

Laville, H., Turon, J.-L., Texier, J.-P., Raynal, J.-P., Delpech, F., Paquereau, M.-M., Prat, F. & Debénath, A. 1983. Histoire paléoclimatique de l'Aquitaine et du Golfe de Gascogne au Pléistocène supérieur depuis le dernier interglaciaire. *Bulletin de l'Institut Géologique du Bassin de l'Aquitaine, Bordeaux* 34: 219–241.

Laville, H., Raynal, J.-P. & Texier, J.-P. 1986. Le dernier interglaciaire et le cycle climatique würmien dans le sud-ouest et le Massif Central Français. *Bulletin de l'Association Française pour l'Etude du Quaternaire* 1986(1–2): 35–46.

Lawson, A. 1978. A hand-axe from Little Cressingham. *East Anglian Archaeology* 8: 1–8.

Lee, P.C. 1991. Biology and behaviour in human evolution. *Cambridge Archaeological Journal* 1: 207–226.

Lee, R.B. & DeVore, I. 1968. *Man the Hunter*. Chicago: Aldine.

Lefèbvre, A. 1969. Le gisement préhistorique de Coquelles (village). *Septentrion* 1: 8–14.

Lenoir, M. 1983. *Le Paléolithique des Basses Vallées de la Dordogne et de la Garonne*. Doctoral dissertation, University of Bordeaux I.

Lenoir, M. 1986. Un mode d'obtention de la retouche 'Quina' dans le Moustérien de Combe-Grenal (Domme, Dordogne). *Bulletin de la Société Anthropologique du Sud-ouest* 21:153–160.

Lenoir, M. 1990. Le passage du Paléolithique moyen au Paléolithique supérieur dans les basses vallées de la Dordogne et de la Garonne. In:

Paléolithique Moyen Récent et Paléolithique Supérieur Ancien en Europe (ed. C. Farizy). Nemours: Mémoires du Musée de Préhistoire d'Ile de France No. 3, pp. 215–221.

Leroi-Gourhan, A. 1958. Etude des restes humains fossiles provenant des grottes d'Arcy-sur-Cure. *Annales de Paléontologie (Vertébrés)* **44**, 97–140.

Leroi-Gourhan, A. 1976. Les structures d'habitat au Paléolithique supérieur. In: *La Préhistoire Française* (ed. H. de Lumley). Paris: Centre National de la Recherche Scientifique, pp. 656–663.

Leroi-Gourhan, A. & Brézillon, M. 1972. *Fouilles de Pincevent: essai d'analyse ethnographique d'un habitat magdalénien*. Paris: Centre National de la Recherche Scientifique (Gallia Préhistoire Supplement 7).

Leroi-Gourhan, A. & Leroi-Gourhan, Arl., 1964. Chronologie des grottes d'Arcy-sur-Cure (Yonne). *Gallia Préhistoire* **7**, 1–64.

Leroi-Gourhan, Arl. 1975. The flowers found with Shanidar IV, a Neanderthal burial in Iraq. *Science* **190**: 562–564.

Leroi-Gourhan, Arl. 1984. La place du Néandertalien de St-Césaire dans la chronologie würmienne. *Bulletin de la Société Préhistorique Française* **81**: 196–198.

Leroi-Gourhan, Arl. 1988 Le passage Moustérien-Chatelperronnien à Arcy-sur-Cure. *Bulletin de la Société Préhistorique Française* **85**: 102–104.

Leroi-Gourhan, Arl. & Renault-Miskovsky, J. 1977. La palynologie appliquée à l'archéologie: méthodes, limites et resultats. In: *Approche Ecologique de l'Homme Fossile* (ed. H. Laville & J. Renault-Miskovsky). Paris: Association Française pour l'Etude du Quaternaire, pp. 35–49.

Leroyer, C. 1986. Les gisements Castelperroniens de Quinçay et de Saint-Césaire: quelques comparaisons préliminaires des études palynologiques. In: *Préhistoire de Poitou-Charentes: Problèmes Actuels* (ed. B. Vandermeersch). Paris: Editions du Comité des Travaux Historiques et Scientifiques, pp. 125–134.

Leroyer, C. 1988. Des occupations castelperroniennes et aurignaciennes dans leur cadre chrono-climatique. In: *L'Homme de Néandertal*, Vol. 8: *La Mutation* (ed. M. Otte). Liège: Etudes et Recherches Archéologiques de l'Université de Liège, pp. 103–108.

Leroyer, C. 1990. Nouvelles données palynologiques sur le passage Paléolithique moyen-Paléolithique supérieur. In: *Paléolithique Moyen Récent et Paléolithique Supérieur Ancien en Europe* (ed. C. Farizy). Nemours: Mémoires du Musée de Préhistoire d'Ile de France No. 3, pp. 49–52.

Leroyer, C. & Leroi-Gourhan, A. 1983. Problèmes de chronologie: le castelperronien et l'aurignacien. *Bulletin de la Société Préhistorique Française* **80**: 41–44.

Le Tensorer, J.-M. 1969. Le Moustérien de las Pélénos (Lot-et-Garonne): étude statistique. *Bulletin de la Société Préhistorique Française* **66**: 232–236.

Le Tensorer, J.-M. 1973. Les industries moustériennes du Plateau Baillard (Lot-et-Garonne). *Bulletin de la Société Préhistorique Française* **70**: 73–79.

Le Tensorer, J.-M. 1976. Les civilisations du Paléolithique moyen dans les Causses et le Lot. In: *La Préhistoire Française* (ed. H. de Lumley). Paris: Centre National de la Recherche Scientifique, pp. 1027–1030.

Le Tensorer, J.-M. 1978. Le Moustérien type Quina et son évolution dans le sud de la France. *Bulletin de la Société Préhistorique Française* **75**: 141–149.

Le Tensorer, J.-M. 1981. *Le Paléolithique de l'Agenais*. Paris: Centre National de la Recherche Scientifique (Institut du Quaternaire, Université de Bordeaux I, Cahiers du Quaternaire No. 3).

Lévêque, F. 1987. Les gisements castelperroniennes de Quinçay et de Saint-Césaire: comparaisons préliminaires: stratigraphie et industries. In: *Préhistoire de Poitou-Charentes: Problèmes actuels* (ed. B. Vandermeersch). Paris: Editions du Comité des Travaux Historiques et Scientifiques, pp. 91–98.

Lévêque, F. 1993. Les données du gisement de Saint-Césaire et la transition Paléolithique moyen/supérieur en Poitou-Charentes. In: *El Origen del Hombre Moderno en el Suroeste de Europa* (ed. V. Cabrera Valdés). Madrid: Universidad Nacional de Educacion a Distancia, pp. 263–286.

Lévêque, F., & Vandermeersch, B., 1980. Découverte de restes humains dans un niveau castelperronien à Saint-Césaire (Charente-Maritime). *Comptes Rendus de l'Académie des Sciences de Paris*, series 2, **291**: 187–189.

Lévêque, F., Backer, A.M. & Gilbaud, M. 1993. *Context of a Late Neandertal: implications of multi-disciplinary research for the transition to Upper Paleolithic adaptations at Saint-Césaire, Charente-Maritime, France*. Madison (Wisconsin): Prehistory Press (Monographs in World Archaeology No. 16).

Levine, M.A. 1983. Mortality models and the interpretation of horse population structure. In: *Hunter-Gatherer Economy in Prehistory: a European perspective* (ed. G.N. Bailey). Cambridge: Cambridge University Press, pp. 23–46.

Lhomme, V. & Freneix, S. 1993. Un coquillage de bivalve du Maastrictien-Paléocene *Glyptoactis (Baluchicardia)* sp. dans la couche inférieure du gisement moustérien de 'Chez-Pourré-Chez-Comte' (Corrèze). *Bulletin de la Société Préhistorique Française* **90**: 303–306.

Lieberman, P. 1989. The origins of some aspects of human language and cognition. In: *The Human Revolution: behavioural and biological perspectives on the origins of modern humans* (eds P. Mellars & C. Stringer). Princeton: Princeton University Press, pp. 391–414.

Lieberman, P. 1991. *Uniquely Human: the evolution of speech, thought, and selfless behavior*. Cambridge (Mass.): Harvard University Press.

Lieberman, P. & Crelin, E.S. 1971. On the speech of Neanderthal man. *Linguistic Inquiry* **2**: 203–222.

LIGA members 1991a. Report of the 1st discussion group: the last interglacial in high latitudes of the northern hemisphere: terrestrial and marine evidence. *Quaternary International* **10–12**: 9–28.

LIGA members 1991b. Report of the 2nd discussion group: interrelationships and linkages between the land, atmosphere and oceans during the last interglacial. *Quaternary International* **10–12**: 29–48.

Lindly, J.M. & Clark, G.A. 1990. Symbolism and modern human origins. *Current Anthropology* **31**: 233–261.

Lock, A. 1993. Human language development and object manipulation: their relation in ontogeny and its possible relevance for phylogenetic questions. In: *Tools, Language and Cognition in Human Evolution* (eds K.R. Gibson & T. Ingold). Cambridge: Cambridge University Press, pp. 279–299.

Lorius, C.J. 1989. Polar ice cores and climate. In: *Climate and Geo-Sciences: a challenge for science in the 21st century* (eds A. Berger, S. Schneider & J.-C. Duplessy). Dordrecht: Kluwer, pp. 77–103.

Lorius, C.J., Jouzel, J., Ritz, C., Merlivat, L., Barkov, N.I., Korotkevich, Y.S. & Kotlyakov, V.M. 1985. A 150,000-year climatic record from Antarctic ice. *Nature* **316**: 591–596.

Lovejoy, C.O. 1981. The origin of man. *Science* **211**: 341–350.

Lucotte, G. 1989. The evidence for the paternal ancestry of modern humans: evidence from a Y-chromosome specific sequence polymorphic DNA probe. In: *The Human Revolution: behavioural and biological perspectives on the origins of modern humans* (eds P. Mellars & C. Stringer). Princeton: Princeton University Press, pp. 39–46.

Lumley, H. de (ed.) 1969a. *Une Cabane Acheuléenne dans la Grotte du Lazaret (Nice)*. Paris: Mémoires de la Société Préhistorique Française 7.

Lumley, H. de 1969b. A Palaeolithic camp at Nice. *Scientific American* **220**(5) (May 1969): 42–50.

Lumley, H. de (ed.) 1972. *La Grotte de l'Hortus (Valflaunès, Hérault)*. Marseilles: Université de Provence (Etudes Quaternaires, Mémoire 1).

Lumley, H. de (ed.) 1976. *La Préhistoire Française* (2 vols). Paris: Centre National de la Recherche Scientifique.

Lumley, H. de & Boone, Y. 1976a. Les structures d'habitat au Paléolithique inférieur. In: *La Préhistoire Française* (ed. H. de Lumley). Paris: Centre National de la Recherche Scientifique, pp. 625–643.

Lumley, H. de & Boone, Y. 1976b. Les structures d'habitat au Paléolithique moyen. In: *La Préhistoire Française* (ed. H. de Lumley). Paris: Centre National de la Recherche Scientifique, pp. 644–655.

Lumley, H. de, Pillard, B, & Pillard F. 1969. L'habitat et les activités de l'homme du Lazaret. In: *Une Cabane Acheuléenne dans la Grotte du Lazaret (Nice)* (ed. H. de Lumley). Paris: Mémoires de la Société Préhistorique Française 7, pp. 183–222.

Lumley, H. de, Lumley, M.-A. de, Brandi, R., Guerrier, F. & Pillard, B. 1972. Haltes et campements de chasseurs Néandertaliens dans la Grotte de l'Hortus. In: *La Grotte de l'Hortus (Valflaunès, Hérault)* (ed. H. de Lumley). Marseilles: Université de Provence (Etudes Quaternaires, Mémoire 1), pp. 527–624.

Maddison, D.R. 1991. African origin of human mitochondrial DNA reexamined. *Systematic Zoology* 40: 355–363.

Madelaine, S. 1990. Inventory of Middle Palaeolithic faunal assemblages in the National Museum of Prehistory, Les Eyzies (unpublished manuscript).

Mania, D. & Mania, U. 1988. Deliberate engravings on bone artefacts of *Homo erectus*. *Rock Art Research* 5: 91–107.

Marks, A.E. 1988. The curation of stone tools during the Upper Pleistocene: a view from the Central Negev, Israel. In: *Upper Pleistocene Prehistory of Western Eurasia* (eds H. Dibble & A. Montet-White). Philadelphia: University of Pennsylvania, University Museum Monographs No. 54, pp. 275–286.

Marks, A.E. 1989. Early Mousterian settlement patterns in the Central Negev, Israel: their social and economic implications. In: *L'Homme de Néandertal*, Vol. 6: *La Subsistence* (ed. M. Otte). Liège: Etudes et Recherches Archéologiques de l'Université de Liège, pp. 115–127.

Marks, A.E. 1993. The early Upper Paleolithic: the view from the Levant. In: *Before Lascaux: the complex record of the early Upper Paleolithic* (eds H. Knecht, A. Pike-Tay & R. White). Boca Raton: CRC Press, pp. 5–22.

Marks, A.E. & Ferring, C.R. 1988. The early Upper Paleolithic of the Levant. In: *The Early Upper Paleolithic: Evidence from Europe and the Near East* (ed J.F. Hoffecker & C.A. Wolf), pp. 43–72. Oxford: British Archaeological Reports International Series 437.

Marshack, A. 1972. *The Roots of Civilization*. New York: McGraw Hill.

Marshack, A. 1981. On Paleolithic ochre and the early uses of color and symbol. *Current Anthropology* 22: 188–191.

Marshack, A. 1988. The Neanderthals and the human capacity for symbolic thought: cognitive and problem-solving aspects of Mousterian symbol. In: *L'Homme de Néandertal*, Vol. 5: *La Pensée* (ed. M. Otte). Liège: Etudes et Recherches Archéologiques de l'Université de Liège, pp. 57–92.

Marshack, A. 1989. Evolution of the human capacity: the symbolic evidence. *Yearbook of Physical Anthropology* 32: 1–34.

Marshack, A. 1990. Early hominid symbol and evolution of the human capacity. In: *The Emergence of Modern Humans: an archaeological perspective* (ed. P. Mellars). Edinburgh: Edinburgh University Press, pp. 457–498.

Martinson, D.C., Pisias, N.G., Hays, J.D., Imbrie, J., Moore, T.C. Jr. & Shackleton, N.J. 1987. Age dating and the orbital theory of the ice ages: development of a high-resolution 0 to 300,000-year chronostratigraphy. *Quaternary Research* 27: 1–29.

Mazière, G. & Raynal, J.-P. 1976. Les civilisations du Paléolithique Moyen en Limousin. In: *La Préhistoire Française* (ed. H. de Lumley). Paris: Centre National de la Recherche Scientifique, pp. 1077–1084

McBurney, C.B.M. 1950. The geographical study of the older Palaeolithic stages in Europe. *Proceedings of the Prehistoric Society* 16: 163–183.

McBurney, C.B.M. 1960. *The Stone Age of Northern Africa*. Harmondsworth: Penguin.

McCown, T.D. & Keith, A. 1939. *The Stone Age of Mount Carmel*, Vol 2: *The Fossil Human Remains from the Levalloiso-Mousterian*. Oxford: Clarendon Press.

McGhee, R. 1977. Ivory for the Sea Woman: the symbolic attributes of a prehistoric technology. *Canadian Journal of Archaeology* 1: 141–149.

McGrew, W.C. 1992. *Chimpanzee Material Culture: implications for human evolution*. Cambridge: Cambridge University Press.

McIntyre, A., Ruddiman, W.F. & Jantzen, R. 1972. Southward penetrations of the North Atlantic Polar Front: faunal and floral evidence of large-scale surface water movements over the last 225,000 years. *Deep-Sea Research* 19: 61–77.

McIntyre, A., Bé, A.W.H., Hays, J.D., Gardner, J.V., Lozano, J.A., Molfino, B., Prell, W., Thierstein, H.R., Crowley, T., Imbrie, J., Kellogg, T., Kip, N., & Ruddiman, W.F. 1975. Thermal and oceanic structures of the Atlantic through a glacial-interglacial cycle. In: *Proceedings of the WMO Symposium on Long-term Climatic Fluctuations*. Geneva: World Meteorological Organization, pp. 75–80.

Meignen, L. 1982. Supports d'outils chauffés au paléolithique moyen. *Studia Praehistorica Belgica* 2: 111–117.

Meignen, L. 1988. Un exemple de comportement technologique différentiel selon les matières premières: Marillac, couches 9 et 10. In: *L'Homme de Néandertal*, Vol. 4: *La Technique* (ed. M. Otte). Liège: Etudes et Recherches Archéologiques de l'Université de Liège, pp. 71–79.

Meignen, L. (ed.) 1993. L'Abri des Canalettes: un habitat moustérien sur les grands Causses (Nant, Aveyron): fouilles 1980–1986. Paris: Centre National de la Recherche Scientifique (Centre de Recherches Archéologiques, Monograph 10).

Meignen, L. & Bar-Yosef, O. 1988. Variabilité technologique au Proche Orient: l'exemple de Kébara. In: *L'Homme de Néandertal*, Vol. 4: *La Technique* (ed. M. Otte). Liège: Etudes et Recherches Archéologiques de l'Université de Liège, pp. 81–95.

Meignen, L. & Coularou, J. 1981. *Le gisement paléolithique moyen – La Roquette (Conqueyrac, Gard) : I – Etude archéologique*. Valbonne: Centre de Recherches Archéologiques, Notes Internes 26: 1–19.

Meignen, L. & Vandermeersch, B. 1986. Le gisement de Marillac (Charente) couches 9 et 10: caractéristiques des outillages, économie des matières premières. In: *Préhistoire de Poitou-Charentes: Problèmes Actuels* (ed. B. Vandermeersch). Paris: Editions du Comité des Travaux Historiques et Scientifiques, Ministère de l'Education Nationale, pp. 135–144.

Meignen, L., Chech, M. & Vandermeersch, B. 1977. Le gisement moustérien d'Arténac à Saint-Mary (Charente). *Gallia-Préhistoire* 20: 281–291.

Mellars, P.A. 1964. The Middle Palaeolithic surface artifacts at Kokkinopilos. *Proceedings of the Prehistoric Society* 30: 229–244.

Mellars, P.A. 1965. Sequence and development of Mousterian traditions in south-western France. *Nature* 205: 626–627.

Mellars, P.A. 1967. *The Mousterian Succession in South-west France*. Doctoral dissertation, University of Cambridge.

Mellars, P.A. 1969. The chronology of Mousterian industries in the Périgord region of south-west France. *Proceedings of the Prehistoric Society* 35: 134–171.

Mellars, P.A. 1970. Some comments on the notion of 'functional variability' in stone-tool assemblages. *World Archaeology* 2: 74–90.

Mellars, P.A. 1973. The character of the Middle-Upper Palaeolithic transition in south-west France. In: *The Explanation of Culture Change: Models in Prehistory* (ed. C. Renfrew). London: Duckworth, pp. 255–276.

Mellars, P.A. 1976. Settlement patterns and industrial variability in the British Mesolithic. In: *Problems in Social and Economic Archaeology* (eds G. de G. Sieveking, I.H. Longworth & K.E. Wilson). London: Duckworth, pp. 375–400.

Mellars, P.A. 1982. On the Middle/Upper Palaeolithic transition: a reply to White. *Current Anthropology* 23: 238–240.

Mellars, P.A. 1985. The ecological basis of social complexity in the Upper Paleolithic of southwestern France. In: *Prehistoric Hunter-Gatherers: the emergence of cultural complexity* (eds T.D. Price & J.A. Brown). Orlando: Academic Press, pp. 271–297.

Mellars, P.A. 1986a. A new chronology for the French Mousterian period. *Nature* 322: 410–411.

Mellars, P.A. 1986b. Dating and correlating the French Mousterian: reply. *Nature* 324: 113–114.

Mellars, P.A. 1988. The chronology of the south-west French Mousterian: a review of the current debate. In: *L'Homme de Néandertal*, Vol. 4: *La Technique* (ed. M. Otte). Liège: Etudes et Recherches Archéologiques de l'Université de Liège, pp. 97–120.

Mellars, P.A. 1989a. Major issues in the emergence of modern humans. *Current Anthropology* 30, 349–385.

Mellars, P.A. 1989b.Technological changes across the Middle-Upper Palaeolithic transition: technological, social, and cognitive perspectives. In: *The Human Revolution: behavioural and biological perspectives on the origins of modern humans* (eds P. Mellars & C. Stringer). Princeton: Princeton University Press, pp. 338–365.

Mellars, P.A. 1989c. Chronologie du Moustérien du sud-ouest de la France: actualisation du débat. *L'Anthropologie* **94**: 1–18.

Mellars, P.A. 1990a. Comment on radiocarbon-accelerator dating of Roc de Combe samples. *Archaeometry* **32**, 101–102.

Mellars, P.A. (ed.) 1990b. *The Emergence of Modern Humans: an Archaeological Perspective*. Edinburgh: Edinburgh University Press.

Mellars, P.A. 1991. Cognitive changes and the emergence of modern humans in Europe. *Cambridge Archaeological Journal* **1**: 63–76.

Mellars, P.A. 1992a. Technological change in the Mousterian of southwest France. In: *The Middle Paleolithic: adaptation, behavior, and variability* (eds H.L. Dibble & P.A. Mellars). Philadelphia: University of Pennsylvania, University Museum Monographs No. 72, pp. 29–43.

Mellars, P.A. 1992b. Archaeology and the population-dispersal hypothesis of modern human origins in Europe. In: *The Origin of Modern Humans and the Impact of Chronometric Dating* (eds M.J. Aitken, C.B. Stringer & P.A. Mellars). London: Royal Society (Philosophical Transactions of the Royal Society, series B, **337**, no. 1280), pp. 225–234.

Mellars, P.A. 1992c. Archaeology and modern human origins in Europe. *Proceedings of the British Academy* **82**: 1–35.

Mellars, P.A. 1994. The European evidence (review feature on Stringer & Gamble 1993). *Cambridge Archaeological Journal* **4**: 103–104.

Mellars, P.A. & Gibson, K. (eds) 1995. *Modelling the Early Human Mind*. Cambridge: McDonald Institute for Archaeological Research (in press).

Mellars, P.A. & Grün, R. 1991. A comparison of the electron spin resonance and thermoluminescence dating methods: the results of ESR dating at Le Moustier (France). *Cambridge Archaeological Journal* **1**: 269–276.

Mellars, P.A. & Stringer, C.B. (eds) 1989. *The Human Revolution: Behavioural and Biological Perspectives on the Origins of Modern Humans*. Princeton: Princeton University Press.

Mellars, P.A. & Tixier, J. 1989. Radiocarbon-accelerator dating of Ksar 'Aqil (Lebanon) and the chronology of the Upper Palaeolithic sequence in the Middle East. *Antiquity* **63**, 761–768.

Mellars, P.A., Bricker, H.M., Gowlett, J.A.J. & Hedges, R.E.M. 1987. Radiocarbon accelerator dating of French Upper Palaeolithic sites. *Current Anthropology* **28**, 128–133.

Mercier, N. & Valladas, H. 1993. Contribution des méthodes de datation par le carbone 14 et de la thermoluminescence à la chronologie de la transition du Paléolithique moyen au Paléolithique supérieur. In: *El Origen del Hombre Moderno en el Suroeste de Europa* (ed. V. Cabrera Valdés). Madrid: Universidad Nacional de Educacion a Distancia, pp. 47–60.

Mercier, N., Valladas, H., Joron, J.-L., Reyss, J.-L., Lévêque, F. & Vandermeersch, B. 1991. Thermoluminescence dating of the late Neanderthal remains from Saint-Césaire. *Nature* **351**: 737–739.

Michel, D. 1990. Les foyers du Paléolithique inférieur du nord Cotentin. In: *Nature et Fonction des Foyers Préhistoriques* (eds M. Olive & Y. Taborin). Nemours: Mémoires du Musée de Préhistoire d'Ile de France No. 2, pp. 131–134.

Mithen, S.J. 1989. To hunt or to paint: animals and art in the Upper Palaeolithic. *Man* (N.S.) **23**: 671–695.

Mithen, S.J. 1993. Simulating mammoth hunting and extinction: implications for the late Pleistocene of the Central Russian Plain. In: *Hunting and Animal Exploitation in the Later Palaeolithic and Mesolithic of Eurasia* (eds G.L. Peterkin, H.M. Bricker & P. Mellars). Archeological Papers of the American Anthropological Association No. 4, pp. 163–178.

Mithen, S.J. 1994. Technology and society during the Middle Pleistocene: hominid group size, social learning and industrial variability. *Cambridge Archaeological Journal* **4**: 3–32.

Moen, A. 1973. *Wildlife Ecology*. San Francisco: Freeman.

Monnier, G. 1992. *Form, function and typology: a case study of Mousterian of Acheulian tradition backed knives*. Master's Dissertation, University of Cambridge.

Morala, A. 1980. *Observations sur le Périgordien, l'Aurignacien et leurs Matières Premières Lithiques en Haut Agenais*. Dissertation, Ecole des Hautes Etudes en Sciences Sociales, Toulouse.

Morawski, W. 1976. Middle Palaeolithic flint assemblages from Piekary IIa site. *Swiatowit* **34**: 139–146.

Mortillet, G. de. 1869. Essai d'une classification des cavernes et des stations sous abri fondée sur les produits de l'industrie humaine. *Matériaux pour l'Histoire Primitive et Naturel de l'Homme* **5**: 172–179.

Mortillet, G. de. 1883. *Le Préhistorique: Antiquité de l'Homme*. Paris: Bibliothèque des Sciences Contemporaines.

Mountain, J.L., Lin, A.A., Bowcock, A.M. & Cavalli-Sforza, L.L. 1992. Evolution of modern humans: evidence from nuclear DNA polymorhisms. In: *The Origin of Modern Humans and the Impact of Chronometric Dating* (eds M.J. Aitken, C.B. Stringer & P.A. Mellars). London: Royal Society (Philosophical Transactions of the Royal Society, series B, **337**, no. 1280), pp. 159–166.

Movius, H.L. 1950. A wooden spear of third interglacial age from lower Saxony. *Southwestern Journal of Anthropology* **6**: 139–142.

Movius, H.L. 1953. Mousterian cave of Teshik-Tash, southeastern Uzbekistan. *Bulletin of the American School of Prehistoric Research* **17**: 11–71.

Movius, H.L. 1975. Summary of the stratigraphic sequence. In: *Excavation of the Abri Pataud, Les Eyzies (Dordogne)* (ed. H.L. Movius). Cambridge (Mass.): Peabody Museum, Harvard University, pp. 7–18.

Movius, H.L., David, N., Bricker, H.M. & Clay, R.B. 1968. *The Analysis of Certain Major Classes of Upper Palaeolithic Tools*. Cambridge (Mass.): Peabody Museum, Harvard University. (American School of Prehistoric Research Bulletin 26).

Munday, F. 1976. Intersite variability in the Mousterian occupation of the Avdat/Aqev area. In: *Prehistory and Palaeoenvironments in the Central Negev, Israel*, Vol. 1 (ed. A. Marks). Dallas: Southern Methodist University Press, pp. 113–140.

Mussi, M. 1990. Le peuplement de l'Italie à la fin du Paléolithique moyen et au début du Paléolithique supérieur. In: *Paléolithique Moyen Récent et Paléolithique Supérieur Ancien en Europe* (ed. C. Farizy). Nemours: Mémoires du Musée de Préhistoire d'Ile de France No. 3, pp. 251–262.

Newcomer, M.H. 1971. Some quantitative experiments in handaxe manufacture. *World Archaeology* **3**: 85–93.

Niederlender, A., Lacam, R., Cadiergues, Dr. & Bordes, F. 1956. Le gisement moustérien du Mas-Viel (Lot). *L'Anthropologie* **60**: 209–235.

Nitecki, M.H. & Nitecki, D.V. 1994. *Origins of Anatomically Modern Humans*. New York: Plenum Press.

Oakley, K.P., Andrews, P., Keeley, L.H. & Clark, J.D. 1977. A reappraisal of the Clacton Spear point. *Proceedings of the Prehistoric Society* **43**: 1–12.

O'Connell, J.F., Hawkes, K. & Blurton-Jones, N. 1988a. Hadza hunting, butchering, and bone transport and their archaeological implications. *Journal of Anthropological Research* **44**: 114–161.

O'Connell, J.F., Hawkes, K. & Blurton-Jones, N. 1988b. Hadza scavenging: implications for Plio-Pleistocene hominid subsistence. *Current Anthropology* **29**: 356–363.

Ohnuma, K. & Bergman, C.A. 1990. A technological analysis of the Upper Palaeolithic levels (XXV-VI) of Ksar Akil, Lebanon. In: *The Emergence of Modern Humans: an archaeological perspective* (ed. P. Mellars). Edinburgh: Edinburgh University Press, 56–80.

Oliva, M. 1993. The Aurignacian in Moravia. In: *Before Lascaux: the complex record of the early Upper Paleolithic* (eds H. Knecht, A. Pike-Tay & R. White). Boca Raton: CRC Press, pp. 37–56.

Olive, M. & Taborin, Y. (eds) 1989. *Nature et Fonction des Foyers Préhistoriques*. Nemours: Mémoires du Musée de Préhistoire d'Ile de France 2.

Orquera, L.A. 1984. Specialization and the Middle/Upper Paleolithic transition. *Current Anthropology* **25**: 73–98.

Otte, M. 1979. *Le Paléolithique Supérieur Ancien*

en Belgique. Brussels: Musées Royaux d'Art et d'Histoire, Monographies d'Archéologie Nationale No. 5.

Otte, M. 1990. From the Middle to the Upper Palaeolithic: the nature of the transition. In: *The Emergence of Modern Humans: an archaeological perspective* (ed. P. Mellars). Edinburgh: Edinburgh University Press, 438–456.

Otte, M., Evrard, J.-M. & Mathis, A. 1988. Interprétation d'un habitat au Paléolithique moyen: la grotte de Sclayn, Belgique. In: *Upper Pleistocene Prehistory of Western Eurasia* (eds H. Dibble & A. Montet-White). Philadelphia: University of Pennsylvania, University Museum Monographs No. 54, pp. 95–124.

Paquereau, M.M. 1969. Etude palynologique du Würm I du Pech de l'Azé (Dordogne). *Quaternaria* 11: 227–235.

Paquereau, M.M. 1975. Le Würmien ancien en Périgord: Etude palynologique. *Quaternaria* 18: 67–159.

Paquereau, M.M. 1979. Quelques types de flores tardi-glaciaires dans le sud-ouest de la France. In: *La Fin des Temps Glaciaires en Europe* (ed. D. de Sonneville-Bordes). Paris: Centre National de la Recherche Scientifique, pp. 151–157.

Paquereau, M.M. 1984. Etude palynologique du gisement de la Ferrassie (Dordogne). In: *Le Grand Abri de la Ferrassie: fouilles 1968–1973* (ed. H. Delporte). Paris: Institut de Paléontologie Humaine (Université de Provence, Etudes Quaternaires, Mémoire no. 7), pp. 51–59.

Parker, S.T. 1990. Why big brains are so rare: energy costs of intelligence and brain size in anthropoid primates. In: *'Language' and Intelligence in Monkeys and Apes: comparative developmental perspectives* (eds S.T. Parker & K.R. Gibson). Cambridge: Cambridge University Press, pp. 129–154.

Parker, S.T. & Gibson, K.R. 1979. A developmental model for the evolution of language and intelligence in early hominids. *The Behavioural and Brain Sciences* 2: 367–408.

Parker, S.T. & Gibson, K.R. (eds) 1990. *'Language' and Intelligence in Monkeys and Apes: comparative developmental perspectives*. Cambridge: Cambridge University Press.

Parker, S.T. & Milbraith, C. 1993. Higher intelligence, propositional language, and culture as adaptations for planning. In: *Tools, Language and Cognition in Human Evolution* (eds K.R. Gibson & T. Ingold). Cambridge: Cambridge University Press, pp. 314–333.

Parks, D.A. & Rendell, H.M. 1992. Thermoluminescence dating and geochemistry of loessic deposits in southeast England. *Journal of Quaternary Science* 7: 99–107.

Passingham, R.E. 1975. Changes in the size and organization of the brain in Man and his ancestors. *Brain, Behavior, and Evolution* 11: 73–90.

Passingham, R.E. 1989. The origins of human intelligence. In: *Human Origins* (ed. R. Durant). London: Clarendon Press, pp. 123–136.

Paterne, M., Guichard, F., Labeyrie, J., Gillot, P.Y. & Duplessy, J.C. 1986. Tyrrhenian sea tephrochronology of the oxygen isotope record for the past 60,000 years. *Marine Geology* 72: 259–285.

Paterson, T.T. & Tebbutt, C.F. 1947. Studies in the Palaeolithic succession in England, No. III: Palaeoliths from St. Neots, Huntingdonshire. *Proceedings of the Prehistoric Society* 13: 37–46.

Patou-Mathis, M. 1993. Taphonomic and paleoethnographic study of the fauna associated with the Neandertal of Saint-Césaire. In: *Context of a Late Neandertal: implications of multidisciplinary research for the transition to Upper Paleolithic adaptations at Saint-Césaire, Charente-Maritime, France*. (eds F. Lévêque, A.M. Backer & M. Gilbaud). Madison (Wisconsin): Prehistory Press (Monographs in World Archaeology No. 16), pp. 79–102

Păunescu, A. 1988. Structures d'habitat moustériennes mises au jour dans l'établissement de Ripiceni-Izvor (Roumanie) et quelques considérations concernant le type d'habitat Paléolithique moyen de l'est des Carpates. In: *L'Homme de Néandertal*, Vol. 6: *La Subsistence* (ed. M. Otte). Liège: Etudes et Recherches Archéologiques de l'Université de Liège, pp. 127–144.

Pelegrin, J. 1986. *Technologie Lithique: une méthode appliquée à l'étude de deux séries du Périgordien ancien: Roc de Combe couche 8, La Côte niveau III*. Doctoral dissertation, University of Paris X.

Pelegrin, J. 1990. Observations technologiques sur quelques séries du Châtelperronien et du MTA

B du sud-ouest de la France: une hypothèse d'évolution. In: *Paléolithique Moyen Récent et Paléolithique Supérieur Ancien en Europe* (ed. C. Farizy). Nemours: Mémoires du Musée de Préhistoire d'Ile de France No. 3, pp. 195–201.

Perkins, D. & Daly, P. 1968. A hunters' village in Neolithic Turkey. *Scientific American* **219**(5): 97–106.

Perlès, C. 977. *Préhistoire du Feu*. Paris: Masson.

Perpère, M. 1989. Les frontières de débitage Levallois: typométrie des éclats. *L'Anthropologie* **95**: 837–850.

Peterkin, G.L. 1993. Lithic and organic hunting technology in the French Upper Palaeolithic. In: *Hunting and Animal Exploitation in the Later Palaeolithic and Mesolithic of Eurasia* (eds G.L. Peterkin, H.M. Bricker & P. Mellars). Archeological Papers of the American Anthropological Association No. 4, pp. 49–68.

Peterkin, G.L., Bricker, H.M. & Mellars, P. (eds) 1993. *Hunting and Animal Exploitation in the Later Palaeolithic and Mesolithic of Eurasia*. Archeological Papers of the American Anthropological Association No. 4.

Pettitt, P.B. 1992. Reduction models and lithic variability in the Middle Palaeolithic of south west France. *Lithics* **13**: 11–31.

Pettitt, P.B. (Forthcoming) *Tool Reduction Models, Primary Flaking, and Lithic Assemblage Variability in the Middle Palaeolithic of South-West France*. Doctoral dissertation, University of Cambridge.

Peyrony, D. 1920. Le Moustérien – ses faciès. *Comptes Rendus de l'Association Française pour l'Avancement des Sciences*, 44th Session (Strasbourg): 496–497.

Peyrony, D. 1921. Les moustériens inhumaient-ils leurs morts? *Bulletin de la Société Historique et Archéologique du Périgord* (1921): 132–139.

Peyrony, D. 1930. Le Moustier: ses gisements, ses industries, ses couches géologiques. *Revue Anthropologique* **40**: 48–76, 155–176.

Peyrony, D. 1932. La station préhistorique de la Gare de Couze ou de Saint-Sulpice-des-Magnats (Commune de Lalinde, Dordogne). *Bulletin de la Société Historique et Archéologique du Périgord* (1932): 1–23.

Peyrony, D. 1934. La Ferrassie: Moustérien, Périgordien, Aurignacien. *Préhistoire* **3**: 1–92.

Peyrony, D. 1939. Le Comte Begouën en Périgord. In: *Mélanges de Préhistoire et d'Anthropologie offerts par ses Collègues, Amis et Disciples au Professeur Comte H. Begouën*. Toulouse, Editions du Museum, pp. 235–241.

Peyrony, D. 1943. Combe Capelle. *Bulletin de la Société Préhistorique Française* **40**: 243–257.

Peyrony, D. 1948. *Le Périgord Préhistorique: essai de géographie humaine*. Périgueux: Société Historique et Archéologique du Périgord.

Pfeiffer, J.E. 1982. *The Creative Explosion: an enquiry into the origins of art and religion*. New York: Harper & Row.

Piaget, J. 1960. *The Psychology of Intelligence*. Totowa: Littlefield, Adams & Co.

Piaget, J. 1970. *Genetic Epistemology*. New York: Columbia University Press.

Pigeot, N. 1991. Réflexions sur l'histoire technique de l'homme: de l'évolution cognitive a l'évolution culturelle. *Paléo* **3**: 167–200.

Pike-Tay, A. 1991. *Red Deer Hunting in the Upper Paleolithic of Southwest France: a study in seasonality*. Oxford: Tempus Reparatum. British Archaeological Reports International Series S569.

Pike-Tay, A. 1993. Hunting in the Upper Perigordian: a matter of strategy or expedience? In: *Before Lascaux: the complex record of the early Upper Paleolithic* (eds H. Knecht, A. Pike-Tay & R. White). Boca Raton: CRC Press, pp. 85–100.

Plisson, H. 1988. Technologie et tracéologie des outils lithiques moustériens en Union Soviétique: les travaux de V.E. Shchelinskii. In: *L'Homme de Néandertal*, Vol. 4: *La Technique* (ed. M. Otte). Liège: Etudes et Recherches Archéologiques de l'Université de Liège, pp. 121–168.

Pons, A., Campy, M. & Guiot, J. 1989. The last climatic cycle in France: the diversity of records. *Quaternary International* **3–4**: 49–55.

Poplin, F. 1988. Aux origines néandertaliennes de l'art. Matière, forme, symétries. Contribution d'une galène et d'un oursin fossile taillé de Merry-sur-Yonne (France). In: *L'Homme de Néandertal*, Vol. 5: *La Pensée* (ed. M. Otte). Liège:

Etudes et Recherches Archéologiques de l'Université de Liège, pp. 109–116.

Potts, R. 1988. *Hominid Activities at Olduvai*. New York: Aldine de Gruyter.

Pradel, L. 1954. *Les Gisements Paléolithiques de Fontmaure*. Toulouse: Suppléments aux Annales de la Faculté des Lettres de Toulouse.

Pradel, L. 1963. Les enseignements ethnographiques du Moustérien de Fontmaure. *Bulletin de la Société des Amis de Grand-Pressigny* (1963): 1–8.

Pradel, L, & Pradel, J.H. 1970. La station paléolithique de Fontmaure, commune de Vélleches (Vienne). *L'Anthropologie* 74: 481–526.

Pradel, L. & Tourenq, C. 1972. Choix de matériaux par les Paléolithiques de Fontmaure et essais de fragmentation dynamique. *Bulletin de la Société Préhistorique Française* 69: 12.

Prat, F. & Suire, C. 1971. Remarques sur les cerfs contemporains des deux premiers stades würmiens. *Bulletin de la Société Préhistorique Française* 68: 75–79.

Ragir, S. 1985. Retarded development: the evolutionary mechanism underlying the emergence of language. *The Journal of Mind and Behavior* 6: 451–468.

Raynal, J.-P. 1988. Paléoenvironnements et chronostratigraphie du Paléolithique moyen dans le Massif Central Français: implications culturelles. *L'Homme de Néandertal*, Vol. 2: *L'Environnement* (ed. M. Otte). Liège: Etudes et Recherches Archéologiques de l'Université de Liège, pp. 113–145.

Raynal, J.-P. & Decroix, C. 1987. L'abri de Baume-Vallée (Haute-Loire, France), site moustérien de moyenne montagne dans son contexte régional. In: *Hommage à l'Abbé Jean Roche*. Porto: Arqueologia, pp. 17–41.

Raynal, J.-P. & Guadelli, J.-L. 1990. Milieux physiques et biologiques: quel changements entre 60 et 30,000 ans à l'ouest de l'Europe? In: *Paléolithique Moyen Récent et Paléolithique Supérieur Ancien en Europe* (ed. C. Farizy). Nemours: Mémoires du Musée de Préhistoire d'Ile de France No. 3, pp. 53–61.

Raynal, J.-P. & Huxtable, J. 1989. Premières datations par thermoluminescence du Moustérien charentien du Velay (Massif Central, France). *Comptes Rendus de l'Académie des Sciences de Paris* II, 309: 157–162.

Reille, M. & de Beaulieu, J.-L. 1990. Pollen analysis of a long upper Pleistocene continental sequence in a Velay maar (Massif Central, France). *Palaeogeography, Palaeoclimatology, Palaeoecology* 80: 35–48.

Renault-Miskovsky, J. & Leroi-Gourhan, A. 1981. Palynologie et archéologie: nouveaux résultats du Paléolithique supérieur au Mésolithique. *Bulletin de l'Association Française pour l'Etude du Quaternaire* 3/4: 121–128.

Renfrew, A.C. 1969. Trade and culture process in European prehistory. *Current Anthropology* 10: 151–169.

Renfrew, A.C. 1987. *Archaeology and Language: the puzzle of Indo-European origins*. London: Jonathan Cape.

Révillion, S. 1989. Le débitage du gisement paléolithique moyen de Seclin (Nord). In: *Paléolithique et Mésolithique du Nord de la France: nouvelles recherches* (ed. A. Tuffreau). Lille: Centre d'Etudes et de Recherches Préhistorique (CERP), Université des Sciences et Techniques de Lille Flandres Artois, pp. 79–90.

Révillion, S. 1993. Question typologique à propos des industries laminaires du Paléolithique moyen de Seclin (Nord) et de Saint-Germain-des-Vaux/Port-Racine (Manche): lames Levallois ou lames non Levallois? *Bulletin de la Société Préhistorique Française* 90: 269–273.

Ricard, J.-L. 1980. Le Moustérien de tradition acheuléenne de la Croix-Guémard (Deux-Sèvres). *Bulletin de la Société Préhistorique Française* 77: 306–316.

Rigaud, J.-P. 1969. Gisements paléolithiques de plein air en Sarladais. *Bulletin de la Société Préhistorique Française* 66: 319–334.

Rigaud, J.-P. 1982. *Le Paléolithique en Périgord: les données du Sud-Ouest Sarladais et leurs implications*. Doctoral dissertation, University of Bordeaux I.

Rigaud, J.-P. (ed.) 1988. *La Grotte Vaufrey: paléoenvironnement, chronologie, activités humaines*. Paris: Mémoires de la Société Préhistorique Française 19.

Rigaud, J.-P. 1989. From the Middle to the Upper Paleolithic: transition or convergence? In: *The Emergence of Modern Humans: biocultural adaptations in the later Pleistocene* (ed. E. Trinkaus).

Cambridge: Cambridge University Press, pp. 142–153.

Rigaud, J.-P. 1993. La transition Paléolithique moyen/Paléolithique supérieur dans le sud-ouest de la France. In: *El Origen del Hombre Moderno en el Suroeste de Europa* (ed. V. Cabrera Valdés). Madrid: Universidad Nacional de Educacion a Distancia, pp. 117–126.

Rigaud, J.-P. & Geneste, J.-M. 1988. L'utilisation de l'espace dans la Grotte Vaufrey. In: *La Grotte Vaufrey: paléoenvironnement, chronologie, activités humaines* (ed. J.-P. Rigaud). Mémoires de la Société Préhistorique Française 19, pp. 593–611.

Roche, J. 1971. Stratigraphie de la grotte du Placard (fouilles 1958–1968). *Mémoires de la Société Archéologique et Historique de la Charente*, 1971: 253–259.

Rodseth, L., Wrangham, R.W., Harrigan, A.M. & Smuts, B.B. 1991. The human community as a primate society. *Current Anthropology* 32: 221–254.

Roe, D.A. 1981. *The Lower and Middle Palaeolithic Periods in Britain*. London: Routledge & Kegan Paul.

Roebroeks, W. 1986. Archaeology and Middle Pleistocene stratigraphy: the case of Maastricht-Belvédère (The Netherlands). In: *Chronostratigraphie et Faciès Culturels du Paléolithique Inférieur et Moyen dans l'Europe du Nord-Ouest* (eds A. Tuffreau & J. Sommé). Paris: Association Française pour l'Etude du Quaternaire, pp. 81–88.

Roebroeks, W. (ed.) 1988. *From Finds Scatters to Early Hominid Behaviour: a study of Middle Palaeolithic riverside settlements at Maastricht-Belvédère (The Netherlands)*. University of Leiden: Analecta Praehistorica Leidensia 21.

Roebroeks, W., Kolen, J. & Rensink, E. 1988. Planning depth, anticipation and the organization of Middle Palaeolithic technology: the 'archaic natives' meet Eve's descendants. *Helinium* 28: 17–34.

Rolland, N. 1972. Etude archéométrique de l'industrie moustérienne de la grotte de l'Hortus. In: *La Grotte de l'Hortus (Valflaunès, Hérault)*. Marseilles: Université de Provence (Etudes Quaternaires, Mémoire 1), pp. 489–508.

Rolland, N. 1976. *The Antecedents and Emergence of the Middle Palaeolithic Industrial Complex in Western Europe*. Doctoral dissertation, University of Cambridge.

Rolland, N. 1977. New aspects of Middle Palaeolithic variability in western Europe. *Nature* 266: 251–252.

Rolland, N. 1981. The interpretation of Middle Palaolithic variability. *Man* 16: 15–42.

Rolland, N. 1988a. Variabilité et classification: nouvelles données sur le 'complexe moustérien'. In: *L'Homme de Néandertal*, Vol. 4: *La Technique* (ed. M. Otte). Liège: Etudes et Recherches Archéologiques de l'Université de Liège, pp. 169–183.

Rolland, N. 1988b. Observations on some Middle Paleolithic time series in southern France. In: *Upper Pleistocene Prehistory of Western Eurasia* (eds H. Dibble & A. Montet-White). Philadelphia: University of Pennsylvania, University Museum Monographs No. 54, pp. 161–180.

Rolland, N. 1990. Middle Palaeolithic socio-economic formations in western Eurasia: an exploratory survey. In: *The Emergence of Modern Humans: an archaeological perspective* (ed. P. Mellars). Edinburgh: Edinburgh University Press, pp. 347–388.

Rolland, N. & Dibble, H. 1990. A new synthesis of Middle Paleolithic variability. *American Antiquity* 55: 480–499.

Ruddiman, W.F. & McIntyre, A. 1976. Northeast Atlantic paleoclimatic changes over the past 600,000 years. *Geological Society of America Memoir* 145: 111–146.

Russell, J. 1995. Development and evolution of the symbolic function: the role of working memory. In: *Modelling the Early Human Mind* (eds P. Mellars & K. Gibson). Cambridge: McDonald Institute for Archaeological Research (in press).

Rust, A. 1950. *Die Höhlenfunde von Jabrud (Syrien)*. Neumünster: Karl Wacholtz.

Ruvolo, M., Zehr, S., von Dornum, M., Pan, D., Chang, B. & Lin, J. 1993. Mitochondrial COII sequences and modern human origins. *Molecular Biology and Evolution* 10: 1115–1135.

Sackett, J.R. 1966. Quantitative analysis of Upper

Paleolithic stone tools. *American Anthropologist* **68** (no. 2 part 2): 356–394.

Sackett, J.R. 1968. Method and theory of Upper Paleolithic archaeology in southwestern France. In: *New Perspectives in Archaeology* (eds S.R. Binford & L.R. Binford). Chicago: Aldine, pp. 61–83.

Sackett, J.R. 1973. Style, function, and artifact variability in Palaeolithic assemblages. In: *The Explanation of Culture Change* (ed. C. Renfrew). London: Duckworth, pp. 317–325.

Sackett, J.R. 1977. The meaning of style in archaeology: a general model. *American Antiquity* **42**: 369–380.

Sackett, J.R. 1982. Approaches to style in lithic archaeology. *Journal of Anthropological Archaeology* **1**: 59–112.

Sackett, J.R. 1986a. Isochrestism and style: A clarification. *Journal of Anthropological Archaeology* **5**: 266–277.

Sackett, J.R. 1986b. Style, function, and assemblage variability: a reply to Binford. *American Antiquity* **51**: 628–634.

Sackett, J.R. 1988. The Mousterian and its aftermath: a view from the Upper Paleolithic. In: *Upper Pleistocene Prehistory of Western Eurasia* (eds H. Dibble & A. Montet-White).Philadelphia: University of Pennsylvania, University Museum Monographs No. 54, pp. 413–426.

Sancetta, C., Imbrie, J. & Kipp, N.G. 1973. Climatic record of the past 130,000 years in north Atlantic deep-sea core V23–82: correlation with the terrestrial record. *Quaternary Research* **3**: 110–116.

Sánchez Gõni, M.F. 1991. On the last glaciation and the interstadials during the Solutrean. *Current Anthropology* **4**: 573–575.

Sánchez Gõni, M.F. 1993. The identification of European Upper Palaeolithic interstadials from cave sequences. *Palynology* **17**: 1–22.

Schick, K.D. & Toth, N.T. 1993. *Making Silent Stones Speak: human evolution and the dawn of technology*. London: Weidenfeld & Nicolson.

Schlanger, N. 1989. *Trimming flakes and changing forms: lithic technology, tool design and early Paleolithic behaviour*. M.Phil. Dissertation, Department of Archaeology, Cambridge University.

Schlanger, N. 1994. *Flintknapping at the Belvédère: archaeological, technological and psychological investigations at the early Palaeolithic site of Maastricht-Belvédère (Limburg, the Netherlands)*. Doctoral dissertation, University of Cambridge.

Scott, K. 1980. Two hunting episodes of Middle Palaeolithic age at La Cotte de St Brelade, Jersey (Channel Islands). *World Archaeology* **12**: 137–152.

Scott, K. 1986. The bone assemblages of layers 3 and 6. In: *La Cotte de St Brelade 1961–1978: excavations by C.B.M. McBurney* (eds P.Callow & J.M. Cornford). Norwich: Geo Books, pp. 159–183.

Scott, K. 1989. Mammoth bones modified by humans: evidence from La Cotte de St Brelade, Jersey, Channel Islands. In: *Bone Modification* (eds R. Bonnischen & M.H. Sorg). Orono: Center for the Study of the First Americans, pp. 335–346.

Sejrup, H.P. & Larsen, E. 1991. Eemian-early Weichselian N-S temperature gradients; North Atlantic-NW Europe. *Quaternary International* **10–12**: 161–166.

Semenov, S.A. 1964. *Prehistoric Technology: an experimental study of the oldest tools and artefacts from traces of manufacture and use*. London: Cory, Adams & Mackay.

Sept, J.M. 1992. Was there no place like home? A new perspective on early hominid archaeological sites from the mapping of chimpanzee nests. *Current Anthropology* **33**: 187–207.

Séronie-Vivien, M. 1972. *Contribution à l'Etude du Sénonien en Aquitaine Septentrionale: ses stratotypes: Coniacien, Santonien, Campanien*. Paris: Centre National de la Recherche Scientifique.

Séronie-Vivien, M. & Séronie-Vivien, M.R. 1987. *Les Silex du Mézozoïque Nord-Aquitain: approche géologique de l'étude des silex pour servir à la recherche préhistorique*. Bulletin de la Société Linnéenne de Bordeaux 25 (Supplement).

Shackleton, N.J. 1969. The last interglacial in the marine and terrestrial records. *Proceedings of the Royal Society of London*, B, **174**: 135–154.

Shackleton, N.J. 1977. The oxygen isotope stratigraphic record of the Late Pleistocene. *Philosophical Transactions of the Royal Society of London*, B, **280**: 169–182.

Shackleton, N.J. 1987. Oxygen isotopes, ice volume and sea level. *Quaternary Science Reviews* 6: 183–190.

Shackleton, N.J. & Opdyke, N.D. 1973. Oxygen isotope and palaeomagnetic stratigraphy of equatorial Pacific core V28–238: oxygen isotope temperatures and ice volumes on a 10^5 and 10^6 year scale. *Quaternary Research* 3: 39–55.

Shackleton, N.J., Hall, M.A., Line, J. & Cang Shuxi 1983. Carbon isotope data in core V19–30 confirm reduced carbon dioxide concentration in the ice age atmosphere. *Nature* 306: 319–322.

Shackley, M.L. 1977. The *bout coupé* handaxe as a typological marker for the British Mousterian industries. In: *Stone Tools as Cultural Markers: change, evolution and complexity* (ed. R.V.S. Wright). Canberra: Australian Institute of Aboriginal Studies.

Shea, J.J. 1989 A functional study of the lithic industries associated with hominid fossils in the Kebara and Qafzeh caves, Israel. In: *The Human Revolution: behavioural and biological perspectives on the origins of modern humans* (eds P. Mellars & C. Stringer). Princeton: Princeton University Press, pp. 611–625.

Shea, J.J. 1993. Lithic use-wear evidence for hunting by Neandertals and early modern humans from the Levantine Mousterian. In: *Hunting and Animal Exploitation in the Later Palaeolithic and Mesolithic of Eurasia* (eds G.L. Peterkin, H.M. Bricker & P. Mellars). Archeological Papers of the American Anthropological Association No. 4, pp. 189–198.

Simek, J. 1988. Analyse de la répartition spatiale des vestiges de la couche VIII de la grotte Vaufrey. In: *La Grotte Vaufrey: paléoenvironnement, chronologie, activités humaines* (ed. J.-P. Rigaud). Mémoires de la Société Préhistorique Française 19, pp. 569–592.

Simpson, I.M. & West, R.G. 1958. On the stratigraphy and palaeobotany of a Late-Pleistocene organic deposit at Chelford, Cheshire. *New Phytologist* 57: 239–250.

Sireix, M. & Bordes, F. 1972. Le Moustérien de Chinchon (Gironde). *Bulletin de la Société Préhistorique Française* 69: 324–336.

Slott-Moller, R. 1990. La Faune. In: *Les Chasseurs d'Aurochs de La Borde: un site du Paléolithique moyen (Livernon, Lot)* (eds J. Jaubert, M. Lorblanchet, H. Laville, R. Slott-Moller, A. Turq & J.-P. Brugal) Paris: Maison des Sciences de l'Homme (Documents d'Archéologie Française 27), pp. 33–68.

Smith, F.H. 1983. A behavioral interpretation of changes in craniofacial morphology across the archaic/modern *Homo sapiens* transition. In: *The Mousterian Legacy* (ed. E. Trinkaus). Oxford: British Archaeological Reports International Series S164, pp. 141–163.

Smith, F.H. 1984. Fossil hominids from the Upper Pleistocene of Central Europe and the origin of modern Europeans. In: *The Origins of Modern Humans: A World Survey of the Fossil Evidence* (ed F.H. Smith & F. Spencer), pp. 137–210. New York: Alan R. Liss.

Smith, F.H. 1991. The Neandertals: evolutionary dead ends or ancestors of modern people? *Journal of Anthropological Research* 47, 219–238.

Smith, F.H. 1994. Samples, species and speculations in the study of modern human origins. In: *Origins of Anatomically Modern Humans* (eds M.H. Nitecki & D.V. Nitecki). New York: Plenum Press, pp. 227–249.

Smith, F.H. & Paquette, S.P. 1989. The adaptive basis of Neandertal facial form, with some thoughts on the nature of modern human origins. In: *The Emergence of Modern Humans: biocultural adaptations in the later Pleistocene* (ed. E. Trinkaus). Cambridge: Cambridge University Press, pp. 181–210.

Smith, F.H., Simek, J.F., & Harrill, M.S. 1989. Geographic variation in supraorbital torus reduction during the later Pleistocene (c. 80,000–15,000 BP). In: *The Human Revolution: behavioural and biological perspectives on the origins of modern humans* (eds P. Mellars & C. Stringer). Princeton: Princeton University Press, pp. 172–193.

Smuts, B.B., Cheney, D.L., Seyfarth, R.M., Wrangham, R.W. & Struhsaker, T.T. (eds) 1987. *Primate Societies*. Chicago: University of Chicago Press.

Soffer, O. 1985a. *The Upper Paleolithic of the Central Russian Plain*. Orlando: Academic Press.

Soffer, O. 1985b. Patterns of intensification as seen from the Upper Paleolithic of the Central Russian Plain. In: *Prehistoric Hunter-Gatherers: the emergence of cultural complexity* (eds T.D. Price & J.A. Brown). Orlando: Academic Press, pp. 235–270.

Soffer, O. 1989. The Middle to Upper Palaeolithic transition on the Russian Plain. In: *The Human Revolution: behavioural and biological perspectives on the origins of modern humans* (eds P. Mellars & C. Stringer). Princeton: Princeton University Press, pp. 714–742.

Soffer, O. 1992. Social transformations at the Middle to Upper Paleolithic transition: the implications of the European record. In: *Continuity or Replacement? Controversies in Homo sapiens evolution* (eds G.Bräuer & F.H. Smith). Rotterdam: Balkema, pp. 247–260.

Soffer, O. 1994. Ancestral lifeways in Eurasia – the Middle and Upper Paleolithic records. In: *Origins of Anatomically Modern Humans* (eds M.H. Nitecki & D.V. Nitecki). New York: Plenum Press, pp. 101–109.

Solecki, R.L. 1992. More on hafted projectile points in the Mousterian. *Journal of Field Archaeology* **19**: 207–212.

Sonneville-Bordes, D. de 1960. *Le Paléolithique Supérieur en Périgord*. Bordeaux: Delmas.

Sonneville-Bordes, D. de 1969. Les industries moustériennes de l'abri Caminade-Est, commune de La Canéda (Dordogne). *Bulletin de la Société Préhistorique Française* **66**: 293–310.

Sonneville-Bordes, D. de 1989. Foyers paléolithiques en Périgord. In: *Nature et Fonction des Foyers Préhistoriques* (eds M. Olive & Y. Taborin). Nemours: Mémoires du Musée de Préhistoire d'Ile de France No. 2, pp. 225–238.

Spaulding, A.C. 1953. Statistical techniques for the discovery of artifact types. *American Antiquity* 18: 305–313.

Speth, J.D. 1983. *Bison Kills and Bone Counts: decision making by ancient hunters*. Chicago: University of Chicago Press.

Speth, J.D. 1987. Early hominid subsistence strategies in seasonal habitats. *Journal of Archaeological Science* **14**: 13–29.

Speth, J.D. 1990. Seasonality, resource stress, and food sharing in so-called 'egalitarian' foraging societies. *Journal of Anthropological Archaeology* **9**: 148–188.

Speth, J.D. & Spielmann, K.A. 1983. Energy source,

protein metabolism, and hunter-gatherer subsistence strategies. *Journal of Anthropological Archaeology* **2**: 1–31.

Spiess, A.E. 1979. *Reindeer and Caribou Hunters: an archaeological study*. New York: Academic Press.

Spurrell, F.C.J. 1880. On the discovery of the place where Palaeolithic implements were made at Crayford. *Quarterly Journal of the Geological Society* **36**: 544–548.

Standen, V. & Foley, R.A. (eds) 1989. *Comparative Socioecology*. Oxford: Blackwell.

Stiner, M.C. 1990. The use of mortality patterns in archaeological studies of hominid predatory adaptations. *Journal of Anthropological Archaeology* **9**: 305–351.

Stiner, M.C. 1991a. Food procurement and transport by human and non-human predators. *Journal of Archaeological Science* **18**: 455–482.

Stiner, M.C. 1991b. An interspecific perspective on the emergence of the modern human predatory niche. In: *Human Predators and Prey Mortality*, (ed. M.C. Stiner). Boulder: Westview Press.

Stiner, M.C. 1991c. The faunal remains at Grotta Guattari: a taphonomic perspective. *Current Anthropology* **32**: 103–117.

Stiner, M.C. 1992. Overlapping species 'choice' by Italian Upper Pleistocene predators. *Current Anthropology* **33**: 433–451.

Stiner, M.C. 1993a. Modern human origins – faunal perspectives. *Annual Reviews of Anthropology* **22**: 55–82.

Stiner, M.C. 1993b. Small animal exploitation and its relation to hunting, scavenging, and gathering in the Italian Mousterian. In: *Hunting and Animal Exploitation in the Later Palaeolithic and Mesolithic of Eurasia* (eds G.L. Peterkin, H.M. Bricker & P. Mellars). Archeological Papers of the American Anthropological Association No. 4, pp. 107–126.

Stiner, M.C. & Kuhn, S.L. 1992. Subsistence, technology, and adaptive variation in Middle Paleolithic Italy. *American Anthropologist* **94**: 306–339.

Stoneking, M., & Cann, R.L., 1989. African origin of human mitochondrial DNA. In: *The Human Revolution: behavioural and biological perspectives on the origins of modern humans* (eds P. Mellars &

C. Stringer). Princeton: Princeton University Press, pp. 17–30.

Stoneking, M., Sherry, S.T., Redd, A.J. & Vigilant, L. 1992. New approaches to dating suggest a recent age for the human DNA ancestor. In: *The Origin of Modern Humans and the Impact of Chronometric Dating* (eds M.J. Aitken, C.B. Stringer & P.A. Mellars). London: Royal Society (Philosophical Transactions of the Royal Society, series B, **337**, no. 1280), pp. 167–176.

Stringer, C.B. 1990. The emergence of modern humans. *Scientific American* **263**: 68–74.

Stringer, C.B. 1992. Reconstructing recent human evolution. In: *The Origin of Modern Humans and the Impact of Chronometric Dating* (eds M.J. Aitken, C.B. Stringer & P.A. Mellars). London: Royal Society (Philosophical Transactions of the Royal Society, series B, **337**, no. 1280), pp. 217–224.

Stringer, C.B. 1994. Out of Africa – a personal history. In: *Origins of Anatomically Modern Humans* (eds M.H. Nitecki & D.V. Nitecki). New York: Plenum Press, pp. 149–172.

Stringer, C.B. & Andrews, P., 1988. Genetic and fossil evidence for the origin of modern humans. *Science* **239**: 1263–1268.

Stringer, C.B. & Gamble, C. 1993. *In Search of the Neanderthals: solving the puzzle of human origins*. London: Thames & Hudson.

Stringer, C.B. & Gamble, C. 1994. Reply: Confronting the Neanderthals. *Cambridge Archaeological Journal* **4**: 112–119.

Stringer, C.B., Hublin, J.-J., & Vandermeersch, B. 1984. The origin of anatomically modern humans in Western Europe. In: *The Origins of Modern Humans: A World Survey of the Fossil Evidence* (eds F.H. Smith & F. Spencer). New York: Alan R. Liss, pp. 51–135.

Stuart, A.J. 1982. *Pleistocene Vertebrates in the British Isles*. London: Longman.

Svoboda, J. 1993. The complex origin of the Upper Paleolithic in the Czech and Slovak Republics. In: *Before Lacaux: the complex record of the early Upper Paleolithic* (eds H. Knecht, A. Pike-Tay & R. White). Boca Raton: CRC Press, pp. 23–36.

Taborin, Y. 1990. Les prémices de la parure. In: *Paléolithique Moyen Récent et Paléolithique Supérieur Ancien en Europe* (ed. C. Farizy). Nemours: Mémoires du Musée de Préhistoire d'Ile de France No. 3, pp. 335–344.

Taborin, Y. 1993. Shells of the French Aurignacian and Perigordian. In: *Before Lacaux: the complex record of the early Upper Paleolithic* (eds H. Knecht, A. Pike-Tay & R. White). Boca Raton: CRC Press, pp. 211–229.

Tavoso, A. 1984. Réflexion sur l'économie des matières premières au Moustérien. *Bulletin de la Société Préhistorique Française* **81**: 79–82.

Tavoso, A. 1987a. *Les Premiers Chasseurs de la Vère*. Marseilles: Université de Provence, Groupement d'Etudes de Recherches Préhistoriques.

Tavoso, A. 1987b. Le Moustérien de la Grotte Tournal. *Cypsela* **6**: 161–174.

Taylor, K.C., Hammer, C.U., Alley, R.B., Clausen, H.B., Dahl-Jensen, D., Gow, A.J., Gundestrup, N.S., Kipfstuhl, J., Moore, J.C. & Waddington, E.D. 1993. Electrical conductivity measurements from the GISP2 and GRIP Greenland ice cores. *Nature* **366**: 549–552.

Templeton, A.R. 1993. The 'Eve' hypothesis: a genetic critique and reanalysis. *American Anthropologist* **95**: 51–72.

Texier, J.-P. 1982. *Les Formations Superficielles du Bassin de l'Isle*. Paris: Centre National de la Recherche Scientifique. (Institut du Quaternaire, Université de Bordeaux I, Cahiers du Quaternaire 4).

Texier, J.-P. 1990. Bilan sur les paléoenvironnements des Moustériens charentiens d'Europe occidentale d'après les données de la géologie. In: *Les Moustériens Charentiens* (Proceedings of International Colloquium, Brive, August 1990), pp. 15–19.

Thorne, A.G. & Wolpoff, M.H. 1992. The multiregional evolution of humans. *Scientific American* **266**: 28–33.

Tixier, J. 1978 *Méthode pour l'Etude des Outillages Lithiques: notice sur les travaux scientifiques de Jacques Tixier presentée en vue de grade de Docteur es Lettres*. University of Paris X.

Tixier, J., Inizan, M.L. & Roche, H. 1980. *Préhistoire de la Pierre Taillée I: Terminologie et technologie*. Antibes: Cercle de Recherches et d'Etudes Préhistoriques.

Tooby, J. & DeVore, I. 1987. The reconstruction of

hominid behavioral evolution through strategic modelling. In: *The Evolution of Human Behavior: primate models* (ed. W.G. Kinzey). Albany: State University of New York Press, pp. 183–237.

Trinkaus, E. 1983. *The Shanidar Neandertals*. New York: Academic Press.

Trinkaus, E. (ed.) 1989a. *The Emergence of Modern Humans: biocultural adaptations in the later Pleistocene*. Cambridge: Cambridge University Press.

Trinkaus, E. 1989b. The Upper Pleistocene transition. In: *The Emergence of Modern Humans: biocultural adaptations in the later Pleistocene* (ed. E. Trinkaus). Cambridge: Cambridge University Press, pp. 42–66.

Trinkaus, E. & Shipman, P. 1993. *The Neandertals: changing the image of mankind*. London: Jonathan Cape.

Tuffreau, A. 1971. Quelques aspects du Paléolithique ancien et moyen dans le nord de la France (Nord et Pas-de-Calais) *Bulletin Spécial de la Société de Préhistoire du Nord* **8**: 1–99.

Tuffreau, A. 1979. Le gisement moustérien du château d'eau de Corbehem (Pas-de-Calais). *Gallia-Préhistoire* **2**: 371–389.

Tuffreau, A. (ed.) 1989a. *Paléolithique et Mésolithique du Nord de la France: nouvelles recherches*. Lille: Centre d'Etudes et de Recherches Préhistorique (CERP), Université des Sciences et Techniques de Lille Flandres Artois.

Tuffreau, A. 1989b. Le gisement paléolithique moyen de Champvoisy (Marne). In: *Paléolithique et Mésolithique du Nord de la France: nouvelles recherches* (ed. A. Tuffreau). Lille: Centre d'Etudes et de Recherches Préhistorique (CERP), Université des Sciences et Techniques de Lille Flandres Artois, pp. 69–78.

Tuffreau, A. 1992. Middle Paleolithic settlement in northern France. In: *The Middle Paleolithic: adaptation, behavior, and variability* (eds H.L. Dibble & P.A. Mellars). Philadelphia: University of Pennsylvania, University Museum Monographs No. 72, pp. 59–74.

Tuffreau, A. (ed.) 1993. *Riencourt-les-Bapaume (Pas-de-Calais): un Gisement du Paléolithique Moyen*. Paris: Maison des Sciences de l'Homme (Documents d'Archéologie Française no. 37).

Tuffreau, A. & Marcy, J.-L. 1988. Synthèse des données archéologiques. In: *Le Gisement Paléolithique Moyen de Biache-Saint-Vaast (Pas-de-Calais)*. Vol. I: *Stratigraphie, environnement, études archéologiques* (eds A. Tuffreau & J. Sommé). Paris: Mémoires de la Société Préhistorique Française 21, pp. 301–307.

Tuffreau, A. & Sommé, J. (eds) 1986. *Chronostratigraphie et Faciés Culturels du Paléolithique Inférieur et Moyen dans l'Europe du Nord-Ouest*. Paris: Association Française pour l'Etude du Quaternaire.

Tuffreau, A. & Sommé, J. 1988. *Le Gisement Paléolithique Moyen de Biache-Saint-Vaast (Pas-de-Calais)*. Vol. I: *Stratigraphie, environnement, études archéologiques*. Mémoires de la Société Préhistorique Française 21.

Tuffreau, A. & Zuate y Zuber, J. 1975. La terrasse fluviatile de Bagarre (Etaples, Pas-de-Calais) et ses industries: note préliminaire. *Bulletin de la Société Préhistorique Française* **72**: 229–235.

Tuffreau, A., Révillion, S., Sommé, J., Aitken, M., Huxtable, J. & Leroi-Gourhan, A. 1985. Le gisement paléolithique moyen de Seclin (Nord, France). *Archäologisches Korrespondenzblatt* **15**: 131–138.

Turner, A. 1989. Sample selection, schlepp effects and scavenging: the implications of partial recovery for interpretations of the terrestrial mammal assemblage from Klasies River Mouth. *Journal of Archaeological Science* **16**: 1–11.

Turner, C. 1985. Problems and pitfalls in the application of palynology to Pleistocene archaeological sites in western Europe. In: *Palynologie Archéologique*, (eds J. Renault-Miskovsky, Bui-Thai-Mai & M. Girard). Paris: Centre National de la Recherche Scientifique, pp. 347–373.

Turner, C. & Hannon, G.E. 1988. Vegetational evidence for late Quaternary climatic changes in southwest Europe in relation to the influence of the North Atlantic ocean. *Philosophical Transactions of the Royal Society of London*, series B, **318**: 451–485.

Turon, J.-L. 1984. Direct land/sea correlations in the last interglacial complex. *Nature* **309**: 673–676.

Turq, A. 1977a. Le complexe d'habitat paléolithique du Plateau Cabrol. *Bulletin de la Société Préhistorique Française* **74**: 489–504.

Turq, A. 1977b. Première approche sur le Paléolithique moyen du gisement des Ardailloux, commune de Soturac (Lot). *Bulletin de la Société des Etudes du Lot* **98**: 222–242.

Turq, A. 1978. A propos de deux sites moustériens de plein air du Fumelois (Lot-et-Garonne). *Bulletin de la Société Préhistorique Française* **75**: 460–471.

Turq, A. 1988a. Le Paléolithique inférieur et moyen en Haut-Agenais: état des recherches. *Revue de l'Agenais* **115**: 83–112.

Turq, A. 1988b. Le Moustérien de type Quina du Roc de Marsal à Campagne (Dordogne): contexte stratigraphique, analyse lithologique et technologique. *Documents d'Archéologie Perigourdine* (A.D.R.A.P.) **3**: 5–30.

Turq, A. 1989a. Exploitation des matières premières lithiques et occupation du sol: l'exemple du Moustérien entre Dordogne et Lot. In: *Variations des Paléomilieux et Peuplement Préhistorique*, (ed. H. Laville). Paris: Centre National de la Recherche Scientifique, pp. 179–204.

Turq, A. 1989b. Approche technologique et économique du faciès Moustérien de type Quina: étude préliminaire. *Bulletin de la Société Préhistorique Française* **86**: 244–256.

Turq, A. 1989c. Le squelette de l'enfant du Roc de Marsal: les données de la fouille. *Paléo* 1: 47–54.

Turq, A. 1990. Exploitation du milieu minéral: technologie, économie et circulation du silex. In: *Les Chasseurs d'Aurochs de La Borde: un site du Paléolithique moyen (Livernon, Lot)* (eds J. Jaubert, M. Lorblanchet, H. Laville, R. Slott-Moller, A. Turq & J.-P. Brugal). Paris: Maison des Sciences de l'Homme (Documents d'Archéologie Française 27), pp. 103–115.

Turq, A. 1992a. Raw material and technological studies of the Quina Mousterian in Périgord. In: *The Middle Paleolithic: adaptation, behavior, and variability* (eds H.L. Dibble & P.A. Mellars). Philadelphia: University of Pennsylvania, University Museum Monographs No. 72, pp. 75–85

Turq, A. 1992b. *Le Paléolithique Inférieur et Moyen Entre les Vallées de la Dordogne et du Lot.* Doctoral dissertation, University of Bordeaux I.

Turq, A. 1993. L'approvisionnement en matières premières lithiques au Moustérien et au début du Paléolithique supérieur dans le Nord-est du Bassin Aquitain (France). In: *El Origen del Hombre Moderno en el Suroeste de Europa* (ed. V. Cabrera Valdés). Madrid: Universidad Nacional de Educacion a Distancia, pp. 315–326.

Turq, A. & Dolse, P. 1988. Le site moustérien de Tour-de-Faure, Lot. *Bulletin de la Société des Etudes du Lot* **109**:189–219.

Turq, A., Geneste, J.-M, Jaubert, J., Lenoir, M. & Meignen, L. 1990. Les Moustériens charentiens du sud-ouest et du Languedoc oriental: approche technologique et variabilité géographique. In: *Les Moustériens Charentiens* (Proceedings of International Colloquium, Brive, August 1990), pp. 53–64.

Valensi, L. 1960. De l'origine des silex protomagdaléniens de l'Abri Pataud, Les Eyzies (Dordogne). *Bulletin de la Société Préhistorique Française* **57**: 80–84.

Valladas, H. 1985. *Datation par Thermoluminescence de Gisements Moustériens du Sud de la France.* Doctoral Dissertation, Université Pierre et Marie Curie (Paris VI).

Valladas, H., Geneste, J.-M., Joron, J.-L. & Chadelle, J.-P. 1986. Thermoluminescence dating of Le Moustier (Dordogne, France). *Nature* **322**: 452–454.

Valladas, H., Chadelle, J.-P., Geneste, J.-M., Joron, J.-L., Meignen, L. & Texier, P.-J. 1987. Datations par la thermoluminescence de gisements moustériens du sud de la France. *L'Anthropologie* **91**: 211–226.

Valoch, K. 1983. L'origine des différents technocomplexes du Paléolithique supérieur en Morave. In: *Aurignacien et Gravettien en Europe* Vol. 2, (ed. L. Bánesz & J.K. Kozlowski). Liège: Etudes et Recherches Archéologiques de l'Université de Liège 13, pp. 371–378.

Valoch, K. 1990. La Morave il y a 40,000 ans. In *Paléolithique Moyen Récent et Paléolithique Supérieur Ancien en Europe* (ed. C. Farizy). Nemours: Mémoires du Musée de Préhistoire d'Ile de France No. 3, pp. 115–124.

Van den Brink, F.H. 1967. *A Field Guide to the Mammals of Britain and Europe.* London: Collins.

Van der Hammen, T., Maarleveld, G.C., Vogel, J.C. & Zagwijn, W.H. 1967. Stratigraphy, climatic succession and radiocarbon dating of the last glacial of the Netherlands. *Geologie en Mijnbouw* **46**: 79–95.

Vandermeersch, B. 1965. Position stratigraphique et chronologie relative des restes humains du Paléolithique moyen du sud-ouest de la France. *Annales de Paléontologie* (Vertébrés) 51: 69–126.

Vandermeersch, B. 1970. Une sépulture moustérienne avec offrandes découverte dans la grotte de Qafzeh. *Comptes Rendus de l'Académie des Sciences de Paris* **270**: 298–301.

Vandermeersch, B. 1976. Les sépultures néanderthaliennes. In: *La Préhistoire Française* (ed. H. de Lumley). Paris: Centre National de la Recherche Scientifique, pp. 725–727.

Vandermeersch, B. 1981. *Les Hommes Fossiles de Qafzeh (Israël)*. Paris: Centre National de la Recherche Scientifique.

Vandermeersch, B. 1989. The evolution of modern humans: recent evidence from southwest Asia. In: *The Human Revolution: behavioural and biological perspectives on the origins of modern humans* (eds P. Mellars & C. Stringer). Princeton: Princeton University Press, pp. 155–164.

Vandermeersch, B. 1993a. Le Proche Orient et l'Europe: continuité ou discontinuité. In: *El Origen del Hombre Moderno en el Suroeste de Europa* (ed. V. Cabrera Valdés). Madrid: Universidad Nacional de Educacion a Distancia, pp. 361–372.

Vandermeersch, B. 1993b. Appendix: was the Saint-Césaire discovery a burial? In: *Context of a Late Neandertal: implications of multidisciplinary research for the transition to Upper Paleolithic adaptations at Saint-Césaire, Charente-Maritime, France* (eds F. Lévêque, A.M. Backer & M. Gilbaud). Madison (Wisconsin): Prehistory Press (Monographs in World Archaeology No. 16), pp. 129–131.

Van Peer, P. 1991. Interassemblage variability and Levallois styles: the case of the northern African Middle Palaeolithic. *Journal of Anthropological Archaeology* **10**: 107–151.

Van Peer, P. 1992. *The Levallois Reduction Strategy*. Madison (Wisconsin): Prehistory Press (Monographs in World Archaeology 13).

Van Vliet-Lanoë, B. 1989. Dynamics and extent of the Weichselian permafrost in western Europe (substage 5e to stage 1). *Quaternary International* **3–4**: 109–113.

Van Vliet-Lanoë, B. 1990. Le pédocomplexe de Warneton: où en est-on? Bilan paléopédologique et micromorphologique. *Quaternaire* **1**: 65–76.

Van Vliet-Lanoë, B., Valdès, A. & Cliquet, D. 1993. Position stratigraphique des industries à lames du Paléolithique moyen en Europe occidentale. In: *Riencourt-les-Bapaume (Pas-de-Calais): un Gisement du Paléolithique Moyen* (ed. A. Tuffreau). Paris: Maison des Sciences de l'Homme, pp. 104–106.

Villa, P. 1977. Sols et niveaux d'habitat du Paléolithique inférieur en Europe et au Proche Orient. *Quaternaria* **19**: 107–134.

Villa, P. 1983. *Terra Amata and the Middle Pleistocene archaeological record of southern France*. University of California Publications in Anthropology **13**: 1–303.

Villeneuve, L. & Farizy, C. 1989. Les témoins de combustion du gisement moustérien de Champlost (Yonne). In: *Nature et Fonction des Foyers Préhistoriques* (eds M. Olive & Y. Taborin). Mémoires du Musée de Préhistoire d'Ile de France No. 2, pp. 135–140.

Vogel, J.C. & Van der Hammen, T. 1967. The Denekamp and Paudorf interstadials. *Geologie en Mijnbouw* **46**: 188–194.

Wainscoat, J.S., Hill, A.V.S., Thein, S.L., Flint, J., Chapman, J.C., Weatherall, D.F., Clegg, J.B. & Higgs, D.R. 1989. Geographic distribution of alpha- and beta-globin gene cluster polymorphisms. In: *The Human Revolution: behavioural and biological perspectives on the origins of modern humans* (eds P. Mellars & C. Stringer). Princeton: Princeton University Press, pp. 39–46.

Washburn, S.L. & Lancaster, C.S. 1968. The evolution of hunting. In: *Man the Hunter* (eds R.B. Lee & I. DeVore). Chicago: Aldine, pp. 293–303.

Watts, W.A. 1988. Europe. In: *Vegetation History* (eds B. Huntley & T. Webb III). Dordrecht: Kluwer, pp. 155–192.

West, R.G. 1977. *Pleistocene Geology and Biology*. London: Longman.

West, R.G., Dickson, C.A., Catt, J.A., Weir, A.H. & Sparks, B.W. 1974. Late Pleistocene deposits at Wretton, Norfolk II: Devensian deposits. *Philosophical Transactions of the Royal Society*, series B, **267**: 337–420.

Whallon, R. 1989. Elements of cultural change in the later Palaeolithic. In: *The Human Revolution: behavioural and biological perspectives on the origins of modern humans* (eds P. Mellars & C. Stringer). Princeton: Princeton University Press, pp. 433–454.

White, C. & Peterson, N. 1969. Ethnographic interpretation of the prehistory of western Arnhem Land. *Southwestern Journal of Anthropology* **25**: 45–67.

White, R. 1982. Rethinking the Middle/Upper Paleolithic transition. *Current Anthropology* **23**: 169–192.

White, R. 1985. *Upper Paleolithic Land Use in the Périgord: a topographic approach to subsistence and settlement.* British Archaeological Reports International Series S253.

White, R. 1989. Production complexity and standardization in early Aurignacian bead and pendant manufacture: evolutionary implications. In: *The Human Revolution: behavioural and biological perspectives on the origins of modern humans* (eds P. Mellars & C. Stringer). Princeton: Princeton University Press, pp. 366–390.

White, R. 1993. Technological and social dimensions of 'Aurignacian-age' body ornaments across Europe. In: *Before Lascaux: the complex record of the early Upper Paleolithic* (eds H. Knecht, A. Pike-Tay & R. White). Boca Raton: CRC Press, pp. 277–300.

Wiessner, P. 1983. Style and social information in Kalahari San projectile points. *American Antiquity* **48**: 253–276.

Wiessner, P. 1984. Reconsidering the behavioral basis for style: a case study among the Kalahari San. *Journal of Anthropological Archaeology* **3**: 190–234.

Wijmstra, T.A., Young, R. & Witte, H.J.L. 1990. An evaluation of the climatic conditions during the Late Quaternary in northern Greece by means of multivariate analysis of palynological data and comparison with recent phytosociological and climatic data. *Geologie en Mijnbouw* **69**: 243–251.

Wilson, A.C. & Cann, R.L. 1992. The recent African genesis of humans. *Scientific American* **266**(4): 22–27.

Wilson, J.F. 1975. The last glacial environment at the Abri Pataud. In: *Excavation of the Abri Pataud, Les Eyzies (Dordogne)* (ed. H.L. Movius). Cambridge (Mass.): Peabody Museum, Harvard University, pp. 75–186.

Wind, J., Pulleybank, E.G., de Grolier, E. & Bichakjian (eds) 1989. *Studies in Language Origins*, Vol. 1. Amsterdam: John Benjamins.

Wintle, A.G. 1990. A review of current research on TL dating of loess. *Quaternary Science Reviews* **9**: 385–397.

Wintle, A.G., Shackleton, N.J. & Lautridou, J.P. 1984. Thermoluminescence dating of periods of loess deposition and soil formation in Normandy. *Nature* **310**: 491–493.

Wobst, M. 1974. Boundary conditions for Palaeolithic social systems: a simulation approach. *American Antiquity* **39**: 147–178.

Wobst, M. 1976. Locational relationships in Paleolithic society. *Journal of Human Evolution* **5**: 49–58.

Woillard, G.M. 1978. Grande Pile peat bog: a continuous pollen record for the last 140,000 years. *Quaternary Research* **9**: 1–21.

Woillard, G.M. & Mook, W.G. 1982. Carbon-14 dates at Grande Pile: correlation of land and sea chronologies. Science **215**: 159–161.

Wolpoff, M.H. 1980. *Palaeoanthropology.* New York: Knopf.

Wolpoff, M.H. 1989. Multiregional evolution: the fossil alternative to Eden. In: *The Human Revolution: behavioural and biological perspectives on the origins of modern humans* (eds P. Mellars & C. Stringer). Princeton: Princeton University Press, pp. 62–108.

Wolpoff, M.H. 1992. Theories of modern human origins. In: *Continuity or Replacement: Controversies in* Homo sapiens *evolution* (eds. G. Bräuer & F.H. Smith). Rotterdam: Balkema, pp. 25–64.

Wolpoff, M.H., Thorne, A.G., Smith, F.H., Frayer, D.W., & Pope, G.G. 1994. Multiregional evolution: a world-wide source for modern human populations. In: *Origins of Anatomically Modern*

Humans (eds M.H. Nitecki & D.V. Nitecki). New York: Plenum Press, pp. 175–199.

Wrangham, R.W. 1987a. Evolution of social structure. In: *Primate Societies* (eds B.B. Smuts, D.L. Cheney, R.M. Seyfarth, R.W. Wrangham & T.T. Struhsaker). Chicago: University of Chicago Press, pp. 282–296.

Wrangham, R.W. 1987b. The significance of African apes for reconstructing human social evolution. In: *The Evolution of Human Behavior: primate models* (ed. W.G. Kinzey). Albany: State University of New York Press, pp. 51–71.

Wreschner, E. 1980. Red ochre and human evolution: a case for discussion. *Current Anthropology* 21: 631–644.

Wymer, J.J. 1968. *Lower Palaeolithic Archaeology in Britain, as represented in the Thames Valley*. London: John Baker.

Wynn, T. 1979. The intelligence of later Acheulean hominids. *Man* 14: 379–91.

Wynn, T. 1985. Piaget, stone tools and the evolution of human intelligence. *World Archaeology* 17: 32–43.

Wynn, T. 1989. *The Evolution of Spatial Competence*. Urbana: University of Illinois Press.

Wynn, T. 1991. Tools, grammar and the archaeology of cognition. *Cambridge Archaeological Journal* 1: 191–206.

Wynn, T. 1993. Layers of thinking in tool behavior. In: *Tools, Language and Cognition in Human Evolution* (eds K.R. Gibson & T. Ingold). Cambridge: Cambridge University Press, pp. 389–405.

Wynn, T. & Tierson, F. 1990. Regional comparison of the shapes of later Acheulian handaxes. *American Anthropologist* 92: 73–84.

Zagwijn, W.H. 1961. Vegetation, climate and radiocarbon datings in the Late Pleistocene of the Netherlands. Part I: Eemian and Early Weichselian. *Mededelingen Geologisches Stichting* (New Series) 4: 15–45.

Zagwijn, W.H. 1990. Vegetation and climate during warmer intervals in the late Pleistocene of western and central Europe. *Quaternary International* 3/4: 57–67.

Zubrow, E. 1989. The demographic modeling of Neanderthal extinction. In: *The Human Revolution: behavioural and biological perspectives on the origins of modern humans* (eds P. Mellars & C. Stringer). Princeton: Princeton University Press, pp. 212–231.

Index of Sites ———————————————————

Sites with many references are given subheadings

Abri Audi, 187
Abri Blanchard, 189, 246, 396, 397, 398
Abri Brouillaud, 145, 146–7, 245, 246
Abri Caminade, 90, 185, 186, 192, 246, 325, 327, 353
Abri Cellier, 397
Abri Chadourne, 102, 180, 185–6, 190, 192, 325, 327, 348, 353, 363
Abri du Chasseur, 184, 185, 186, 192
Abri Olha, 132
Abri Pataud, 201, 364, 365, 400, 409
Aldène, 302, 303, 364
Amud, 3
L'Arbreda, 409
Arcy-sur-Cure
    absolute age measurements, 409
    bone artefacts and animal-tooth pendants, 416
    fossils, imported, 371
    hut plans, 313
    industries
        Châtelperronian stone tools, 414
        racloir-reduction, 102
    pollen sequences, 48–9
    population interactions, 412, 414, 416
    see also Grotte du Bison; Grotte de l'Hyène; Grotte du Renne
Les Ardailloux, 146–7, 266
Aveyron, 263

Bacho Kiro, 409
Bagarre, 64
Barbe Cornio, 158, 160, 266
Baume les Peyrards, 299, 304
Baume Vallée, 188, 247
Baume-Bonne, 302, 303, 364
Belcayre, 397
Bergerac region, 253, 257, 259, 265
Biache-Saint-Vaast
    dating, 3, 4

Ferrassie-type industries, 341, 354
    paving, 303
    raw materials, 101, 340
    tool technology, 88
        racloirs, 101, 103, 113, 114–15
        recurrent techniques, 67–9, 70
Bilzingsleben, 372, 373
Bisitun, 105, 107
Bison, Grotte du Bison, 299, 301, 309
Blanchard *see* Abri Blanchard
Bocksteinschmiede, 125, 373, 374, 375
Boker Tachtit, 419
La Borde, 232–5
    faunal assemblages, 223, 229–41, 332
        behavioural interpretations, 236–44
    industries, 63, 132, 242, 244, 263
    location map, 234
    raw materials, 148
Border Cave, 402
Bournemouth, 129
Le Breuil, 303
Brugas, 188
Buffebale, 258
La Burlade, 145, 146–7, 266

Caminade *see* Abri Caminade
Campsegret, 159, 265
Les Canalettes, 287–92, 309, 310, 311
Carrière Thomasson, 267, 332
Castanet, 396
El Castillo, 131, 409, 410
La Cavaille, 187
Le Cérisier, 303
Chadourne *see* Abri Chadourne
La Chaise, 127, 353, 372
Champlost, 101, 102, 299, 341
Champvoisy, 102, 340
La Chapelle-aux-Saints
    burial practices, 375–6

General Index

Names of sites are located in the Index of Sites

Acheulian industries, 35, 48, 121, 124, 136, 264,
    281, 311, 404
Africa
    Koobi Fora, 311
    Olduvai, 311, 381
    origin of modern humans, 402, 404
Alaska
    Nunamiut, 321
    reindeer, 321, 324
    Tuluaqmiut, 324
Alpes Maritimes Department, 281
Amersfoot interstadial, 13, 15, 19, 20-1, 27
Amudian industries, 77
Angoumian deposits, 145
animals *see* faunal exploitation
Armeria, 39
art and decoration, 396-8
Artemisia, 19, 25
Asinipodian Mousterian, 199
Asinus, steppe ass, 196
âteliers de taille, 167, 386
Aterian industries, 174
Atlantic coast, 53, 257
Atlantic, North
    cold/warm currents, 22, 24
    sea-bed sediments, 14-15
    sea-surface temperatures, 21, 23
Atlantic Plain, 247, 262
Auel, J., *Clan of the Cave Bear*, 1, 411
Aurignacian industries, 49, 166, 167, 264, 301
    chronology, 393, 409-10
    distribution, 406
    interstratification with Châtelperronian, 416
    vs Mousterian, reindeer, 201
    sea shells, use and export, 399
    stone and bone artefacts, 394-9
    transition, MP to UP, 393-400
aurochs, 188-9
    see also bovids

Australia, 321, 335, 366
Auvézère river, 255
aven deposits, 235, 237

backed knives, 120-2, 134, 171, 173, 326
Bajocian/Bathonian outcrop, 258, 266
becs, 122
Bergerac flint, 144, 167-9
Beune valley, 246, 248
bifacial tools, 124-36, 382-3
Binford models, 315-32
    social organization, 357-60, 364-5
bison, 188-9
    see also bovids; faunal assemblages
Black Sea coast, Russia, 398
blade technology, 77-87, 134, 412
Bølling–Allerød interstadial, 20
Bond cycles, 27
bone and ivory artefacts, 371-5, 395, 396, 397, 400,
    416
    decorated animal teeth, 373, 374, 375
    incised bones, 372, 395
    perforated bones and teeth, 373, 395, 396
Bordes, F
    Tale of Two Caves, 292
    taxonomy of flake tools,170, 180-3, 315
        evaluation, 180-3, 315
Bos primigenius (aurochs), 188-9
    see also bovids
Boulou valley, 248
bovids, 48, 52-3, 194-240
    butchery sites, 119, 138, 238, 310, 332
    Denticulate and Quina Mousterian, Combe
        Grenal, 330
    estimated age distributions, 236
    frequencies, 37, 195, 197, 200
    hunting sites, 231-6
        Combe Grenal, 207, 208, 240-1
        Coudoulous, 235-6

orbital forcing, 10
ornaments *see* decorative or symbolic artefacts
oxygen-isotope ratios
 deep sea sediment patterns, 9–10
 GRIP and GISP2 cores, Greenland, 25, 26
 Mediterranean, 27
 see also climate

paving, 301–3, 364
 Aldène, 303
 cobble paving, 302–3, 364
 paved hearths, 296–7, 299
 Solvieux, 303
perçoirs, 122
perforated objects, 373–5
Permotriassic limestone, 141
phallic carvings, 398
Piaget, J, on cognitive development, 384–5
piercers, 122
pigments
 manganese dioxide, 363, 369–70
 ochre, 363, 369
 use, 369–71
pits, 305, 306, 307, 364
planes (*rabots*), 122
plant food, processing, 119, 138, 351
Poland
 Piekary, 78
 principal stadials and interstadials, 19
pollen sequences
 Arcy-sur-Cure, 49
 Combe Grenal, 36, 40
 La Grande Pile, 15, 18
 Le Moustier, 47
 Les Echets, 14–18
 Padul, 14
 Pech de l'Azé II, 43
 Valle di Castiglione, 14
population dispersal, 401–18
 interactions, 411–18
population fluctuations, 345–8
post-holes, 308, 364
Poterium, 39
primates
 chimpanzee, use of pigment, 370
 planning abilities, 390
 social patterns, 311, 356–7
processualist approaches, 7
procurement and distribution of raw materials,
 141–68
Pyrenees, 231, 232, 257, 262, 347
 Pyrenean foothills, 229
Pyrenees and Cantabria, Vasconian industries,
 130–2, 174, 320, 344, 354–5

quarry sites *see* extraction sites
quartzite

artefacts, 289, 291
chopping tools, 243–4
vs flint tools, 139
see also raw materials
Quina Mousterian industries, 73–7, 130, 157, 171,
 262–3, 293
 functional interpretation, 325, 349–51
 limace forms, 174, 176
 non-Levallois flakes, 327
 Quina-type retouch, 99, 175, 178, 179
 spatial and chronological distribution, 183–90,
 352–3
 taxonomy, 181–3
 tool-type frequencies, 173

rabots, 122
racloirs, 96–110
 forms, 104–10, 114
 bulbar face, 97
 déjeté, 121–2
 double-edged, 97
 intensive reduction, Dibble hypothesis,
 104–10
 mean lengths, 102, 328
 pointed forms, 110–17
 taxonomy, 170
 with thinned back, 97
 transverse, 98, 105, 109, 179
 hafting procedures, 114, 115–16
 index, 171–2
 multi-dimensional-scaling analysis, 180
 raw materials, 136–40
 resharpening, 100, 103, 107, 157
 spatial distribution, 274–6, 293–4
raw materials, 141–86, 250–1
 chaîne opératoire approach, 57, 58–60, 154
 jasper, 144, 168
 procurement and distribution, 141–68
 quartz, quartzite, 139, 243–4
 artefacts, 289
 distribution, 291
 social and mobility implications, 161–8
Rébières valley, 248
red deer, 211, 212, 213, 220, 223, 229
 frequencies, 37, 195, 197, 200
 Italy, 229
 processing patterns, 211–16
 teeth, ratio to postcranial bones, 209–10
reduction/resharpening models, 99–110, 112–17,
 120, 332–41
reindeer, 201, 211, 215, 231, 329, 387
 Alaska, 321, 324
 Combe Grenal, 204–6
 Denticulate and Quina Mousterian, Combe
 Grenal, 330
 frequencies, 37, 195, 197, 200, 201
 migration routes, 247, 400